Medicinal Plants

Shankar Gopal Joshi

Oxford & IBH Publishing Co. Pvt. Ltd.
New Delhi
(*A Unit of* CBS Publishers & Distributors Pvt Ltd)

CBS

CBS Publishers & Distributors Pvt Ltd

New Delhi • Bengaluru • Chennai • Kochi • Kolkata • Mumbai
Bhopal • Hyderabad • Jharkhand • Nagpur • Patna • Pune • Uttarakhand

Medicinal Plants

ISBN-13: 978-81-204-1414-3
ISBN-10: 81-204-1414-4

First Edition: 2000
CBS Reprint: 2018, 2021

OXFORD & IBH
New Delhi
(A Unit of CBS Publishers & Distributors Pvt Ltd)

Published by Satish Kumar Jain and Produced by Varun Jain for

CBS Publishers & Distributors Pvt Ltd
4819/XI Prahlad Street, 24 Ansari Road, Daryaganj, New Delhi 110 002, India.
Ph: 011-23289259, 23266861, 23266867 Fax: 011-23243014 Website: www.cbspd.com
e-mail: delhi@cbspd.com; cbspubs@airtelmail.in.

Corporate Office: 204 FIE, Industrial Area, Patparganj, Delhi 110 092, India
Ph: 011-4934 4934 Fax: 011-4934 4935 e-mail: publishing@cbspd.com; publicity@cbspd.com

Branches

- **Bengaluru:** Seema House 2975, 17th Cross, K.R. Road, Banasankari 2nd Stage, Bengaluru 560 070, Karnataka
  Ph: +91-80-26771678/79 Fax: +91-80-26771680 e-mail: bangalore@cbspd.com
- **Chennai:** 7, Subbaraya Street, Shenoy Nagar, Chennai 600 030, Tamil Nadu
  Ph: +91-44-26680620, 26681266 Fax: +91-44-42032115 e-mail: chennai@cbspd.com
- **Kochi:** 42/1325, 1326, Power House Road, Opp KSEB, Kochi 682 018, Kerala, India
  Ph: +91-484-4059061-65,67 Fax: +91-484-4059065 e-mail: kochi@cbspd.com
- **Kolkata:** 6/B, Ground Floor, Rameswar Shaw Road, Kolkata-700014 (West Bengal), India
  Ph: +91-33-2289-1126, 2289-1127, 2289-1128 e-mail: kolkata@cbspd.com
- **Mumbai:** 83-C, Dr E Moses Road, Worli, Mumbai-400018, Maharashtra
  Ph: +91-22-24902340/41 Fax: +91-22-24902342 e-mail: mumbai@cbspd.com

Representatives

• Bhopal	0-8319310552	• Hyderabad	0-9885175004	• Jharkhand	0-9811541605
• Nagpur	0-9421945513	• Patna	0-9334159340	• Pune	0-9623451994
• Uttarakhand	0-9716462459				

Printed at Chaman Enterprises, Daryaganj, New Delhi, India

Dedicated to

Dr. Apte Ganesha Dattatraya (1892-1967)

and

Vd. Vartak Shankar Gangadhar (1908-1984)

FOREWORD

I am glad to introduce Dr. S.G. Joshi, writer of this treatise on Ayurvedic Materia Medica for the medical fraternity in general and Ayurvedic practitioners in particular. The source of his erudiction in Āyurveda is his maternal uncle Vd. S.G. Vartak, who happens to be my guru also, and is a recognised authority on Āyurvedic Philosophy and Therapeutics. Dr. Joshi took full advantage of his association with Vd. Vartak. He got interested in Homoeopathy as he got acquainted with Late Dr. G.D. Apte, M.B.B.S., an Allopathic doctor who primarily practiced Homoeopathy. Dr. Apte was also a great recognized authority in Āyurveda and was chosen as the president of their Conference. Dr. Joshi got enough knowledge about Homoeopathy and Āyurveda.

Since Dr. Joshi got introduced to Homoeopathy he started practice in Homoeopathy and continued his study in Āyurveda. As for his understanding of Āyurveda Dr. Joshi by his rational and logical thinking came to the conclusion that Āyurveda is a self-contained complete system of medicine having its eight essential branches. It does not require any help from Modern knowledge of medical sciences for its therapeutical results and laboratory analysis to prove its merit.

The philosophical base of Āyurveda is the concept of "Panca-Tānmātrā". The Panca-mahābhuta which are the basic constituents of the universe and also of the human body which is part and parcel of each other. Caraka in his Sanhitā states that constituents of human body are six. 'Ṣad Dhātvātamaka Puruṣa'. Ṣad-Dhātu constitute the anatomical structure of the body or Śarīra. Cetanā Dhātu or life fource or Prāṇa-Śakti is an integrated part of the living body or Jīvita Śarīra.

Vāta-Pitta-Kapha are derived from Pancamahābhuta. Cetanā Dhātu is derived from 'Ātman' and is a dynamic principle. 'Ata' is to go and 'Ana' to add. This in fact is the definition of Life or 'Āyu-Jīvita.

Āyurveda is not merely a science of therapeutics but is the science of Life. It deals with all aspects of healthy living. It lays more emphasis on promotion of positive health and prevention of disease.

Fundamental principles of Āyurvedic philosophy are Pancamahābhuta sidhānta, Tridoṣa sidhānta, Rasavīryavipāka sidhānta.

According to Pancamahābhuta sidhānta, the entire material world originates from five mahābhuta. Everything in this world living and non-living is composed of Panca-mahābhuta. Characteristic properties of each mahābhuta are definite and different from one another. In human body the five mahābhuta are expressed in terms of Doṣa, Dhātu and Mala. Doṣa are concerned with the functions of body, Dhātu with the formation of body and Mala are substances formed in the body which are partly utilised in the body and partly excreated.

The diagnostic methodology of Ayurveda is based on 'Tri-Doṣa' concept of 'Vāta-Pitta-Kapha.' Tridoṣa represent a broad based generalisation of life processes and vital activities of body and form the basis for the maintenance of positive health and diagnosis, as well as treatment.

All the Doṣa have all five mahābhuta in their composition but are dominated by one or more of them. Their locations are fixed. Aggravation of Doṣa results in certain symptoms.

Doṣa are three in number, namely Vāta, Pitta and Kapha. Location of Doṣa and their functions are fixed.

Even in health, every individual is endowed with all the three Doṣa. But each individual has his own steady state of equilibrium of the three Doṣa. This state is unstable by nature and is disturbed by internal and external forces. Even due to change of seasons of the year Doṣa undergo certain changes. Doṣa become excited and imbalanced and confer a predisposition to or actually cause morbidity or disease in the person. Abnormalities or maladies are due to the imbalance of Tridoṣa. Thus disease or roga is the condition of altered intrinsic balance of Tridosa in the patient.

The treatment or 'Cikitsā' is based on clinical approach of Roga-Nidana and Roga-Cikitsā in terms of disturbance of intrinsic individual balance of Tridoṣa. In Nidān-pancaka the five aspects are (1) Cause (2) Pūrva rūpa (3) Rūpa (4) Upaṣaya and (5) Samprāpti. In Roganidāna we are to acertain which of the three Doṣa are aggravated or decreased and disturbed the intrinsic individual balance and the site or localisation of Prakupita Doṣa i.e. Sthāna Sanṣraya. Sthān-śrota-duṣti is the disease or Vyādhi or Roga. This is Roga-samprāpti or pathogenesis of disease.

Duṣta Doṣa travel from Koṣṭha to Śākhā and then to Marma. These channels are called Vyādhi-Mārga.

Cikitsā or treatment is the use of proper herbal medicine and Pancakarma treatment to reverse the pathogenesis i.e. Samprāpti-bhanga and lead Duṣta-dhātu from Marma, Madhyamarga to Śākhā, Kapha to Koṣṭha, Pitta to intestine and Vāta to Bṛhat-āntra. Duṣtadoṣa are first ameliorated by proper digestive medicine, then led to the eliminative Śākhā. The deficient Dhātu is nourished while increased Dhātu is eliminated and the individual balance is restored.

Rasa having common Guṇa with the Doṣa increase that Doṣa. Rasa having opposite Guṇa to Doṣa decrease Doṣa. Thus Madhura, Amla and Lavaṇa decrease Vāta but increase Kapha. Ticta, Katu and Kaṣāya decrease Kapha and increase Vāta. Dr. Joshi has given details about properties of different Guṇa and interrelationships of Rasa, Guṇa, Vīrya and Vipāka with the three Doṣa. This information is useful while treating the patient. Dravya-Guṇa-Rasa Śāstra determines the medicinal use of the plant. Plants have their curative properties through their Rasa, Guṇa, Vīrya, Vipāka and Prabhāva.

To understand and appreciate the meaning of the technical terms used in Āyurveda it will be appropriate to explain in some detail these terms and their interrelationship.

Doṣa are three namely Vāta, Pitta and Kapha. Their properties are as follows:

Vāta : Structure of Vāta is Ākāśa and Vāyu. Enthusiasm, inspiration and expiration, voluntary actions, circulation of rasa, rakta etc. throughout the body and discharge of excretory products out of body are the functions to Vāta.

Pitta : Structure of Pitta is teja. Digestion, heat-production, hunger, thirst, softness and suppleness of body, lustre, cheerfulness and intelligence are the main functions of Pitta.

Kapha : Structure of Kapha is Āpa and Pṛthvi. Unctuousness of body, smooth working of joints, general stability of body, strength, forbearance, courage and greedlessness are properties of Kapha.

Rasa : Rasāyate āsvādyate iti rasaḥ.

Rasa is that which is determined by tongue. There are six Rasa or tastes viz. Madhur (Sweet), Amla (sour), Lavaṇa (saline), Kaṭu (pungent), Tikta (bitter), Kaṣāya (astrifungent). The effect of different tastes are on increasing or reducing dosa.

Relationship between Rasa and Tridosha.

Dosa :	Vāta	Pitta	Kapha
Madhura	Amel.	Amel.	Agg.
Amla	Amel.	Agg.	Agg.
Lavana	Amel.	Agg.	Agg.
Kaṭu	Agg.	Agg.	Amel.
Tikta	Agg.	Amel.	Amel.
Kaṣāya	Agg.	Amel.	Amel.

Guṇa : Dravyam guṇena karma kurute. Samavāyi tu niscestaḥ.

Guṇa are twenty in number. Their action is stated below.

(1) Gurū : Yasya dravyasya bṛhaṇe karmāṇi śaktiaḥ sa gurūaḥ.
(2) Laghu : Langhane laghuaḥ.
(3) Śīta : Stambhane himaḥ.
(4) Ūṣṇa : Swedane ūṣṇaḥ.
(5) Snigdha : Yasya kledane śaktiaḥ sa snigdhaḥ
(6) Rūkṣa : Yasya śoṣane śaktiah sa rūkṣaḥ.
(7) Manda : Yasya śamane śaktiaḥ sa mandaḥ.
(8) Tīkṣṇa : Yasya śodhane śaktiaḥ sa tīkṣṇaḥ.
(9) Sthira : Yasya dhāraṇe śaktiaḥ sa stiraḥ.
(10) Sara : Yasya preraṇe śaktiaḥ sa saraḥ.
(11) Mṛdu : Ślathane mṛduaḥ.
(12) Kaṭhiṇa : Yasya dṛdhikaraṇe śaktiaḥ sa kaṭhiṇaḥ.
(13) Picchila : Yasya lepane śaktiaḥ sa picchilaḥ.
(14) Viṣada : Yasya salane śaktiah sa viṣadaḥ.
(15) Ślakṣṇa : Yasya ropaṇe śaktiaḥ sa ślakṣṇaḥ.
(16) Khara : Yasya lekhane śaktiaḥ sa kharaḥ.
(17) Sthula : Yasya savarṇe śaktiah sa sthulaḥ.
(18) Sukṣma : Yasya vivaraṇe śaktiah sa sukṣmaḥ.
(19) Sāndra : Yasya prasādane śaktiah sa sāndraḥ.
(20) Drava : Yasya vilodane śaktiah sa dravaḥ.

Relationship between Guna and Tridosha.

	Vāta	Pitta	Kapha.	
Gurū :	Amel.	-	Agg.	Agnimāndya.
Laghu :	Agg.	-	Amel.	Agnidīpana, Vraṇa-ropaṇa.
Śīta :	Agg.	Amel.	Agg.	Rakta-Stambhana, Dāhaśamana.
Ūṣṇa :	Amel.	Agg.	Amel.	Agnidīpana, Āmapacana.
Snigdha :	Amel.	Amel.	Agg.	Balya, Snehana, Varnaprāsadana.
Rūḳsa :	Agg.	-	Amel.	Balakṣaya, Śuṡkatā.
Manda :	Amel.	Amel.	Agg.	
Tīkṣṇa :	-	-	-	Śodhana, Lekhana.
Sthira :	Amel.	Amel.	Agg.	Mala-Mūtra-Sweda-Stambhana.
Sara :	-	-	-	Doṣa-Śodhana. Anulomana.
Mrdu :	Amel.	Amel.	Agg.	Śaithilyakara.
Kaṭhiṇa :	-	-	Amel.	Malāvarodha.
Picchila :	Amel.	Amel.	Agg.	Balya, Vṛṣya, Brnhana.
Viśada :	Agg.	-	Amel.	Vraṇa Śodhana-ropaṇa.
Slakṣṇa :	Amel.	-	-	Vraṇaropaṇa, Malamūtrapravartana.
Khara :	Agg.	-	Amel.	Lekhana.
Sthūla :	Amel.	-	Agg.	
Sūkṣma :	Agg.	-	Amel.	Swedapravartana.
Sāndra :	Amel.	-	Agg.	
Drava :	Amel.	-	Agg.	Swedapravartana.

Vīrya : Navīryam kurute kincit sarva vīryakṛtaḥ kriyaḥ.
Ūṣṇaśītaguṇot karsat tatra vīrya dvidhā smṛtam.

Vīrya is the therapeutic efficacy of drug. They are two, Ūṣṇa and Śīta. Śīta Vīrya is good for health and is 'para'. It diminishes secretions, stabilises excretory functions, stops haemorrhages and increases vigour and vitality, aggravates Vāta and Kapha but subdues Pitta. Ūṣṇa vīrya leads to storing up of energy, easy digestion, diapphoresis, emesis, thirst and fatigue, subdues Vāta and Kapha but excites Pitta.

Relationship between Rasa and Virya.

Madhura, tikta and kaṣāya rasa are increasingly Śīta Vīrya.
Kaṭu, amla and lavaṇa rasa are increasingly Ūṣṇa Vīrya.

Relationship between Vīrya and Tridoṣa.

Doṣa :	Vata	Pitta	Kapha
Śīta :	Agg.	Amel.	Agg.
Ūṣṇa :	Amel.	Agg.	Amel.

Vipāka : Parināma lakṣṇo vipākaḥs.

Vipāka are three, Madhura, Kaṭu **and Amla.**

Madhura Vipāka is good for health and is considered 'para'.

Relationship between Rasa and Vipaka.

Madhura and lavana rasa yield madhura vipāka.

Amla rasa yield amla vipāka.

Kaṭu, tikta and kaṣāya rasa yield kaṭu vipaka.

Exceptions : Pippali and Śunthi have kaṭu rasa but madhura Vipāka. Guduchi has tikta Rasa and madhura Vipāka. Haritaki has kaṣāya Rasa and madhura Vipāka.

Relationship between Vipāka and Tridoṣa.

	Vata	Pitta	Kapha
Madhura	Amel.	Amel.	Agg.
Amla	Amel.	Agg.	Amel.
Kaṭu	Agg.	-	Amel.

Prabhava : Viṣesaḥ karmanam caiva prabhāvaḥ tasya sa smṛtaḥ.

Dr. Joshi has devoted himself to Ayurvedic Materia Medica or 'Dravya-Guna Sastra' and after deep study in its principles 'Rasa-Vīrya-Vipāka' Sidhānta. The book is specially useful to Ayurvedic teachers, practitioners and research workers.

It is authentic and uptodate in respect of Botanical information and the line drawings of given plant species.

Lastly I feel it my duty to offer my sincere thanks to my friend Dr. Joshi for such a basic contribution. I specially appreciate his tenacity to study and grasp the Principles of Āyurveda and Dravya-Guṇa-Śāstra in particular.

Vd. M.G. Wadalakar
Ex. - Principal,
Tilaka College of Āyurveda, Pune 11.

PREFACE

India was one of the foremost developed countries in ancient times. Learned persons of vedic culture were quite aware regarding unimaginable obligation of plants for the very sustenance of animal life. Though not scientifically proved at that time, they knew that the air we breath remains pleasant by surrounding plants. There are a number of verses in ancient literature depicting this generosity of vegetable kingdom. They have also realised that there is no conduct of life where the plant kingdom does not make its contribution like food, fuel, shelter, fiber, fodder and medicine. No wonder that many such plants species have been revered as God.

From time immemorial many medicinal plants are well-known in this country. Use of herbal medicine can be traced to the remote past. One of the oldest treaties in the world is RIG-VED [4500 B.C.-1000 B.C.] where healing properties of some herbs are mentioned in the form of sonnets, which were often recited in religious rituals. Later on a special faculty was developed known as AYURVEDA, mostly dealing with human philosophy of health including utilization of medicinal plants for restoring normal physical fitness.

The further advancement of this process was materialization of SUSRUTA-SAMHITA and CHARAKA-SAMHITA [1000 B.C.] which incorporates comprehensive chapters on the therapeutic use of various plant species. These treaties deal with about 700 drugs, some of these are not indigenous to Indian subcontinent. The invaluable knowledge of medicine was composed in lyrical sutras (Sonnets) which often reveal valuable information in few words.

Interestingly enough there used to be regular rapport between sages of various communities interested in miraculous effects of herbal medicine. There are records in ancient scripts regarding periodic conferences, seminars and also workshops in selected areas where exchange of knowledge was often manifested. Even it was mentioned that women scholars like Maitrai, Gargi contributed some knowledge about medicinal plants and their maintenance. During the glorious days of Buddha philosophy (600 B.C.- 400 A.D.) there was friendly mingling of therapeutic values of plants, through religious norms were different. Scholars practicing medicine were quite familiar with wild medicinal plants growing in jungles. Often they do cultivate some selected plant species for procuring fresh and authentic material for preparing drugs. Such drugs when administered by expert and learned physician were sure to get desire results.

Later on number of new plant species were added to the then known pharmacopeia of indigenous medicinal plants. Attempts were also made to identify the required medicinal plant species by observing their exomorphic characters. For example plant species like Bala (Sida acuta Burm.); Ati-bala (Sida rhombifolia Linn.); Nag bala (Sida spinosa Linn.) and Bhu bala (Sida veronicaefolia Lam.) were named and grouped together under one class. In modern classification all these belong to Hibiscus(Malvaceae) family. Indeed Indian scholars do have realised the importance of correct identity of plants and that the knowledge of diagnostic external morphological characters are essential requirements for that purpose. But they could not imagine the use of Binomial system of naming the plants and thereby fixing absolute identity of the concerned plant material. Instead of that they have started giving various names to the same plant species by considering broad features like shape, color of flowers, smell, taste, along with local names based on plant profile and mythological folklores in various regional languages. For example Krisna Tulsi (*Ocimum sanctum* L.) have 52 different names in various Indian dialects. Such trend of thinking became prelude to more confusion while undertaking any research project on indigenous Indian plants.

From 12th century A.D. Arabs and Turkish tribal empires imbibed with fervour of Islam, conquered 3\4 of Spain and started sriking Eastern frontiers of Europe. More or less during the same period warring empires of Moghals and Tarters invaded Indian subcontinent through the narrow corridor of Afganistan. Non-believers of Islam became slaves and faced extreme cruel treatment. The scholerly good work began by the Greeks in Europe and Hindu culture in India was mostly lost and forgotten during the prolonged period. This era is often recorded as dark-age in the world history. The hostile eclips started its withdrawal during the 16th century in Europe and in India after establishment of British empire during 19th century.

There are however, some learned moderate scholers like Avicenna, Galen from Basara and Bagdad who had realised the importance of logical nature of thinking manifested by Indian scholers and their contribution in the technological developments in the field of various sciences including mathematics. From vanquished Indian culture Arabian scholers acquired valuable knowledge of healing human ailments by using herbal medicines. They also knew way of living healthy life professed by Ayurveda.

During fifteen century A.D. all Europian countries became united under the banner of crusaders and totally driven out Moslem dynasties along with their autocratic culture. Perhaps this may be called as the beginning of the initiation period of renaissance for Europian countries.

In Indian subcontinent, however, due to waves of Moslem invesions and internal rivalry among the feudal princely states, progress of science and culture were mostly in nonexistance. There were, however, eminent scholers in the services of Moslem emperors, Maratha, Rajput and other southern rulers dynasties of feudal states. Among these there were also some experts acquainted with the art of healing human ailments. These Pandits, Vaidyas, Hakims and knowledgeable persons of Siddha way of treating patients were mostly using herbal medicines derived from indigenous plant species. Their method of practising is mostly based on the texts of Ayurveda, Unani and Siddha; versions of traditional ancient knowledge. Large number of tribal communites living nomad life were quite acquainted with the wild plants known for their miraculous healing properties. Urban scholers learnt various new wild plants discovered by tribal Vaidoos showing effective remedies for some chronic diseases.

Modern scientists are now critically studying specialised faculty of medico-botany by using sophisticated equipment and accepted scientific parameters.

The systematic and scientific studies on traditional use of local herbal medicine was initiated after India was forced to become prestigious jewel in British Crown. Even before the British dominence over south-eastern countries of Asian continent, Portuguese, Dutch, French had already started establishing the colonial rule in small pockets of Western India and coastal area of Malbar. Europian travelers, business persons, adventurers, explorers were all amazed to observe the richness in Biodiversity of Indian flora, fauna and mineral wealth. They had meticulously documented Natural resources of Indian subcontinent in the form treaties, books and research papers in journals. Garcia Da Orta (1490-1570) was a surgeon with the Portuguese armed forces. He had recorded reputed 45 plant species in Indian traditional medicine. Among the narration were Coconut, Areca, Borassus, Dhatura, Cinnamomum, Cardamine etc. The documented record is also supplimented with several local names and some commercial names known to Europian pharmacists. Heinrich van Rheeds (1637-1692) and his contemporary colleagues of various European nationalities added more than 450 plant species with their botanical description and line drawings. The old plant names given by Europian botanist were often referd the habit, habitat, morphological pecularities, known medicinal use and supply folklores. For determining the status and identity of certain plant specimen along with the mostly accepted names 3 or more epithats were also supplimented. This arrangement however, makes process of identification more clumsy.

In 1753 Carolus Linnaeus (1707-1770) the great botanist started using binomial system of naming the plants and animals. The system was recognised throughout the world. The binomial system revolutionised the floristic and faunal system of classification for its easy applied use.

Now most of the plant species have their internationally approved botanical names. International rules of Botanical nomenclature confirmed by Botanical congress during their deliberations serves as guideline for plant taxonomists. At present their is no ambiguity about the correct identity of plants. Our Mango fruit tree may have several different names in various languages and countries but no body will differ regarding its authanticate valid name *Mangifera indica* L. throughout the world. Of course varities of cultivars are often appended as and when required.

During 2nd half of nineteen century and first half of 20th century floristic studies were metaculously undertaken by regional circles of Botanical Survey of India, Botany Departments of premier Universities and renowned research laboratories. Now we have more than 20,000 botanically identified species of flowering plants and ferns with detailed description, accurate line drawing and voucher herbarium specimens.

Medico scientists practicing allopathy and research minded Vaidyas, Hakims have contributed valuable knowledge regarding efficacy of reputed medicinal plants, indigenous to India. Among allopathic practitioners and Botany scholers, following are well-known stalwarts: Watt; Dymock; Chopra; Dutt; Kirtikar and Basu; Nadkarni; Nayar and Chopra. Ayurveda, Unani and Sidha practitioner have also contributed and given valuable information in regional language. In Maharashtra, Pade, Desai, Ogle, Sawant have published books giving comprehensive information. They have always stressed the need of more research in Ayurveda.

There are a large number of big and small booklets on medicinal plants of India. Beside these there are innumerable research papers and project reports giving various virtues of medicinal plants and importance of traditional system of medicine.

It is interesting to note that while we are entering in the 21st century, 50 years after Independance 70 per cent of our population is still dwelling in villages and living a substandard life. In remote villages health problem is tackled by using herbal medicine prescribed by Vaidya or medicinal man or woman from the tribal communities. Even after a period of 2000 years Ayurvedic system is still proving useful for these needy persons.

For raising Ayurvedic system of herbal medicine to world standard, it is very essential that each component or the factor of the system should be critically studied and made perfect. Starting from correct identification of plant material; no adulteration of any substitute, critical pharmacopial and pharmalogical research should be thoroughly conducted. After completing these formalities then only the drug should be authenticated by the high power committee of experts.

Nowadays there is volume of literature on Ayurvedic system of herbal medicine scattered in various languages and states of the Indian subcontinent. It is however very difficult to get access to particular component required for research work. I congratulate the author for his strenuous efforts to bring all these essential factors together for practical use. The present publication is based on the results of ten years of study of various components of Ayurvedic system of medicine. The endevour requires library work, discussions with stalwart Vaidya and Hakim, research scholers of Botany and Biochemistry which help to fulfil the desired project.

The scientific names are kept as upto date as possible. When name other than the usual one has been adopted, the one which the newer name is based has also been given. Popular local names are given for easy conformation of plant species. Sanskrit names are provided for the benefit of Ayurveda scholers. English names designate the importance of plant in commercial market. The main text include habit, habitat of the plant, its chemical composition, Ayurvedic properties with Sanskrit shloka in English as well as in Devanagari script. Action of medication and other uses as mentioned in Ayurveda and in Homoeopathic treatment has been also narrated. Therapeutical values based on traditional observations and modern results of research have been mentioned.

I am sure that this voluminous work on Ayurved will meet required needs and will be of great use to research minded persons in Ayurveda. In my opinion, the author has done a service to the cause of glorifying our Ayurvedic system of traditional medicine.

Dr. V. D. Vartak Ph.D.
Ex. - Director M.A.C.S.,
Pune 4.

ACKNOWLEDGEMENT

For Ayurvedic information, I received encouragement and guidance from one Vaidya from Pune who allowed me to use his library from time to time. I am also thankful to the Authorities of Fergusson College and MACS Libraries, Pune for giving access to me to refer the books.

Botanical part of this book is critically checked by Dr. V. D. Vartak, an Ex-Director of Maharashtra Association for the Cultivation of Science, Pune and Member of Programme Advisory Committee of Botanical Survey of India and Zoological Survey of India, Department of Environment & Forestry, Central Government of India. He also assisted me in selecting figures for the plants icluded.

In the initial efforts, I am indebted to a number of my friends Shri B. G. Nitsure for computer work, Dr. G. G. Muzumdar, Shri N. G. Gokhle, Shri Diwakar and Shri Alhad Panawalkar for figures; Dr. Paranjpe Prakash for photographs for Coverpage and Mrs. S. S. Joshi, Mrs. Surekha Panawalkar and Dr. Mrs. S. A. Chiplunkar for library reference, without whose encouragement and support this book would not have been completed. Shri Vyankatesha Mandke of V. Graphics & Arts arranged all the D.T.P. work and Shri Ganu of Rational Printers took the responsibility of Printing and Binding. I sincerely thank them for their keen interest in bringing out this book.

S.G. JOSHI

CONTENTS

KEY TO TRANSLITERATION

Sanskrit	Transliteration Key	Sanskrit	Transliteration Key
अ	A, a	ट	ṭa
आ	Ā, ā	ठ	ṭha
इ	I, i	ड	ḍa
ई	I, ī	ढ	ḍha
उ	U, u	ण	ṇa
ऊ	Ū, ū	त	ta
ऋ	Ṛ, ṛ	थ	tha
-	Ṛ, ṛ	द	da
ए	E, e	ध	dha
-	Ē, ē	न	na
ऐ	Ai, ai	प	pa
ओ	O, o	फ	pha
-	Ō, ō	ब	ba
औ	Ou/ou	भ	bha
अं	aṃ/aṁ	म	ma
अः	aḥ	य	ya
क	ka	र	ra
ख	kha	ल	la
ग	ga	व	va
घ	gha	श	śa
ङ	ṅa	ष	ṣa
च	ca	स	sa
छ	cha	ह	ha
ज	ja	ळ	ḷa
झ	jha	क्ष	kṣa
ञ	ña	ज्ञ	dna

INTRODUCTION

Plants are used for their medicinal properties since ancient times. A number of books on Indian Medicinal Plants have been Published so far by Government Agencies, Leading Publishers, Institutions and Research Bodies. A special feature of this book is that for each plant information about its Rasa, Guṇa, Vīrya and Vipāka is given. To substantiate this useful data authentic Sanskrit Slokas are given from Standard Nighantus both in Devanagari and Roman Scripts using transliteration marks, explaining the use of the particular plant for the treatment.

Arrangement :

Information regarding 420 medicinal plants which are in use in Ayurvedic practice; belonging to 106 different families is included in this book. I have adopted family as a basis for grouping of these plants. Families are presented in Alphabetical order. Species of plants belonging to each family are also presented in alphabetical order of their Botanical names.

In dealing with each plant, after its Botanical name, all available synonyms are mentioned. This is followed by names popularly used in English and Indian languages. Many of the plants have a number of different names in Sanskrit. These names are mentioned.

To help to identify the particular plant a line drawing with the characteristic botanical features is given below information about its Habit and Habitat.

Available information about the chemical composition of the plants is given.

Only certain parts of the plant are generally employed in medicine and these are mentioned.

Side effects of allopathic drugs have scared people all over the world and there is a consorted effort to find an alternative therapeuptic method. As drugs of vegetable origin preponderate in Ayurvedic recipes, Ayurveda is increasingly attracting attention of the medical scientists of the advanced countries. Hence, Ayurvedic therapeutics is given special attention.

Under Ayurvedic properties, for each plant material, information regarding its Rasa, Guṇa, Vīrya, Vipāka is given.

Sanskrita Slokas from relevent Nighantus describing these properties are first given in Devanagary script followed by its rendering in Roman script using Internationally recognized transliteration markings.

This is followed by giving its action and uses according to Ayurvedic therapeutics.

Information about its use in general therapeutical practice is then given.

Some of the species dealt with are also in use in Homeopathy. In such cases their Homeopathic therapeutics is described.

Index for Botanical names, Index for Sanskrit names and a combined Index for English and Indian names are given.

The material is drawn from authentic texts. I have taken great care in drawing information from various sources to avoid any mis-statements. The possibility of some errors of omission or commission cannot be ruled out. Suggestions for corrections are welcome and will be attended to for future publications.

Botanical identification of plants is necesssary owing to confusion in nomenclature especially in case of Sanskrit names where same name appears for more than one plant. This difficulty can be solved if Botanical name of the plant according to modern classification and information about its habit and habitat together with a line drawing is given.

Salient features of this book are : (1) There is a revision of Families of plants. In this book plants are grouped under revised families. (2) To facilitate identification of plants line drawings of all plants with their characteristic Botanical features are presented. (3) To help workers in Ayurvedic research, information about Rasa, Vīrya, Guṇa and Vipāka characteristics for plants together with proper Shlokas describing them is given. This will help understand the therapeutic use of the plants. (4) Some of the plants are used in Homeopathy. Their therapeutic use is given.

The present compilation is intended to be a good ready reference work for Teachers, Students, and Research workers in Medicine, both traditional and Ayurvedic, also in Botany, Agriculture, Forestry, Pharmacy, Pharmacology and Sidha workers.

Family : I - Acanthaceae.

Adhatoda vassica Nees.

SYNONYMS : *Adennanthera vasika, Adhatoda zeylanica* Medic., *Justicia adhatoda* Linn.

ENGLISH : Malabarnut, Vasaka.

INDIAN : Adullasaa, Wasaakaa, Adosh, Arusha, Rus, Bansa.

SANSKṚTA : Vāsaka, Vāsikā, Vṛṣa, Sinhāsya, Vansa, Vājidanta, Sinhī, Bhiṣaṅgmātā, Pancamukhi, Mātrukā, Sinhaparṇa, Arusak, Vris, Sinhamukhi, Adarsahaḥ.

HABIT : Small tree or large shrub.

HABITAT : All over India. Commonly cultivated as a hedge plant, often grows wild near human inhabitations.

CHEMICAL COMPOSITION : *A. vasica* is a source of quinozoline alkaloids—vasicolin, adhatodine, vacicolinone, and anisotine. It contains betaine, vasicinone and new alkaloid vasicine (1%); in addition B-sitosterol and tritriacontane in different parts. Leaves, visicine (0.79%); and flowering tops, 0.47%. Peganine (vasinine)—the chief active principle in quinazoline alkaloids yield from different leaf samples from India, 0.54 to 1.11 % dry wt. basis while in foreign samples it is high as 2.18% Adhatodic acid. Alcoholic extract of leaves is useful as hypotensive, bronchodilator, respiratory stimulent, hypoglycemic and artispos-modic. Oil from leaves, flowers and roots is active against Tubercule bacilli. Essential oil from leaves is bronchodilator, also vasicinone and ephedrine, potentiated; anti-insect and juvenile hormone mimicking activity.

Parts used : Roots, leaves and flowers.

Ayurvedic properties :
Rasa - Tikta-kaṣāya. Vīrya - Śīta.
Guṇa - Laghu, Rūkṣa. Vipāka - Kaṭu.

वासको वातकृत्स्वर्य कफपित्तास्त्रनाशनः।
तित्तरस्तुवरको हृद्यो लघु शीतरस्तृडर्तिकृत्।
श्वासकासज्वरच्छर्दि मेघकुष्ठक्षयापहः।। (भा.प्र.)

Vāsako vātakṛtsvaryam kaphapittasranāśanaḥ,
Tiktastuvarako hṛdyo laghu śitastsudartikrt,
Svāsakāsajvaracchardi meghakuṣṭhaksyāpahaḥ.
(Bhāvaprakāṣa)

वासयां विद्यमानायां आशायां जीवितस्यच।
रक्तपित्ती क्षयी कासी श्वासी किमवसीदति।। (वृन्दमाधव)

Vāsāyām vidyamānāyām āśāyām jīvitasyaca,
Raktapitti ksayī kāsī śvāsī kimavasīdati.
(Vrndamādhava)

Actions/Uses : Kaphaghna, Pittaghna, Vātakar.

Therapeutics :
Vāsakā is a reputed remedy for all sorts of cough and cold, bronchitis and other respiratory disorders due to its expectorant action. It is the main constituent of cough syrup "Adulsa syrup". Plant is bitter, astringent, diuretic, antispasmodic, expectorant and alterative. It cures vomiting, thirst, dermatosis, jaundice, fever, phthisis and haematemesis. It is particularly useful in fevers associated with bilious and respiratory troubles and also in piles. Roots are expectorant and mild bronchial antiseptic; given in intermittent fever,

pulmonary and catarrhal affections. Leaves and roots are hypoglycemic. Juice of leaves relieves cough by its soothing action on nerves and by liquefying sputum. Fresh leaf juice is beneficial in haemoptysis and menorrhagia. Leaves are useful against Ranikhet disease virus. In haemoptysis it is a grand remedy. Leaves and wood ashes mixed with honey are used for cough and asthma. Juice of leaves mixed with juice of Feronia limonica cures nose bleeding. A preparation of leaves in clarified butter is used for glandular tumors. Crude extract is more useful for respiratory ailments. Flowers and roots with ginger are given in ague, rheumatism, constipation, asthama, chronic bronchitis and other chest affections. Shoots are used in liver enlargement. It is one of the constituents of drug "Geriforte" used against senile pruritus and as antifatigue. 'Justica rubrum or Rakta vasaka' is not so common as *A. vasica* but is more largely found in Coochabehar and Darjeeling. It is used in all those complaints and ailments in which generally *A. vasica* is used. But its usefulness and efficacy are more sure and certain where there is more blood with the cough and where there is more expectoration or bloody vomiting in tuberculosis. A. bedddomei C. B. Clarke is a variety found in Kerala which is more powerful and active than *A. vasica* Nees. and is used as antiemetic, antibechic, haemostatic, particularly in haemorrhages.

HOMEOPATHIC USE : Homeopathic preparation from extract of leaves is used for colds coryza, coughs, pneumonia, spitting of blood, fever, jaundice, catarrh, whooping cough and bronchitis.

Andrographis paniculata Nees.

SYNONYMS : *Andrographis subspathulata Clarke., Justicia paniculata Burm.*

HABIT : An annual erect, branched glabrous herb.

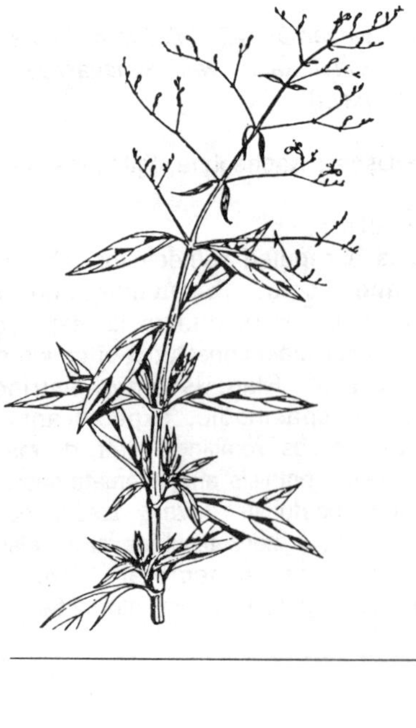

HABITAT : Common among undergrowths in forests areas and also cultivated in private compounds.

ENGLISH : Creat, Chiretta.

INDIAN : Kalmegha, Kiraayat-oli, Mahatita.

SANSKṚTA : Bhūnimba, Kirāta, Yavatikta, Ṡvetatikta, Ṡvetakunha, Ṡankhini, Sūksmapuṣpa, Tiktaphala, Yaṡasvani, Yavi, Kālamegha, Kalpanātha, Mahātikta.

CHEMICAL COMPOSITION : Pigments, andrographin and panicolin from roots. Leaves, β-sitosterol glucoside and unknown compounds. Andrographolide, panicolide, polyphenols, caffeic and chlorogenic acids. Roots, flavones andrographin and panicolin and a-sitosterol. Plant, diterpene glucoside, neoandrographolide. This plant better known as Kalmegha from the bitter substance Kalmeghin which it contains, is different from the well-known Kadekiraaeet or chiraeeta of ancient Indian medicine. Kalmegh increased biliary flow and liver weight in rats.

Parts used : All parts.

Ayurvedic properties :

Rasa - Tikta. Vīrya - Ūṣṇa.
Guṇa - Laghu, Rūkṣa, Tīksṇa. Vipāka - Kaṭu.

भूनिम्बो वातलस्तिक्तो व्रणरोपणकारकः।
सरः शीतः पथ्यकरो लघु रुक्षस्तृषापहः।।
कफंपित्तंज्वरं कुष्ठं कण्डूं शोफं कृमींस्तथा।
सन्निपातज्वरं दाहं शूलं मेहं व्रणं तथा।।
श्वासं कासं च प्रदरं शोषं चार्शोऽरुचि जयेत्। (नि.र.)

Bhunimbo vātalastikto vranaropaṇa kārakaḥ,
Saraḥ śitaḥ pathyakaro laghu rukṣastṛsāpahaḥ.
Kapham pittam jvaram kuṣṭham kṇandum
śopham kṛmimimstathā,
Sannipatājvaram dāham śūlam meham
vraṇam tathā.
Śvāsam kāsam ca pradaram śoṣam
cārśo arucim jayet.
(Nighanturatnākara)

किरातः सारको रुक्षः शीतलस्तिक्तको लघुः।
सन्निपातज्वर श्वास कफपित्तास्त्रदाहनुत्।।
कासशोफतृषाकुष्ठज्वरव्रणकृमिप्रणत्। (भा. प्र.)

Kirātah sārako rukṣaḥ śitalastiktako laghuḥ,
Sannipātajvarasvāsakaphapittāsradāhanut.
Kāsaśophatṛṣākuṣṭhajvaravranakṛmipranut.
(Bhāvaprakāṣa)

Actions/Uses : Kapha-Pittaghna, Recana, Yakṛtottejaka, Śothahara, Dīpana, Krmighna, Śvedakāri, Jvaraghna.

Therapeutics :

Kirata plant is considered to be useful for children and not so much for older persons. It is a specific for all types of fevers especially intermittent fevers and overcomes Sannipaata type of fever. It is laxative, dry, cooling, bitter, light, tonic and anthelmintic. It is useful in constipation, colic, dysentery and dyspepsia, strangulation of intestitine, in spleen complaints and debility. It is a domestic medicine for flatulence and diarrhoea of children. Used in torpidity of liver, neuralgia and convalescence after fever and in advanced stages of dysentery. Powdered plant mixed with sarsoo oil applied in itching. It is the chief constituent of an Ayurvedic drug "SG-1 Switraadilepa" for dermatological diseases. Roots and leaves are tonic, stomachic, antipyretic alterative, anthelmintic, febrifuge and cholagogue. Root combined with pepper and aloes is a tonic, stimulant and gentle aperient and valuable as a remedy in the treatment of several forms of dyspepsia and in torpidity of alimentary canal. Used as curative or preventive in snake venom poisoning.

Barleria prionitis Linn.

INDIAN : Koraanti-piwalli, Katsareya.

SANSKṚTA : Saireyaka, Kurantaka, Vajradanti, Karunta.

HABIT : A much branched prickly shrub, with 2-3 spines in axil of each leaf. Yellow colored flowers.

HABITAT : Occurs thoughout the hotter parts of India. A common undershrub occasionally found wild but generally cultivated as a hedge plant or for its ornamental flowers.

CHEMICAL COMPOSITION : Plant contains β-sitosterol. Flowers, scutellarein-7-rhamnosylglucoside. Leaves and stems showed the presence of five iridoid glucosides; three of them are iridoids, barlerin & Ac-barlerin.

Parts used : All parts, especially leaves.

Ayurvedic properties :
Rasa - Tikta, Madhura. Vīrya - Ūṣṇa.
Guṇa - Laghu. Vipāka - Kaṭu.

सैरेयः कुष्ठवाताकफकण्डुविषापहः।
तिक्तोष्णो मधुरोऽनम्ला सुस्निग्धः केशरंजनः।।
(भा. प्र.)

Saireyaḥ kuṣṭhavātāsrakaphakandu viṣāpahaḥ, Tiktoṣṇo madhuro anamlaḥ susnigdhaḥ keśaranjanaḥ.
(Bhāvaprakaṣa)

Actions/Uses : Kaphaghna, Vātaghna, Śothahara, Vedanāsthāpan.

Therapeutics :

Saireyaka has antiseptic properties; its decoction is used in dropsy to wash the body. It is especially useful for children. Roots are febrifuge. A decoction of roots is used as mouthwash in toothache and paste is applied to disperse boils and glandular swellings. Leaves are chewed to relieve toothache. Leaf juice is used in urinary and paralytic affections, and stomach disorders; applied to lacerated sole. Juice of leaf administered in a little honey or sugar in catarrhal affections of children, which is accompanied by fever and much phlegm. Tribals apply infusion of leaves to soles of feet to harden them to tolerate extreme heat, cold and rough soil. Crushed fresh leaves applied to wounds; also used for rheumatic pain and itch; juice with coconut oil applied on pimples. Leaves and flowering tops are valued as a diuretic and are rich in soluble potassium salts. Juice of fresh bark is diaphoretic and expectorant and used in anasaraca. Dried bark powder with honey is used in cough and whooping cough.

Gendarussa vulgaris Nees.

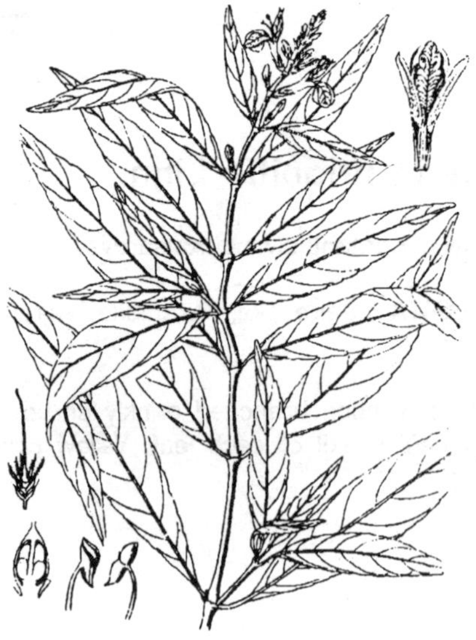

SYNONYMS : *Justicia gendarussa* Burm.

INDIAN : Adullasaa-kaalla, Udisanbhalu, Nilinargandi, Bakas, Tao.

SANSKṚTA : Nīlanirgudi, Nīlamanjari, Kṛisṇa nirguda.

HABIT : An erect branched shrub. Flowers creamish-white.

HABITAT : Native of China. Cultivated as a hedge plant and often found wild as an escape from cultivation.

CHEMICAL COMPOSITION : Roots yield B-sitosterol. Leaves B-sitosterol and an alkaloid.

Parts used : Leaves.

Ayurvedic properties :
Rasa - Tikta, Kaṣāya. Vīrya - Ūṣṇa, Śīta.
Guṇa - Laghu. Vipāka - Kaṭu.

नीलानिर्गुडिका तिक्ता रुक्षा चोष्णा कटुः स्मृता।।
आघ्मानवानं प्रदरं क्रासंशोथं कफं हरेत्।
वातनाश करीप्रोक्तामुनिभिर्गुणपार गैः।।
(निघण्टु रत्नाकर)

Nīlānirgudikā tiktā rūksā coṣṇā katuḥ smṛtā,
Ādhmānavānam pradaram kāsamśotham kapham haret.
Vātanāsa karīproktā munibhirguṇapara gaiḥ.
(Nighantu Ratnākar)

Actions/Uses : Vātakāraka, Svedakāri, Vāmaka, Recana, Tīvra, Jvaraghna, Selsmānisāraka.

Therapeutics :

Nīlanirgudi is pungent, bitter, hot and dry. It is a strong medicine not to be used for children or old persons freely as it causes vomiting and diarrhoea. It is used for lung diseases. Useful in dyspepsia; tympanitis, inflammations, eye diseases and as febrifuge, emetic, emmenagogue and diaphoretic in bronchitis and also for vaginal discharges. Root boiled in milk is used in chronic indigestion, dysentery, rheumatism and fever. Leaves are antiperiodic, alterative and insecticide. Juice of leaves, in coughs of children, also efficacious in colic of children. 2 to 4 leaves with ash of Achyranthus aspera and honey are used in acute cough. Leaves and tender shoots are diaphoretic and are given in the form of decoction in chronic rheumatism. Infusion of leaves given internally in cephalagia, hemiplegia and facial paralysis. Juice of leaves for earache. Leaves made into paste and applied on fractured or dislocated bones. Bark is a good emetic.

Hygrophilia spinosa T. And.

SYNONYMS : *Asteracantha longifolia* Nees.; *Bahel schulli* Ham. *Hygrophila auriculata* Heine.

INDIAN : Kollasundaa, Taalimkhaanaa, Gokshura.

SANSKṚTA : Kokilākṣa, Ikṣūra, Śrungāli, Śrunkhalā, Tailakantaka, Śhrugālaghanti, Vajrakantaka, Pikekṣaṇa.

HABIT : An erect stout herb. Stems thickened at nodes. Axillary branches reduced into sharp spines.

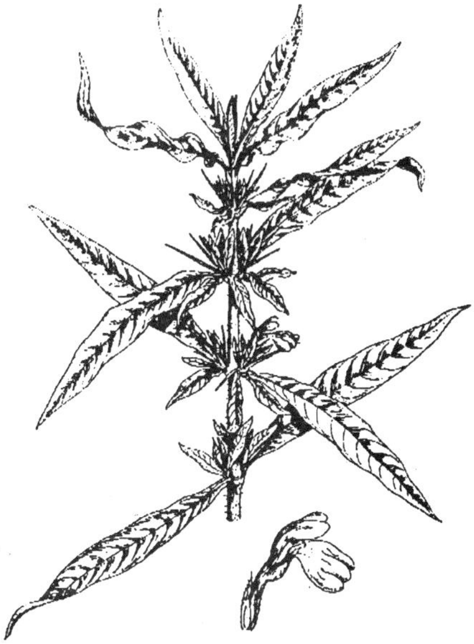

HABITAT : A common herb throughout India around ponds, on sides of rivers and in marshy areas.

CHEMICAL COMPOSITION : Diuretic properties of seeds are due to large amount of mucilage and potassium salts. Seeds contain 23% of a yellow semi-drying oil; they also contain diastase, lipase and protease. Xylose and uronic acids from oil of seeds; lupenol, stigmasterol and straight-chain hydrocarbons from lipid extract of plant. Palmitic (7.2), stearic (0.8), oleic (11.9) and linoleic (80.1%) acids in seed oil.

Parts used : Roots, seeds, salts and other parts.

Ayurvedic properties :
Rasa - Madhura. Vīrya - Śīta.
Guṇa - Gurū, Snigdha, Picchila.
Vipāka - Madhura.

कोकिलाक्षस्तु मधुरः शीतः पित्ताश्मरिप्रणुत्।
वृष्यः कफहरो बल्यो रुच्यः संतर्पण परः।।
(रा. नि.)

Kokilākṣatu madhuraḥ śītaḥ pittāṣmaripranut,
Vṛṣaḥ kaphaharo balyo rucyaḥ santarpaṇa paraḥ.

(Rājanighaṇtu)

Actions/Uses : Pittaghna, Vātaghna.

Therapeutics :

Kokilāksa is a reputed remedy for arthritis. It is sour, bitter, cold, aphrodisiac, roborant and demulcent. It promotes strenghth and appetite, cures oedema, ascites, thirst, bladder stones, eye diseases and dysentery. Root is a cooling bitter, tonic, diuretic, demulcent and refrigerant. Decoction of roots is diuretic. Seeds given in gonorrhoea and with milk and sugar in spermatorrhoea. Seeds are tonic and aphrodisiac. Leaves, roots and seeds are diuretic and are employed for jaundice, dropsy, rheumatism, anasaraca and diseases of urinogenital tract. Plant is useful in cancer and tubercular fistula and juice in anaemia. Leaves and roots alongwith flowers of Stuea frondosa taken in leucorrhoea. It is a source of Ayurvedic drug "Kokilaksha"; Unani drug, Talimkhana; and a Siddha drug "Neermulli". Herb ingredient of indigenous drug "Speman" having anabolic-cum androgen-like activity.

Justicia procumbens Linn.

SYNONYMS : *Rostellularia procumbens* Nees.

INDIAN : Ghatipittapaapadaa, Karambal, Kalmashi, Pitpapada.

SANSKṚTA : Parpaṭa, Pittapāpadā.

HABIT : A procumbent, diffuse, slender, branching annual.

HABITAT : A common herb among the grasses and along the road-sides, Konkan, Western Ghats of Maharashtra and Madras; coast of Madras to Travancore. The plant is eaten in some parts of Maharashtra.

CHEMICAL COMPOSITION : Leaves contain an essential oil which is antifungal; Justicidins A & B (major), diphyllin and lignan neojusticin.

Parts used : All parts.

Ayurvedic properties :
Rasa - Katu, Tikta. Vīrya - Ūṣṇa.
Guṇa - Laghu, Rūkṣa. Vipāka - Kaṭu.

Actions/Uses : Vātakara, Pittahara, Kaphahara, Vamakanibandha, Jvaraghna, Svedakāri, Mūtrajanana. Properties are similar to those of Rungia repens Nees.

Therapeutics :

Pittapapāda is considered as a laxative, diaphoretic, diuretic, alterative, expectorant, anthelmintic and febrifuge. An infusion of herb is given in asthma, cough, rheumatism, backache, plethora, flatulence and lumbago. Decoction with some bitter principle is used in bilious fever; after perspiration, burning of body is relieved, and after urination and one or two diarrhoic stools inflammation and pain of liver is reduced. Juice of leaves is squeezed into the eye in cases of opthalmia. Leaf decoction is used in curvature and diseases of bones. Properties and uses are the same as Fumaria oaeviflora. Used as a substitute for true Pithpapara (Fumaria officinalis) and for Fumaria parviflora.

Rhinacanthus nasutus Kurz.

SYNONYMS : *Rhinacanthus communis* Nees.; *Justicia nasuta* Linn. *Hygrophila auriculata* Heine.

HABIT : An annual or biennial undershrub.

HABITAT : Cultivated in gardens in the southern States. All over Maharashtra, especially in Konkan and Mahableshwar.

INDIAN : Gajakarnni, Yuthakaparnni, Palak juhi.

SANSKṚTA : Yuthakaparṇi, Juthikaparṇi.

CHEMICAL COMPOSITION : Roots and bark contain Rhinocanthin which is similar to Chrysophanic acid.

Parts used : Root, leaves, seeds and bark.

Ayurvedic properties :
Rasa - Kaṭu, Tikta. Vīrya - Ūṣṇa.
Guṇa - Laghu, Rūkṣa. Vipāka - Madhura.

गजकर्णी स्वादुपाका तिक्ता चोष्णा कफापहा।
शीतज्वरं च वातं च नाशयेदिति कीर्तिता।।
कन्दोऽस्या पाण्डुशोधार्शः कृमिगुल्मोदरा पहा।
आनाहवातं प्लीहां च ग्रहणीमपि नाशयेत्।।
(निघण्टु रत्नाकर)

Gajakarṇi svādupāka tiktā coṣṇā kaphāpahā,

Śītajvaram ca vātam ca nāsayediti kīrtitā.
Kando asyā pāndusodharśaḥ kṛmigulmodarā,
Ānāhavātam plīhām ca grahaṇīmapi nāṣayet.
(Nighanturatnākara)

Actions/Uses : Kaphavātasamaka, Raktaśodhaka, Uttejaka, Vājikara, Krimighna, Visagna, Kuṣṭhagna, Dadrughna.

Therapeutics :

Yuthakaparṇi possesses antiseptic properties. Roots boiled with milk is used as aphrodisiac. Root powdered and made into a paste with lime-juice useful for eczema, ring-worm and Dhobie itch. Juice of leaves is efficaceous when applied to eczema. Leaves crushed with salt are externally applied for ringworm. Leaves when chewed have a pungent taste and are used against cancer. Root, leaves and seeds are useful remedies for ringworm and other skin diseases.

Rungia repens Nees.

SYNONYMS : *Justicia repens* Linn.

INDIAN : Ghatipitta; Paarpaatha. Kharmor.

SANSKṚTA : Parpatha.

HABIT : A sub-erect or diffuse perennial herb.

HABITAT : A rare herb in shady places throughout the warmer parts of India.

CHEMICAL COMPOSITION : Flowers, Luteolin-7-glycoside and delphinidin, besides luteolin and another flavone - chrysoeriol.

Parts used : All parts

Ayurvedic properties :
Rasa - Tikta. Vīrya - Śīta.
Guṇa - Laghu. Vipāka - Kaṭu.

पर्पट, शीतलस्तिक्त, संग्राही वात कोपनः।
लघुः पाके च कटुको हरेत्पित कफत्पित कफज्वरान्।।
रक्त दोषा रुचिदाह ग्लानीभ्रममदान्जयेत्।
प्रमेहषमितृड्रक्तपित्तानां च विनाशकः।।
अस्यशाकं तु संग्राहि शीत वातकरं लघु।
तिक्तं रक्तरुजं पित्तज्वरं तृष्णांच नाशयेत्।।
कफं भ्रमं च दाहं च नाशयेदिति कीर्तितम्। (नि. र.)

Parpatah śītalastiktaḥ sangrāhi vātakopanaḥ,
Laghuḥ pāke ca kaṭuko haretpitta kaphatpitta kaphajvarān.
Raktadoṣa rucidaha glānibhramamadānjayet,
Pramehavamitṛnraktapittānam ca vināśakaḥ.
Asyaśākam tu sangrāhi śītavātakaram laghu,
Tiktam raktarujam pittajvaram tṛsnam ca nāṣayet.
Kapham bhramam ca dāham ca nāṣayediti kīrtitam. (Nighanturatnakāra)

Actions/Uses : Vātakara, Kaphapittaghna, Grāhi, Krimighna, Jvaraghna, Svedakāri, Mūtrala.

Therapeutics :

Parpatha is diuretic and vermifuge and is given in snake bite. Plant dried and pulverized and given in fevers and cough and is considered to be a vermifuge. Fresh leaves, bruised and mixed with castor oil are applied to scalp in cases of tinea capitis.

Family : II - Adiantaceae.

Adiantum philippense Linn.

SYNONYMS : *Adiantum lunulatum* Burm.; *Pteris lunulata* Retz.

ENGLISH : Walking maidenhair fern.

INDIAN : Kalijhant, Hansapadi, Ghodyachi pawale.

SANSKṚTA : Hansapādi, Mayūrahāśikhā, Nilakanthasikha, Hangsavati.

HABIT : A graceful fern. A small rhizomatous herb.

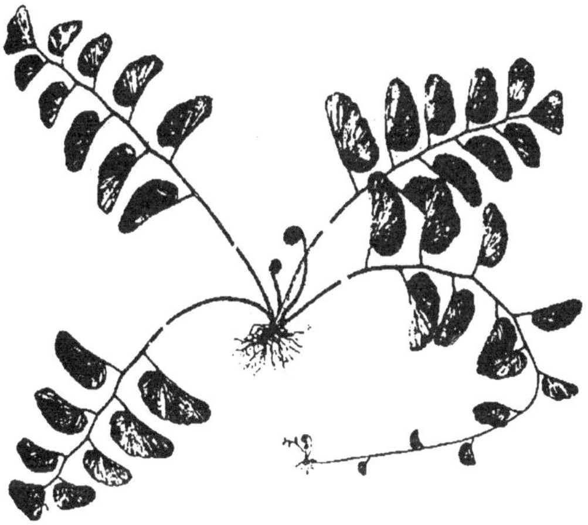

HABIṬAT : Throughout N. India in moist places. S. India very general on western side in planes.

CHEMICAL COMPOSITION : Chlorophyll-degradation products and higher carotenoids.

Parts used : Whole plant.

Ayurvedic properties :
Rasa - Kaṣāya. Virya - Śīta.
Guṇa - Gurū. Vipāka - Madhura.

हंसपादी गरुः शीता हन्ति रक्तविषव्रणान्।
विसर्प दाहातीसारलूता भूताग्निरोहिणीः ।।
(भा. प्र.)

Hansapādi guruḥ śītā hanti raktaviṣavraṇān,
Visarpadāhatīsāralūta bhūtāgnirohiṇiḥ.
(Bhvāprakaṣa)

मयूराहशिखा प्रोक्ता सहस्त्रांहिमधुच्छदा।
नीलकण्ठ शिखा लघ्वी पित्तश्लेष्मातिसारजित्।।
(भा. प्र.)

Mayūrāhaśikhā proktā sahasrāhimadhucchdā,
Nīlakanthaśikhā laghvi pittaslesmātisārajit.
(Bhāvaprakāṣa)

मयुराह्रा शिखा शीता कशाषा कटुपाकिनी।
लघ्वी पित्तकफौ रक्तमतिसारं विनाशयेत्।।
(कै. नि.)

Mayūrāha sikhā sita kasaya katupakini,
Laghvi pittakaphau raktamatisaram vinasayet.
(Kaiyadeva)

Actions/Uses : Kaphapittaśāmaka, Dāhapraśamana, Viṣaghna, Vranaropaṇa, Visarpa, Stambhana, Mūtrala.

Therapeutics :

Hansapādi is pungent and alexiteric. Plant is used in blood diseases, burning sensation, epileptic fits, dysentery, febrile affections and ulcers. Also useful in atrophy, muscular pain, emaciation or cachexy, diarrhoea and rabies and as a cooling lotion in cases of erysepelas. With Asparagus racemosus it is used in gonorrhoea. Roots used in fever and erysipelas. Rhizome is prescribed for strangury and fever due to elephantiasis. A decoction of rhizome is given in throat affections and used for febile conditions of children. The fronds and fruits are used in leprosy, fever and erysipelas. The fronds, made into plaster are applied to chronic gouty and other swellings and also in chronic tumours.

Family : III - Alangiaceae.

Alangium salvifolium Wang.

SYNONYMS : *Alangium decapetalum* Lamk.; *Alangium lamarckii* Thw.; *Grewia salvifolia* Linn.; *Alangium tomentosum* Lam.

EINGLISH : Sage-leaved alangium.

INDIAN : Ankola, Akola.

SANSKṚTA : Ankota, Dīrghakila, Gandhapuṣpa, Tāmraphala, Vanavāsi, Guptasneha, Ankola, Angulipatra, Dṛdhamūlaka, Śubhrapatra, Reci, Dherā, Kantaki, Kotara, Gudhapatra, Nikocaka, Madana, Mallikā, Kolaka, Lambakarṇa, Rocana, Vijnyāna, Tailagarbha, Bhūsūtā, Hundikā, Soedhanam.

HABIT : A deciduous, rambling shrub or tree.

HABITAT : A rare tree in costal areas but grows vigorously in forests of S. India and found throughout the drier parts of India.

CHEMICAL COMPOSITION : The plant contains amorphous alkaloid alangine A and B, alangicine, marckine and marckidine; emetin, demethylcephaeline, cephaeline, tubulosine, and psychotrine. In small doses alkaloid reduces blood pressure temperorily, depresses the heart and produces irregular respiration, increases peristaltic movement of the intestines. Alkaloid extract of leaves is mild adrenolytic, antispasmodic, hypotensive and anticholinesterase activity. Total alkaloids of seeds show a sustained and prolonged hypertensive effect at lower and hypotensive effect at higher doses.

Parts used : Root, root-bark, leaves, fruits and seeds.

Ayurvedic properties :
Rasa - Tikta, Katu, Kaṣāya. Vīrya - Ūṣṇa.
Guṇa- Laghu, Snigdha, Tīkṣṇa, Sara.
Vipāka - Kaṭu.

अङ्कोटकः कटुस्तीष्णः स्निग्धोष्णस्तुवरो लघुः।
रेचनः कृमिशूलाम शोफग्रह विषापहः।।
विसर्पकफपित्तास्त्रमूषिकाहिविषापहा।
तत्फलं शीतलं स्वादु श्लेष्मघ्नं बृंहणं गुरु।।
बल्यं विरेचनं वातपित्तदाहक्षयांस्त्रजित्। (भा. प्र.)

Aṅkodakah kaṭustīksnah snigdhoṣṇastuvaro laghuḥ,
Recanaḥ kṛmisūlāma śophagraha viṣāpaḥa.
Visarpa kaphapittāsra mūṣikāhiviṣāpahā,
Tatphalam śītalam svāduḥ sleṣmaghnam bṛnhaṇam guru,
Balyam virecanam vātapittadāhakṣāyasrajit.
(Bhāvaprakāṣa)

Actions/Uses : Pittaghna, Vātaghna, Kapha-vāta-pitta-sāmaka, Viṣagha, Krimighna.

Therapeutics :
Ankoṭ is a reputed single drug for the treatment of rabies. Various parts are used in enlargement of spleen, dropsy, anasarca, colic pain, stomach-ache, prolapsus ani and fistula ani, cholera, phthisis, bronchitis and snakebite. Root is astringent, emolient, anthelmintic, diuretic and purgative, useful in fever and skin diseases. Root paste or decoction cures skin diseases. Root bark is administered internally and externally in cases of rabid dog-bites, also as an antidote for other poisonous bites. Bark extract is taken orally for lowering B.P. It is

useful in diarrhoea, simple continued fevers, worms, colic, hemopathy and inflammations and is a reputed remedy for leprosy and other skindiseases and syphylis. Oil of the root-bark is a useful external application in acute rheumatism. Leaves are used as poultice in rheumatic pains. Fruit is laxative, antiphlegmatic and tonic, useful in the treament of burning sensation of body, emaciation, haemorrhages and morbidity. It is also a preventive and cure for eye ailments. Extract of flowers and fruit is used externally to cure eye sores. Seeds are cooling and tonic used in treatment of haemorrhages and as a cure for boil.

Family : IV - Alliaceae.

Allium cepa Linn.

ENGLISH : Red onion.

INDIAN : Taambadaa Kaandaa, Pyaaja.

SANSKṚTA : Palāndu, Tīkṣṇagandha, Rocana, Śudrapriya, Kandarpa, Ulli, Durgandha, Pharada, Mukhadūṣaṇa, Krimighna, Rudrasangyaka, Mukhagandhaka, Bahupatra, Viṣvagandha, Yavaneṣta, Śikhādhara, Dīpana, Durbalāsya, Śikhakanda, Sukanda, Śikhamaṇi, Śiroratna, Sikhāmūla, Varhiṇī.

HABIT : A bulbous biennial or perennial sub-erect herbs.

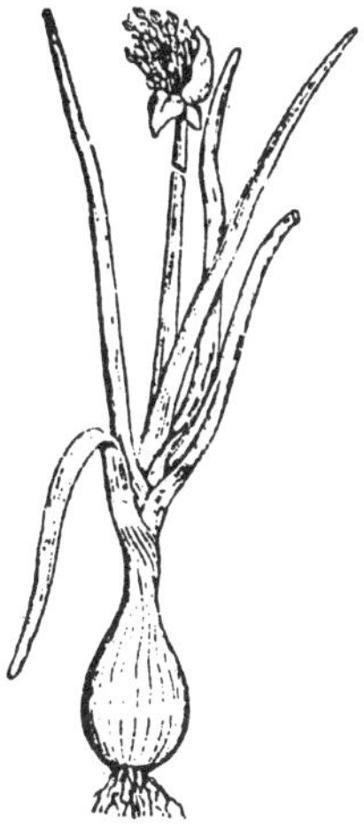

HABITAT : Extensively cultivated all over India.

CHEMICAL COMPOSITION : Pure quercetin from onion skin. Kaemferol, quercetin, diosgenin, flavonoid and phenolic compounds and sugar components. Essential oil and organic sulphides. Catechol and protocatechuic acid. A heart stimulant, increases pulse volume and frequency of systolic pressure and coronary flow, stimulates intestinal smooth musculature and uterus, promotes bile production and reduces blood sugar. Allyl propyl disulphide is chief constituent of crude oil.

Parts used : Bulb

Ayurvedic properties :
Rasa - Madhura, Kaṭu. Vīrya - Ūṣṇa.
Guṇa - Gurū, Snigdha, Tīkṣṇa.
Vipāka - Madhura.

पलाण्डुस्तु गुणैर्ज्ञेयो रसोनसदृशो बुधैः।
स्वादु पाके रसेनोष्णः कफकृन्नातिपित्तलः।।
हरते केवलं वातं बलवीर्यकरो गुरुः।
(भा. प्र.)

Palāṇdustu guṇaurjneyo rasonasadṛśo budhaiḥ,
Svādu pāke rasenoṣṇaḥ kaphakṛmnāti pittalaḥ.
Harate kevalam vātam balavīryakaro guruḥ.
(Bhāvaprakāṣa)

पलाण्डुः कटुको बल्यः कफपित्तहरो गुरुः।
वृष्यश्च रोचनः स्निग्धो वान्ति दोषविनाशनः।।
(रा. नि.)

Palāṇduḥ kaṭuko balyaḥ kaphapittaharo guruḥ,
Vṛṣyascya rocanaḥ snigdho vānti doṣavināśanaḥ.
(Rājanighantu)

Actions/Uses : Vātasāmaka, Kaphapittavardhaka, Tvagrogahara, Vedanāsthāpana, Dīpana, Pācanī, Balya,

Medhya, Rasāyana, Rocana, Anulomana, Śoṣaghna, Kuṣṭaghna, Śophaghna, Arśhoghna, Yakṛtottejaka, Svaraprada, Cakṣuśya, Vṛṣya, Varnakara, Bhaghnasandhānakara, Hṛdrogahara, Ajīrṇanāśana, Jvaraghna, Krīmighna, Gulmahara, Kukṣiśulahara, Vibandaghna, Kāsaghna, Svāsaghna.

Therapeutics :

Bulbs of palānaṇdu are stimulant, diuretic, expectorant, aphrodisiac, emmenagogue; useful in flatulence, haemorrhoids, dysentery, colic, jaundice, splenopathy, pneumonia, asthma and bronchitis. They are also used in eye troubles, lumbago, epilepsy, paralysis, tumours, leucoderma and skin diseases. With salt, it is domestic remedy for colic and scurvy. In raw state it is diuretic; roasted or fresh, used as polutice to boils, bruises and wounds. Applied to navel in dysentery and in body heat. Eaten with black pepper, in malarial fever. Decoction very useful in strangury and extreme heat sensation. It is useful for diabetes. Regular use of onion, 50 gms. per day reduces insulin requirement of a diabetic patient from 40 to 20 units a day. Continuous consumption of onion 50 gms.per day for 5 months decreased serum cholesterol in healthy human.

HOMEOPATHIC USE : Allium cepa, better known as Cepa is a Homeopathic remedy. Coryza profuse, watery and acrid nasal discharge; with profuse bland lachrymation is typical symptom. Acute catarrhal inflammation of mucus membranes with increased secretion; eyes burning, biting smarting as from smoke, must rub them are other symptoms of note.

Allium sativum Linn.

SYNONYMS : *Allium ophioscordon* Don., *Porrum sativum* Reichenb.

HABIT : A hard perennial sub-erect herb with bulbs made up of cloves.

HABITAT : Wildly cultivated in India.

ENGLISH : Garlic.

INDIAN : Lasoonna.

SANSKṚTA : Laśuṇa, Rasonā,Ugragandha, Ariṣṭa, Yavaneṣta, Mahauṣadha, Bhūtaghna, Āmlakānta, Rasāyanavara, Jugupsita, Mlencchakandaka.

CHEMICAL COMPOSITION : Garlic contains protein 6.3; fat, 0.1; and minerals 1.0 g/100gm. Thiamine,0.06; riboflavin,0.23; niacin 0.4 and vitamin C 13.0 mg/100 gm. Folic acid 6.15; and iodine 0.07 mg./100 gm. and amino acids leucine, methionine etc. Garlic bulbs contain a mixture of polysaccharides containing peptic acid, a D-galactan and a fructan compound. Essential oil contains allyl propyl disulphide, diallyl disulphide and two more sulphur compounds. Allicin, allisatin I and allisatin II. Blood cholesterol was significantly decreased in all human subjects after two months ingestion of garlic.

Parts used : Bulb and oil.

Ayurvedic properties :
Rasa - Kaṭu and all other except Amla.
Vīrya - Ūṣṇa.
Guṇa- Snigdha, Tīkṣṇa, Picchila, Gurū, Sara.

Vipāka - Kaṭu.

स्निग्धोष्णतीक्ष्णः कटुपिच्छिलश्च।
गुरुः सरः स्वादुरसश्च बल्यः।
वृष्यश्च मेधास्वरवर्णचक्षु-
र्भग्नास्थिसन्धानकरो रसानेः॥
हृद्रोग जीर्णज्वरकुक्षिशूल -
विबन्धगुल्मारुचिकासशोषान्।
दुर्नामकुष्ठानल सादजन्तु -
समीरण श्वासक फांश्च हन्ति॥ (सु. सूं. ४६)

Snigdhoṣṇatikṣṇaḥ kaṭupicchilaśca
Guruḥ saraḥ svādurasaśca balyaḥ,
Vṛṣyaśca medhāsvaravarṇacakṣur,
Bhagnāsthisandhānakaro rasonaḥ.
Hṛdrogajīrṇajvarakukśi,
Śulavibandhagulmārucikāsaśosān,
Durnāmkuṣṭhānalasādajantu,
Samiraṇasvāsakaphāmsca hanti.
(Su.S.46)

कटुकश्चापि मूलेषु तिक्तः पत्रेषु संस्थितः।
नाळे कषाय उद्दिष्टो नाळाग्रे लवणः स्मृतः।
बीजे तु मधुरः प्रोक्तो रसरत्तद्गुणवेदिभिः।
रसोनो बृंहणो वृष्यः स्निग्धोष्णेः पाचनः सरः॥
पञ्चभिश्च रसैर्युक्तो रसेनाम्लेन वर्जितः।
तस्माद्रं सोन इत्युक्तो द्रव्याणां गुणवेदिभिः॥ (भा. प्र.)

Kaṭukaścāpi mūleṣu tiktaḥ patreṣu samsthitaḥ,
Nāḷe kaṣāya uddiṣto nāḷāgre lavaṇaḥ smṛtaḥ.
Bīje tu madhuraḥ prokto rasastad guṇavedibhiḥ,
Rasono bṛmhaṇo vṛṣyaḥ snigdhoṣṇaḥ pācanaḥ saraḥ
Pañcabhiśca rasairyukto rasenāmlena varjitaḥ,
Tasmādrasona ityuto dravyāṇām guṇavedibhiḥ.
(Bhāvaprakāṣa)

Actions/Uses : Kaphaghna, Vātaghna, Pittakara, Yakṛttottejaka, Malānulomana, Raktapittaprakopī, Dīpana, Pācana, Śūlagna, Krimighna, Kandughna.

Therapeutics :

Laśūṇa is a stimulant, diaphoretic, expectorant, diuretic, and tonic. It is a very useful drug for rheumatism and catarrhal conditions and is used as an effective long term preventive treatment for all rheumatic and catarrhal conditions; rubefacient when applied externally. It is an anthelmintic and emmenagogue. Bulb is carminative, aphrodisiac, expectorant, stimulant; used in fevers, coughs, in intermittent fever. Juice used as rubefacient in skin diseases and as eardrops in earache; useful in atonic dyspepsia, flatulence and colic. Raw garlic decreased glucose, total cholesterol, phospholipids, triglycerides etc. in healthy individuals. Garlic is applied as resolvent to indolent tumors. Liniment beneficial in infantile convulsions and other spasmodic affection; in gout, sciatica, etc. Internally given with common salt in nervous diseases, headache etc. Juice applied to bruises and sprains, used to relieve earache and to allay otorrhoea. Fresh garlic inhibited mammary tumor in females.

HOMEOPATHIC USE : Allium sativum is a Homeopathic remedy. It is useful for dyspepsia of patients who eat a great deal more, especially meat. Sensitive to cold; constant rattling of mucus in bronchi; cough in the morning after leaving bedroom, with mucous expectoration, which is tenacious and difficult to raise; fetid expectoration; darting pain in chest; eruption in vagina and on breasts and vulva during menses are the chief guiding symptoms.

Family : V - Amaranthaceae.

Achyranthes aspera Linn.

ENGLISH : Prickly-chaff flower.

INDIAN : Aaghaadaa, Chirchira, Latjira.

SANSKṚTA : Apāmārga, Śikhari, Adhāhśalya, Mayūraka, Durgrahā, Pratyakapuspa, Markaṭī, Kharamanjarī, Kiṇihī, Kaṭī, Kāṇḍakaṇṭa, Śaikharika, Durabhigrahā, Vaśirā, Markaṭa-pippalī, Kṣavaka, Panktikaṇtaka,Mālākaṇṭa, Kubjaka, Aghaṭa.

HABIT : An erect or procumbent, annual or perennial herb with woody base.

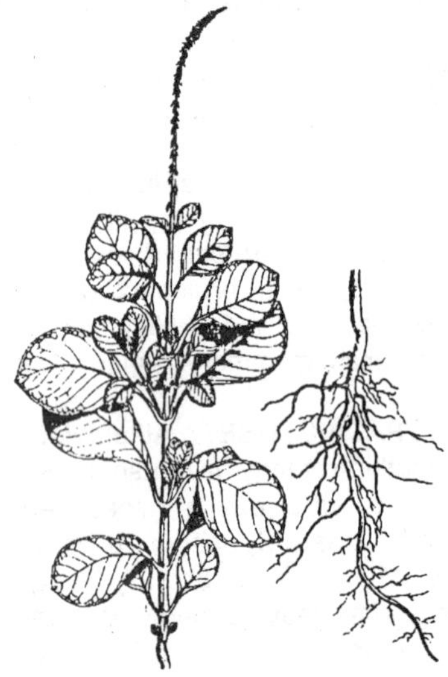

HABITAT : Throughout India up to 3000 ft. Common weed along roadsides, waste land and also as an undergrowth along forest borders during rainy season.

CHEMICAL COMPOSITION : Chemical analysis shows that seeds contain hentriacontane and saponin, oleanolic acid. Alkaloid Achyranthine identified as betaine lowers B.P.; depresses heart; causes vasodilation; increased respiration; spasmodic on rectal muscle; has a diuretic and purgative action. Plant ash contains large amount of potash. Root contains ecdysterone, ecdtsone and inokosterone. Seeds contain saponin A & B and amino acids.

Parts used : Herb, leaves, seeds and roots.

Ayurvedic properties :
Rasa - Kaṭu, Tikta. Vīrya - Uṣṇa.
Guna - Laghu, Rūkṣa, Tikṣṇa. Vipāka - Katu.

अपामार्गः सरस्तीष्णो दीपनस्तिक्तकः कटुः।
पाचनो रोचंश्छर्दिकफ मेदोऽनिलापहा।।
निहन्ति हृद्रुजाध्मानार्शः कण्डूशूलोदरापर्चा। (भा. प्र.)

Apāmārgah sarastīkṣno dīpanastiktakah kaṭuḥ,
Pācano rocanścchardikaphamedo anilāpaḥ,
Nihanti hṛdrujādhmānārśahkaṇḍuśūlodarāpacī.
(Bhāvaprakāṣa)

अपामार्गस्तु तिक्तोष्णः कटुश्च कफनाशनः।
अर्शकण्डूदराध्मानो रक्तहृद् ग्राहि वान्तिकृत्।।
(ध.नि.रा.नि.)

Apāmārgastu tiktoṣṇaḥ kaṭuśca kaphanāsanaḥ,
Arśakaṇḍudarāmhgni raktahṛd grāhivāntikrt.
(Rājanighaṇtu)

Actions/Uses : Kaphaghna, Vātaghna, Pittakar, Śothahara, Vedanāsthāpan.

Therapeutics :

Apāmārga is bitter, acrid, pungent, pectoral, purgative, cardiotonic, astringent, carminative, diuretic, alleviative of kapha and vaata and is useful as an errhine. It cures hiccough, ascites, enlargement of cervical glands, piles, and is used in dropsy and urinary diseases, boils, skin eruptions, pruritus, anorexia, colic and snakebite. Various parts of the herb are used in atrophy, cachexy, rheumatism, scabies, labor complaints and blindness of cattle. Decoction of herb is useful in pneumonia, cough, kidney stone. In large doses acts as ecbolic. Root is given in stomach pain; and paste applied to remove opacity of cornea and to wounds as haemostatic. Used in abdominal tumor. In Tahiti, roots are used in mouth sores, toothache and syphilitic sores; powder in leprosy; decoction and paste as antifertility agent; extract in menstrual disorders. Infusion of roots is astringent. Leaves are used as a cure for gonorrhoea and excessive perspiration. Leaf extract called "Achyrol" is used for leprosy and juice in eczema and leprosy. In Philippines it is used as a diuretic, and sap for treating cataract; heated sap is valued in tetanus. In Tonga used in wounds; juice as an antidote to insect bite; used in ear and eye trouble. Seeds used as emetic, in hydrophobia. Seed paste applied to insect-bite. Leaf, seed and twigs used in bronchitis affections. Powdered seeds soaked in butter-milk given in billiousness. Flowers ground and mixed with curds and sugar given in menorrhagia. Flower tops employed in treatment of rabies. Pollen is suspected to cause allergy. Herb is used for Zn processing in Ayurvedic preparations. It is one of the constituent of indigenous drug "Cystone" useful in urinary tract infection. Tablets made from herb paste with fruits of Piper longum is used to cure effects of bite of mad dog. "Surasaadi tailum", Aaviltolaadi bhasmam, Suvarnnamuktaadi gulika, Jaatyaadi tailam, Ardhavilvam kashaayam are important formulations using the drug.

Aerva lanata Juss.

SYNONYMS : *Achyranthus lanata* Linn.; *Aerva floribunda* Wight., *Celosia lanata* Linn.

INDIAN : Kupuri madhuri, Astambeta.

SANSKṚTA : Gorakṣagañjā, Astamabayda, Ādānapākī, Śatakābhedī, Aśhmabheda, Bhadrā, Pāṣāṇabheda.

HABIT : An erect or prostrate, hoary-tomentose perennial herb.

HABITAT : Common weed along forest borders; throughout the plains of India ascending to 3000 ft. in the hills. Every where in Konkan in rainy season.

CHEMICAL COMPOSITION : α-Amyrin, campesterol, β-sitosterol, its palmitate, chrysin and four flavonoid glucosides from heartwood.

Parts used : Roots, flowers and leaves.

Ayurvedic properties :
Rasa - Tikta, Kaṣāya. Vīrya - Ūṣṇa.
Guṇa - Laghu, Tīkṣṇa. Vipāka - Kaṭu.

गोरक्षगञ्जा तुवरा सतिक्ता
लघ्वी च तीक्ष्णा परमुष्णवीर्या।
कफपित्त्हमूत्रविरेचनीया
प्रभावतोऽप्यश्मरिनाशनी स्यात्।। (द्र. गु. वि.)

Gōrakśagañjā tuvarā satiktā
Laghvī ca tīkṣṇā paramuṣṇavīryā,
Kaphapittahṛṃn mūtravirecanīyā,
Prabhāvātoapyāsmarināṣanī syāt.
(Dravyaguṇa vijnāna)

Actions/Uses : Snehana, Mūtrajanana, Vedanāsthāpan, Ashmarighna, Kāsahara.

Therapeutics :

Gorakṣaganja is cooling, diuretic, lithontriptic and is used in haemetemesis, diabetes and lithiasis. Root is diuretic and demulcent. Inflammation of kidney is ameliorated by a decoction of roots and urination is increased hence it is used for kidney stone. It is used as painkiller in the treatment of headache and for cough. Dry leaves and flowers are smoked during asthma. It is used as an anthelmintic, in haematemesis and diabetes and for lithiasis; also used against swellings and cutaneous affections; in white urine, diarrhoea, cholera, dysentery and in snake-bite. Flowers useful in kidney stone and in gonorrhoea. The action is similar to that of Achyranthus aspera.

Amaranthus spinosus Linn.

ENGLISH : Prickly amaranthus.

INDIAN : Kate matha, Tandullajaa.

SANSKṚTA : Taṇḍulīya, Māriṣa, Taṇḍūleraka, Meghanāda, Kāndera, Mandīra.

HABIT : An erect branched spinous, evergreen, annual or perennial herb.

HABITAT : Common and abundant weed in waste lands and in cultivated fields. Grows abundantly in Malabara and West Bengal.

CHEMICAL COMPOSITION : Higher alkanes and their methyl derivatives, higher aliphatic alcohols, acids and esters, amino acids, β-sitosterol, stigmasterol, campesterol, cholesterol, α-spinasterol, α-spinasterol octacosanoate, glycosides of a-spinasterol and oleanolic acid. Chemical analysis of leaves and stems gives hentriacontane and a-spinasterol. Root a-spinosterol and mixture of saponins composed of oleanolic acid, D-glucose and D-glucuronic acid; also a-spinosterol octacosanoate and a saponin.

Parts used : Roots and all parts.

Ayurvedic properties :
Rasa - Madhura. Vīrya - Śīta.
Guṇa - Gurū, Laghu, Rūkṣa. Vipāka - Madhura.

तण्डुलीयो लघुः शीतो रक्षः पित्तकफास्त्रजित्।
सृष्टमूत्रमलो रुच्यो दीपनो विषहारकः।।
(भा. प्र.)

Taṇḍulīyo laghuḥ śīto rūkṣaḥ pittakaphāsrajit,
Sṛṣṭamūtramalo rucyo dīpano viṣahārakaḥ.
(Bhāvaprakāṣa)

मारिषो मधुरः शीतो विष्टम्भी पित्तनुद्गुरुः।
वातश्लेष्मकरो रक्तपित्तनुद् विषमाग्निजित्।।
(राजनिघण्टु)

Māriṣo madhuraḥ śīto viṣṭambhī pittanudguruḥ,
Vātaśleṣmakaro raktapittanud viṣamāgnijit.
(Rājanighantu)

Actions/Uses : Pittakaphaśāmaka, Śītala, Dīpana, Rucya, Mūtrajanana, Snehana, Stanyajanana, Vedanāsthāpana.

Therapeutics :
Taṇḍulīya is light, cooling, acrid, carminative,

diuretic and laxative and used for strengthening uterus and ameliorating its pains. Herb is febrifuge and suitable food for patients attacked by fever. It is used for snake-bite. It is also used against burning sensation, dyspepsia, haemophilic conditions. Decoction of roots is used in gonorrhoea. It causes profuse urination and pains are ameliorated and pus is stopped. It is suitable for first two stages of gonorrhoea. Use of A. aspera and Glycyrrhiza glabra along with this decoction is recommended. Decoction of roots is widely used for menorrhoea and metrorrhagia. It reduces uterine pains and stops bleeding. Dry fruit of Phyllanthus embelica, bark of Saraca indica and Berberis aristata are usually employed simultaneously. *A. spinosus* is used for leucorrhoea. It is more effective when given with Balsamodendron myrrha. It is useful to avoid habitual tendency for abortion. During pregnancy, when given for three or four days at the period of regular menses abortion is avoided. It is used to increase flow of milk. Extract of plant is spasmogenic. In Ghana, decoction of herb is used as mouth-wash for tooth-ache; poultice mixed with native soap used as whitlow remedy. Leaf infusion is diuretic. Leaves in the form of poultice are applied to abscess, bubos, wounds and burns to promote suppuration and discharge of pus. Roots and leaves boiled and given to children as laxative, and applied as emollient, poultice to abscesses, boils and burns. Leaves used as enema, in stomach troubles, in curing piles and leprosy. Used in vomiting. Pollen extract useful in allergic asthma or allergic rhinitis.

Celosia argentea Linn. var. cristata Kuntze.

SYNONYMS : *Celosia cristata* Linn.

HABIT : An erect annual herb about three feet in height.

HABITAT : Common weed in cultivated fields, hedges, river banks, along slopes of forest areas, road sides and among grasses in open sunny areas.

ENGLISH : Cockís comb.

INDIAN : Kurdu, Kombadaa, Laal Murgaa, Huldi-murga.

SANSKṚTA : Mūrvā, Śītavāra, Vitunnaka, Mayūr-śikhā.

CHEMICAL COMPOSITION : It contains hyaluronic acid, celosianin betanin and isocelosianin.

Parts used : Seeds.

Ayurvedic properties :
Rasa - Madhura, Tikta. Virya - Śita.
Guṇa - Gurū, Sara. Vipāka - Madhura.

मूर्वा सरा गुरुः स्वादुस्तिक्ता पित्तास्त्रमेहनुत्।
त्रिदोष तृष्णाहृरोगकण्डू कुष्ठज्वरापहा।। (द्रव्यगुण)

Mūrvā sarā guruḥ svādustiktā pittāsramehanut,
Tridoṣa tṛṣṇā hṛdroga kandu kuṣṭha jvārapaha.

(Dravyaguṇa)

Actions/Uses : Śītala, Snehana, Paustika, Mūtrakruchha, Sangrahaṇī.

Therapeutics :

Mūrvā or Śitavāra is useful in dyscrasia and polyuria; pacifies deranged tridosa, allays thirst and is beneficial in heart diseases, pruritus, dermatois and fever. Extract of plant is antiprotozoal and spasmolytic. Seeds are diuretic, aphrodisic, useful in seminal diseases, diarrhoea, blood diseases and mouth sores, for clearing the vision, and for diseases of the eye. Tender leaves are used as vegetable and is known in Bengal as useful for inducing sleep. Seeds are used as a nourishing food. Powdered seed with sugar and milk used as aphrodisic tonic. Decoction of seeds useful for diarrhoea and seed powder with sugar for retention of urine. Flowers are astringent and are used in diarrhoea. Seeds are demulcent and useful in couhg, painful micturition and dysentery.

Family : VI - Amaryllidaceae.

Crinum asiaticum Linn.

SYNONYMS : *Crinum toxicarium* Roxb.; *Crinum bracteatum* Willd.

ENGLISH : Poison bulb.

INDIAN : Naagadawanna, Kanwal, Pindar, Naginkapatta.

SANSKṚTA : Nāgadananī, Sudarśana, Somavalli, Cakrāngī, Macchaka, Madhuparṇikā, Vatsadani, Kandali, Saptācakrakam, Viṣakandara, Viṣamandala, Viṣṇukanda, Kandari.

HABIT : A sub-erect tuberous herb.

HABITAT : Grows in water-logged places especially in Konkan and Bengal. Occurs throughout tropical India, wild or cultivated in gardens.

CHEMICAL COMPOSITION : Fruit yields alkaloid hamayne; bark, sterols and triterpines. Root contains alkaloids narcissine (lycorine) and crinamine.

Parts used : Bulb or tuber, root and leaves.

Ayurvedic properties :
Rasa - Madhura, Tikta. Vīrya - Ūṣṇa.
Guṇa - Laghu, Rukṣa. Vipāka - Madhura.

नागदमनी स्वादुतिक्तोष्णा कफशोफास्त्रवातजित्।
(स्व.)

Nāgadamanī svādutiktoṣṇā kaphaśophāsravātajit. (Swa.)

प्रोक्ता नागदमन्युष्णा तिक्ता लघ्वी रुचिप्रदा।
कोष्ठशुद्धिकरी तीष्णा कटुका योनिदोषजित्।।
लूतां सर्पविषं वातंकफं वान्ति कृमीन्व्रणम्।
मूत्रकृच्छ्रं चोदरंच जालगर्दभकं तथा।।
त्रिदोषं च प्रमेहं च कासं कण्ठं रुजं तथा।।
शूलं गुल्मं रक्तदोषं ज्वरं सर्पविषाणिच।।
आधानं ग्रहपीडां च नाशयेदिती कीर्तिता। (निघण्टु रत्नाकर)

*Proktā nāgadamanyuṣṇā tikta laghvī rucipradā,
Koṣṭhaśudhikarī tīkṣṇā kaṭukā yonidoṣajit.*

Lutām sarpaviṣam vātam kapham vāntim kṛminvraṇam,
Mūtrakṛccham codaram ca jalagardhabhakam tathā.
Tridosanca pramehanca kasam kantham rujam tatha,
Sulam gulmam raktadosam jvaram sarpavisanica.
Adhmanam grahapidanca nasayediti kirtita.
(Nighaṇṭuratnakāra)

Therapeutics :

Nāgadamanī is useful in early stages of inflammation of bronchii. It can be safely administered to children for vomiting of cough. Tuber is pungent, bitterr, vulnerary, laxative, aphrodisiac; useful in bronchitis and diseases of chest and lungs, gonorrhoea, defective vision, diseases of spleen, urinary concretions, lumbago, and anuria. In full doses it is emetic and in small doses nauseant and diaphoretic. It is a dependable emetic which is moderate in action. Vomiting is immigeate yet without fatigue after it or without any side effects such as tenesmus or diarrhoea. Bulb is bitter, used as tonic, laxative, expectorant. Useful in biliousness, strangury and other urinary troubles. Fresh root is emetic, nauseant, diaphoretic. Leaves are used as external application on swellings after applying castor oil to them. Swelling is quickly removed, arrests formation of pus. Useful in whitlow and other skin diseases. Leaves as expectorant, applied to skin diseases and to reduce inflammation. Juice of leaves is used for earache. Seeds are used as tonic, purgative, diuretic and emmenagogue.

Crinum defixum Ker-Gawl.

SYNONYMS : *Amaryllis zeylanicum* Linn. *Crinum asiaticum* Roxb.; *Crinum encifolium* Roxb.; *Crinum latifolium* Linn., *Crinum roxburghii* Dalz.

HABIT : A perennial herb with ovoid bulbs.

HABITAT : Common in water-logged stony places.

ENGLISH : Poison bulb.

INDIAN : Naagadawanna, No.2; Sukhadarshana, Sudarshana.

SANSKṚTA : Dadhyālī, Sudarṣana,

CHEMICAL COMPOSITION : Bulb contains a toxic principle. Leaves devoid of toxic principle. Lycorine from bulbs and seed.

Parts used : Bulbs and leaves.

Ayurvedic properties :
Rasa - Madhura, Tikta. Vīrya - Ūṣṇa.
Guṇa - Rūkṣa, Tīkṣṇa. Vipāka - Madhura.

दध्याली स्वादुतिक्तोष्णा कफशोथास्रवातजित्। (कै. नि.)

Dadhyālī svādutiktoṣṇā kaphaśothāsravātajit. (Kaiyadeva)

Actions/Uses : Kaphavātaśāmaka, Śothahara, Raktavikārghna.

Therapeutics :

Leaves and root of Sudarsana are a good

substitue for ipecacuanha. They act without griping, purging or any other distressing symptoms. Bulb is nauseant, emollient, emetic, diaphoretic; used for the treatment of burns, whitlow, and carbuncle. Medicinal properties and uses are the same as *C. asiaticum* Linn. Anti-T.B. activity of lycorine in this variety is established.

Family : VII - Anacardiaceae.

Anacardium occidentale Linn.

SYNONYMS : *Cassuvium pomiferum.*

ENGLISH : Cashew nut, East Indian Almond.

INDIAN : Kaaju.

SANSKṚTA : Vātāda, Kājūtaka, Agnikṛta, Aruṣkara, Snigdhapitaphala, Gucchapuṣpa, Kajuta, Parvati, Pṛthakbīja, Rujakara, Śophahara, Veṇāmra, Vṛukkabīja, Vṛkaphala, Vṛktapatra.

HABIT : A small, spreading evergreen tree below 5 m.; fruit kidney shaped.

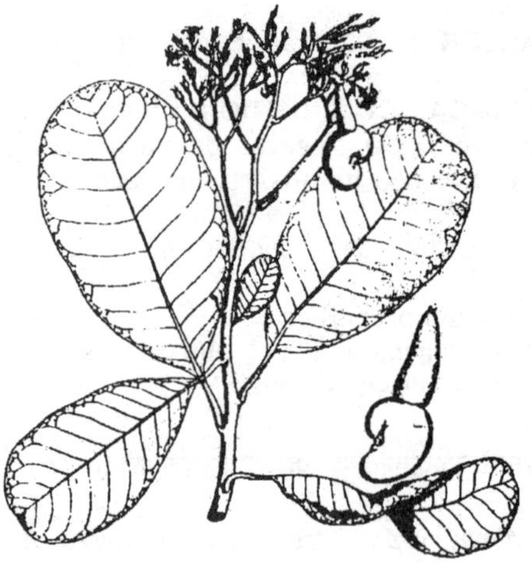

HABITAT : Native to tropical America and naturalized in the warmer parts of India, and cultivated in the coastal districts of India, especially in the West Coast; Ratnagiri, Karnatak, Goa, Kerala and Tamilnaadu.

CHEMICAL COMPOSITION : Cashewnut shell yields gum, shell liquid and shell oil. Liquid-free nutshells give syringic and gallic acids and (+)-gallocatechin. Defatted shell yields narimgenin and prunin-6″-O-p-coumarate. Fruit contains anacardic acid, a phenol cardol. Essential amino acids present in cashewnut protein are arginine, 10.4; histidine, 2.08; lysine, 4.64; phenylalanine, 4.32; methionine, 1.44; threonine, 3.2; tryptophan, 1.76; leucine, 8.16; isoleucine, 5.12; and valine, 5.76 g/16 g N. Alinine, aspartic acid, cystine, glutamic acid, glycine, serine and tyrosine are also present. Unsaponifiable matter of lipids from cashewnuts contain squalene, cycloartenol, 24-methylene cycloartanol, β-amyrin, β-sitosterol and campesterol. Husk contains D-catechin, gallic, caffeic and quinic acids. Ethyl gallate, hyperoside, methyl gallate, leucocyanidine and leucodelphinidin.

Parts used : Fruits, seeds, bark and oil.

Ayurvedic properties :
Rasa - Madhura. Vīrya - Ūṣṇa.
Guṇa - Laghu. Vipāka - Madhura.

काजूतकस्तु तुवरो मधुरोष्णो लघुः स्मृतः।
धातुवृद्धिकरो वातकफगुल्मोदरज्वरान्।।
कृमिव्रणाग्निमान्धानि कुष्ठा च श्वेतकुष्ठकम्।
संग्रहण्यर्शानाहान्नाशयेदिति कीर्तितः।।
(नि. र.)

Kājutakastu tuvaro madhuroṣṇo laghuḥ smṛtaḥ,
Dhātuvṛddhikaro vātakaphagulmodarajvarān,
Kṛmivraṇāgnimāndyāni kuṣṭha ca śvetakuṣṭhakam,
Sangrahaṇyārśānāhannāsayediti kīrtitaḥ.
(Nighaṇṭuratnākara)

वाताद उष्णः सुस्निग्धो वातघ्नः शुक्रकृत् गुरुः।
वातादमज्जा मधुरो वृष्यः पित्तानिलापहः।
स्निग्धोष्णः कफकृन्नेष्टो रक्तपित्तविकारिणाम्।। (भावप्रकाश)

Vātāda uṣṇaḥ susnigdho vātaghnaḥ śukrakṛt guruḥ,
Vātādamajjā madhuro vṛṣyaḥ pittānilāpahaḥ,
Snigdhoṣṇaḥ kaphakṛnneṣto raktapittāvikarıṇām.
(Bhāvaprakāṣā)

Actions/Uses : Vājikar, Mūtrala, Vātagna, Vṛṣya.

Therapeutics :

Roots have purgative properties while bark is alterative. Bark and leaves are useful in ondontalgia andulitis. Extract of bark is anti-hypertensive; when given orally lowers blood sugar. It has proved efficient in certain malarial fevers not yielding to treatment by Quinine. Bark exudes a pale yellow gum. It is used as a substitute for gum arabic and has insecticidal properties for which it is used in pharmaceuticals. Tar or acrid oil from bark is alterative and astringent, irritant and vesicant and is applied in leprosy, corns, ulcers and as counter irritant and rubefacient. Fruit is acrid, sweet, hot, antidiarrhoeal, aphrodisiac and anthelmintic; cures vaata and kapha and is used as a counter irritant. Spirit distilled from the fruit is locally rubefacient. Kernel is nutritious, demulcent and emollient. Cashewnut shell extract is molluscicide for snails and tonic for fish. Shell yields gum, a liquid and an oil. The liquid is vesicant, a source of naturally occuring phenols from which several pesticides, drugs are prepared. Shell oil, a dark brown viscous oil is mosquito larvicide. Oil is mild purgative and used in treatment of hook-worm; also for cracks on soles of feet, warts, corns, leporous sores. The fresh juice of shell is acrid and corrosive. Resinous juice of cashewnut is useful in mental derangement, sexual debility, nervous prostration following seminal emission, morning sickness in pregnancy, palpitation of heart and rheumatism pericarditis and loss of memory as a sequel to small pox. Kernel good food for weak patients suffering from incessant and chronic vomiting. The nuts are roasted to obtain the kernels, which are edible and of agreeable flavour. The fruit or cashew apple is used to make a strongly flavoured liquor called "fenny", valued as diuretic, having healthy effect on kidneys and advanced cases of cholera. It contains Vit. C, 261.5 mg. and Vit. E, 210 mg.

HOMEOPATHIC USES : It acts powerfully on the skin causing erysipelas, blisters and swelling. The juice is used locally as an application to corns, warts, hard excrescences, ringworms and obstinate ulcers. It causes weakness of memory and mind. The main symptoms are general paralytic state; tongue painfully swollen; vesicular eruption, especially on face; itching almost intolerable; umbilicated vesicles as in small pox; the erysipelas spreads from left to right. It cures erysipelas spreading from right to left. *Rhus t.* cures cases spreading left to right.

Buchanania lanzan Spreng.

SYNONYMS : *Buchanania latifolia* Roxb.

ENGLISH : Almondi tree.

INDIAN : Chaarholli, Chironji, Piyal, Achar.

SANSKṚTA : Cāra, Priyāla, Piyala, Cāroḷī, Ciraujī, Cirolī, Karakā, Bahalvalkala, Capavata, Caramamidi, Kharaskandha, Lalana, Rajadana, Snehabija, Svaraskanda, Tapasapriya.

HABIT : An evergreen tree with a straight, cylindrical trunk.

HABITAT : Common throughout India in dry or moist deciduous forests or in semi evergreen forests.

CHEMICAL COMPOSITION : Albumimpids, mucilage, oil. Amino acids, linoleic, myristic, oleic, palmitic and stearic acids. Vitamins; gallotannins, triterpenoids, saponins and reducing sugars. Leaves contain kaempferol glucoside, quercetin. Bark contain alkaloids and 13.4% tannin.

Parts used : Fruit, leaves, bark, oil from kernel.

Ayurvedic properties :
Rasa - Madhura. Vīrya - Śīta.
Guṇa - Snigdha, Gurū, Sara. Vipāka - Madhura.

चारः पित्त कफास्रघ्नः तत् फलं मधुरं गुरु।
स्निग्धं सरं मरुत्पित्तदाहज्वर तृषापहम्।।
प्रियालमज्ञा मधुरो वृष्यः पित्तानिलापहः।
हृद्योतिदुर्जरः स्निग्धो विष्टम्भी चामवर्द्धनः।।
(भावप्रकाश)

Cārah pittakaphāsraghnaḥ tatphalam madhuramguru,
Snigdham saram marutpittadāha jvaratṛṣāpaham,
Priyālamajjā madhuro vṛṣyaḥ pittānilāpaḥ,
Hṛdyotidurjaraḥ snigdho viṣṭambhī cāmavarddhanaḥ.
(Bhāvaprakāṣa)

Actions/Uses : Vātaghna, Pittaghna, Śothahara.

Therapeutics :

Various plant parts of Priyāla are used in inflammation of gums, in fever, burns, dysuria, cholera, phthisis, bronchitis and asthma. Extract of plant is anticancer. Root is acrid, removes kapha and biliousness. Leaves are valued as tonic and cardiotonic; crushed or powdered are applied to wounds. Fruit is sour, sweet, demulcent, alterative, fattening, laxative, cooling and aphrodiasic; cures vaata, biliousness, fevers, thirst, ulcers and blood diseases. Seeds are heating. Oil from kernel causes kapha; removes vaata and biliousness; is used in skin diseases caused by blood defects; ointment used to cure itch and blemishes from face; also applied to glandular swellings of neck; kernels as brain tonic. Gum is used in intercostal pains. Gum in cow's milk is used for rheumatic pains.

Mangifera indica Linn.

ENGLISH : Mango.

INDIAN : Aambaa, Aamra, Aam, Amb.

SANSKṚT : Āmra, Cūta, Rasāḷa, Sahakāra, Pikavallabha, Kāmaśara, Madhudūta, Amva,Modākhya, Kāmavallabbha, Kāmāṅga, Kīreṣṭa, Mādhavadruma, Bhṛṅgābhīṣṭa, Sīdhurasa, Madhulī, Kokilotsava, Priyāmbū, Vasantadūta, Amlaphala, Manmathālaya, Madhvāvāsa, Sumadana, Pikarāga, Nṛpapriya, Parapuṣṭamahotsava, Manojnya, Vanapuṣpotsava, Atisaurabha, Vasantapādapa, Śareṣṭa, Supatha, Kānta, Madirāsava, Manmatha, Moda, Śāmataila, Caitravṛkṣa, Śukapriya, Phalaśreṣtha, Keśavāyudha, Madhukarapriya.

HABIT : Middle sized evergreen tree.

HABITAT : A common tree all over India cultivated for its delicious fruits.

CHEMICAL COMPOSITION : Leaves contain two flavonoids, two phenolic compounds, glucose, galactose, arbinose, xylose, rhamnose and tannin. Flowers yield 0.04% of an essential oil. Pannicles give ethyl galate. Mangiferin has been isolated from bark. Roots give mangiferin, friedelin and b-sitosterol. Fruit contains vitamins A, C and D. Mangiferrin has cardiotonic and diuretic properties. The extract of leaves, bark and stems and unripe fruit exhibit moderate anti-bacterial activity. In ripe mango anti-fungal microorganisms are found. Seed kernel has an astringent taste and is free from toxic principles and contains amino acids.

Parts used : Fruit, seeds, root, leaves and bark.

Ayurvedic properties :
Rasa - Kaṣāya (unripe), Madhura (ripe).
Vīrya - Śīta.
Guṇa - Laghu, Rūkṣa. Vipāka - Kaṭu.

आम्रः कषायाम्लरसः सुगन्धिः
कंठामयघ्नोग्निकरश्च वालः।
पित्तप्रकोपानिलरक्तदोष -
प्रदः पटुत्वादिरुचि प्रदश्च।।
(राजनिघण्टु)

Āmraḥ kaṣāyāmlarasaḥ sugandhih
Kaṇṭhāmayaghnognikaraśca vālaḥ,
Pittaprakopānilaraktadoṣapradaḥ
Paṭutvādirūcipradaśca.
(Rājanighaṇṭu)

आम्रं बालं कषायाम्लं रुच्यं मारुतपित्तकृत्।
पक्वं तु मधुरं वृष्यं स्निग्धं बल सुखप्रदम्।।
गुरुवात हरं हृद्यं वर्ण्यं शीतमपित्तलम्।।
(भा. नि.)

Āmram bālam kaṣāyāmlam rucyam mārutapittakṛt,
Pakvam tu madhuram vṛṣyam snigdham bala sukhapradam,
Guruvātaharam hṛdyam varnyam śītamapittalam.
(Bhāvaprakāśa)

Actions/Uses : Rucya, Hṛdya, Grāhī, Balaprada, Vātala, Pittakar, Atisārahara, Pramehahara, Vṛṣya, Śukravardhaka, Vātahara, Varṇya.

Therapeutics :

Root and bark of Āamra are acrid, cooling, astringent. Leaves are acrid, astringent, cure vaata, pitta and kapha and are used in scorpion-sting. Decoctions of leaves, bark, gum, flowers, fruits and seeds are used in medicines. Flowers are cooling and astringent, improve taste and appetite; cause vaata; cure leucorrhoea, and are good in dysentery, bronchitis, biliousness, urinary discharge. Dried flowers are given for diarrhoea, chronic dysentery, catarrh of the bladder and gleet. Unripe fruit is acrid, acid, astringent, tasty, stomachic and antiscorbutic, cures vaata, kapha and tridosha and biliousness. Useful in throat troubles, ulcers and vaginal troubles, in opthalmia and eruptions. Ripe fruit is laxative, nourishing and invigourating, diuretic, astringent, diaphoretic, and refrigerant, useful in haemorrhages from intestines, uterus, lungs. Rind of fruit is astringent, stimulant tonic in debility of stomach. Seed is used in asthma and its powder as anthelmintic. Kernel is astringent, anthelmintic and used in diarrhoea, in haemorrhage. Its juice if snuffed can stop bleeding. Bark is astringent, used in uterine haemorrhage and other discharges, diarrhoea, in diphtheria and rheumatism. It has a tonic action on the mucous membrane. Gum is used as a substitute for gum arabic, in dressings for cracked feet and for scabies, also considered anti-syphilitic.

Semecarpus anacardium Linn. f.

SYNONYMS : *Anacardium orientale* Linn.

HABIT : Moderate-sized dioecious, deciduous tree.

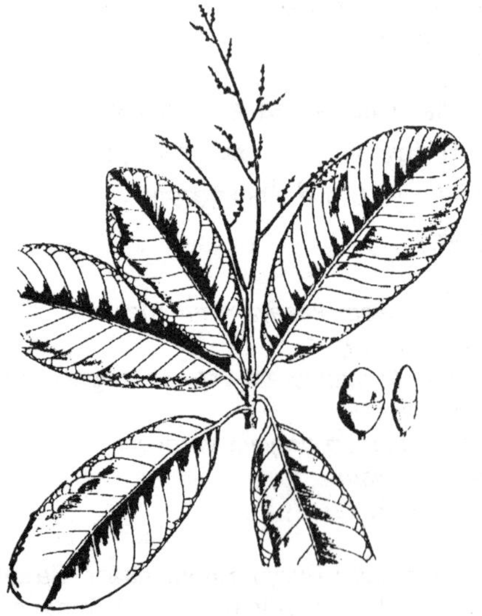

HABITAT : Common in dry or moist deciduous forests.

ENGLISH : Marking nut, Oriental cashee.

INDIAN : Bibbaa, Bhallaataka, Bhela, Bhilawa.

SANSKRT : Bhallātaka, Arūṣkara, Agnika, Śophakṛuta, Agnimukha, Anala, Vātāri, Mahātīkṣṇa, Sphotabījaka, Kṛmighna, Tailabija, Vīrataru, Dhanurvṛkṣa, Vanhi, Kṣatakṛta, Nījapādapa, Arśohita, Bhallatamu, Pṛthakabīja, Sphotahetu.

CHEMICAL COMPOSITION : Anacardic acid, cardol, catechol, anacardol and fixed oil; semicarpol, bhilawanol.

Parts used : Fruit, oil, godambi.

Ayurvedic properties :
Rasa - Madhura, Kaṣāya. Vīrya - Ūṣṇa.
Guṇa - Laghu, Snigdha, Tīkṣṇa.
Vipāka - Madhura.

भल्लातकः कटुस्तिक्तः कषायोष्णः क्रिमीन् जयेत्।
कफवातोदरानाहमेहदुर्नामनाशनः।।
(राजनिघण्टु)

Bhallātakaḥ kaṭustiktaḥ kaṣāyoṣṇaḥ krimīn jayet,
Kaphavātodarānāha mehadurnāma nāsanaḥ.
(Rājanighaṇṭu)

भल्लातकः कषायोष्णःशुक्रलो मधुरो लघुः।
वातश्लेष्मोदरानाह कुष्ठार्शो ग्रहणीगदान्।।

हन्ति गुल्मज्वरश्वित्रवह्निमांद्य कृमिव्रणान्।।
(भा. प्र.)
कफजो न स रोगोऽस्ति न विबंधोऽस्ति कश्चनः।
यं न भल्लातकं हन्यात् शीघ्रमग्निबलप्रदम्।। (च.चि.१)

*Bhallātakaḥ kaṣāyoṣṇaḥ śukralo madhuro laghuḥ.
Vātaślesmodarānāha kuṣṭhārso grahanīgadān,
Hanti gulmajvaraśvitravahnimāndya krmivranān.
(Bhāvaprakāśa.)*

Actions/Uses : Vātaghna, Kaphaghna, Pittakar.

Therapeutics :

Bibba or Bhallātaka is a pre-eminent Ayurvedic medicine. Its uses are many and varied. Fruit is acrid, hot, sweetish; aphrodisiac, anthelmintic; stays looseness of bowels; removes vaata, kapha and is considered beneficial in ascites, tumours and warts, acute rheumatism, asthma, neuralgia, piles, dysentery, fevers, loss of appetite, urinary discharges, epilepsy and psoriasis. It heals ulcers. Juice of pericarp and the oil are powerful counterirritant and vesicant. Oil is powerful antiseptic and cholagogue. Nut is a gastro-intestinal irritant when taken by mouth. Ripe fruits are stimulant, digestive, nervine and escharotic. Nut bruised and applied to Os Uteri to procure abortion; also given as a vermifuge. Oil from the nuts is vesicant and is used externally in rheumatism and leprous nodules. Gum from the nuts is used in scrofulous, venareal and leprous affections and nervous debility. Kernel is nutritive, appetiser, digestive and carminative. It is a good cardiac tonic and a general respiratory stimulant.

HOMEOPATHIC USES : Anacardium exerts a depressing influence on the system, not only as to the mind but as to the body as 'well. It produces weakness of memory. The emotional mind is very much disturbed; propensity to swear, penchant for profanity; has many mental aberrations; is clairaudient hungry all the time, better by eating but worst after it.

Spondias pinnata Kurz.

SYNONYMS : *Spondias mangifera* Willd.; *Mangifera pinnata* Linn.; *Spondias acuminata* Roxb, *non Gamble.*, *Spondias macrophylla* Wall. *ex Hook. f.*

HABIT : A middle sized deciduous glabrous tree below 15 m.

HABITAT : Common tree all over, in wild as well as in cultivation.

ENGLISH : Bile tree, Wild mango, Indian hog-plum.

INDIAN : Aambaadaa, Dhoraaambaa, Aamraataka, Jungli aam.

SANSKRT : Āmrataka, Kapipriya, Pittavṛkṣa, Markātamra, Pitanaka, Advagabhoga, Madhuramlaka.

CHEMICAL COMPOSITION : β-Amyrin and oeanolic acid from fruits. Glycine, cystine, serine, alanine and leucine in fruits. Lignoceric acid, β-sitosterol and its glucoside in aerial parts.

Parts used : Dried fruits. Leaves and Bark.

Ayurvedic properties :
Rasa - Madhura, Amla, Kaṣāya.
Vīrya - Śīta, Ūṣṇa.

Guṇa - Snigdha. Vipāka - Madhura.

आम्रातकं कषायाम्लमामहृत् कण्ठहर्षणम्।
पक्वम् तु मधुराम्लादयम् स्निग्धं पित्तकफापहम्।।
(राजनिघण्टु)

Āmrātakam kaṣāyāmlamāmahṛt kaṇṭhaharṣaṇam,
Pakvam tu madhurāmlādayam snigdham pittakaphopaham.
(Rājanighaṇṭu)

Actions/Uses : Grāhī, Raktapittapraśamana, Kaphaśāmaka.

Therapeutics :

Āmrataka cures vaata and promotes appetite. Root is useful in regulating menstruation. Leaves are tasty, appetising, astringent. Juice of leaves is used in earache. Leaves and bark are aromatic, astringent and refrigerent; useful in dysentery and diarrhoea and given to prevent vomiting. Paste of bark is rubbed on in both articular and muscular rheumatism. Unripe fruit is acrid, refrigerant, tonic, aphrodisiac, and astringent, cures vaata, biliousness, ulcers, burning sensations, phthisis and blood complaints. Fruit is antiscorbutic and the pulp astringent, used in bilious dyspepsia. Decoction of wood is used in gonorrhoea and leucorrhoea. Gum exuded from the tree is demulcent.

Family : VIII - Annonaceae.

Annona reticulata Linn.

ENGLISH : Bullockís heart, Custard apple.

INDIAN : Raamaphalla.

SANSKṚTA : Rāmaphala, Rāmaphalāṣa, Kṛisṇabījam, Lavali, Lavanī, Mṛduphalam.

HABIT : A small deciduous or semi-deciduous tree. Fruits sub-globose, yellowish-red when ripe, edible with peculiar aroma.

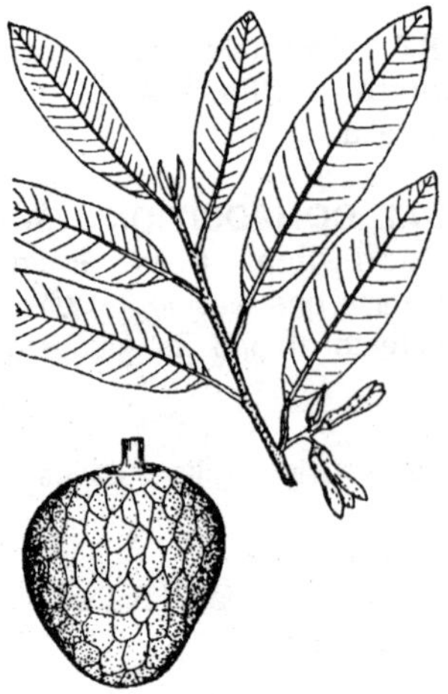

HABITAT : Native to tropical America, particularly the West Indies. Naturalized and common cultivated in India; occasionally found wild.

CHEMICAL COMPOSITION : Root bark contains an alkaloid liriodenine and oxoushinsunine, anonine, norushinsunine, michelalbine and a new alkaloid reticulin containing 2 phenolic hydroxyl groups. Anonaine showed hypotensive activity in mice and rabbits. The edible portion contains per 100 g protein, 1.4; fat, 0.2; minerals 0.7 g.; calcium, 5.0; phosphorus, 10.0; iron,0.6; riboflvin, 0.07; vitamin C, 5.0; and niacin, 0.6 mg; carotene, 67 ug.

Parts used : Bark, fruit, seeds and leaves.

Ayurvedic properties :
Rasa - Madhura. Vīrya - Śīta.
Guṇa - Snigdha. Vipāka - Madhura.

पक्वं रामफलं स्वादु मधुरं कफवातलम्।
अम्लं रुचिकरं दाहक्षुत्पित्त श्रमनुत्परम्।।
(निघण्टु रत्नाकर)

Pakvam rāmaphalam svādu madhuram kaphavātalam,
Amlam rucikaram dāhakṣutpittañjamanutparm.
(Nighaṇṭuratnākara)

Actions/Uses : Kaphakara, Vātavardhaka, Virecana, Grāhī, Rucya.

Therapeutics :

Rāmaphala root is drastic purgative. Seed, leaf, stem and root are incecticidal. Leaves are insecticide, anthelmintic and externally useful as suppurant and are used against inflammatory tumors. Extract of leaves and stems is inotropic, positive chronotropic and spasmolytic activities. Bark is a powerful astringent and used as a tonic, vermifuge and antidysenteric. Fruit is anthelmintic and antidysenteric. Unripe and dried fruits are used in diarrhoea and bloody dysentery. Seed is astringent and vermifugal. Seed kernel is highly poisonous. Fruit pulp is sweet and edible but its paste is used to kill lice on cattle.

Annona squamosa Linn.

ENGLISH : Custard apple, Sugar apple, Sweet sop.

INDIAN : Sitaaphalla, Sharifah.

SANSKṚTA : Sitāphala, Gandhagātra, Kṛsṇabījam, Suda, Śhubha, Paruṣaphalam, Guṭiā, Ganḍa.

HABIT : A large, evergreen, stagglin shrub or small tree. Fruit globose, yellowish-green when ripe, edible, with peculiar aroma. Seeds brownish black.

HABITAT : A common cultivated tree in gardens. Sometimes found wild as an escape from cultivation.

CHEMICAL COMPOSITION : Roots contain alkaloids anonine, michelalbine, oxoushinsuine (liriodenine), L(+)-reticuline, analobine; also diazepine, sqamolone. Root and bark alkaloids corydine, isocorydine, anonaine, glaucine; β-sitosterol; camphor, borneol and a new monoterpinoid which is laxative. Tender leaves and stem contain anonaine, soemerine, norcorydine, corydine, norlaureline, isocorydine, norisocorydine and glaucine. Edible portion contain per 100 g, protein, 1.6; fat, 0.4; minerals, 0.9 g; calcium, 17.0; phosphorus, 47.0; iron, 1.5; thiamine, 0.07; riboflavin, 0.17; niacin, 1.3; and vitamin C, 37 mg. Following aminoacids alanine, β-alanine, cystine, glutamine, serine, y-aminobutyric acid, pipecolic acid and traces of lysine, histidine, arginine and asparagine.

Parts used : Fruit, leaves, roots and seeds.

Ayurvedic properties :
Rasa - Madhura. Vīrya - Śīta.
Guṇa - Laghu, Snigdha. Vipāka - Madhura.

सीताफलं तु मधुरं शीतं हृद्यं बलप्रदम्।
वातलं कफकृत् स्वादु पुष्टिकृत् पित्तनाशनम्।। (नि. र.)

Sītaphalam tu madhuram sitam hṛdyam balapradam,
Vātalam kaphakṛt svādu puṣṭikṛt pittanāsanam.
(Nighaṇṭuratnākara)

Actions/Uses : Vātaghna, Pittaghna, Vātakara, Kaphakara, Jvaraghna, Śothahara.

Therapeutics :

Sitaphal plant is cardiotonic and invigorating and

is used in cold, rheumatism, cancre, syphilis, carbuncle and puerperal fever. Root is diuretic and a drastic purgative, and is given in acute dysentery. Leaf is a stimulant, anti-spasmodic, sudorific, anthelmintic and insecticidal. Leaf tea is used to alleviate fever. Leaf extract is insecticide. Aquious and alcoholic extracts are spasmogenic and spasmolytic and oxytocic; also cardiorespiratory and anticancer. Poultice of leaves with salt is used as a cataplasm over boils and ulcers to induce suppuration and to relieve pains and swellings. Leaves and fruits are used against tumors. Ripe fruit is used for dysentery; made into paste with betel leaves is applied to tumour to hasten suppuration. Unripe fruit is given in diarrhoea, dysentery and atonic dyspepsia. Seeds are irritant to conjuctiva and os uterus and abortifacient. Powdered seeds along with leaves of Plumbago zeylanica are used for abortion. Seeds, fruits and leaves are insecticide, fish poison, and used to remove lice in head. Bark is powerful astringent and tonic. Bark is used in diarrhoea.

Family : IX - Apiaceae.

Anethum sowa Roxb. ex Flem.

SYNONYMS : *Anethum graveolens* Linn. *var. sowa Roxb.*, *Anethum sowa* Kurz., *Peucedanum graveolens* Linn., *Peucedanum sowa* Roxb. *Anethum graveolens DC.*; *Peucedanum graveolens* Benth.

HABIT : An annual, erect, glabrous herb.

HABITAT : Cultivated all over India and used as vegetable.

ENGLISH : Indian dill, Sowa.

INDIAN : Shepoo, Baallanta shepa.

SANSKRTA : Śatapuṣpā, Miṣreyā, Śatāhvā, Śatapatrikā, Suvākapuṣpī, Miśi, Chatrā, Śiphā, Kāravī,Yonidoṣaghnī, Māghadhī, Vajrapuṣpā, Ghoṣā, Pītikā, Puṣpāhvā, Ahichatra, Sanghātapatrikā, Gandhādhikā, Śataprasūnā, Bahulā, Vanapuṣpā, Bhūripuṣpā, Madhurasā, Paṇa, Avākpuṣpī, Cetikā, Sūkṣmapatrikā, Madhurā, Sanhatapatri, Kṣetrā, Sitachatrā.

CHEMICAL COMPOSITION : The aerial parts are aromatic but most of the volatile oil is contained in the fruit (the so called seed). Seeds yield 3-3.5% essential oil and contains myristicin and apiol. Tripetroselinin, petroselinicdiolein and dipetroselinicolein from seed oil. Vicenin from fruits. A new xanthone glycoside- dillanoside- from fruit. Carvone, Dihydro carvone, D-limonene, Phellandrene, Dillapiol.

Parts used : Leaves, fruits, seeds and seed oil.

Ayurvedic properties :
Rasa - Kaṭu, Tikta. Vīrya - Ūṣṇa.
Guṇa - Laghu, Tīkṣṇa, Rūkṣa. Vipāka - Kaṭu.

मिश्रेया :- विशेषाद् योनिशूलनुत्।
अग्निमांघहरी हृद्या बद्धविट्कृमिशुक्रहृत्।।
रुक्षोष्णा पाचनी कासकृमिश्लेष्मानिलान्हरेत्। (भा. प्र.)

Misreya :-
Viśeṣād yoniśūlanut, agnimāndyaharī hṛdyā,

Badhaviṭ kṛmiśukrahṛt.
Rūkṣoṣṇa pācanī kāsakṛmiślesmānilān haret.
(Bhāvaprakāṣa)

शतपुष्पा लघुस्तीक्ष्णा पित्तकृद्दपिनी कटुः।
उष्णा ज्वरानिलश्लेष्मव्रणशूलाक्षि दोषहृत्।।
(भा. प्र.)

Ṡatapuṣpā lghustīkṣṇā pittakṛt dīpani kaṭuḥ,
Uṣṇa jvarānilaśleṣmavraṇaśūlākṣi dośahṛt.
(Bhāvaprakāṣa)

Actions/Uses : Pittakara, Vedanāsthapan, Dīpana, Ṡothahara, Ārtavagamanakara, Stanya, Jvaraghna, Kṛmighna, Mūtrala, Pācana, Vraṇapācana.

Therapeutics :

Dried ripe fruits are acrid, bitter, digestive, carminative, stimulant, stomachic, anthelmintic, diuretic, emmenagogue and galactagogue. In children's complaints, they are used in flatulence, disordered digestion, colic and intestinal worms. It is an excellent remedy mostly given in the form of Dillwater or with limewater. An infusion is given as cordial drink to women after confinement. Fruits are also useful for ammenorrhoea, dysmenorrhoea, cough, asthma, bronchitis, fever, cardiac debility, ulcers, spermatorrhoea, gleet and syphilis. Extract of fruit is hypoglycaemic and spasmolytic. Oil of seeds is antifungal and is administered as drops on sugar. It is a stimulant for uterus.

Apium graveolens Linn.

SYNONYMS : *Apium graveolens var. dulce (Mill.) Pers., Apium graveolens var. rapaceum (Mill.) Gaudic. Carum roxburghianum Sprague.*

HABIT : An erect, annual or biennial herb with very small seeds.

HABITAT : Native to Europe and naturalized and cultivated at the foot of Himalayas and outlying hills in Punjab and U.P.

ENGLISH : Celery, Celeriac, Black cumin.

INDIAN : Owaa, Ajamodaa.

SANSKṚTA : Ajamoda, Bastamoda, Kharāśva, Mayūri, Dīpya, Locamostaka.

CHEMICAL COMPOSITION : Seeds yield 3% essential oil containing d-limonene; sedanenolide and 3-n-Buphthalide. Myristicic acid, 8-hydroxy-5-methoxypsoralen and umbelliferone from seeds. Fruits yield a number of furocoumarins. Glucoside apin in the essential oil contracts gravid and vaginal uterus.

Parts used : Fruits and roots

Ayurvedic properties :
Rasa - Kaṭu, Tikta. Vīrya - Ūṣṇa.
Guṇa - Tīkṣṇa, Laghu, Snigdha. Vipāka - Kaṭu.

अजमोदा कटुस्तीक्ष्णा दीपनी कफवातनुत्।
उष्णा विदाहिनी हृद्या वृष्या बलकरी लघुः।।
नेत्रामयकफच्छदि हिक्काबस्ती रुजो हरेत्।
(भा. प्र.)

Ajamodā kaṭustikṣṇā dīpanī kaphavātanut,
Uṣṇā vidāhinī hṛdyā vṛṣyā balakarī laghuḥ,
Netrāmayakaphacchardi hikkabasti rujo haret.

(Bhāvaprakāsa)

अजमोदा तु शूलघ्नी तिक्तोष्णा कफवातजित्।
हिक्काध्मानारुचिर्हन्ती कृमिजित् वन्हि दीपनी।।
(ध. नि.)

Ajamodā tu śulāghnī tiktoṣṇā kaphavātajit,
Hikkādhmānārucirhantī kṛmijit vanhidīpanī.
(Dhanvantari Nighaṇṭu)

Actions/Uses : Kaphaghna, Vātaghna, Vidānhī, Hṛidya, Vṛṣya, Balya, Netrarogahara, Kṛmighna, Chardināśana, Hikkāśamana, Vastirogasamanakar.

Therapeutics :

Ajamodā herb is antioxidant and is a known preventive of rheumatism and gout. It is used as tonic, carminative, diuretic and emmenagogue. Roots show antibacterial properties, and are used as alterative, diuretic and given in ansarca and colic. Seeds are stimulant, cordial, tonic, aphrodisiac carminative, diuretic and emmenagogue. They are used for flavouring food products and as antispasmodic in bronchitis and asthma and to some extent for liver and spleen complaints. Decoction of seeds is a popular household remedy for rheumatism. Essential oil from seeds called celery oil is tranquilizing and anticonvulsions. It is deobstruent and resolvent and used internally as pectoral and tonic and carminative adjunct to purgatives; also as diuretic, emmenagogue, lithontriptic and alexipharmic.

HOMEOPATHIC USES : Symptoms of the drug are unpleasant feeling at stomach, with belchings tasting celery; heartburn, spitting of food; gone feeling at pit of stomach lasting for hours and partially relieved by eating. Urticaria, always appearing with shuddering. Profuse discharge from granulating ulcers intense constriction over sternum, with drawing feeling through to back on lying down.

Carum carvi Linn.

HABIT : A perennial or biennial small herb.

HABITAT : Wild in the north Himalayan regions. Cultivated in the plains as a cold season crop.

ENGLISH : Caraway.

INDIAN : Shahaajire.

SANSKṚTA : Jaraṇā, Kṛṣṇajīraka, Kṛsna, Bahugandhā, Bhedinī, Kāśmīra, Jīraka, Nīlakaṇā, Kāli, Vāntiśodhinī, Hṛdya, Varṣā, Kṛṣṇājājī, Kālapeśikā, Suṣavi.

CHEMICAL COMPOSITION : Fruit or so called seeds contain volatile oil (3-7%), fatty oil (18-23%), protein (21-23%), abscisic acid; yield 2% essential oil (Caraway oil) containing 18% aldehydes; Trans- and cis- carveol, d-limonene. Seeds contain flavons and lipids umbelliferone, scopoletin and herniarin. Lipids from roots contain glyceryl esters, β-sitosterol. α-pinene, camphene, β-pinene, myrcene, limpnine, myristicine etc. in essential oil.

Parts used : Fruits and roots.

Ayurvedic properties :
Rasa - Kaṭu. Vīrya - Ūṣṇa.
Guṇa - Laghu, Rūkṣa. Vipāka - Kaṭu.

जरणा कटुरुष्णा च कफशोफनिकृन्तनी।
रुच्या जीर्णज्वरघ्नी च चक्षुष्या ग्राहिणी परा।।
(ध. नि.)

Jaraṇā kaṭuruṣṇā ca kaphaśophanīkṛntanī,
Rucyā jīrṇajvaraghnī ca cakṣuṣyā grāhiṇī parā.
(Dhanvantari nighaṇṭu)

Actions/Uses : Kaphaghna, Vātaghna, Rocana, Dīpana, Pācana, Medhya, Vṛṣya, Balya, Stanyajanaka, Hṛdrogaghna, Śothahara, Garbhāṣayaśodhanī, Jīrṇajvaraghna, Ajirṇaghna.

Therapeutics :

Kṛsnajīraka is carminative, stimulent, stomachic, lactagogue and spasmolytic. It is used in children's ailments, flatulence and stomachic detangements. A caraway bath is recommended for painful swelling of womb, a polutice for painful and protruding piles. Various plant parts are used in venereal sores, syphilis, constipation, cholera, prolapsus ani and fistula ani. Essential oil from seed is antibacterial and is useful in hemorrhoids.

Centella asiatica (Linn.) Urban.

SYNONYMS : *Hydrocotyle asiatica* Linn.

ENGLISH : Indian Pennywort.

INDIAN : Ekapaani, Kaariwanna, Mandukaparni, Karinga, Brahma-manduki, Khulakhudi.

SANSKṚTA : Maṅḍūkaparṇi, Maṇḍūki, Bramhamaṇḍukī.

HABIT : A prostrate perennial faintly aromatic creeping herb, rooting at the nodes.

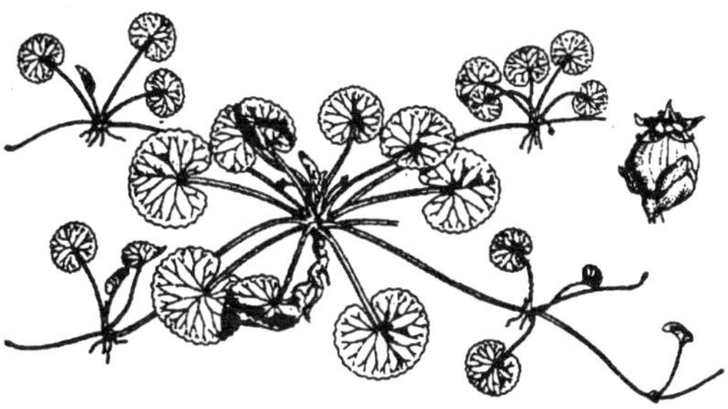

HABITAT : Common in moist grounds and along the sides of cultivated fields.

CHEMICAL COMPOSITION : Fresh leaves contain a glucoside asiaticoside and asiatic acid. Vellarine, pectic acid and resin present in roots and leaves. Asiatocoside is active in treatment of leprosy. Asiaticoside and oxy-asiaticoside employed in treatment of certain types of tuberculosis. The plant also contains ascorbic acid in a concentration of 13.8 mg. per 100 g.

Parts used : Whole plant and leaves.

Ayurvedic properties :
Rasa - Tikta , Kaṣāya, Madhura. Vīrya - Śīta.
Guṇa - Laghu, Sara. Vipāka - Madhura.

रक्तपित्तरहान्याहु र्हधानी सुलघुनि च।
कुष्ठमेहज्वरश्वासकासारुचिहरानि च।।
कषाया तु हिता पित्ते स्वादुपाकरसा हिमा।
लघ्वी मण्डूकपर्णी तु ....।।
(सुश्रुतसंहिता)

Raktapittaharānyāhurhṛdyāni sulaghuni ca.
Kuṣṭhameha jvaraśvāsa kāsaruci harāni ca.
Kaṣāyātu hitā pitte svādupākarasā hima,
Laghvī Maṇḍūkaparṇī tu .. .
(Su. Śu.)

Actions/Uses : Kaphaghna, Vātaghna, Tridoṣāsamaka, Medhya, Āyuṣya, Smṛtiprada, Agnivardhaka, Hṛidya, Balya, Raktapittaghna, Kaṇḍuhara, Mehagna, Jvaraghna, Śvāsaghna,

Śothahara, Arucighna, Viṣaghna, Vraṇaśodhana, Kāsaghna, Stanyaśodhana, Mūtrajanana, Sangrāhaka, Rasāyana.

Therapeutics :

Mandūkaparṇī is a medhya drug. It maintains youthful vigour and strength, improves receptive and retentive capacity of mind. It is a nervine and cardio-tonic, astringent and diuretic. It is useful in dermatosis, anaemia, diabetes, cough, dysponea, emaciation, and insanity. Plant is useful as alterative and tonic and in diseases of skin; leprosy, for nerves and blood. Etherial extract of plant is antiprotozoal and spasmolytic. Plant tablets administered orally to mentally retarded children for 12 weeks showed very significant increase in both general ability and behavioural pattern. Leaves are taken as tonic and for improving memory. Useful in syphilitic skin diseases both internally and externally. Juice of leaves is useful in cataract and other eye troubles; also given in fevers; in diarrhoea among children. Leaves are used to cure severe headache.

HOMEOPATHIC USES : The skin and the female generative organs are the chief seat of action of the drug; though the liver, the nerves and mucous membranes are also powerfully acted upon. It has cured a great variety of skin affections such as acne, eczema, pemphigus, lupus, copper colored eruption, papular eruption on face and intolerable itching in various places. It is employed in ulceration of womb, eczema, elephantiasis, ascariasis and in granular cervicitis.

Conium maculatum Linn.

SYNONYMS : *Coriandrum maculatum Roth* Germ., *Conium major* Bauh.

HABIT : A herb.

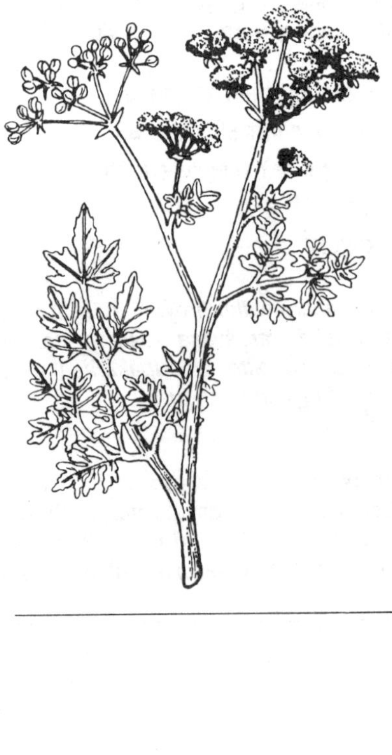

HABITAT : Europe and temperate Asia; common in England.

ENGLISH : Hemlock.

INDIAN : Kurdumana, Jeeraa-kaalaa, Visha jeeraa.

SANSKṚTA : Sukarana.

CHEMICAL COMPOSITION : Fruits contain volatile alkaloids, d-coniine, conhydrine, hesperidine, N-Me-coniine and y-coniceine. Ripe seeds yield coumarines, bergapten and xanthotoxin.

Parts used : Leaves.

Ayurvedic properties :

Rasa - Kaṭu, Tikta. Vīrya - Ūṣṇa.
Guṇa - Gurū. Vipāka - Kaṭu.

Actions/Uses : Viṣakta, Avasādaka, Vṛsya, Vedanāśamaka.

Therapeutics :

Plant causes poisoning in sheep. Alkaloids are poisonous and produce paralysis of motor nerves terminations and stimulation followed by depression of central nervous system, cause nausea and vomiting. Used as sedative, anodyne. Is of particular service in all spasmodic, afections such as chorea, epilepsy, acute mania. Also used with advantage in whooping cough. Neurotic in painful affections of skin and aphrodisiac. Plant extract is a constituent of medical preparation for treating obstruction of lymphatic system. Mother tincture beneficial in prevention of immature cataract.

HOMEOPATHIC USES : Primary action is rigidity, condensation and contraction of fibres, with swelling of glands and diminution of senses. It has cured cases from simple weakness of vitality to complete paralysis. It is anti-scofulous medicine reducing swollen glands, especially when very hard. It has checked, if not cured, cancers of the breast, lip and stomach. It has a specific action on the female breast, dissipating its engorgements and tumors and relieving its pains. Caries and ulcers; hysteria and apoplexy, fainting fits are other significant symptoms.

Coriandrum sativum Linn.

HABIT : An annual glaberous herb.

HABITAT : Common in cultivated fields. Cultivated for the leaves which are used for salads and other cooking purposes and also for fruit Dhane.

ENGLISH : Coriander.

INDIAN : Dhane, Kothimbira.

SANSKṚTA : Dhānyakam, Chatrā, Dhanika, Hṛudyā, Kastumburu, Dhānā, Vitunnaka, Dhaniyaka, Dhānya, Dhenukā, Dhānaya, Allaka, Sugandhī, Sūkṣmapatra, Chatradhānya, Śākayogya, Veṣanā, Nṛtyakuṇḍaka, Janapriya, Bījadhānya, Vedhaka, Challidhānya.

CHEMICAL COMPOSITION : Aflatoxins B1 & B2 are found in samples of coriander. Leaves a rich source of vitamin C and of carotene. Oxalic acid and Ca content of leaves are 0.012% and 0.172%. Fruits contain β-sitosterol, chlorogenic and caffeic acids, rutin, umbelliferone and scopoletin. Seeds, quercetin 3-O-caffeyl glucoside and kaempferol 3-glucoside. Seeds contain 19-21% fatty oil, and an essential oil which causes irritation when in contact with skin for a long time. Seed oil contains a-pinene, limonene, β-phellandrene, linalool, borneol, citronellol, thymol, geraniol, linalys acetate, geranyl acetate, caryophellene oxide, elemol and methylheptenone.

Parts used : Fruit, green parts and oil.

Ayurvedic properties :
Rasa - Kaṣāya, Tikta, Madhura, Kaṭu.
Vīrya - Ūṣṇa.
Guṇa - Laghu, Snigdha. Vipāka - Madhura.

धान्यकं कासतृड्छर्दीज्वरहृच्चक्षुषो हितम्।
कषायं तिक्तमधुरं हृद्यं रोचनदीपनम्।।
(ध. नि.)

Dhānyakam kāsatṛnḍ chardījvarahṛccakṣuṣo hitam,
kaṣāyam tiktamadhuram hṛdyam rocanadīpanam.
(Dhanvantarī nighaṇṭu)

धान्यकं मधुरं शीतं कषायं पित्तनाशनम्।
ज्वरकासतृषाच्छर्दि कफहारि च दीपनम्।।
(राजनिघण्टु)

Dhānyakam madhuram śītam kaṣāyam pittanāśanam,
Jvarakāsatṛṣācchrdi kaphahāri ca dīpanam.
(Rājanighaṇṭu)

Actions/Uses : Vātaghna, Pittaghna, Kaphaghna, Dīpana, Mūtraviracanīya, Pipāsāghna, Dāhapraṣamana, Vāyunaśi, Abhiṣyandapraśama.

Therapeutics :

Dhānyakam is aromatic, stimulant, tonic, carminative, stomachic, antibilious, refrigerant, diuretic and aphrodisiac. Leaves are useful as carminative and antibilious. Decoction is used in sore throat and catarrh; also to wash eyes in conjuctivitis. Fruits prevent griping. Seed extract beneficial as antibacterial and sedative, used in lotions and shampoos. Useful with castor oil in rheumatism. Roasted seeds useful in dyspepsia. A poultice is applied in chronic ulcers and cabuncles. Various plant parts are used in spleen complaints, sores, venereal sores and syphilis.

Cuminum cyminum Linn.

HABIT : A small slender annual herb.

HABITAT : Indigenous to the Upper Nile. Cultivated along the Mediterranean coast and in Punjab and South India.

ENGLISH : Cummin, Caraway.

INDIAN : Jire, Safed jeera.

SANSKṚTA : Śubhrajīraka, Jīraka, Dīrghajīraka, Kaṇajīraka, Ajājī, Gaurajīraka, Dīrghaka, Sitājājī, Kaṇāhvā, Kuncika, Kaṇajīrṇā, Śuklājājī, Dīpya, Hrasvānga.

CHEMICAL COMPOSITION : Fruits contain apigenin-7-O-glucoside and luteolin-7-O-glucosides, an essential oil and fixed oil. Cuminin and flavons are chief constituents. Alcoholic extract of seeds showed 100% antifertility effect in early pregnancy in rats.

Parts used : Seeds.

Ayurvedic properties :
Rasa - Tikta, Kaṭu. Vīrya - Ūṣṇa.
Guṇa - Laghu, Snigdha. Vipāka - Kaṭu.

शुभ्र जीरं कटू ग्राहि पाचनं दीपनं लघू।
किञ्चिदुष्णं च मधुरं चक्षुष्यं कृमिकृन्मतम्।।
गर्भाशयशुद्धिकरं (रुक्षं) बल्य सुगन्धिकम्।
तिक्तं वमिक्षयाध्मानं वातं कुष्ठं विषं ज्वरम्।।
अरोचकं रक्तदोषं अतिसारं कृर्मास्तथा।
पित्तं च गुल्मरोगं च नाशयेदिति कीर्तितम्।। (नि. र.)

Śubhrajīram kaṭū grāhi pācanam dīpanam laghu,
Kiṅciduṣnam ca madhuram caksuṣyam kṛmikrnmatam.
Garbhāśayaśudhikaram rūkṣyam balyam sugandhikam,
Tiktam vamikṣayādhmānam vātam kuṣṭham viṣam jvaram.
Arocakam raktadoṣam atisāram kṛmīnstathā,
Pittam ca gulmarogam ca nāśayeti kīrtitam.
(Nighaṇṭuratnākara)

जीरकः कटुरुष्णश्च वातहृत् दीपनः परः।
गुल्माध्मानातिसारघ्नो ग्रहणी क्रिमिहृत् परः।।
(राजनिघण्टु)

Jīrakaḥ kaṭuruṣṇaśca vātahṛt dīpanaḥ paraḥ,
Gulmādhmānātisāraghno grahanī krimihṛt paraḥ.
(Rājanighaṇṭu)

Actions/Uses : Kaphaghna, Vātaghna, Pittakar, Mūtravirajaniya, Dāhapraśamana, Vedanāsthapana.

Therapeutics :
Jīraka fruit is carminative, aromatic, stomachic, diuretic, stimulant, galactagogue and astringent, cooling in effect, useful in hoarseness, dyspepsia, chronic diarrhoea and in chronic fever. It increases appetite and strength. It is used for flatuence, vomiting, diarrhoea and indigestion. In skin diseases, paste is applied which reduces itching and pain. Also used for scorpion bite. Etherial extract is spasmolitic and hypotensive.

Daucus carrota Linn. var sativa DC.

HABIT : A bristly annual or biennial herb.

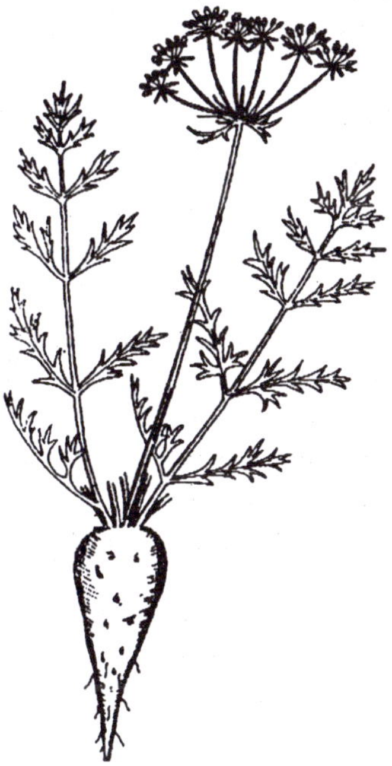

HABITAT : Cultivated throughout India.

ENGLISH : Carrot.

INDIAN : Gaajara.

SANSKRTA : Garjara, Gṛunjana, Gunjaka, Nārangavarṇaka, Svādumūla, Raktakanda, Marjara, Arangaka, Varṇaka, Pītaka, Pītamūla, Supīta, Pinḍamūla, Śikhamulama.

CHEMICAL COMPOSITION : Root contains carotin, hydrocarotin, sugar, pectin, malic acid, lignin, albumin, salts and a volatile oil. β-carotene; daucic acid, pyrrolidine. Oil contains a-pinene, nopinene, sabinene, dipentene, p-thymol, linalool, geraniol, bergamottin, β-bisabolene. Aqueous extract of seeds showed spasmodic action on smooth muscles of ileum, trachea and rectus. Aqueous suspension of essential oil produced a transient fall of blood pressure.

Parts used : Roota and seeds.

Ayurvedic properties :
Rasa - Madhura, Kaṣāya, Tikta. Vīrya - Ūṣṇa.
Guṇa - Laghu, Tīkṣṇa, Vidahi.
Vipāka - Madhura.

गाजरं मधुरं तीक्ष्णं तिक्तोष्णं दीपनं लघु।
संग्राहि रक्तपित्तार्शो ग्रहणीकफ वातजित्।।
(भा. प्र.)

Gājaram madhuram tīkṣṇam tiktoṣṇam dīpanam laghu,
Sangrāhi raktapittārśograhaṇīkaphavātajit.
(Bhāvaprakāṣa)

गुन्जनं मधुरं रुच्यं किञ्चत् कटु कफापहम्।
आध्मानक्रिमिशूलघ्नं दाहपित्ततृषापहम्।।
(राजनिघण्टु)

Gṛnjanam madhuram rucyab kuñcit kaṭu kaphāpaham,
Ādhmānakrimiśūlaghnam dāhapittatṛṣāpaham.
(Rājanighaṇṭu)

Actions/Uses : Vātaghna, Pittaghna, Sangrāhī, Dīpana, Hṛdya, Vṛṣya, Arśoghna, Balya, Mūtrajanana, Garbhāśayauttejaka.

Therapeutics :

Garjara is diuretic, deobstruent and stimulant. It has a beneficial influence on kidneys and dropsy and prevents the brick-dust sediment sometimes found in urine. It is an active and valuable remedy in the treatment of dropsy, retention of urine, gravel and affections of bladder. Used in leprosy, piles, tumors. Considered to be a blood purifier, and antiseptic. They cleanse blood and are recommended in chronic diarrhoea. Carrot decoction is beneficial in biliousness and jaundice. An ointment with lard is useful in burns and scalds; used with leaves as poultice for oozing sores and ulcers. Seeds are carminative, aromatic and stimulant and are used as aphrodisiac and nervinve tonic, given in uterine pain and also for producing abortion. They are useful in diseases of kidney and dropsy.

Ferula alliacea Boiss.

INDIAN : Hingu; Hinga; Abushaaheri Hinga.

SANSKṚTA : Hingu, Balhika.

HABIT : A small tree.

HABITAT : Iran and Afghanistan.

CHEMICAL COMPOSITION : Essential oil. Ferulin. Fruit contains a mixture of furocoumarin, byakangelicin, isopimpinelin and feruli (a mixture of phellopterin and byakangelicol.

Parts used : Niryaasa

Ayurvedic properties :

Rasa - Kaṭu, Tikta. Vīrya - Ūṣṇa.
Guṇa - Tīkṣṇa. Vipāka - Kaṭu.

हिंगु वातकफानाहशुलघ्नं पित्तकोपनम्।
(वा. सू. अं. ६))

Hingu vātakaphānāhaśulaghnam pittakopanam.
(Va.Sū.)

हिंगुनियसि श्छेदनीय दीपनीया
नुलोमिक वातकफप्रशमनानाम् श्रेष्ठः।
(च. सू. २५)

Hinguniryāsa śchhedañiya dīpanīya,
Anulomikavātakaphapraśnānām śresṭhaḥ.
(C.Sū. 25)

Actions/Uses : Dīpana, Pācana, Vāyunāśaka, Āanulomika, Kṛmighna, Chedanīya, Śleṣmāhara, Āmāśayauttejaka.

Therapeutics :

Gum-resin is used for scorpion sting and is an intestinal antiseptic, carminative. Useful in hysteria and epilepsy. Latex of *F. alliacea* Boiss, alongwith cow's milk and ashes of seeds of palash (Butea monosperma) is a moderate aborticide. It is appetizer, good for digestion; stimulant for stomach and intestines, deflatulent, carminative, nerve tonic, stimulant for uterus; antispasmodic and useful for typhoid fever and removing cough and bad breath. Useful in chest infections, asthma, whooping cough, dry cough etc. (Anulomika, chhedaneeya). Used in hysteria, epilepsy and scorpion-sting and also as intestinal antiseptic and carminative.

Ferula assa-foetida Linn.

SYNONYMS : *Ferula foetida* Regel.

ENGLISH : Asafoetida; Devil's Dung.

INDIAN : Hinga.

SANSKṚTA : Hinguka, Sahatravedhī, Jatuka, Bāhlika, Jantughna, Ūgragandha, Rakṣoghna, Dīpta, Supānga.

HABIT : Tree with large thick roots with few fibres, of a blackish color outside and very white within. Oleo-gum-resin is drawn from the roots and stems. The gum resin is distinguished by its powerfully alliaceous odor. Imported into India.

HABITAT : Iran, Eastern Persia and western Afghanistan. Cultivated in nortwestern parts of India.

CHEMICAL COMPOSITION : Luteolin and its compound. Essential oil, ferulic acid, organic sulphur compound, pinene, isobutyl propanyl disulphide and umbeliferone.

Parts used : Root, stem, leaves and gum-resin.

Ayurvedic properties :
Rasa - Kaṭu, Tikta. Vīrya - Ūṣṇa.
Guṇa - Tikṣṇa. Vipāka - Kaṭu.

हृद्यं हिङ्गु कटूष्णञ्च क्रिमिवातकफापहम्।
विवन्धाध्मानशूलघ्नं चक्षुष्यं गुल्मानाशनम्।।
(राजनिघण्टु)

Hṛdyam hiṅgu kaṭūṣṇānca krimivātākaphāpaham,
Vivandhādhmanaśūlaghnam cakṣuṣyam gulmānāśanam.
(Rājanighaṇṭu)

Actions/Uses : Kaphghna, Vātaghna, Dīpana, Pācana, Vāyunāṣaka, Ānulomika.

Therapeutics :
Hinguka leaves are anthelmintic, carminative and diaphoretic. Stem is a brain and liver tonic. Root is antipyretic. Gum resin is stimulant, antispasmodic and expectorent. Useful in infantile convulsions, croup and flatulent colic also in asthma, whooping cough, chronic bronchitis, hysteria, epilepsy and cholera. As enema, stimulates intestine, resperatory tracts and nervous system. Oleo-gum-resin is used for medical purposes. Uses same as *F. alliacea.* Hinguka and its ether extracted essential oil is protective against fat-induced increase in plasma fibrinogen and decrease in coagulation

time. Diluted with vinegar and taken, acts as abortifacient. Used in spleen, madness, pain, hemiplegia, epilepsy, convulsions, cramps, colic and rinderpest. Oleo-gum resin used alongwith others in herbal compound "Vidangadi Yoga", very effective antifertility drug.

HOMEOPATHIC USES : Asafoetida is a Homeopathic remedy especially useful in two classes of diseases. First, in nervous diseases developing a perfect type of hysteria; it acts upon the muscular fibres, producing a reverse teristaltic action in oesophagus and intestines. Thus, it causes a sensation as though a ball started in the stomach and rose into the throat. It produces a bursting feeling, upwards, as though in the abdomen were coming out at the mouth as found in colic in hysteria. The second action is upon bones. It produces periosteal inflammations, resulting in ulcers, especially upon the shin bones. Many of the discharges of Asafoetida are fetid.

Ferula narthex Boiss.

ENGLISH : Tibetian assafoetida.

INDIAN : Hinga, Anjudan.

SANSKṚTA : Hingu, Sahastravedhī, Jatuka, Bahlika, Ūgragandha, Ūgravirya, Rakṣoghna, Jantughna, Dīpta, Supānga.

HABIT : A perennial odorous herb.

HABITAT : Baltistan and Astor.

CHEMICAL COMPOSITION : Essential oil umbelliferon.

Parts used : Niryaasa.

Ayurvedic properties :
Rasa - Kaṭu. Vīrya - Ūṣṇa.
Guṇa - Laghu, Snigdha, Tīkṣṇa. Vipāka - Kaṭu.

सहस्त्रवेधि जतुकं दाहिकं हिंगुरामठम्।
हिंगूष्णं पाचनं रुच्यं तीक्ष्णं वातबलासहृत्।।
शूलगुल्मोदरानाहकृमिघ्नं पित्तवर्धनम्।
स्त्रीपुष्पजनबल्यं मूर्च्छापस्मारहृत् परम्।।
(भा. प्र.)

Sahasravedhi jatukam dāhikam hingurāmaṭham,
Hingūṣṇam pācanam rucyam tīkṣṇam vātabalāsahṛt.
Śūlagulmodarānāha kṛmighnam pittavardhanam,
Strīpuṣpajana balyam Mūrcchāpasmārahṛt param.
(Bhāvaprakāṣa)

Actions/Uses : Kaphaghna, Vātaghna.

Therapeutics :

Hingu or Assafoetida is stimulant, powerful antispasmodic and expectorant. It is also antiseptic, anthelmintic, nervine stimulant and emmenagogue; useful in asthma, whooping cough, flatuent colic, and in pneumonia and bronchitis of children. Leaves are diaphoretic and carminative. Essential oil is antibacterial against several pathogenic and nonpathogenic bacteria.

Foeniculum vulgare Mill.

SYNONYMS : *Foeniculum capillaceus* Gilb.

ENGLISH : Fennel, Indian sweet fennel.

INDIAN : Baddishepa, Saunf, Sonp, Shĕpu.

SANSKṚTA : Miśreyā, Madhurikā, Madhurā, Chatrā, Mādhavī, Kāravi, Bahupuṣpāvhā, Ahichatrā, Śatapuṣpa, Śatāhvā, Miśi, Ghoṣā, Pītikā, Vajrapuṣpā, Śataprasūna, Sanhatapatrī, Suvākpuṣpi, Yonidoṣaghni, Sanghātapatrikā, Ceṭikā, Māgadhā, Paṇā.

HABIT : A stout, glabrous aromatic biennial or perennial herb.

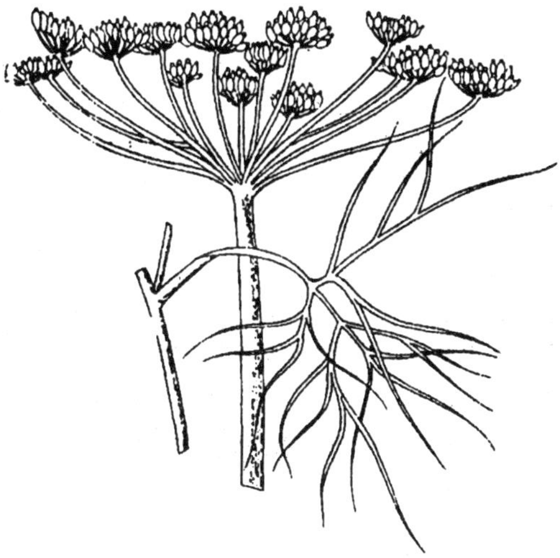

HABITAT : Native of Southern Europe and Asia. Extensively cultivated during cold season in India.

CHEMICAL COMPOSITION : Constituents of plant are fatty oil, protein, anethose, fenchone, anisalehyde, feniculin, nelumboside; flavons, abscisic acid. Essential oil from fruit anethole, xanthotoxin, bergapten, also psoralen, scoparone, vanillin, a-pinene, camphene, limonene, α-phellandrene, fenchone, methylchavicol.

Parts used : Fruit, leaves, root and oil from fruits.

Ayurvedic properties :
Rasa - Madhura, Tikta, Kaṭu.
Vīrya - Ūṣṇa, Śīta.
Guṇa - Laghu, Snigdha, Tīkṣṇa, Sara.
Vipāka - Kaṭu.

Actions/Uses : Vātaghna, Pittaghna, Dīpani, Vrṣya, Jvaraghna, Stanyavardhinī, Rucya, Netravranahara, Vranaghna, Dāhaśamani, Śūlahara, Āmadoṣaghna, Triṣṇāśamani.

मिश्रेया मधुरा स्निग्धा कटुः कफहरापरा।
वातपित्तोत्थदोषघ्नी प्लीहजन्तुविनाशनी।।
(राजनिघण्टु)

Miśreyā madhurā snigdhā kaṭuh kaphaharāparā,
Vātapittotthadoṣaghnī plihajantuvināśanī.
(Rājanighaṇṭu)

तिक्ता स्वादुर्हिमा वृष्या दुर्नामक्षयजित्मिशि।
क्षतक्षीणहिता बल्या वातपितास्त्रदोषजित्।।
(ध. नि.)

Tiktā svādurhimā vṛṣyā durnāmakṣayajit miśi,
Kṣatakṣīnahita balyā vātapittāsradoṣajit.
(Dhanvantari nighaṇṭu)

Therapeutics :

Miśreyā leaves are diuretic and improve eyesight. Leaves along with those of Lycopersicon species are boiled and given in gonorrhoea. Seeds are stimulant, aromatic, appetizer, diuretic, emmenagogue, carminative, spasmolytic, galactagogue and stomachic. Tea made from bruised seeds help detoxify body by using it in bath or in a vaporizer to increase release of toxic wastes; as a lotion rarely used for inflammed eye. Dried fruit is useful in chest, spleen and kidney troubles. Aqueous extract is used as enema for infants for expulsion of flatus. Hot infusion helps to increase lacteal secretion and stimulates sweating. With other ingredients used in antiasthmatic drug. Oil from seed is vermicide and antimicrobial.

Family : X-Apocynaceae.

Alstonia scholaris R. Br.

SYNONYMS : *Echites scholaris* Linn., *Alstonia cuneata* Wall.

ENGLISH : Dita Bark, Devil's Tree.

INDIAN : Satwina, Saptaparnna, Datyuni, Chhatium.

SANSKṚTA : Saptaparṇā, Raktachada, Viśālatvak, Śālmalīpatrika, Viṣamachada, Brihatvak, Ayugmapatra, Madagandha, Surabhi, Śārada, Uchairika, Gucchapuṣpaka, Pṛthakpatra, Yugmaparṇa, Bahucchada, Sudīrgha, Ayugmacchada.

HABIT : A large buttressed evergreen tree.

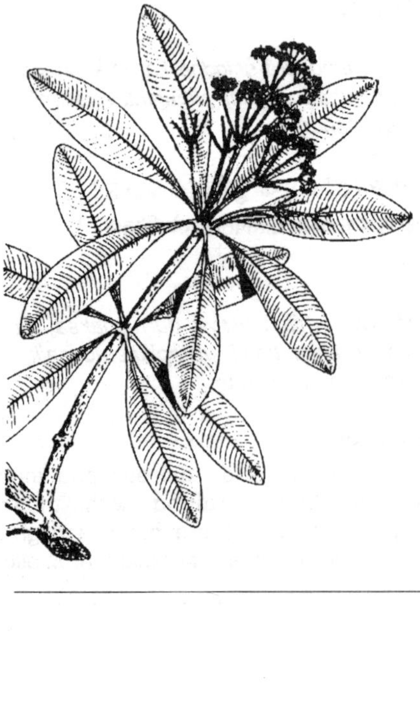

HABITAT : Common in deciduous and evergreen forests.

CHEMICAL COMPOSITION : Several alkaloids of various skeletal patterns have been isolated from various parts. The total alkaloidal content is 0.16-0.27 per cent. Bark contains rhazine; echitamine; glucoside of venoterpine, triterpenes. Leaves and flowers contain alkaloids picrinine, strictamine, tetrahydroalstonine etc. Root bark contains a-amyrin, lupeol, stigmasterol and a-sitosterol. Picrinine showed CNS depressant action.

Parts used : Leaves, bark, wood and milky juice.

Ayurvedic properties :
Rasa - Tikta, Kaṣāya, Kaṭu. Vīrya - Ūṣṇa.
Guṇa - Laghu, Snigdha. Vipāka - Kaṭu.

सप्तपर्णो व्रणश्लेष्मवातकुष्ठास्त्रजन्तुजित्।
दीपनः श्वासगुल्मघ्नः स्निग्धोष्णस्तुवरः सरः॥
(भा. प्र.)

Saptaparṇo vraṇaślesmavātakuṣṭhātrajantujit, dīpanaḥ śvāsagulmaghnaḥ snigdhoṣṇastuvaraḥ saraḥ.
(Bhāvaprakāṣa)

सत्यपर्णस्तु तिक्तोष्णः त्रिदोषघ्नश्च दीपनः।
मदगन्धों निरुग्धोयं व्रणरक्तामयक्रिमीन्॥
(राजनिघण्टु)

Saptaparṇastu tiksnoṣṇaḥ tridoṣaghnaśca dīpanaḥ,
Madagandho nirugdhoyam vraṇaraktamayakrimin.
(Rājanighaṇṭu)

Actions/Uses : Kaphaghna, Vātaghna, Hṛdya, Dīpanī, Vranaghna, Raktamayaghna, Kṛmighna, Kuṣṭhaghna, Jirnasvararaghna, Svāsaghna, Stanyajanana, Gulmaghna, Grahaṇīrogahara.

Therapeutics :

Tender leaves in the form of polutice are good for ulcers with foul discharge. The juice of leaves with that of ginger is given after confinement. Decoction of leaves is used in beriberi and in congestion of liver. Bruised leaves boiled in oil are given internally in dropsy. Bark is stimulant, carminative, stomachic, bitter tonic, astringent, expectorant, alterative, anthelmintic and galactogogue. It cures gastro-intestinal troubles. It is a mild febrifuge, antiperiodic, useful in malaria, like quinine and in diarhhoea and dysentery, also for snake-bite. It is used in skin diseases. It has proved valuable in chronic diarrhoea and in advanced stages of dysentery. It is an efficient remedy in debility after fevers and other exhausting diseases. It is used in heart diseases, asthma and to stop bleeding of wounds. Bruised and boiled in cotton seed oil, the bark is applied to the ear for deafness. Fresh bark juice with milk is administered in leprosy and dyspepsia. Stem bark and leaves of Vitex negundo are used as poultice in body-ache and joint pains. Sap, gum and roots are used in tumors and cancer. Various plant parts are used in headache, pain in legs and loins, rheumatism, haemoplegia; cholera, bronchitis, phthisis and asthma, pleurisy, pneumonia and lactation complaints. Milky juice or latex is applied to ulcers, sores, toothache, tumours and in rheumatic pain. Etherial extract of bark is hypotensive and anticancer.

HOMEOPATHIC USES : It causes debility, low fever, often with diarrhoea, and when pushed, rigors, sweat, etc., purging, cramps and vertigo. Clinically it has been used in chills and fever, in diarrhoea of malarial origin and for chronic intermittent fevers suppressed by quinine. It paralyzes peripheral motor nerve filaments as a property of Ditaine which it contains.

Holarrhena antidysenterica Wall.

SYNONYMS : *Holarrhena pubescens (Buch.-Ham.)* Wallich.

HABIT : A deciduous laticiferous shrub or small tree.

HABITAT : Common in tropical parts of India.

ENGLISH : Conessi bark tree, Tellicherry, Kurchi.

INDIAN : Kudaa-pandharaa, kutaja, Kurchi, Kureya, Karra, Kodaga, Dola-kuda.

SANSKṚTA : Kuṭaja, Vatsaka, Kalinga, Indravṛkṣa, Yavaphala, Śakra, Prāvṛuṣya, Vṛukṣaka, Girimallikā, Varatikta.

CHEMICAL COMPOSITION : Bark contains a large number of alkaloids, the chief amongst them are conessine, nor-conessine, conesimine, iso-conessimine, kurchine, conimine, conamine, conarrhimine, conessidine, conkurchine, holarrhenine, holarrhimine, holarrhine, kurchicine and lettocine. In addition to the alkaloids it contains gum, resin and tannin. A triterpene alcohol, lupeol and β-sitosterol from the bark. Seeds contain a drying oil. Latex contains caoutchouc and two resinols. It is useful in the synthesis of steroid hormones.

Parts used : Bark, seeds and leaves.

Ayurvedic properties :
Rasa - Kaṭu, Kaṣāya, Tikta. Vīrya - Śīta.
Guṇa - Laghu, Rūkṣa. Vipāka - Kaṭu.

कुटजः कटुको रुक्षो दीपनस्तुवरो हिमः।
अर्शोतिसारपितास्त्र कफ तृष्णामकुष्ठजित्।।
(भा. प्र.)

Kūṭajaḥ kaṭuko rukṣo dīpanastuvaro himaḥ,
Arśotisārapittāsrakaphatṛsṇāmkuṣṭhajit.
(Bhāvaprakāṣa)

कूटजः कटुतिक्तोष्णः कषायश्चातिसारजित्।
तत्रासितोस्त्रपित्तघ्नः त्वग् दोषार्शोनिकृन्तनः।।
(रा. नि.)

Kūtajaḥ kaṭutiktoṣṇaḥ kaṣāyaścātisārajit,
Tatrāsitosrapittaghnaḥ tvag doṣārśonikṛntanaḥ.
(Rājanighaṇṭu)

Actions/Uses : Kaphaghna, Pittaghna, Vātakar, Tridoṣaghna, Dīpana, Sangrāhi, Arśahara, Jvaraghna, Atisāraghna, Raktatisāraghna, Kuṣṭhaghna, Visarpahara, Dāhasamana, Śulaghna.

Therapeutics :
Kūṭaja is used in anaemia, colic pain, diarrhoea, haematuria, menorrhagia, obsteric conditions, spermatorrhoea, spenomegaly. It is astringent and tonic. A single remedy for chronic dysentery and diarrhoea. As extract or decoction it is used in the treatment of amoebic dysentery and diarrhoea. Root-bark is bitter, astringent, antiperiodic, cooling and carminative, cures piles, diarrhoea, haemorrhage, indigestion and skin diseases. It is useful in heart diseases, fever and vomiting. Leaves are used in chronic bronchitis and applied to boils and ulcers. Bark is astringent, anthelmintic, antidysenteric, stomachic, febrifuge and tonic. It is used in the treatment of amoebic dysentery and diarrhoea. It is a good remedy for colitis. Its paste is applied in pruritus. It is given either alone or with other astringent drugs in piles, colic, dyspepsia, chest affections and diuresis. It is used for diseases of skin and spleen. Dried and ground, it is rubbed over the body in dropsy. Seeds are astringent, febrifuge, useful in dysentery, diarrhoea, bleeding piles and intestinal worms.

Holarrhena antidysenterica Willd.

SYNONYMS : *Wrightia antidysenterica J. Grah.; Holarrhena antidysenterica (Roth) A. DC.; Echites antidysenterica Heyne.; Holarrhena pubescens (Buch.-Ham.) Wall. ex DC.*

INDIAN : Naagakuda; Kudaa-pandharaa.

SANSKṚTA : Indrayava.

HABIT : A small tree. Seeds are known as "Kadawaa Indrajava."

HABITAT : Common and abundant in open areas of the jungles. Local people use the leaves for making biddies.

CHEMICAL COMPOSITION : Bark contains L-quebrachitol; alkaloids dihydroisoconessimine and 3 a-aminoconan-5-ene; base kurcholessine; 7a-OH-conessine and holonamine. Holantosines A and B.

Parts used : Whole plant, root-bark, bark and seeds.

Ayurvedic properties :
Rasa - Kaṭu, Kaṣāya, Tikta. Vīrya - Śīta.
Guṇa - Laghu, Rūkṣa. Vipāka - Kaṭu.

इन्द्रयवं त्रिदोषघ्नं संग्राहि कटुशीतलम्।
ज्वरातिसाररक्तार्शः कृमिवीसर्पकुष्ठनुत्।
दीपनं गुदकीलास्त्रवातास्त्रश्लेष्मशूलनुत्।।
(भा. प्र.)

Indrayavam tridoṣaghnam sangrāhi kaṭuśītalam,
Jvarātisāraraktārśaḥ kṛmivīsarpakuṣṭhanut,
Dīpanm gudakīlasravātāsraślesmaśūlanut.
(Bhāvaprakāṣa)

Actions/Uses : Kaphaghna, Pittaghna, Vātakar, Tridoṣaghna, Dīpana, Sangrāhi, Arśahara, Jvaraghna, Atisāraghna, Raktātisāraghna, Kuṣṭaghna, Visarpahara, Dāhaśamana, Śūlaghna.

Therapeutics :

Various parts of Indrayava are used for cold, colic, anaemia, constipation, spleen complaints, epilepsy, spermatorrhoea, labor complaints and dog-bite. Root-bark is very useful for dysentery with bloody stools. Fresh bark as paste in sour butter-milk reduces fever and frequency of stools as well as blood in stools. Bark-paste is applied for rheumatic inflammation. Bark is used in dropsy and dysentery. Pounded bark is used in stomach disorders, in abdominal and glandular tumors. Gargling with decoction of bark reduces tooth-ache. Fruit extract is antiprotozoal, anticancer and hypoglycaemic. Seeds are astringent, febrifuge, useful in diarrhoea and intenstinal worms and for eczema. They are useful in regulating menstruation. Bleeding of gums and pus formation and bad breath are treated by applying powder of seeds. Paste is applied to head in headache. Decoction is taken internally in dysentery and fevers. It is an ingredient of compositions for giardiasis. Their extract is hypotensive. Vegetable of tender pods is useful to eradicate worms. Powder of pod is applied on affected part in snake-bite to reduce burning pain and swelling.
Uses are similar to *H. antidysenterica* Wall.

Ichnocarpus frutescens R. Br.

SYNONYMS : *Apocynum frutescens* Linn.; *Echiles frutescens.*

ENGLISH : Black creeper.

INDIAN : Kaatebhowari, Sareevaa, Shyaamalataa, Krishnasarava, Bhadraa, Kalli-dudhi.

SANSKṚTA : Sārivā, Kṛṣṇsārivā, Paravalli, Śyāmalatā.

HABIT : A large evergreen, laticiferous twining shrub.

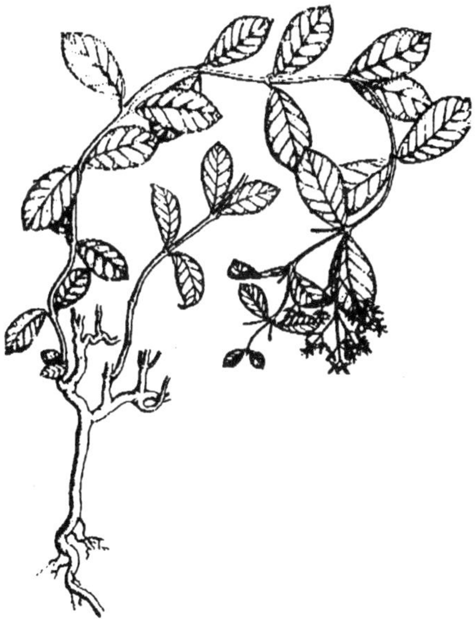

HABITAT : More or less throughout India. Common in forest areas.

CHEMICAL COMPOSITION : Leaves contain flavons and phenolic acids. Stem contains a new triterpne glucoside.

Parts used : Roots and leaves.

Ayurvedic properties :
Rasa - Madhur. Vīrya - Śīta.
Guṇa - Gurū, Snigdha. Vipāka -Madhur.

सारिवा द्वे तु मधुरा कफवातास्त्रनाशिनी।
कुष्ठकण्डूज्वरहरो मेहदुर्गन्धीनाशने।।
कृष्णमूली तु संमाहिशिशिरा कफवातजित्।
तृष्णारुचिप्रशमनी रक्तपित्तहरा स्मृता।। (राजनिघण्टु)

Sārivā dve tu madhurā kaphavātāsranāśinī,
Kuṣṭhakaṇḍūjvaraharā mehadurgandhināśane.
Kṛṣṇamulī tu sanmāhiśiśirā kaphavātajit,
Tṛṣṇā ruci praśamanī raktapittaharā smṛtā.
(Rājanighaṇṭu)

Actions/Uses : Kaphahara, Vātahara, Grāhī, Stanyaśodhana.

Therapeutics :

Kṛṣṇasārivā plant is used for bleeding gums, convulsions, cough, delirium, dysentery, glossitis, haematuria, measles and night blindness. It is beneficial in anorexia, leucorrhoea, syphilis, urinary calculi and rheumatism. Root is alerative, tonic, diuretic and diaphoretic, and is used as a substitute for sarsaparila. Decoction of leaves and stalks is used for fever and skin eruptions. Uses and Ayurvedic description are similar to Hemidesmus indicus.

Nerium indicum Mill.

SYNONYMS : *Nerium odorum* Soland.

ENGLISH : Sweet scented oleander, Indian oleander.

INDIAN : Kannhera, Dhavi-kaneri, Karber.

SANSKṚTA : Karavīra, Aśvamāraka, Hayamāraka, Śvetapuṣpa, Laguda, Raktapuṣpa, Pratihāsa, Śatakumbha, Angulīpatra, Caṇḍaka, Kunda, Mahavīra, Hayaghna.

HABIT : Low shrub with small canopy.

HABITAT : Cultivated in Public parks and home gardens.

CHEMICAL COMPOSITION : Dambonitol from leaves and stems. Ursolinic acid, oleandrin and neriodorin, plumericin and odorin. Root, bark and seeds contain toxic principles neriodorin, nerioderin and karabin. They are a powerful cardiac poison and act on the heart in a somewhat similar manner as digitalis. Leaves, flowers and bark show carditonic potency and diuretic activity. Leaves contain a compound giving reactions described for rutin.

Parts used : Root, root-bark and leaves.

Ayurvedic properties :
Rasa - Kaṭu, Tikta, Kaṣāya. Vīrya - Ūṣṇa.
Guṇa - Laghu, Rūkṣa, Tīkṣṇa. Vipāka - Kaṭu.

करवीर द्वयं तिक्तं कषायं कटुकं च तत्।
व्रणलाघवकृन्नेत्र कोपकुष्ठव्रणापहं।
वीर्योष्णं कृमिकण्डुघ्नं भक्षितं विषवन्मतम्।।
(भा. प्र.)

Karavīradvayam tiktam kṣāyam kaṭukam ca tat,
Vraṇalāghavakṛnnetrakopakuṣṭhvraṇāpaham,
Viryoṣṇam kṛmikaṇḍughnam bhakṣitam

viṣavanmatam.
(Bhāvaprākaṣa)

करवीरः कटुस्तीक्ष्णः कुष्ठकण्डूतिनाशनः।
व्रणार्त्तिविषविस्फोटशमनोऽश्वमृतिपदः।।
(राजनिघण्टु)

Karavīraḥ kaṭustīksṇaḥ kuṣṭhkaṇḍūtināśanaḥ,
Vraṇārttiviṣavisphoṭaśamanośvamṛtipradaḥ.
(Rājanighaṇṭu)

Actions/Uses : Kaphaghna, Vātaghna, Viṣavata, Kuṣṭhahara, Jvaraghna, Kaṇḍūghna, Kṛmighna, Netrakopavraṇahara.

Therapeutics :

Karavīra is a poisonous plant. Leaves and root-bark are used as medicine. Root is a powerful resolvent and attenuant, and used externally; beaten into a paste with water and applied to chancres and ulcers on penis. It is also employed as abortifacient. Decoction of roots is used to reduce swellings. Oil prepared from root-bark is used in skin dieases of a scaly nature, and in leprosy. It acts like digitalis on heart and has similarity with strophanthus. It is useful in heart diseases where it acts as a diuretic. It is useful in skin diseases. There is much similarity in the action of this remedy and Thevetia peruviana.

Nerium oleander Linn.

Note : *N. indicum* and *N. oleander* are commonly treated as separate species. However some consider them nonspecific.

ENGLISH : Rose Laurel, Oleander, Rose bay.

HABIT : An evergreen glabrous shrub cultivated in gardens and as ornamental plant. Thrives best in moist ground.

HABITAT : Native of Mediterranean region. South Europe, North Africa and West Asia.

CHEMICAL COMPOSITION : Leaves contain nerin, oleandrin and folinerin; neritaloside, urechitoxin, adynerin, B-sitosterol, ursolic and oleanolic acids, neriaside, oleosides A,B,C,D,E and F. Oleandrin showed digitalis-like action on heart and emetic action in experimental animals. Bark yields glucoside cortenerin; stimulates diuresis.

Therapeutics :

The bark and flowers possess cardio-tonic properties similar to the leaves. Leaves are used in cutaneous eruption. A decoction of leaves is used to destroy maggots infesting wounds.

HOMEOPATHIC USES : It is known as Oleander and was proved by Hahnemann. Plant is poisonous. In animals poisoned by it, heart becomes paralysed, first the auricles, then the entire heart. Palpitation of the heart, anxiety, insomnia, and unconsciousness are some of the symptoms. It has violent pulsation of the carotids and violent vertigo, which has been noted in patients of high blood pressure. It is indicated in the obtuse continued stitch in the sternum; dull drawing pain over the heart, more violent when stooping, and continued during expiration. It acts profoundly on the digestive system and is a great remedy for indigestion, especially when the stools consist undigested food. It is indicated in sensitiveness of the molars and teeth and sharp drawing toothache in the left second molar. It will be found to be if not a complete remedy yet an indispensable intermediate remedy in some kinds of mental derangements, e.g. absence of mind, certain kinds of painless paralysis, and some external head affections. It is the front rank of remedies affecting the scalp, more particularly the back part. It has great sensitiveness of the skin of the whole body; general itching. Voluptuous dreams with emission of semen.

Plumeria rubra Linn. var. acutifolia Bailey.

SYNONYMS : *Plumeria acuminata* Ait.; *Plumeria acutifolia* Poir.

ENGLISH : Pagoda Tree.

INDIAN : Khairachaafaa, Chaafaa-paandharaa, Gglainchi.

SANSKṚTA : Śvetacampā, Kṣīracampaka.

HABIT : An evergreen or partly deciduous tree.

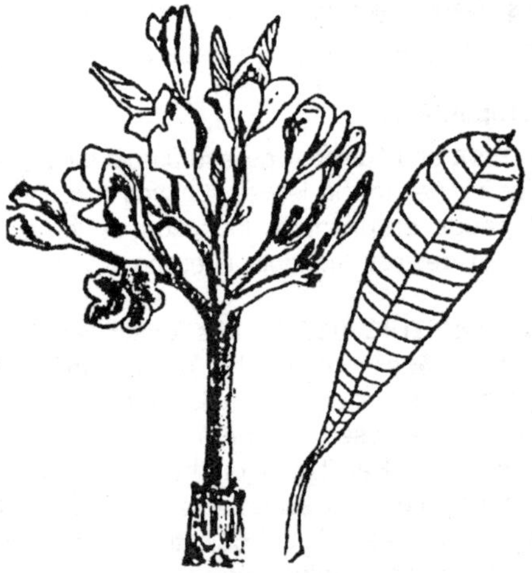

HABITAT : A native of Mexico. Cultivated as an ornamental plant in gardens and also as a hedge plant.

CHEMICAL COMPOSITION : Bark contains fulvopluimerin, plumieride, amyrin and luoeol. It also contains a bitter glucoside, an essential oil and plumeric acid.

Parts used : Root, root-bark, bark, leaves, branches, flowers, and latex.

Ayurvedic properties :
Rasa - Kaṭu, Tikta, Kaṣāya. Vīrya - Ūṣṇa.
Guṇa - Sara, Gurū, Snigdha. Vipāka - Kaṭu.

श्वेतस्तु चम्पकः प्रोक्तः सरस्तिक्तः कटुः स्मृतः।
तुवरोष्णः कुष्ठकण्डूव्रणशूलकफापहः।।
वातं चोदररोगं च आध्मानं चैव नाशयेत्।
(निघण्टु रत्नाकर)

Śvetastu ca campakaḥ proktah sarastiktaḥ kaṭuḥ smṛtaḥ,
Tuvaroṣṇaḥ kuṣṭha kaṇḍū vraṇa śūla kaphāpaḥ.
Vātam codararogam ca ādhmānam caiva nāśayet.
(Nighaṇṭuratnākar)

Actions/Uses : Tīvrarecana, Mūtrajanana, Śothaghna, Vātahara, Jvaraghna, Kaṇḍūghna, Niyatakalika jvarapratibandhaka.

Therapeutics :

Plant is febrifuge and used in cholera and indigestion. Roots are demulcent, alterative, tonic, diuretic, diaphoretic, and a powerful cathartic. They are used as a substitute for Hemidesmus Indicus. Root-bark is a strong purgative, diuretic, anti-inflammatory, anti-itch, febrifuge and antiherpetic, useful in gonorrhoea and venereal sores, and is a cure for intermittent fevers. Leaves are febrifuge and their paste is applied as a polutice to indolent swellings. Bark has a stimulating action and are antidiarrhoeal. A decoction is used as a purgative, emmenagogue and febrifuge. It is also given for dropsical affections and venereal affections and is a powerful antiherpatic. It is given with coconut, ghee and rice in diarrhoea. Branches are used to induce abortion. Flowers are contraceptive. Milky juice or latex is purgative and employed as a rubefacient in rheumatism.

Rauwolfia serpentina Benth. ex Kurz.

SYNONYMS : *Ophioxylon serpentinum.*

ENGLISH : Rauvolfia, Serpentine.

INDIAN: Sarpagandhaa, Hadaki, Chandrabhaga, Chota-Chanda, Patalgarud.

SANSKṚTA : Sarpagandhā, Dhavalaviṭapa, Candramāra, Sarpāngī, Bhujangāksī, Nākulī, Viṣanāśinī, Surasā, Nāgasugandhā, Gandhanākulī, Nakuleṣṭhā, Cundrika.

HABIT : A small erect or sub-erect perennial shrub, generally 15-45 cm. high, but growing upto 90 cm. under favourable conditions.

HABITAT : Indigenous to moist deciduous forests of S.E. Asia. Common in open areas among bushes in Konkan, W. and E. Ghats and other parts of India.

CHEMICAL COMPOSITION : More than 50 alkaloids have been reported from this plant. Root yields alkaloids ajmaline, ajmalinine, ajmalicine, yohimbine, alloyohimbine, isoyohimbine, r-yohimbine, chandrine, desrpidine, isoajmaline, papaverine, corynanthine, raunatine, rauvolfinine, rauwolscine, reserpiline, reserpine, rescinnamine, reserpinine, reserpoxidine, sarpagine, serpinine, serpentine, serpentinine and amorphous bases. Alkaloid rauwoline decreases heart rate and intravenous injection reduces blood pressure and increases tone of small intestines and decreases peristaltic contractions. Alkaloid reserpin has a very marked hypnotic effect and lowers blood pressure.

Parts used : Root and leaves.

Ayurvedic properties :

Rasa - Tikta, Kaṭu.	Vīrya - Ūṣṇa.
Guṇa - Rūkṣa.	Vipāka - Kaṭu.

सर्पगन्धातितिक्तोष्णा रुक्षा कटुविपाकिनी।
पित्तवृद्धिकरी रुच्या शूलप्रशमनी सरा।।
कफवातहरा निद्राप्रदा हृदवसादिनी।
कामावसादिनी चैव हन्ति शूलज्वरक्रिमीन्।।
अनिद्रां भूतमुन्मादमपस्मारं भ्रमं तथा।
अग्निमान्घं विषं रक्तवाताधिक्यं व्यपोहति।।
(भा. प्र.)

Sarpagandhā atitiksnoṣṇā rukṣā kaṭuvipākinī,
Pittavṛdhikari rucyā śūlapraśamanī sarā,
Kaphavātaharā nidrāpradā hṛdvasādinī,
Kāmāvasādinī caiva hanti śūla jvarakrmīn.
Anidrām bhūtamunmādam apasmāram bhramam tathā,
Agnimāndyam viṣam raktavāta adhikyam vyapohati.
(Bhāvaprakāṣa)

Actions/Uses : Kaphaghna, Vātaghna, Pittakar, Pauṣtika, Nidrajanana, Kṛmighna, Garbhāśaya-uttejaka, Jvaraghna, Viṣahara.

Therapeutics :

In India, Sarpagandhā has been employed for centuries for the relief of various central nervous system disorders, both psychic and motor, including anxiety states, excitement, maniacal behaviour associated with psychosis, schizophrania, insanity, insomnia and epilepsy. Root is bitter tonic, hypnotic, sedative, specific for insanity, reduces blood pressure. It is a remedy in painful affections of the bowels. Extract of roots is used for the treatment of intestinal disorders, particularly diarrhoea and dysentery and also as anthelmintic. Roots stimulate uterine contraction and are used in child-birth in difficult cases. Decoction of roots is employed in labors to increase uterine contractions. Leaves are bitter, stimulant for uterus, nutritive, anthelmintic, and febrifuge. Juice of leaves is used for removal of opacities of cornea. Its action is similar to Plumbogo zeylanica. Ajmaloon, a drug from Rauvolfia serpentina has been found to be highly effective for the treatment of hypertension.

Tabernaemontana coronaria R. Br.

SYNONYMS : *Ervatamia coronaria Stapf.; Ervatamia divaricata (Linn.) Alston.; Tabernaemontana divaricata (L) R.Br. ex R. & S. Ervatamia divericata (Linn.) Burkill.; Nerium divaricatum Linn. Tabernaemontana divaricata (L) Roem & Scult.*

ENGLISH : Waxflower, East Indian rosebay.

INDIAN : Tagara, Chandni.

SANSKRTA : Tagar, Ananta, Nandīvṛkṣa, Nadyāvarta, Viṣṇupriya, Tārāvata.

HABIT : A low evergreen shrub with small canopy.

HABITAT : Indigenous to India and cultivated in gardens for its scented flowers.

CHEMICAL COMPOSITION : Two new alkaloids. Leaves contained coronaridine, voacristine, tabernaemontanine and dregamine. Bark of stem and root contains a crystaline substance and alkaloids tabernaemontanine and coronarine.

Parts used : Root, root-bark, leaves, stem-bark, flowers, wood, and milky juice.

Ayurvedic properties :
Rasa - Tikta, Kaṣāya. Vīrya - Ūṣṇa.
Guṇa - Laghu, Śīta. Vipāka - Kaṭu.

तगरं शीतलं तिक्तं दृष्टिदोषविनाशनम्।
विषार्त्तिशमनं पथ्यं भूतोन्मादभयापहम्।।
(राजनिघण्टु)

Tagaram śītalam tiktam dṛṣṭidoṣa vināśanam,
Viṣārttiśāmanam pathyam
bhūtonmādabhayāpaham.
(Rājanighaṇṭu)

Actions/Uses : Śītala, Jvaraghna, Vedanāsthāpana, Śāmaka, Garbhāśaya-uttejaka, Vraṇaropaṇa.

Therapeutics :

Tagar or Anant plant is febrifuge, pain-killer and stimulant for uterus. Root is acrid, bitter and used as local anodyne. For tooth-ache root is chewed and gum is applied to wounds which acts as refrigerant and helps granulation in healing. For eradication of "Aarasheeche ful" the root is rubbed on stone in lemon juice and applied in the eyes as ointment. Action of the root is on brain and nerves which results in activating body. The root is rubbed with water to make a paste which is applied to body during puerperal fever. Root-bark is anthelmintic. Juice of flowers is mixed with oil to alleviate burning sensation. During eye inflammation flowers are crushed and tied over eyes and gum is mixed with oil and applied to scalp. Bark and wood are refrigerant. Milky juice is used for eye diseases and applied to wounds to prevent inflammation.

Tabernaemontana heyneana Wall.

SYNONYMS : *Tabertnemontana alternifolia* Linn.; *Ervatamia heyneana (*Wall.*)* Cooke. *Ervatamia alternifolia (*Linn.*) Almeida.*

ENGLISH : East Indian rose-bay, Wax-flower.

INDIAN : Naagakuda; Kudaa-pandharaa No.2.

SANSKRTA : Śveta-kuṭaja, Nāgakuḍā.

HABIT : A shrub or small deciduous tree.

HABITAT : Common in open forest areas in Konkan and W. Ghats.

CHEMICAL COMPOSITION : Root contained coronaridine, voacangine, ibogamine, 19-oxycoronoridine and pseudoindoxyl of voacangine. Whole plant including roots, contains alkaloid coronaridine; fruit, lupeol-O Ac and β-amyrin-OAc. Fleshy skin contains three alkaloids including heyneanine and coronaidine. Leaves contain 7alkaloid isocoacristine. Methanol extract of leaves showed signficant activity against B16 melanoma in mice. Wood and stem bark yield biogenetically active camptothecine. Coronoridine given orally prevents pregnancy and is cytotoxic against P 388 lymphocytic leukemea. Camptothecine is widely used in China in treatment of various forms cancer.

Parts used : Wood, milky juice, root and stem-bark.

Ayurvedic properties :
Rasa - Tikta, Kaṣāya. Vīrya - Ūṣṇa.
Guṇa - Laghu, Śīta. Vipāka - Kaṭu.

श्वेतस्तु कुट्जस्तिक्तः कटुश्चोष्णोऽ अग्निदीपकः।
पाचकस्तुवरो रुक्षो ग्राहको रक्तदोषहा।।
कुष्ठातिसार पित्तार्शः कफतृट् कृमिहा मतः।
ज्वरं चामं च दाहं च नाशयेदिती कीर्तितः।।
(नि. र.)

Śvetastu kuṭajastikaḥ kaṭuścoṣṇo agnidīpakaḥ,
Pācakastuvaro rūkṣo grāhako raktadoṣahā.
Kuṣṭhātisārapittārśaḥ kaphatṛt kṛmihā mataḥ,
Jvaram cāmam ca dāham ca nāśayedifī kīrtitaḥ.
(Nighaṇṭuratnākara)

Actions/Uses : Śītala, Jvaraghna, Vedanāsthapana, Śāmaka, Garbhāśaya-uttejaka, Vraṇaropaṇa.

Therapeutics :

Uses similar to those of *T. coronaria.* Extracts from root, stem and leaves are anticancer against P 388 lymphocytic leukemia. Root and stem extract also active against L. 1210 lymphocytic leukemia.

Thevetia peruviana (Pers.) Schum.

SYNONYMS : *Cerbera thevetia* Linn.; *Thevetia neriifolia Juss.; Cascabela thevetia* (Linn.) Lippold., *Thevetia peruviana* (Pers.) Merr.

ENGLISH : Yellow oleander, Lucky nut.

INDIAN : Kannhera-piwalli, Kulkephul, Zard kunel, Thivati.

SANSKṚTA : Pītapuṣpa, Karavir-2, Pītakaravīra, Aśvaghna, Divyapuṣpa, Haripriya.
Hapuṣā, Aśvamāraka, Asvāntaka, Śatakuṇḍa.

HABIT : A large evergreen shrub or small tree with bell-shaped yellow flowers.

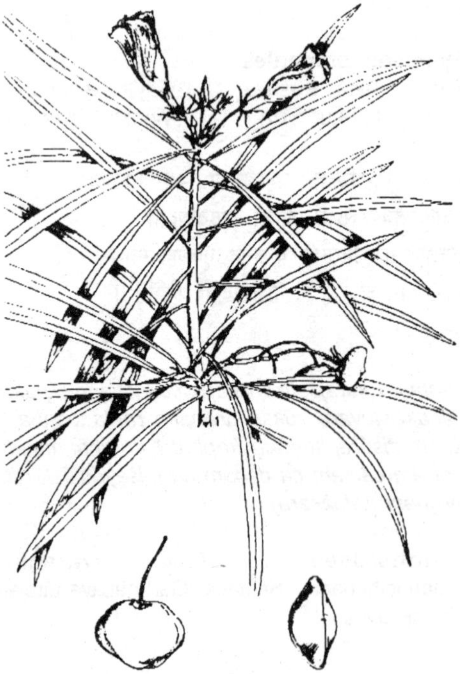

HABITAT : Native of tropical America and West Indies. Commonly cultivated as an ornamental plant in garden in India.

CHEMICAL COMPOSITION : L-(+-) Bornesitol from stems and leaves. Theveside and viridoside and thevefolin from seeds. Epiperuviol acetate, α- and β-amyrin, kaempferol and quercetin from fruit pericarp and flowers. Glucoside thevetin and another thevetoxin. Peruvoside and ruvoside from seeds. Peruvoside or ruvoside has positive inotropic effect followed by cardiac arrest in experimental animals.

Parts used : Root, leaves, bark, kernel, seeds and latex.

Ayurvedic properties :
Rasa - Kaṭu, Tikta, Kaṣāya. Vīrya - Ūṣṇa.
Guṇa - Laghu, Rūkṣa, Tīkṣṇa. Vipāka - Kaṭu.

करवीरः पीतपुष्पा ....।
विशोषात् मूत्रलो हृद्यः शोथहृ प्रकीर्तितः।
(द्रव्यगुणविज्ञान)

Karavīraḥ pītapuṣpaḥ . . . .,
Viśoṣāt mūtralo hṛdyaḥ śothahṛcca prakīrtitaḥ.
(Dravyaguṇa)

Actions/Uses : Jvaraghna, Niyatakālikajvarabandhaka, Dāhajanak, Viṣa, Atisāraka, Vamaka, Hṛdayasanvardhaka.

Therapeutics :

Plant is diuretic, cardiotonic and cures oedema. Roots in the form of plaster are applied to tumours. Leaves are emetic and purgative. Bark is bitter, is used as cathartic, febrifuge, useful in different kinds of intermittent fevers. The action of decoction of bark is strong and hence it is to be used in small quantity. In fevers it is fifteen times stronger than bark of cinchona.

This medicine should be taken only after food. It causes much perspiration, body becomes cold. Hot milk or liquor may be used if there is great fatigue. Apart from its usefulness in fevers action of the plant on heart is more important and is comparable with that of Digitalis purpura. Seeds are poisonous, abortifacient and alterative. They are used as purgative in dropsy and rheumatism. Kernel is acro-narcotic poison. Milky juice is highly poisonous and is used for suicide.

Wrightia tinctoria R. Br.

SYNONYMS : *Wrightia rothii G. Don.; Nerium tinctorium* Roxb.

ENGLISH : Sweet indrajao, Pala indigo-plant.

INDIAN : Kudaa-pandharaa, Mitha-Indrajau, Kaalli kudaee, Shwetakutaja, Kallaakado, Indrajau.

SANSKṚTA : Indrayava, Kṛṣṇakuṭaja, Hayamāraka.

HABIT : A deciduous tree. Flowers white, fragrant.

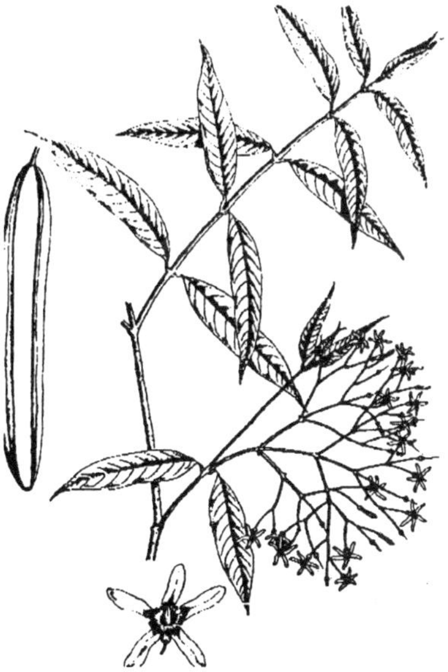

HABITAT : Common in open areas among bushes. Rajputana, Deccan, Konkan, Western ghats.

CHEMICAL COMPOSITION : Indican. Seeds yield 30-49 % fixed oil. β-Sitosterol, β-amyrin and its acetate and lupeol benzoate from the bark.

Parts used : Root, leaves, bark and seed.

Ayurvedic properties :
Rasa - Kaṭu, Tikta, Kaṣāya. Vīrya - Śīta.
Guṇa - Rūkṣa. Vipāka - Kaṭu.

इन्द्रयवा कटुस्तिक्ता शीता कफवातरक्तपित्तहरा।
दाहातिसारशमनी नानाज्वरदोषशूलमूलघ्नी।।
(राजनिघण्टु)

Indrayava kaṭustikta śīta kaphavātaraktapittaharā,
Dāhātisāraśamanī
nānānjvaradoṣaśūlamūlaghnī.
(Rājanighaṇṭu)

Actions/Uses : Kaphahara, Vātahara, Tridoṣahara.

Therapeutics :

Indrayava or Kṛṣṇakuṭaja is astringent, stomachic, tonic and febrifuge. In Karnatak, Tamilnadu and Madras where the tree is named as Jaundice curative tree. Juice of tender leaves is used efficaciously in jaundice. The juice of leaves is also used against serpent-bite. Crushed fresh leaves are filled in cavity of decayed tooth which releaves tooth-ache. If this touches gums or cheeks it causes inflammation of the parts. Leaves and bark are febrifuge, stomachic and tonic. In the Siddha system of medicine it is used in psoriasis and other skin diseases. Bark is tonic. Seeds are astringent, tonic, anthelmintic, antidiarrhoeal, febrifuge and aphrodisiac. They are used in seminal weakness and flatulence. They are known as "Godaa Indrajava". Given in small doses they improve functioning of liver, however they cause vomiting and diarrhoea to a great extent. Uses of bark and seeds are the same as those of Holarrhena antidysenterica.

Family : XI-Araceae.

Acorus calamus Linn.

SYNONYMS : *Acorus gramineous* Soland.

ENGLISH : Sweet Flag.

INDIAN : Wekhanda, Bhadra, Bach.

SANSKṚTA : Vacā, Bhadrā, Ugragandhā, Ugrāvāmani, Ṣaḍgranthā, Vijayā, Golomī, Lomaśā, Śatapatrikā, Kṣūdrapatrā, Rakṣoghnī, Mangalyā.

HABIT : A semi-aquatic, perennial, aromatic herb with creeping rhizomes and tuberous roots.

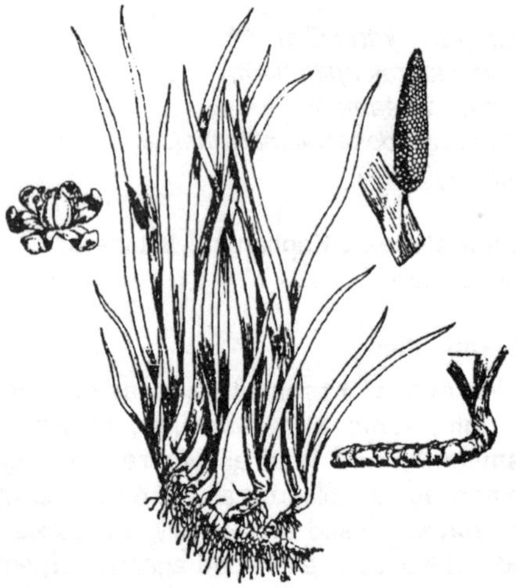

HABITAT : Grows in marshy places; plentiful in Kashmir and Sirmoor, in Manipura and Naga hills.

CHEMICAL COMPOSITION : Essential oil from rhizome, calamus oil has B-asarone as major constituent together with calamen, calamenol and calameon. Dry rhizomes contain a yellow aromatic volatile oil. Glucoside acorin. Leaves contain oxalic acid and calcium. Asarone is a mild sedative, a potent tranquillizer and like chlorpromazine a mild hypotensive and hypothermic substance.

Parts used : Dried rhizome.

Ayurvedic properties :
Rasa - Kaṭu, Tikta. Vīrya - Ūṣṇa.
Guṇa - Laghu, Tīkṣṇa, Sara. Vipāka - Kaṭu.

वचोग्रगन्धा कटुका तिक्तोष्णा वान्तिवहनिकृत्।
विवंन्धाध्मानशूलघ्नी शकृन्मूत्रविशोधिनी।।
अपस्मारककोन्मादभूतजन्त्वनिलान् हरेत्।
(भा. प्र.)

Vacogragandhā kaṭukā tiktoṣṇā vantivahanikṛt
Vibandhadhmānasūlaghnī śakṛnmūtraviśodhinī
Apasmārakaphonmādabhūtanantvanilān haret.
(Bhāvaprakāṣa)

Actions/Uses : Kaphaghna, Vātaghna, Pittakar, Medhya, Kanthya, Kṛmighna, Vibandhaghna, Śūlaghna, Vamanakara, Mūtraśodhanī, Apasmāraghna, Buddhivivardhanī, Medhobalai.

Therapeutics :

Vaca is an important drug capable of improving memory power and intellect. Root is used for fever, cough and toothache. Root extract is one of the constituents of toothpaste. Rhizome is pungent, carminative, emetic, laxative, diuretic, anthelmintic; improves appetite. It is nauseant, expectorant, antiseptic and nervine sedative; useful in treating bronchitis and asthma, in headache, teeth strenthening and in intermittent fevers either by itself or in conjunction with other tonics. It is serviceable in flatulence, flatulent colic, in atonic dyspepsia. It is employed for the cure of ague. Beneficial in diarrhoea and dysentery of children. Powdered rhizome is insecticide, useful against bedbugs, moths, flies. Primitives use it as carminative, and for catarrh, haemorhages and burns, also as nerve stimulant, bitter tonic and anthelmintic. One of the constituents of Ayurvedic drug "Myostal" useful in puerperal patients. Roots and rhizomes are used in various kinds of cancers. Aqueous extract and alcoholic extract of rhizome is hypothermic and hypotensive but not analgesic.

Amorphophallus campanulatus Blume

ENGLISH : Elephant Foot, Telugu potato.

INDIAN : Surannna, Arshoghna, Madan masta.

SANSKṚTA : Sūraṇa, Arśoghna, Kandavardhana, Kanḍūla, Sukanda, Kanda, Kandārha, Sthūlakandaka, Durnāmāri, Suvṛutta, Vātāri, Gudāmayahara, Tīvrakandaka, Kandasūraṇa, Kandāli, Rucyakanda, Vajrakanda, Citrakanda, Surakanda, Surendraka, Gudajaghna.

HABIT : A tuberous, stout herb.

HABITAT : Commonly cultivated in private compounds for the corms, which are cooked and eaten.

CHEMICAL COMPOSITION : Corms contain protein 1.2; fat 0.1; oxalic acid, 1.3 and mineral 0.8 %. Calcium 50.0; phosphorus, 34; iron, 0.6; thiamine; 0.06; riboflavin, 0.07; niacin, 0.7 mg/100 g. carotene 260 ug/100g; vitamin A, 434 I.U. The corms are irritant due the presence of calcium oxalate. Tubers contain protease inhibitors, trypsin and chymotripsin. Corms, triacorane, lupeol, betulinic acid, stigmasterol, triacontane, β-sitosterol and its palmitate and glucose, galctose, rhamnose and xylose.

Parts used : Rhizome.

Ayurvedic properties :
Rasa - Kaṭu, Kaṣāya. Vīrya - Uṣṇa.
Guṇa - Laghu, Rūkṣa, Tīkṣṇa. Vipāka - Kaṭu.

सूरणो दीपनो रुक्षः कषायः कण्डुकृत् कटुः।
विष्टंभि विशदो रुच्यः कफार्शःकृन्तनो लघुः।।
विशेषा दर्शसे पथ्यः प्लीहगुल्मविनाशनः।
सर्वेषां कन्दशाकानां सूरणः श्रेष्ठ उच्यते।।
दद्रुणां कुष्ठिनां रक्तपित्तिनां न हितो हि सः।
सन्धानयोगं सम्प्राप्तः सूरणो गुणवत्तरः।।
(भा. प्र.)

Sūraṇo dīpano rūkṣaḥ kaṣāyaḥ kaṇḍukṛt kaṭuḥ,
Viṣtambhi viśado rucyaḥ kaphārśaḥ kṛntano lghuḥ.
Viśeṣādarśase pathyaḥ plīhagulmavināśanaḥ,
Sarveṣām kandaśākānām sūraṇaḥ śresṭha ucyate.
Dadruṇam kuṣṭhinām raktapittinam na hito hi saḥ,
Sandhānayogam samprāptaḥ sūraṇo guṇavattaraḥ.
(Bhāvaprakāṣa)

Actions/Uses : Kaphavātaśāmaka, Arśoghna, Śothahara, Vedanāsthāpana.

Therapeutics :

Arśhoghna is iritant, acrid, carminative, astringent, tonic, stomachic, restorative, and anthelmintic. It cures piles, colic, abdominal tumours, intestinal worms and obesity and is used in enlargement of spleen, asthma, bronchitis, vomiting and elephantiasis. It is called "Arshoghna" meaning killing piles. It improves action of liver and removes constipation and is useful in leucorrhoea. Useful in piles and dysentery. When fresh acts as an acrid stimulant and expectorant and much used in acute rheumatism. Herb is used in constipation, diarrhoea, cholera, haemorrhages, earache, pain, intercostal neuralgia, pleurisy, pneumonia, and perpueral fever, swelling of throat and difficulty in breathing, enlarged spleen, pimples, septicemia, rinderpest and kala-azar. Root is used in boils and ophthalmia, also as emmenagogue. Tubers are used against abdominal tumors. Vegetable of tuber is useful during intestinal diseases.

Amorphophallus sylvaticus (Roxb) Kunth.

SYNONYMS : *Synantherias sylvatica* (Roxb.) *Schott.; Amorphophallus zeylanicus* Blume.

INDIAN : Suranna-jangali, Wajrakanda, Madanamasta.

SANSKṚTA : Lohita-sūraṇa, Vanasūraṇa, Vajrasūraṇa, Madanamasta, Vajrakanda, Araṇyasūraṇa.

HABIT : Bulbiferous herb.

HABITAT : Tamilnadu state in the Circars, Nilgiris and Coorg. Quite common herb along forest borders.

CHEMICAL COMPOSITION : Contains Calcium oxalate just like onion scales.

Parts used : Rhizome.

Ayurvedic properties :

Rasa - Kaṭu, Kaṣāya. Vīrya - Ūṣṇa.
Guṇa - Laghu, Rūkṣa, Tīkṣṇa. Vipāka - Kaṭu.

लोहितः सुरणः कण्डूकर्ता विष्टम्भको लघुः।
तुवरो रुक्षकटुको रुच्यो विशद्दीपनः।।
पाचनः पित्तलो दाही कृमिवातकफार्शहा ।
श्वासं कासं वमिं शूलं गुल्मं स्थौल्यं च नाशयेत्।।
(निघण्टु रत्नाकर)

Lohitaḥ suraṇaḥ kaṇḍūkartā viṣtambhako laghuḥ,
Tuvaro rukṣakaṭuko rucyo viśaddīpanaḥ.
Pācanaḥ pittalo dāhi kṛmivata kaphārśahā,
Śvāsam kāsam vāmim śūlam gulmam sthailyam ca nāsayet.
(Nighaṇṭuratnākara)

वनसुरण कन्दस्तु विशेषादर्शसां हितः।
गुल्मे स्थौल्ये तथा वाते श्लेष्मवाते हितः परम्।।
(कै. नि.)

Vanasuranakaṇdastu viśeṣadarśam hitaḥ,
Gulme sthaulye tathā vāte śleṣmavāte hitaḥ param. (Kaiyadevanigaṇṭu)

वज्रसुरण को रुच्यः कटुष्णः कृमिनाशनः।
गुल्मशूलादिदोषघ्नः स चारोचकहारकः।।
(शालिग्राम निघण्टु)

Vajrasūraṇako rucyaḥ kaṭuṣṇaḥ kṛmināśanaḥ,
Gulmaśūlādidoṣaghnaḥ sa cārocakahārakaḥ.
(Sāligramanighaṇṭu)

Actions/Uses : Kaphavātaśāmaka, Vājikara.

Therapeutics :

Vanasuraṇa or Madanamasta is an itching variety. It is suspected to be poisonous to human beings and livestock. Bark of corms is cut into pieces dried and woven into a string and sold as "Madanmasta". Seed is a cure for toothache and gland enlargement. Used against tumours. Flour of tuber mixed with milk and sugar is used as aphrodisiac. It stimulates urinary tract, causes itching of penis with erections.

Family : XII - Arecaceae.

Areca catechue Linn.

ENGLISH : Areca palm; Betel nut.

INDIAN : Supari, Pugiphala.

SANSKṚTA : Pūga, Kramuka, Guvāka, Dīrghapādapa, Ghonṭa, Muni, Khapura, Vallataru, Dṛdhavalka, Cikkāṇa, Pophaḷa.

HABIT : A tall palm.

HABITAT : Rocky beach, throughout India and the islands of the Eastern Archipelago.

CHEMICAL COMPOSITION : Several alkaloids belonging to pyridine group. Most important being arecoline; also present are arecaidine, guvacine, isoguvacine. Leucocyanidin from nuts. Aqueous extract of nuts vasoconstractor and adrenaline potentitisor in rats.

Parts used : Seed or kernel, root and leaves.

Ayurvedic properties :
Rasa - Madhura, Kaṣāya. Vīrya - Śīta.
Guṇa - Gurū, Rūkṣa. Vipāka - Kaṭu.

पूगं गुरु हिमं रुक्षं कषायं कफपित्तजित्।
मोहनं दीपनं रुच्यं आस्यवैरस्यनाशनम्।।
आंद्र तद् गुर्वभिष्यन्दि वह्निदृष्टिहर स्मृतम्।
स्विन्नं दोषत्रयच्छेदि दृढमध्य तदुत्तमम्।।
(भा. प्र.)

Pūgam gurū himam rukṣam kaṣāyam kaphapittajit,
Mohanam dīpanam rucyam āsyavairasyanāśanam.
Ārdram tad gurvabhiṣyanda vahnidṛṣṭiharam smṛtam,
Svinnam doṣatrayacchedi dṛdhamadhyam taduttamam.
(Bhāvaprakāṣa)

Actions/Uses : Kaphapittaśāmaka, Tridoṣaśāmaka, Vraṇaropaṇa, Stambhaka, Vikāśī, Rucikara, Dīpana, Śukrastambhana, Āsyavairaśyanāśana.

Therapeutics :

Pūga nut is aphrodisiac, astringent, anthelmintic, nervine tonic, emmenagogue and taenicide; useful in urinary disorders. It is used in veternary medicine as vermifuge for tapeworms. Root is used for liver troubles and its decoction to cure sore lips. Leaves are used for cough and bronchal troubles. Juice of tender leaves is mixed with oil for application in lumbago. Young green shoots are abortifacient in early pregnancy. Bark is deobstruent in flatulence and dropsy and also for choleric affections. Plant is used in rhagades, smallpox, venereal sores, syphilis, cholera and dysentery; also used for fractured bones..

Borassus flabellifer Linn.

SYNONYMS : *Borassus flabelliformis* Roxb.

ENGLISH : Palmyra palm, Brab tree.

INDIAN : Taada.

SANSKṚTA : Tāla, Tāladruma, Patrī, Dīrghaskandha, Cirāyu, Prānśū, Āsavadru, Khala, Dhvajadruma, Tṛṇarāja, Drudhacchada, Madyayonī, Gajabhakṣa, Mahonnata, Śaṭha, Dīrghapatra, Durāruha, Drumeśvara, Śrībīja, Gucchapatra, Dīrghadaṇḍa, Lekhyapatra.

HABIT : A very tall, erect, magnificient dioecious palm.

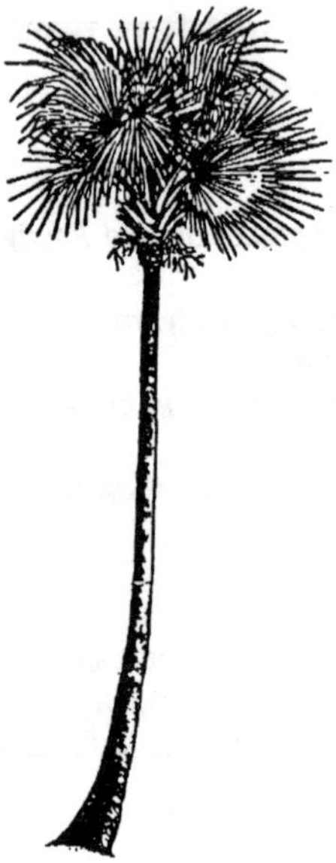

HABITAT : A commonly cultivated tree. Flower stalks are used for preparing liquor.

CHEMICAL COMPOSITION : A good source of sugars and vitamin B complex or biologically available riboflavin.

Parts used : Fruit, leaves and salts.

Ayurvedic properties :

Rasa - Madhura. Vīrya -Śīta.
Guṇa - Gurū, Snigdha. Vipāka - Madhura.

पक्वं तालफलं पित्तरक्तश्लेमविवर्धनम्।
दुर्जरं बहुमूत्रं च तंद्राभिष्यन्दशुक्रदम्।।
तालमज्ञा तु तरुणः किंचिन्यदकरो लघुः।
श्लेष्मलो वातपितघ्नः सस्नेहो मधुरः सरः।।
तालजं तरुणं तोयं अतीव मदकृन्मतम्।
अम्लीभूतं तदा तु स्यात् पित्तकृद्वातदोषहृत्।।
(भा. प्र.)

Pakvam tālaphalam
pittaraktaśleṣmavivardhanam,
Durjaram bahumūtram ca
tandrābhiṣyandaśukradam.
Tālmajjā tu taruṇaḥ kincinmadakari lghuḥ,
Śleṣmalo vātapittaghnaḥ sasneho madhuraḥ saraḥ.
Tālajam taruṇam toyam atīva madakṛnmatam,
Amlībhūtam tadā tu syāt pittakṛd vātadosahṛt.
(Bhāvaprakāṣa)

Actions/Uses : Vātapittaśāmaka, Raktastambhaka, Dāhaśāmaka, Śothahara, Vraṇaropaṇa.

Therapeutics :

Tāla root is cooling, restorative diuretic and anthelmintic, used as a cure for gonorrhoea. Decoction of young root and juice from young terminal buds and leaf stalks is used in gastritis and hiccups. Juice when fresh is diuretic, cooling, stimulant and antiphlogistic. Pulp from unripe fruit is diuretic, demulcent and nutritive. Terminal buds are nutritive and diuretic. Plant is used in heat stroke, headache, earache, nausea and vomiting, epilepsy, convulsions, adenitis, scrofulosacolli, scabies, pain, sores, burns, syphilis, ulcer of palate, spiderlick, menorrhagia and haemorrhage, septicaemia. Powder of burnt bark or decoction is used in dentifrice. Useful as antacid in heart burn. Fruit juice as tonic for asthamatic, anaemic and leprosy patients.

Cocos nucifera Linn.

ENGLISH : Cocoanut.

INDIAN : Naaralla, Naarikela, Maada.

SANSKRTA : Nārikela, Dūdhabīja, Tunga, Dṛdhavṛkṣa, Skandaphala, Dṛdhanīra, Māngalya, Mahāphala, Payodhara, Sadāphala, Pūtodakam, Kurcaśīrṣya, Dakṣinātya, Toyagarbha, Vastuphala, Dīrghaphala, Lanalī.

HABIT : A tall palm.

HABITAT : Cultivated in the hot damp regions of India, especially near the sea.

CHEMICAL COMPOSITION : A mannan and a water-soluble galactomannan from kernel. Enzymes, invertin, oxydase, catalase. Milk, histidine, arginine, lysine, tyrosine, tryptophan, proline, leucine, alanine. Albumin, globulin and prolamine fractions separated.

Parts used : Flowers, root, fruit, oil and ash.

Ayurvedic properties :

Rasa - Madhura. Vīrya - Śīta.
Guṇa - Gurū, Snigdha. Vipāka - Madhura.

नारिकेलं गुरु स्निग्धं पित्तघ्नं स्वादु शीतलम्।
बलमांस प्रदं हृद्यं बृंहणं बस्तिशोधनम्।।
(सु.सु. ४६)

Nārikelam guru snigdham pittaghnam svādu śītalam,
Balamānsapradam hṛdyam bṛnhaṇam bastiśodhanam.
(Su.Śu.46)

विशेषतः कोमलनारिकेलं निहन्ति पित्तज्वरपित्तदोषान्।
तदेव जीर्णं गुरुपित्तकारी विदाही विष्टम्भि मतं भिषग्भिः।
तस्याम्भः शीतलं हृद्यं दीपनं शुक्रलं लघु।
पिपासा पित्तजित् स्वादु बस्तिशुद्धिकरं परम्।।
(भा. प्र.)

Viseṣataḥ komalanārikelam nihanti pittajvarapittadoṣān,
Tadeva jīrṇam gurupittakarī vidāhī viṣṭambhi matam bhiṣagbhiḥ,
Tasyāmbhaḥ śītalam hṛdyam dīpanam śukralam laghuḥ,
Pipāsā pittajit svādu bastiśudhikaram param.
(Bhāvaprakāṣa)

Actions/Uses : Vātaghna, Pittaghna,

Kaphavardhaka. Dāhaśamanī, Keṣya, Vraṇaropaṇa, Anulomanī, Dīpana, Śūlapraśamanī, Raktapittapraśamanī, Vājikara, Hikkāghna, Mūtrala, Mūtravirecani, Brihmana, Ārtavajanani, Balya, Jvaraghna, Hṛdya, Viṣṭambhi, Tarpaṇa, Prinana, Māmsakara, Vṛṣya.

Coconut water : Śīta, Snigdha, Madhura, Dīpani, Hṛdya, Laghu, Vastiśodhana, Śūkrala, Tṛṣṇāasamanī, Dāhaśamanī.

Oil : Brihmana, Balya, Keṣya, Danya, Madhura.

Flower : Raktapittaśamanī, Pramehaśamani, Dīpanī, Somarogaghna, Raktātisāraghna.

Therapeutics :

Nārikela root is diuretic, astringent and used in uterine diseases, gonorrhoea and wounds. Leaves are used to treat abscesses. Shoots and ashes of dry meat used in deep cuts, general debility and swollen womb. Flowers are astringent, mixed with oil and applied to swellings. Water of unripe fruit is cooling, useful in thirst, fever and urinary disorders and is a source of K for cholera patients. Fruit is sweet, tonic, diuretic, laxative, sedative, anthelmintic, stomachic, styptic, and aphrodisiac. They are useful dyspepsia, and burning sensation. Grated meat is used to treat burns and abscesses. Oil is used as a local aplication in alopecia.

Phoenix dactylifera Linn.

HABIT : A medium sized palm.

HABITAT : Cultivated in many parts of India.

ENGLISH : Date palm.

INDIAN : Khajura, Pindakhejur.

SANSKṚTA : Piṇḍa-kharjūra, Kharjūra, Yavaneṣṭā, Kharaskandha, Durāruha, Duṣpradharṣa, Kaṣāya, Haripriya.

CHEMICAL COMPOSITION : Vitamin B and antiscorbutic vitamin; fruit contains vitamin A, B and D. Hemicellulose from pollen grains contained arabinose, galactose, xylose, rhamnose and uronic acid.

Parts used : Fruit, flowers and roots.

Ayurvedic properties :

Rasa	- Madhura.	Virya	- Śita.
Guna	- Gurū, Snigdha.	Vipāka	- Madhura.

पिंडखर्जूरिका वृष्या स्वाद्वी शीता गुरुर्मता।
अग्निमाद्यं कृमींश्चैव करोतीती बुधैर्मता।।
धातुवृद्धिं च तृप्तिंच पुष्टिं चैव करोति हि।
हृद्या बल्या दुर्जराच स्निग्धा पाके रसे मधुः।।
रक्तपित्तं च पित्तं च दाहं श्वासं कफं श्रमम्।
क्षतक्षयं विषं तृष्णां शोषं चैव विनाशयेत्।।
(निघण्टु रत्नाकर)

Pindakharjurikā vṙsyȧ svādvi śīta gururmatā, Agnimandyam kṙminscyaiva karotiti budhairmatā.

Dhátuvrddhim ca trptim ca puśtim caiva karoti hi,
Hŕdyá balyá durjará ca snigdhá pákė rase madhuh.
Raktapittam ca pittam ca dáham śvásam kapham śramam,
Ksátakśayam viśam tŕsnám śosam caiva vináśayet.
(Nighaṇṭurutnakaru)

Actions/Uses : Vātapittaśámaka, Snehana, Vátánulomana, Stambhana.

Therapeutics :

Kharjura fruit is demulcent, expectorant, nutrient, laxative, aphrodisiac; prescribed in asthma, chest complaints and cough; also in fever, gonorrhoea etc. Fresh juice is cooling and laxative. Gum useful in diarrhoea and diseases of genito-urinary system.

Phoenix sylvestris Roxb.

ENGLISH : Date sugar palm, Wild date.

INDIAN : Shindi, Khajuri, Boichand.

SANSKṚTA : Kharijuri, Viṭakharjūra.

HABIT : A medium sized palm.

HABITAT : Common throughout India wild or more often cultivated.

CHEMICAL COMPOSITION : Vitamin content of palm gur is as follows: thiamine, riboflvin, nicotinic acid, and vitamin C. Mineral contents are calcium, phosphorus, iron, sodium and potassium; copper, cobalt, nickel, magnesium and molybdenum are present.

Parts used : Fruit and leaves

Ayurvedic properties :
Rasa - Madhura. Vīrya - Śīta.
Guṇa - Gurū, Snigdha. Vipāka - Madhura.

खुर्जूरी तुवरा ज्ञेयासापक्काम मधुरामता।
तुवराशीतला वृष्याश्लेष्म शुक्रविवर्धिनी।।
लघ्वीच कृमिकृंज्ञेयांवातपित्त मदापहा।
मूर्च्छां मदात्ययंदाहंक्षयंचैव विनाशयेत्।।
(नि. र.)

Kharjūrī tuvarā jneya sāpakvāma madhurā matā,
Tuvarā śītalā vṛṣyā ślesmaśukravivardhinī.
Laghvīca kṛmikṛnjneyām vātapitta madāpahā,
Mūrcchām madātyayam dāham kṣayam caiva vināśayet.
(Nighaṇṭuratnākara)

मधुरं बृंहणं वृष्यं खर्जुरं गुरु शीतलं।
क्षयेऽभिघाते दाहे च वातपित्ते च तद्हितम्।।
(च. सू. २७)

Madhuram bṛmhanam vṛṣyam kharjūram guru śītalam,
Kṣaye abhidhāte dāhe ca vātapitte ca tad hitam.
(Ca.Śú.27)

Actions/Uses : Vātapittaśāmaka, Vṛṣya.

Therapeutics :

Kharijūri or Viṭakharjūra root is used in toothache. Fruit is tonic and restorative. Juice of tree is used as a cooling beverage. Fresh unfermented sap (Nira) is a refreshing sweet drink. After spontaneous fermentaion for 8 to 10 hours it is converted into toddy. It is good source of vitamin B group. Palm gur is more nutritious than cane gur. Kernel made into paste with root of Achryantes aspera, is eaten with betel leaves as remedy for ague.

Family : XIII - Aristolochiaceae.

Aristolochia bracteata Retz.

SYNONYMS : *Aristolochia bracteolata* Lamk.

ENGLISH : Bracteated birthwort, Worm-killer.

INDIAN : Kidamaara, Dhumrapatra, Gandhani, Gaval, Kirmar.

SANSKṚTA : Kīṭamārī, Dhūmrapatra, Nakuli, Ahigandha, Sunanda, Pāniri, Kīḍāmārī, Patra-banga, Gridhrani.

HABIT : A slender, decumbent, glabrous perennial weed in cultivated fields.

HABITAT : In hedges in plains of North India, Andhra Pradesh, Konkan and Maharashtra, Karnataka and Bengal.

CHEMICAL COMPOSITION : Ceryl alcohol, β-sitosterol, aristolochic acid and KCl from leaves. Aristolochic acid, KCl and KNO_3 from roots.

Parts used : Roots, leaves and entire plant.

Ayurvedic properties :
Rasa - Tikta. Vīrya - Ūṣṇa.
Guṇa - Laghu, Rūkṣa, Tīkṣṇa. Vipāka - Kaṭu.

कटुष्णा कीटमारी स्यात् तिक्ता दीपनरेचनी।
कासगुल्मरजोदोष विषर्वा सर्पनाशिनी।।
विषमज्वरजिद्व्रण्या जन्तुघ्नी शोफशूलनुत्।
(स्व)

Kaṭūṣṇā kīṭamārī syāt tiktā dīpanarecanī,
Kāsagulmarajodoṣa viṣarvīsarpanāśinī.
Viṣamjvarajidvraṇyā jantughnī śophaśūlanuta.
(Svayamkṛti)

कीटमारी रसे तिक्ता दन्तकृमिविषापहा। (सो. नि.)

Kīṭamārī rase tiktā dantakṛmiviṣāpahā.
(So.Ni.)

वातश्लेषज्वराहरा सन्ध्यस्थाति प्रसारिणी (शा. नि.)

Vātaśḷeṣmajvaraharā sandhyasthīnī prasāriṇī.
(Shāligrāmanighaṇṭu)

Actions/Uses : Kaphavātaśāmaka, Vraṇaśodhana, Grāhī, Anulomana, Viṣghna, Śothahara, Vedanāsthāpana, Dīpana.

Therapeutics :

Plant is bitter, purgative, emmenagogue and anthelmintic. Root is an emmenagogue and as a dry powder or infusion employed to increase uterine contractions. Decoction of root is used to expel roundworms. Juice of leaves is applied to foul and neglected ulcers. Bruised leaf mixed with castor oil applied to eczema of children's legs.

Aristolochia indica Linn.

SYNONYMS : *Aristolochia lanceolata* Wight.

ENGLISH : Indian birthwort.

INDIAN : Saapasana, Eeshwaree, Kadula, Sapashi, Isharmul.

SANSKṚTA : Gārudī, Isvarī, Isvaramūla, Jatāvallī, Rudrajatā, Raudrijatā, Rudrā, Saumya, Ahigaṇḍa, Sugandhikā, Surabhī, Subahā, Patravallī, Nākulī, Candralatā, Suputrā, Surāhvayā, Netrapuṣkar, Sunanda, Arkamūla, Dhanā, Viṣaghnī, Viṣavega, Karallaka.

HABIT : A perennial climber with greenish white woody stem.

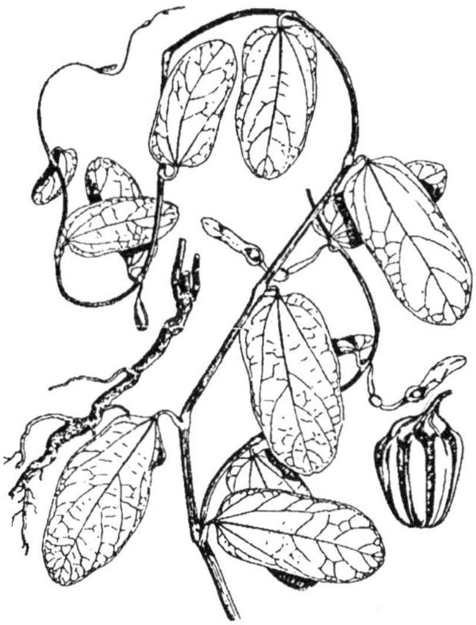

HABITAT : Konkan and Deccan Peninsula; Low hills and plains from Nepal to Chittagong.

CHEMICAL COMPOSITION : Aristolochic acid, aristolochine, a tetracyclic sequiterpene, ishwarane and aristolochene. Essential oil containing carbonyl compounds.

Parts used : Root, rhizome, leaves, stem and seeds.

Ayurvedic properties :
Rasa - Tikta, Kaṭu, Kaṣāya. Vīrya - Ūṣṇa.
Guṇa - Laghu, Rūkṣa. Vipāka - Kaṭu.

मूलं तिक्तमर्तावास्य सुगन्धं च कषायकम्।
विषहारि त्रिदोषघ्न दीपनं पाचनं स्मृतम्।।
कृमिघ्नमगदं प्रोक्तं ज्वरघ्नं कोष्ठशूलजित्।
श्वासकासाङ्गमर्द्दादीन् सन्धिशोफं च नाशयेत्।।
(स्व)

Mūlam tiktamatarvāsya sugandham ca kaṣāyakam,
Viṣahari tridoṣaghnam dīpanam pācanam smṛtam.
Kṛmighnamagadam proktam jvaraghnam koṣṭhaśūlajit,
Śvāsakāsāngamardādīn sandhiśopham ca nāṣayet.
(Svayamkrti)

जटा कटुरसा श्वासकासहृद्रोगनशिनी।
भूतविद्राविनी चैव रक्षसाऽच निवर्हिणी।।
(राजनिघण्टु)

Jaṭā kaṭurasā śvāsakāsahṛdroganāśanī,
Bhūtavidrāviṇī caiva rakṣasānca nivarhiṇī.
(Rājanighaṇṭu)

गारुडी सर्पविषजित् कुष्ठघ्नी कफवातला।।
(हृदयप्रिया)

Gāruḍī sarpaviṣajit kuṣṭhaghnī kaphavātalā.
(Hṛdayapriyā)

Actions/Uses : Vātapittakapha-śāmaka, Kaphavātasodhana, Viṣaghna, Vraṇaśodhana,

Vedanāsthāpana, Ṡothahara.

Therapeutics :

Isvarī has the power to neutralise or resist snake poison. It is purifier of blood and hence useful in skin diseases. It heals wounds and is an appetiser, aphrodisiac, anthelmintic; relieves burning sensation. Root and rhizome are tonic, and used as gastric stimulant, emetic and emmenagogue. They are useful in dropsy and rubbed with honey they are given in white leprosy. Decoction of roots is stimulant and febrifuge and is used in impotency. In combination with black pepper and ginger it is given in bowel complaints of children and in intermittent fever. In powder form it is given in honey for leucoderma and used in snake-bite. Decoction stem is stimulant, tonic and febrifuge. Juice of fresh leaves is used in treating cough of children by inducing vomiting without depression. Juice of leaves as also of bark is chiefly used in bowel complaints of children, cholera and diarrhoea and in intermittent fevers. Seeds are useful in treating inflammations, biliousness and dry cough.

Family : XIV - Asclepiadaceae.

Asclepias curassavica Linn.

ENGLISH : Wild Ipecacunanha.

INDIAN : Kaakatundi, Raktapushpa, Kurki.

SANSKṚTA : Bhāradvāji, Kākatuṇḍi, Raktapuṣpā, Dugdhakṣupa, Vanapicula.

HABIT : An erect, simple or much-branched perennial undershrub.

HABITAT : Native to tropical America. Naturalized in many parts of India, often grown in gardens.

CHEMICAL COMPOSITION : Glucosides asclepiadin and vincetoxin. Asclepin, chief active principle, excellent cardiotonic.

Parts used : Fowers and roots

Ayurvedic properties :
Rasa - Tikta, Kaṣāya. Vīrya - Ūṣṇa.
Guṇa - Laghu, Rūkṣa, Śīta, Tīkṣṇa.
Vipāka - Kaṭu.

भारद्वाजी हिमारुच्या व्रणशस्त्रक्षतापहा।
(राजनिघण्टु)

Bhāradvājī himārucyā vraṇaśastrakṣatāpahā. (Rājanighaṇṭu)

Actions/Uses : Vāmaka, Yakṛtottejaka, Āmāśayottejaka, Pittasrāvī, Svedajanana, Kaphaghna, Ānulomika and Kṛmighna.

Therapeutics :

Bhāradvājī or Kākatuṇḍī is poisonous and is used in phthisis. Whole plant is considered to be emetic, styptic and purgative. It has cardiotonic effect. Root is emetic, cathartic and used as an astringent and a remedy in piles and gonorrhoea. Powder of root mixed with equal quantity of Acorus root is given in chronic ulcers. The root decoction is used against cancer. The juice of leaves is antidysenteric, anthelmintic, sudorific and is used against haemorrhages and gonorrhoea; useful externally in piles and in abdominal pain. The fresh or dried and pulverised leaf is reported to be used against cancer. Various plant parts, as also plant latex are used against warts and cankers.

Calotropis gigantea R. Br.

SYNONYMS : *Asclepias gigantea* Linn.

ENGLISH : Gigantic Swallow Wort.

INDIAN : Ruee, Arka.

SANSKṚTA : Arka, Sūryāśva, Sadāpuṣpī, Ksatakśīrī, Kśīraparṇa, Arkaparṇa, Vikīraṇa, Sūryapuṣpaka, Kṣīrī, Alarka, Puṣpī, Āshpota, Bhānu, Tanuphala, Vasuka, Kāṣṭhila, Pratīyasa, Aharpati, Jambala, Mandāra, Surtapatra, Aharbāndhava, Aharmaṇī, Āsphotaka.

HABIT : A small tree. Flowers lilac or dull white in lateral or terminal panicles of umbellate cymes.

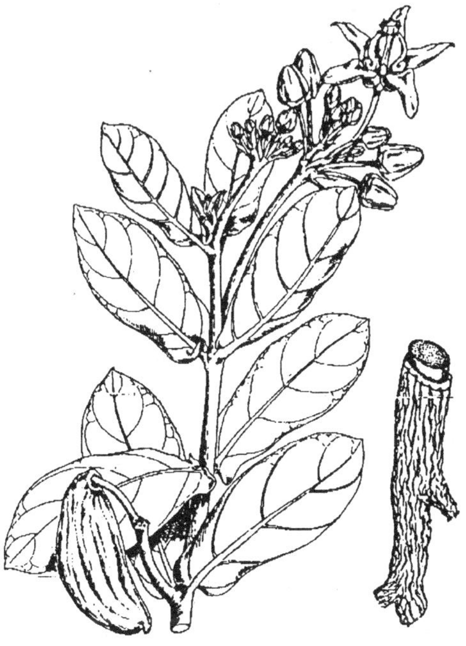

HABITAT : Common throughout India.

CHEMICAL COMPOSITION : A proteolytic enzyme, somewhat similar to papain has been isolated from milky juice.

Parts used : Root, root-bark, leaves, juice of plant and flowers.

Ayurvedic properties :
Rasa - Kaṭu, Tikta, Madhura. Vīrya - Ūṣṇa.
Guṇa - Sara, Snigdha, Laghu. Vipāka - Kaṭu.

अर्कस्तु कटुरुष्णश्च वातजिद्दीपनीयकः।
शोफव्रणहरः कण्डूकष्ठकृमिविनाशनः।।
श्वेतार्कः कटुतिक्तोष्णो मलशोधनकारकः।
मूत्रकृच्छ्रास्त्रशोफार्त्ति व्रणदोषविनाशनः।।
राजार्कः कटुतिक्तोष्णः कफमेदोविषापहः।
वातकुष्ठव्रणान् हन्ति शोफकण्डू विसर्पनुत्।।
(रा नि.)

Arkastu kaṭuruṣṇaśca vātajiddīpanīyakaḥ,
Śophavraṇaharaḥ kaṇḍukuṣṭhakṛimivināśanaḥ.
Śvetārkaḥ katutiktoṣṇo malaśodhana kāralaḥ
Mūtrakṛcchrāsra śophārttivraṇadoṣa vinasānaḥ.
Rajarkaḥ kaṭutiktoṣṇaḥ kapha medo viṣāpahaḥ,
Vāta kuṣṭha vraṇān hanti śopha kaṇḍū visarpanut.
(Rājanighaṇṭu)

Actions/Uses : Kaphaghna, Vātaghna, Pittaprakopī. Kṛmighna, Kuṣṭhaghna, Vraṇaghna, Arśoghna, Gulmaghna, Viṣahara.

Therapeutics :

Arka is purgative, alexipharmic, anthelmintic; cures leprosy, leucoderma, ulçers, tumours, piles, diseases of spleen, liver, and abdomen. Juice is anthelmintic and laxative; cures piles and kapha. Dried and powdered plant is taken with milk acts as a good tonic. Action is similar to digitalis on the heart. Root-bark and juice have emetic, diaphoretic, alterative and purgative properties. It is used in dysentery and as a substitute for Ipecacuantha. It is regarded as a great remedy in syphilitic affections and is called "vegetable mercury". In intermittent fevers it is used as antiperiodic and diaphoretic. It cures asthma and syphilis. In form of paste applied to elephantiasis. Tincture of leaves used in intermittent fevers. Latex is bitter, heating, oleagenous and irritant, used in combination with Euphorbia neriifolia as purgative. Flowers are sweet, bitter, digestive, stomachic, tonic, anthelmintic, analgesic, astringent; cure inflammations, tumours, kapha; and are good in ascites.

Calotropis procera (Ait.) R. Br.

SYNONYMS : *Calotropis procera Ait. f. ssp. hamittonii Ali.; Calotropis procera auct. non Ait.; Calotropis mamiltonii Wight.; Asclepias procera Willd.*

ENGLISH : Gigantic Swallow Wort, Madar.

INDIAN : Akada, Maandaara, Safed-ak.

SANSKṚTA : Alarka, Adityapuṣpika, Dīrghapuṣpa, Arka, Bimbora, Ganarupi, Kaṣṭhila, Raktārka, Sadāpuṣpa.

HABIT : A woody perennial shrub. Flowers pink, spotted with purple.

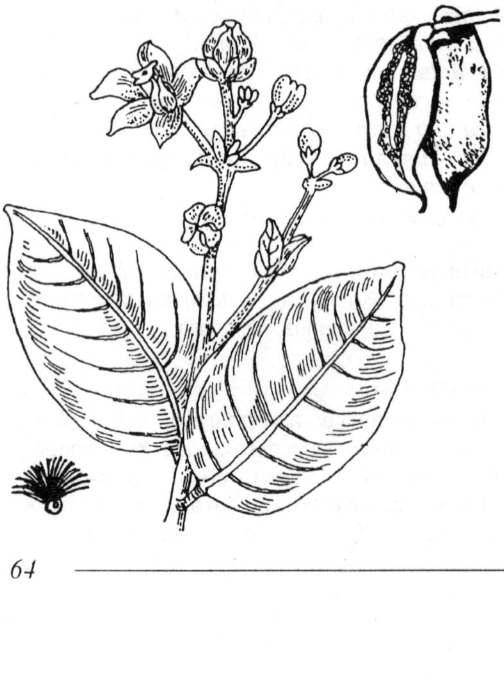

HABITAT : Throughout India in warm dry places.

CHEMICAL COMPOSITION : Leaves and stalks contain calotropin and calotropangenin. Latex contains usharin, calotoxin and calactin. Flowers contain evandin 3-rhamnoglucoside.

Parts used : Root-bark, latex, leaves and flowers.

Ayurvedic properties :
Rasa - Kaṭu, Tikta. Vīrya - Ūṣṇa.
Guṇa - Laghu, Rūksa, Tīkṣṇa. Vipāka - Kaṭu.

अर्कद्वयं सरं वातकुष्ठकण्डू विषव्रणान्।
निहन्ति प्लीहगुल्मार्शः श्लेमोदरशकृत्कृमीन्।।
अळर्ककुसुमं वृष्यं लघुः दीपन पाचनम्।
अरोचक्रः प्रसेकार्शः कासश्वासनिवारणम्।।
रक्तार्कपुष्पं मधुरं सतिक्तं कुष्ठकृमिघ्नं कफनाशनञ्च।
अर्शोविषं हन्ति च रक्तपित्तं संग्राहिगुल्मे श्वयथौ हितंतत्।।
क्षीरमर्कस्य तिक्तोष्णम् स्निग्धम सलवणम् लघु।
कुष्ठ गुल्मोदरहरम् श्रेष्ठमेतद्विरेचनम्।। (भा.प्र.)

Arkadvayam saram
vātakuṣṭhakaṇḍūvisavraṇān,
Nihanti plihagulmārśaḥ śleṣmodarśakṛtkṛmīn,
Aḷārkakusumam vṛṣyam laghu
dīpanapācanam,
Arocakaḥ prasekārśaḥ kāsaśvāsa nivāranam.
Raktārkapuṣpammadhuram satiktam,
Kuṣṭhakṛmighnam kaphanāśanañca,
Arśoviṣam hanti ca raktapittam,
Sangrāhi gulme śvayathau hitam tat.
Kṣīramarkasya tiktoṣṇam snigdham sa
lavaṇam laghu,

Kuṣṭha gulmodara haram śreṣṭhametad virecanam. (Bhāvaprakāṣa)

Actions/Uses : Kaphagna, Vātaghna, Pittaprakopī.

Therapeutics :

Properties and uses of Alārka are the same as of *C. gigantea.* Alarka is an alterative, tonic and diaphoretic, in large doses emetic. Root bark useful for treating chronic cases of dyspepsia, flatulance, constipation, loss of appetite, indigestion and mucus in stools. Leaves used against guinea worms. Flowers useful in asthma. Seed oil is geriatric and tonic. Green copra given in asthma. Plant used in spleen complaints, rheumatism, epilepsy, hemiplegia, sores, smallpox and protracted labour.

HOMEOPATHIC USE : It is used with marked success in treatment of syphilis; also in elephantiasis, leprosy and acute dysentery. Pneumonic phthisis, tuberculosis, increased circulation in the skin, sudorific. Heat in stomach is a good guiding symtom.

Gymnema sylvestre R. Br. ex Schult.

SYNONYMS : *Periploca sylvestris Retz., Asclepis geminata Roxb.*

HABIT : A climbing shrub.

HABITAT : Konkan, Maharashtra and Western Ghats.

ENGLISH : Small Indian Ipecacunaha, Ram's horn.

INDIAN : Medhashingi, Kaawallee, Godmari, Gur-mar.

SANSKRTA : Meṣaṣṛiṅgī, Śṛingī, Ajaṣṛingī, Madhunāśīnī, Śrivṛikṣa, Ghanasringi, Cakṣuhya, Bahulāngaka, Sarpadamstha, Dakṣināvarta, Medravalli, Cakranemī, Varatikta, Mahākālī, Tiktadugdha, Dvitīya, Supucchika, Pādavṛikṣa, Meṣaviṣāṇikā.

CHEMICAL COMPOSITION : Plant bases, choline, betaine etc. Leaves contain an antisaccharin principle gymnemic acid; gymnamine, nonacosane, hentriacontane, tritriacontane and conduritol.

Parts used : Roots and leaves.

Ayurvedic properties :

Rasa - Kaṣāya, Tikta.	Vīrya - Ūṣṇa.
Guṇa - Laghu, Rūkṣa.	Vipāka - Kaṭu.

मेषगृङ्गी रसे तित्का वातलाश्वासकासहृत्।
रुक्षा पाकेकटुः कुष्ठव्रणश्लेष्माक्षिशूलनुत्।।
मेषशृङ्गी फलं तित्कं कुष्ठमेह कफप्रणुेत्।
दीपनं संश्रणं कासक्रिमिव्रणविषापहम्।।
(भा. प्र.)

Meṣaṣṛṅgī rase tiktā vātalāsvāsakāsahṛt.
Rukṣā pāke kaṭuḥ kuṣṭhavraṇa ślesmākṣi śūlanut.
Meṣaṣṛngīphalam tiktam

kuṣṭhamehakāohapranut,
Dīpanam samsraṇam kāsakrmivraṇaviṣāpaham.
(Bhāvaprakāṣa)

Actions/Uses : Kaphavātaśāmaka, Śothahara, Dāhapraśamana, Kāsaghna, Timiraghna, Śvāsaghna, Vraṇaghna, Kuṣṭhaghna, Kṛmighna, Mehaghna.

Therapeutics :

Meṣaṣṛiṅgī is astringent, stomachic, antiperiodic, diuretic, tonic and refrigerant. Root is expectorant and emetic. Root and leaves are used in stomach pain. Extract of leaves is stimulant, diuretic, cardiovascular and hypoglycemic and has purgative action; are used in diabetes, chewed to reduce glycosuria. Leaf powder stimulates heart and the circulatory system, increases secretion of urine and activates uterus.

Hemidesmus indicus R. Br.

SYNONYMS : *Hemidesmus indicus Schult.; Asclepias pseudosarsa var. latifolia Roxb.; Periploca indica Wiild.; Periploca emetica Retz.*

HABIT : A perenial prostrate or slender, laticiferous, twining under shrub.

HABITAT : Upper Gangetic plain, Madhya Pradesh and South India and Konkan.

ENGLISH : Indian Sarsaparila, False sarsaparilla.

INDIAN : Anantamulla, Upalasari, Kapuri, Magrabu, Salsa, Kalisar, Dudvali.

SANSKṚTA : Sārivā, Anantamūla, Gopakanyā, Dādimapatra, Śāradī, Gopī, Dugdhagarbha, Candanagandha, Kṛṣodarī, Gopavallī, Anantā, Utpalasārivā, Nāga-jihvā, Sugandhā, Gopimulam.

CHEMiCAL COMPOSITION : Rutin from leaves. Hexatriacontane, lupeol, its octacosanoate, α-amyrin, β-amyrin, its acertate and sitosterol in roots. Air-dried roots contain essential oil containing methoxy benzaldehyde, sterols and a glucoside; saonin, resin acid and tannins.

Parts used : Roots.

Ayurvedic properties :
Rasa - Madhura, Tikta. Vīrya - Śita.
Guṇa - Gurū, Snigdha. Vipāka - Madhura.

शारिवा युगलं स्वादु स्निग्धं शुक्रकरं गुरु।
अग्निमान्धारुचि श्वासकासामविषनाशनम्।।
दोषत्रयास्त्रप्रदरज्वरातीसार नाशनम्।।
स्वेदनं मूत्रकृद् वल्यं परं वृष्यं रसायनम्।।
औपदंशिकरोगघ्नं सर्वचर्म्मविकारनुत्।
आमवातं वातरक्तं सूतरोगांश्च नाशयेत्।।
(भा. प्र.)

Śārivā yugalam svādu snigdham śukrakaram

guru,
Agnimāndyārucisvāsakāsāmviṣanāśanam,
Doṣatrayāsrapradarajvarātīsāranāśanam.
Svedanam mūtrakṛd balyam param vṛṣyam rasāyanam.
Aupadamśika rogaghnam sarva carmavikāranut,
Āmavātam vātaraktam sūtarogāmśca nāśayet.
(Bhāvaprakāṣa)

Actions/Uses : Vātaśāmaka, Pittaśāmaka, Kāsahara, Śvāsahara, Raktapradaraghna, Raktapittapraśhamanī, Mehanāśinī, Sangrāhaka, Arucihara, Viṣahara, Tridoṣaghna, Jvaraghna, Atisārghna, Kaṇḍūhara, Śukrakara, Kuṣṭhahara.

Therapeutics :

Sārivā root is sweet, bitter, cooling, tonic, aphrodisiac, alterative, astringent, demulcent, antipyretic, antidiarrhdoeal; cures leprosy, leucoderma, itching, skin diseases, fevers, foul odour from the body, loss of appetite, asthma, and bronchitis. Valuable alterative, tonic, demulcent, diaphoretic and diuretic. It is also sudorific. Useful in fever, skin diseases, as a blood purifier, in loss of appetite and disinclination for food and in nutritional disorders. Also syphilis, chronic rheumatism, gravel and other urinary diseases, leucorrhoea, in scorpion sting and snake bite. It is administered in the form of powder, infusion or decoction and as syrup. It is an ingredient of several medicinal preparations. It is used as a substitute for Sarsaparilla.

Oxystelma esculentum R. Br.

SYNONYMS : *Oxystelma secamone (Linn.) Karst.; Periploca secamone Linn.; Periploca esculenta Linn.f.; Asclepias rosea.*

HABIT : A perennial slender laticiferous twining herb with milky juice.

HABITAT : Throughout the plains and lower hills in India, usually near water.

INDIAN : Dudhannee, Dudhialata, Dudlata, Dudhika.

SANSKṚTA : Dugdhika, Dudhiālatā, Grahinī, Svaduparṇi, Tāmramūla.

CHEMICAL COMPOSITION : Not reported

Parts used : Whole plant, roots, milky juice and fruit.

Ayurvedic properties :
Rasa - Kaṭu, Tikta. Vīrya - Ūṣṇa.
Guṇa - Gurū, Rūkṣa. Vipāka - Kaṭu.

दुग्धिका स्वादुपर्णी स्यात्क्षीराविक्षीरिणी तथा।
दुग्धिकोष्णा गुरु रुक्षा वातलागर्भकारिणी।।
स्वादुक्षीरा कटुस्तिक्ता सृष्टमूत्रमलापहा।
स्वादुविष्टंभिनी वृष्या कफकाष्ठकृमिप्रणुत्।।
(भा. प्र.)

Dugdhikā svāduparṇī syātkṣīrā vikṣīriṇī tathā,
Dugdhikoṣṇā gurū rukṣā vātalā garbhakāriṇī.
Svādukṣīrā kaṭustiktā sṛṣṭamūtramalāpahā,
Svāduviṣṭāmbhinī vṛṣyā kaphakoṣṭha

kṛmipraṇut.
(Bhāvaprakāṣa)

Actions/Uses : Mūtral, Vātakar, Kāmoddipaka.

Therapeutics :

Dudhiālatā is hot, dry, bitter, pungent, diuretic, laxative, aphrodisiac, anthelmintic, lactagogue, and indigestible, causes flatulence. It is useful in leucoderma and bronchitis. Decoction of plant is used as gargle in aphthous ulceration of mouth and in sore throat. Roots considered specific for jaundice. Milky sap is used as a wash for ulcers. Fruit is bitter; tonic, expectorant, anthelmintic; juice is used in cough, lecuderma, pain in muscles, gleet and gonorrhoea; given to childrren as an astringent.

Tylophora indica (Burm. f.) Merr.

SYNONYMS : *Tylophora asthmatiça W. & A.; Asclepias asthmatica Linn.f.; Cynanchum indicum Brum.f.*

ENGLISH : Emetic swallow-wort, Country-ipecacuanha.

INDIAN : Pittamaari, Jangli-pikvam, Antmula, Khodiki raasnaa.

SANSKṚTA : Mūlinī, Mūlarārnā, Antamūla, Pittavallī, Antrapācaka.

HABIT : A twining perennial.

HABITAT : Konkan, Karnatak, Bengal and Assam.

CHEMICAL COMPOSITION : Contains three alkaloids A, B and C. Alkaloid B is desmethyl-tylophorine and alkaloid C is desmethyl tylophorinine. Alkaloid A is not identified. It also contains a substance with emetic properties and essential oil. Leaves yield α-amyrin, tylophorine, kaempterol and quercetin.

Parts used : Leaves, root and leaf-ash.

Ayurvedic properties :

Rasa - Tikta.	Vīrya - Ūṣṇa.
Guṇa - Laghu, Rūkṣa.	Vipāka - Kaṭu.

Actions/Uses : Svedajanana, Kaphaghna, Vāmaka, Anulomika, Pittaṣrāvi, Āmapācana.

Therapeutics :

Mūlinī is used as a substitute for Ipecacuanha. It is highly reputed as an alterative and as a purifier of blood and is given in rheumatism. Root is highly effective as alterative and antirheumatic. It is bitter, expectorant in chronic bronchitis and early stages of whooping cough. It is stumulant, reduces lochia, relieves pain in gout when applied locally. Powder of roots is given in dysentery or diarrhoea even in earliest stages and whilst fever is present. It is useful in intermittent malarial fever. Leaves are emetic, diphoretic, expectorant; useful in overloaded states of stomach and other cases requiring use of emetics.

Wattakaka volubilis (Linn.) Stapf.

SYNONYMS : *Dregea volubilis Benth. ex Hook. f.; Marsdenia volubilis T. Cook.; Asclepias volubilis Linn. f.*

INDIAN : Ambri, Dodhi, Nakchhikni, Gharphul, Hirandodi.

SANSKṚTA : Suparṇikā, Madhu-mālati.

HABIT : A large twining shrub.

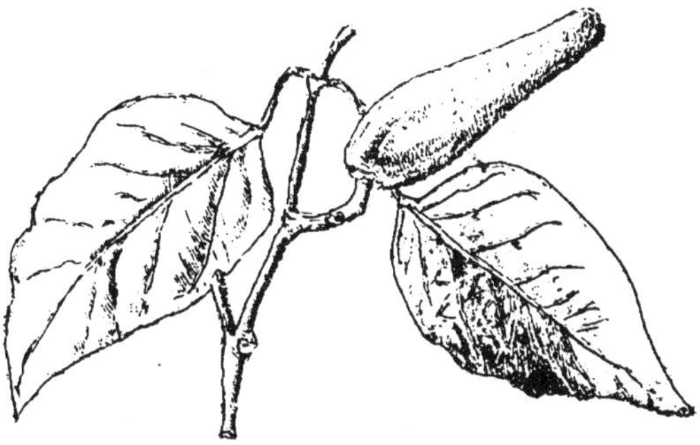

HABITAT : Konkan and Maharashtra, Bengal, Assam.

CHEMICAL COMPOSITION : Glucoside dregein. Drevogenin D from seeds. Steroid along with withaniol and withaferin A. Stems and leaves contain taraxerol, triterenoid, kaempferol, a glycoside of kaempferol and a saponin. Seeds contain a number of pregnane glycosides.

Parts used : Roots, tender shoots, leaves.

Actions/Uses : Vamaka, Kaphanissaraka.

Therapeutics :

Suparṇikā is tonic, cooling, aphrodisiac and expectorant. Cures vaata, biliousness, burning sensation and is used in colds and eye diseases, and to cause sneezing and also in snake bite. Root and tender stalks considered emetic and expectorant. Leaves are used as an application to boils and abscesses.

Family : XV - Asparagaceae.

Asparagus officinalis Linn.

SYNONYMS : *Asparagus adscendens* Roxb.

HABIT : An erect, much-branched, dioecious perennial herb.

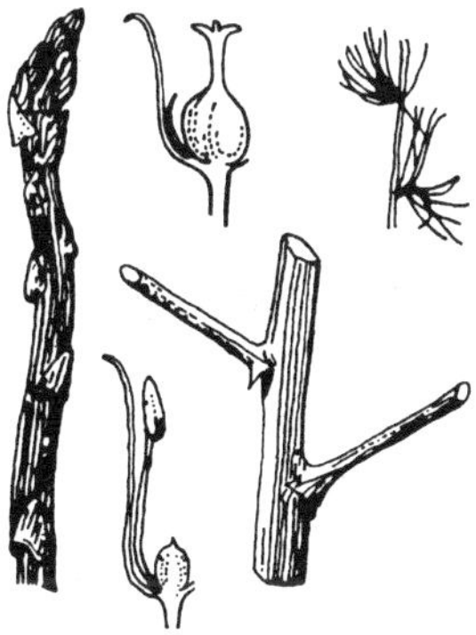

HABITAT : Native to temperate Europe and West Asia. Cutivated in India.

ENGLISH : Asparagus, Sparrowgrass.

INDIAN : Haliyoona, Safed-musallee, Marchuba, Margiyeh, Paragus, Karua, Nakdown.

SANSKṚTA : Muśalī, Haliyuna, Śveta, Taniravittanga, Sakakule, Gholi.

CHEMICAL COMPOSITION : Essential oil, asparagin, tyrosin. Four flavonoids and rutin from leaves. β-sitosterol, sarsasapogenin and nine asparagosides from roots. Asparagusic, dihydroasparagusic and S-acety-dihydroasparagusic acid are the three plant growth inhibitos isolated.

Parts used : Tuberous root or rhizome decorticated.

Ayurvedic properties :

Rasa - Madhura. Vīrya - Śīta.

Guṇa - Gurū, Snigdha. Vipāka - Madhura.

मुशली मधुरा शीता वृष्या पुष्टिबलप्रदा।
पिच्छिला कफदा पित्तदाहश्रमहरा परा।।
(राजनिघण्टु)

Muśalī madhurā śītā vṛṣyā puṣṭibalapradā,
Picchilā kaphadā pittadāhasramaharā parā.
(Rājanighaṇṭu)

Actions/Uses : Vātapittaśāmaka, Kaphavardhaka, Śūkrala, Balya, Brunhaṇa, Rasāyana, Puyamehaghna, Mūtrakṛicchghna.

Therapeutics :

Muśalī or Haliyuna is diuretic, demulcent, tonic, cardiac, sedative, aphrodisiac, laxative. A tincture of the whole plant is used for urinary affections and rheumatism. Roots are more diuretic than shoots. They are recommended in dropsy, enlargement of heart etc. Infusion is used against jaundice and congestive torpor of liver. Root bark taken with milk for vitality and strength. Aspargin stimulates kidneys and imparts a strong smell to urine. A decoction of the fresh or dry fruits is used as contraceptive in Europe. Drug is a mild aperient, tonic, aphrodisiac and sedative. It is given in flatulence, calculous affections, dropsy, rheumatism and chronic gout.

HOMEOPATHIC USE : Asparagus is a Homeopathic remedy. Like other liliaceous plants it acts on the heart, kidneys and dropsical effusions. Severe coryza and nasal catarrh, frequent violent sneezing, copious discharge of tenacious mucus from throat, frequent urination, fulness in chest, violent palpitation are some of the main symptoms.

Asparagus racemosus Willd.

SYNONYMS : *Asparagus javanica Kunth.; Asparagus sarmentosus Graham.; Asparagopsis sarmentosa Dalz. & Gibs.*

INDIAN : Shataawaree, Aswali, Shakakul.

SANSKṚTA : Śatāvarī, Śatamūli, Śītavīrya, Bahusutā, Atirasā, Śatapadī, Indīvarī, Dīpyā, Dvipaśatru, Durbharā, Adharakaṇtikā, Sūkṣmapatrā, Nārāyaṇī, Svadurasā, Bhīrū, Ātmaśakti, Jatāmūlā, Mahodanī, Madhurā, Keśikā, Pīvarī, Viśvākhyā, Vaiṣṇavī, Vāsudevī, Karīyasī, Tejavallī, Bahuputrī.

HABIT : An extensively scandent, much-branched, spinous under-shrub, with tuberous, short rootstock bearing numerous fusiform, succulent tuberous roots.

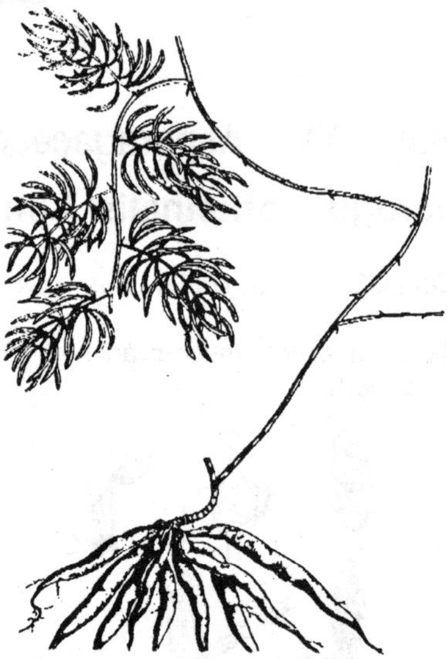

HABITAT : Throughout tropical and subtropical India.

CHEMICAL COMPOSITION : Fresh leaves yield diosgenin. Plant yields shatavarin I to IV. Fruits and flowers contain glycosides of quercetin, rutin and hyperoside; sitosterol, stigmasterol and their glucosides. Root extract caused intitial increase in force and rate of amphibian heart at lower dose and cardiac arrest at higher dose.

Parts used : Roots and leaves.

Ayurvedic properties :
Rasa - Madhura, Tikta. Vīrya - Śīta.
Guṇa - Gurū, Snigdha, Mṛudu.
Vipāka - Madhura.

शतावरी गुरुः शीता तिक्ता स्वाद्वी रसायनी।
मेधाग्निपुष्टिदा स्निग्धा नेत्र्या गुल्मातिसारजित्।।
शुक्लस्तन्यकरी बल्या वातपित्तास्त्रशोफजित्।
महाशतावरी मेध्या हृद्या वृष्या रसायनी।।
शीतवीर्या निहन्त्यर्शोग्रहणीनयनामयान्।
(भा. प्र.)

Śatāvarī guruḥ śītā tiktā svādvī rasāyanī,
Medhāgnipuṣṭidā snigdhā netryā gulmātisārajit.
Suklastanyakari balyā vātapittāsraśophajit,
Mahāśatāvarī medhyā hṛdyā vṛṣyā rasāyanī.
Śītavīryā nihantyarśograhaṇīnayanāmayān.
(Bhāvaprāṣa)

Actions/Uses : Pittaghna, Vātaghna, Kaphakara, Balya, Dīpanā, Vrsya, Vayasthāpana, Vraṇaropaṇa, Vedanāsthāpana, Rasāyana, Yakṛutottejaka, Stanyakarī, Śūlaghna.

Therapeutics :
Śatāvarī is a powerful drug capable of improving memory power, intelligence, and physical strength and maintaining youthfulness. It is sweet, bitter in taste, tonic, aphrodisiac, galactagogue, roborant, diuretic, antidysenteric and demulcent. It increases breast milk, promotes sexual vigour, cures swelling, consumption, diseases due to impurities of blood, diarrhoea, piles, eye diseases. It is a good remedy for vaginal disorders like leucorrhoea, uterine disorders, excess of bleeding and coliky pain. It is a reputed drug for peptic and duodenal ulcers. It is used against ulcerated tongue, pain in limbs or bones, ardor of body, gravel, pleurisy and rinderpest. In Ayurved, it is used in diabetes, jaundice and other urinary disorders. Root is tonic, nutritive, demulcent, diuretic, refrigerant, galactagogue, highly mucilagenous, antidiarrhoeic, antidysenteric, antispasmodic and aphrodiasiac, used as sexual tonics. Roots are used to cure diarrhoea as well as in cases of chronic colic and dysentery. Juice of root taken with milk is useful in gonorrhoea. Tubers are refrigerant, demulcent, diuretic, aphrodisiac, antispasmodic, alterative, antidiarrhoea, antidyentery, galctagogue and are used in fever.

Family : XVI - Asteraceae.

Achillea millefolium Linn.

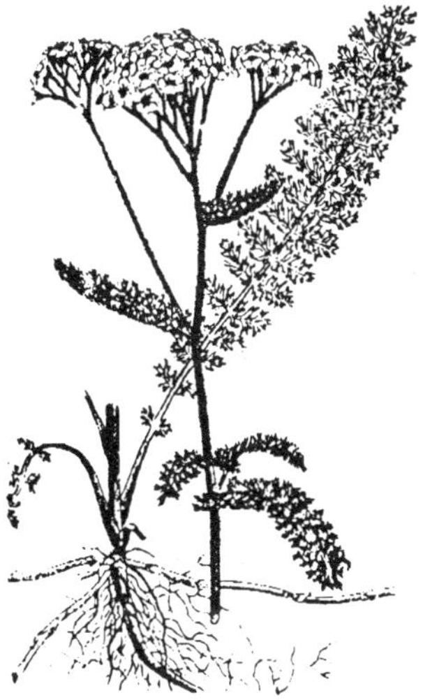

SYNONYMS : *Achillea setacea Waldst. & Kit.*

ENGLISH : Melfoil, Yarrow.

INDIAN : Rojmari, Sahstrapatree, Gandan.

SANSKṚTA : Śvetadurvā, Sahastrapatrī.

HABIT : An erect, slightly aromatic, pubescent, perennial herb.

HABITAT : America and Western Himalayas especially around Simla.

CHEMICAL COMPOSITION : Herb contains an alkaloid achilleine which has haemostatic properties. Herb contains salicylic acid, β-sitosterol and its acetate, inositol, dulcitol, mannitol, betain, choline, trigonekkine, betonicine and stachydrine. Essential oil, HCN-glucoside achillein. Flowers give essential oil containing azulene. Entire American herb contains oil. Leaves contain rutin, apigenin, cosmosin, luteolin and its glucoside.

Parts used : All parts.

Ayurvedic properties :
Rasa - Madhura. Vīrya - Śīta.
Guṇa - Laghu, Snigdha. Vipāka - Madura.

श्वेतदूर्वाऽतिशिशिरा मधुरा वान्तिपित्तजित्।
आमातिसारकासघ्नी रुच्या दाहतृषापहा।।
(राजनिघण्टु)

Śvetadurvā atiśiśirā madhura vāntipittajit,
Āmātisārakāsaghnī rucyā dāhatṛṣāpahā.
(Rājanighaṇṭu)

Actions/Uses : Dīpana, Pācana, Pauṣṭika, Koṣṭhavātapraśamana, Raktasangrāhaka, Kaphaghna, Sankocavikāsapratibandhaka, Uttejaka, Ārtavajanana, Arśoghna, Mūtrajanana,

Therapeutics :

Śvetadurvā or Sahastrapatri is a mild aromatic tonic, diaphoretic, stimulant, antispasmodic, emmenagogue and astringent. It is used in colds, obstructed perspiration and at the commencement of fevers. It is antiseptic, vulnerary and styptic used in flatulent colic, heat burn, hysteria, epilepsy, and rheumatism. It suppresses haemorrhage and profuse mucous discharge. Decoction or fresh juice is applied to cuts, bruises, piles, varicose veins and ulcers. Hot infusion of leaves is a powerful emmenagogue. Leaves and flower heads are stimulant, tonic and carminative.

Ageratum conyzoides Linn.

ENGLISH : Goat weed, White weed.

INDIAN : Osaaddee, Ghaneraosadi.

SANSKṚTA : Viṣamuṣṭī, Osari.

HABIT : A polymorphic, aromatic, erect annual herb.

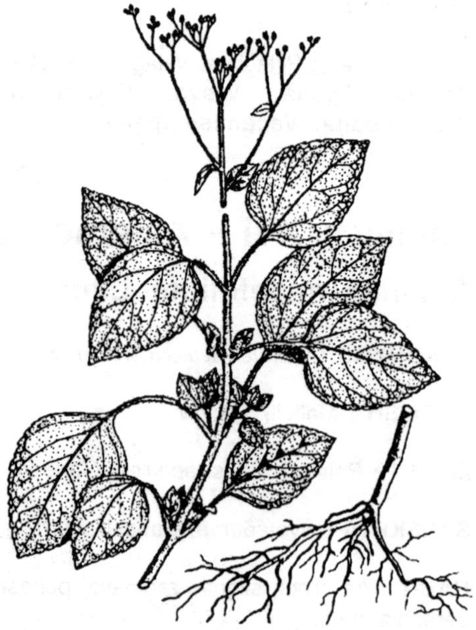

HABITAT : Native to tropical America, naturalized as a weed throughout India.

CHEMICAL COMPOSITION : Phenol, essential oil. Essential oil contains eugenol and powerful nauseating odour. Ageratochromene, caryophyllene, y-cadinene, and 6-demethoxyageratochromene.

Parts used : Roots, leaves and flowers.

Ayurvedic properties :
Rasa - Kaṭu, Tikta. Vīrya - Śīta.
Guṇa- Laghu, Rūkṣa. Vipāka - Kaṭu.

विषमुष्टिः कटुस्तिक्तो दीपनः कफवातहृत्।
कण्ठामयहरो रुच्यो रक्तपित्तार्त्तिदाहकृत्।। (रा. नि.)

Viṣamuṣṭiḥ kaṭustikto dīpanah kaphavātahṛt,
Kaṇṭhāmayaharo rucyo raktapittārttidāhakṛt.
(Rājanighaṇṭu)

Actions/Uses : Dīpana, Kaphavātahara, Rucya, Raktapittaśāmaka, Jvaraghna, Svedajanana.

Therapeutics :

Viṣamuṣṭī or Osari is used internally as a stimulant and tonic. Juice is a good remedy in prolapsus ani. Boiled with oil, applied externally in rheumatism. Decoction or infusion used in diarrhoea and dysentery. Also for colic, rheumatism and fever. Roots are acrid bitter, digestive, appetiser and ophthalmic and are used in dyspepsia, anorexia and purulent ophthalmia, renal and vesical calculi. Leaves are styptic and antidysenteric and are commonly used for haemorrhoids, boils, wounds, sores and in ague. They are also used for fomentation in leprosy and skin diseases, used along with salt as a vulnerary. They are said to prevent titanus. Juice of leaves is antilithic. Flower buds cure cancerous growths.

Anacyclus pyrethrum DC.

SYNONYMS : *Anacyclus officinarum Hayne.; Anthemis pyrethrum DC.*

HABIT : A perennial, procumbent herb.

HABITAT : Native to North Africa, introduced into Europe.

ENGLISH : Spanish pellitory.

INDIAN : Akalakaaraa-kharaa.

SANSKṚTA : Ākārakarabha, Akllakotha, Akalaka.

CHEMICAL COMPOSITION : Roots contain anacyclin, pellitorine, enetriyne alcohol, hydrocarolin, inulin, traces of volatile oil and sesamin. Amides (I, II, III, IV) from roots. Essentail oil, pellitorine or pyrethrin.

Parts used : Flowers and fruit.

Ayurvedic properties :
Rasa - Kaṭu. Vīrya - Ūṣṇa.
Guṇa - Rūkṣa, Tīkṣṇa. Vipāka - Kaṭu.

अकल्लकोष्णो वीर्येण बलकृत् कटुको मतः।
प्रतिश्यायं च शोफं च वातं चैव विनाशयेत्।।
(नि. र.)

Akallakoṣṇo vīryena bakakṛt kaṭuko mataḥ,
Pratiśyāyam ca śopham ca vatam caiva vināśayet.
(Nighaṇṭuratnākara)

Actions/Uses : Kaphaghna, Vātaghna, Krimighna, Uttejaka.

Therapeutics :

Ākārakarabha is cordial, stimulant and sialogogue, used in rheumatism. Roots possess stimulant and rubifacient properties and are used as a masticatory to allay toothache, for gargles and for the preparation of tooth powders. They are regarded as tonic to the nervous system. An infusion of roots is used as a cordial and stimulant and also in certain stages of typhus fever. A decoction of roots is useful for pharyngitis and tonsilitis. It is used in the treatment of paralysis, hemi-plegia, epilepsy and chronic ophthalmia. An indigenous drug FORTEGE, containing A. pyrethrum, Argyrea speciosa Sweet, and Withania somnifera Dunal, is found useful in curing sexual disorders in males.

Artemisia absinthium Linn.

SYNONYMS : *Absinthium officinale* Lam.

HABIT : An aromatic and bitter herb.

HABITAT : Native of Europe, found in Kashmir; cultivated in gardens.

ENGLISH : Wormwood, Abisinthe, Madarwood, Ajenjo.

INDIAN : Wilaayati afasanteena, Serpana.

SANSKṚTA : Dāmara, Indhana, Aphasantīn.

CHEMICAL COMPOSITION : Essential oil known as Absinth or Wormwood oil. Artemisetin, absinthin, artabasin, myrcene, a-pinene, thujyl alcohol and thujyl acetate.

Parts used : All parts.

Ayurvedic properties :

Rasa - Tikta.	Vīrya - Ūṣṇa.
Guṇa - Rūkṣa.	Vipāka - Kaṭu.

Actions/Uses : Sugandhī, Uttejaka, Jvaraghna, Krimighna.

Therapeutics :

Dāmara is tonic, stomachic, febrifuge and anthelmintic. It is useful for inflammation of liver and menstrual disorders. Tincture of the plant is used as tonic, digestive, febrifuge, anthelmintic and for brain concussions. It is a remedy for enfeebled digestions and debility. Flowers are vermifuge, tonoic in intermittent fever. Drug "Afsanteen" used in chronic fever, swellings and inflammation of liver. It is also used as tonic and stimulant.

HOMEOPATHIC USE : Abisinthium is a

Homeopathic remedy. A perfect picture of epileptiform seizure is produced by the drug. Nervous tremors precede attack. Sudden and sever dizziness, delirium with hallucinations and loss of consciousness, a peculiar vertigo on rising, with tendency to fall back are some of the main symptoms.

Artemisia maritima Linn.

SYNONYMS : *Artemisia fragrans (Willd.) Ledeb.; Artemisia spicigera C. Koch.; Artemisia brevifolia Wall.*

ENGLISH : Wormseed, Santonica, Semen sanctum.

INDIAN : Kiramaannee owaa (2).

SANSKṚTA : Cauhāra, Caurāhna, Caurapūrva, Jantunāśana, Kītamari yavānī.

HABIT : A deciduous perennial shrub with much-branched woody rootstock.

HABITAT : Indigenous to Iran, Afganistan and Russia. Grows in W. Himalayas from Kashmir to Kumaon and Nepal.

CHEMICAL COMPOSITION : Essential oil. Santonin, β-santonin, pseudosantonin and a bitter substance artemisin. Bitter substance is appetite stimulating. Erivanine, lumisantonin from flowers.

Ayurvedic properties :
Rasa - Tikta. Vīrya - Ūṣṇa.
Guṇa - Laghu, Rūkṣa, Tīkṣṇa. Vipāka - Kaṭu.

Parts used : All parts and extract.

यवानिका यवानी स्या च्चौहारो जन्तुनाशनः।
चोराहस्तद्गुणः प्रोक्तो विशेषात् कृमिनाशनः।।
(रा. नि.)

Yavanikā yavānī syāccauharo jantunāśanaḥ,
Cauharastad gunaḥ prokto viśeṣāt kṛmināsanaḥ.
(Rājanighaṇṭu)

Actions/Uses : Dīpana, Vedanāsthāpana, Krimighna; Vātānulomana, Yakṛtottejana, Kaphanissāraka, Śvasahara.

Therapeutics :

Cauhāra is used as a deobustruent, stomachic, laxative and tonic. A decoction or infusion of the fresh plant is used in cases of intermittent and remittent fever. Decoction or infusion of leaves is used in ague, intermittent and remittent fevers. Flower heads are anthelmintic. Artemisia maritima varieties contain essential oil varying in quantity and composition.

Artemisia vulgaris auct. (non Linn.)

SYNONYMS : *Artemisia nilagirica (Clarke) Pamp.*

ENGLISH : Felon Herb, Mugwort, Indian wormwood, Flea-bane.

INDIAN : Naagadawanna, Naagadamani, Dhordavana.

SANSKṚTA : Damana, Sūrapaenṇa, Damanaka, Dānta, Vāmana, Barha, Gandhoṭkata, Muni, Puṇḍarīka, Bramhajatā, Jatila, Nāgadamani.

HABIT : A tall, aromatic, pubescent or vilous shrub-like perennial herb.

HABITAT : Konkan and Western Ghats and mountainous districts of India.

CHEMICAL COMPOSITION : Essential oil containing p-cymene, l-linalool, β-thujone, azulene, thujyl alcohol and an unidentified hydrocarbon. Artemisia ketone, a pentacyclic triterpene, thujone, a-amyrin and its acetate, ferneol, stigmasterol and β-sitosterol.

Parts used : Leaves, flowering tops, All parts and salts.

Ayurvedic properties :
Rasa - Tikta, Kaṣāya. Vīrya - Ūṣṇa.
Guṇa - Laghu, Rūkṣa, Tīkṣṇa. Vipāka - Kaṭu.

दमनस्तुवररस्तिक्तो हृद्यः वृष्यः सुंगाधिकः।
ग्रहणी विषकुष्ठास्त्रक्लेदकण्डूत्रिदाष जित्।।
(भा. प्र.)

Damanastuvarastikto hṛdyaḥ vṛṣyaḥ sugandhikaḥ,
Grahaṇīviṣakuṣṭhasrakledakanḍūtridoṣajit.
(Bhāvaprakāṣa)

Actions/Uses : Tridoṣaghna, Vātaghna, Kaphaghna. Dīpana, Pācana, Pittasāraka, Vātānuloman and Kṛmighna.

Therapeutics :

Damana is a valuable stomachic, deobsrtuent, stomachic, emmenagogue, anthelmintic and antispasmodic. Leaves and flowering tops are bitter, astringent, anodyne, diuretic, febrifuge, aphrodisiac and digestive. They are used as an inferior substitute for cinchona in fevers. They are useful in cough, asthma, bronchitis, nervous and spasmodic affections. They are anti-lithic and alexipharmic. A decoction given to children in measles. Externally used as fomentation in skin diseases and ulcers.

Blumea balsamifera DC.

SYNONYMS : *Blumea densiflora Hook f.; Sim. Blumea lacera DC.*

INDIAN : Bhangaruda, Kukundara, Nirmudi, Janglikasni.

SANSKṚTA : Kukundara, Gangāpatrī.

HABIT : An evergreen, robust, tomentose perennial shrub or undershrub or sometimes an erect herb upto 4 m in height.

HABITAT : Subtropical Himalayas, Nepal, Sikkim, Assam and Khasia hills.

CHEMICAL COMPOSITION : Leaves contain 2 quercetin derivatives. The plant is a source of Blumia Camphor.

Parts used : Whole plant, Roots and leaves.

Ayurvedic properties :
Rasa - Tikta, Kaṣāya. Vīrya - Ūṣṇa.
Guṇa - Laghu, Rūkṣa, Tīkṣṇa. Vipāka - Kaṭu.

कुकुन्दरः कटुस्तिक्तो ज्वररक्तकफापहा।
तन्मूलमार्द्रं विक्षित्पं वदने मुखशोषहृत्।।
(भा. प्र.)

Kukundaraḥ kaṭustikto jvararaktakaphāpahā,
Tanmūlamārdram nikṣiptam vadane mukhaśoṣahṛt.
(Bhāvaprakāṣa)

Actions/Uses : Kaphapittaśāmaka. Dīpana, Anulomana, Kṛmighna, Yakrtottejaka.

Therapeutics :

Plant is hot, pungent, bitter, antipyretic, cures fevers, thirst bronchitis, blood diseases. It is liver-tonic, digestive, diuretic and stimulant. Warm infusion is sudorific and decoction expectorant. Leaves are stomachic, hypothermic and given in intestinal diseases such as diarrhoea, cholera, colic, worms, liver disorrders, haemorrhoids and also for bronchitis, cough and leucorrhoea. They are used in the treatment of body-swelling. Extract of leaves is used in chest trouble, headache, dropsy, hypertension and as tranquilizer. Fresh leaf juice is used as eye drops in chronic eye diseases. Plant is a fish poison.

Carthamus tinctorius Linn.

ENGLISH : Safflower.

INDIAN : Karadaee.

SANSKṚTA : Kusumbha, Vyāgrīdala, Kanṭaphala, Vanhiśikha, Kanṭakī, Vastraranjaka.

HABIT : An erect annual herb.

HABITAT : Cultivated throughout India.

CHEMICAL COMPOSITION : Matairesinol monoglucoside, bitter principle of seeds. Acetylene compounds in aerial parts. Yellow pigment safflomin in flowers. Cathamin and neocarthamin; kaempferol and kaempferol glycoside.

Parts used : Whole plant, leaves, seeds, roots, flowers and oil.

Ayurvedic properties :

Rasa - Madhura, Kaṣāya.	Vīrya - Ūṣṇa.
Guṇa - Gurū, Vidahi.	Vipāka - Kaṭu.

कुसुंभो वातळो रुक्षो विदाही कटुकः स्मृतः।
मूत्रकृच्छ्रं कफं रक्तपित्तञ्चैव विनाशयेत्।।
कुसुंभपुष्पं सुस्वादुः त्रिदोषघ्नं च भेदकम्।
कुसुंभ तैलमम्लं स्वादुष्णं गुरु विदध्दि च।।
(भा. प्र.)

Kusumbho vātaḷo rūkṣo vidāhi kaṭukaḥ smṛtaḥ,
Mūtrakṛccham kapham raktapittancaivavināśayet.
Kusumbhapuṣpam susvāduḥ tridoṣaghnam ca bhedakam.
Kusumbha tailamamḷam syāduṣṇam guru vidāhi ca.
(Bhāvaprakāṣa)

Actions/Uses : Kṛmighna, Mūtrakricchaśamana, Raktapittaśamana, Agnikāraka, Rucya, Malamūtraśodhaka, Puṣṭikara, Balahara, Caksusya, Gudarogahara, Madahara.

Therapeutics :

Leaves of Kusumbha are bitter sweet, laxative, appetiser and diuretic and are useful in urrhoea and eye troubles. Flowers are laxative, diaphoretic, stimulant, sedative and emmenagogue, are used in children's complaints and diseases, measles, fevers and eruptive skin complaints. Seeds are tonic, purgative, diuretic and used in rheumatism. Oil is reported to lower the level of blood serum cholesterol, mainly due to the presence of oleic and linoleic acid. It has also been considered useful in arteriosclerosis, atherosclerosis and infant eczema. Charred oil is used for healing sores and in rheumatism.

Centratherum anthelminticum (Willd.) Kuntze.

SYNONYMS : *Vernonia anthelmintica Willd.; Serratula anthelmintica Roxb.; Centratherum anthelminiticum (Linn.) Kuntze.*

ENGLISH : Purple flea bane.

INDIAN : Kaddujire, Kaallejire, Somaraaji.

SANSKṚTA : Āranyajīraka, Bṛhatpallī, Sūkṣmapatrā Vanajīraka, Vanyajīraka, Karajīrī, Agnibīja, Kaṇā, Avalguja, Putiphali, Tictajīraka, Soamrāji, Vakuśi.

HABIT : An tall robust annual herb.

HABITAT : In the Himalayas, Khasi hills. Often cultivated in near villages.

CHEMICAL COMPOSITION : Bark contains a-gallocatechol and a-pyrocatechol. Seed oil, vernosterol. Leaves, abscisic acid, centratherin, germacranolide. Seeds, stigmasterol, stigmastatrienol acetate, veranodalol.

Parts used : Seeds.

Ayurvedic properties :

Rasa - Kaṭu, Tikta. Vīrya - Ūṣṇa.
Guṇa - Laghu, Tīkṣṇa. Vipāka - Kaṭu.

आरण्यजीरकं चोष्णं तुवरं कटुकं मतम्।
स्तम्भवातं कफं चैव व्रणं चैव विनाशयेत्।।
(निघण्टु रत्नाकर)

Āranyajīrakam coṣṇam tuvaram kaṭukam matam,
Stambhavātam kapham caiva vraṇam caiva vināṣayet.
(Nigaṇṭuratnakara)

Actions/Uses : Kaphavātaśāmaka, Kṛmighna, Dīpana, Vāyunāśinī, Kaṭupauṣṭika, Jvaraghna, Mūtrajanana, Stanyajanana, Tvagdoṣahara, Kaṇḍughna.

Therapeutics :

Various plant parts of Āranyajīraka are used in constipation, syphilis and against cholera. Seeds are tonic, stomachic, diuretic, anthelmintic and are used in asthma, skin diseases and scorpion-sting; employed for destroying pediculi.

Echinops echinatus Roxb.

SYNONYMS : *Echinops echinatus DC.*

ENGLISH : Globe-thistle, Camels thistle.

INDIAN : Kaatechumbakaa, Utakanta, Gokru, Utanti.

SANSKṚTA : Utkanṭaka, Kanṭaphala, Raktapuṣpa, Bramhadanḍi, Utati, Karabhādana, Ustrakantha, Vrittaguccha, Śṛgālaśūnakāśana, Kānṭālu.

HABIT : A pubescent annual herb.

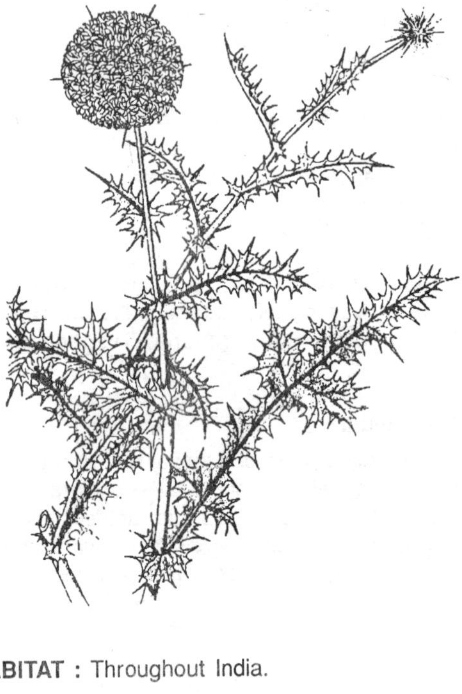

HABITAT : Throughout India.

CHEMICAL COMPOSITION : Hentriacontane, hentriacontanol, β-amyrin and leupeol in plant. Echinopsine in seeds.

Parts used : Whole plant, root, root bark, leaves and fruit.

Ayurvedic properties :
Rasa - Kaṭu, Tikta, Madhura. Vīrya - Ūsna.
Guṇa - Laghu. Vipāka - Katu.

उत्कंटकः कटुस्तिक्तः कफवातहरी लघुः।
तन्मूलपानतः स्त्रीणां शीघ्र प्रसवकारकः॥
शोढल उत्कण्टा रुचिदा चोष्णा तिक्ता वृष्याः प्यूदाहृता।
मूत्रकृच्छ्रं पित्तवातं मेहं तृष्णां च हृद्रुजम्॥
विस्फोटकं तु नाशयति बीजमस्यास्तु शीतलम्।
वृष्यं तृप्तिकरं चैव मधुरं च प्रकीर्तितम्॥
(निघण्टु रत्नाकर)

Utkanṭakaḥ kaṭustikaḥ kaphavātaharī laghuḥ,
Tanmūla pānataḥ strīṇām śīghra prasavakārakaḥ.
Śodhala utkanṭā rucidā coṣṇā tikta vṛṣyaḥ pyūdāhṛtā,
Mūtrakṛccham pittavātam meham tṛṣṇām ca hṛdrujam.
Visphoṭakam tu nāśayati bījamasyāstu śītalam,
Vṛṣyam tṛptikaram caiva madhuram ca prakīrtikam.
(Nighanṭu-adarṣa)

Actions/Uses : General : Rucida, Dīpana, Pauṣṭika, Vṛṣya, Mūtrakṛcchahara, Mehaghna, Tṛṣṇāhara, Raktaśodhana.

Therapeutics :

Utkanṭaka is bitter, alterative, diuretic, nerve tonic; used in hoarse cough, hysteria, dyspepsia, scrofula and ophthalmia. Powder of roots is applied to wounds in cattle to destroy maggots and mixed with acacia gum, applied to hair to destroy lice.

Eclipta alba Hassk.

SYNONYMS : *Eclipta erecta* Linn.; *Eclipta prostrata Roxb.*, *Verbesina alba* Linn.; *Eclipta prostrat* Linn.; *Wiedelia calendulacea.*

INDIAN : Maakaa, Bhrungaraaja, Bhangra, Mochkand, Babri.

SANSKṚTA : Bhṛngara, Bhṛngarāja, Mārkava, Keśarāja, Keśaranjana, Keśya, Piṭṛpriya, Sūryāvarta, Ravipriya, Pittapriya, Ajagara.

HABIT : An erect or prostrate annual succulent herb.

HABITAT : Common weed in moist situations throughout India.

CHEMICAL COMPOSITION : Plant has been found to posses myocardial depressant and hypotensive effect. It is found to be effective in the treatment of infective hepatitis and against liver injury and inflammation. Leaves contain stigmasterol and a-ter-thienylmethanol. Alkaloids ecliptine and nicotine. Sixteen new biogenetically closely related polyacetylenic thiophenes isolated.

Parts used : Leaves and roots.

Ayurvedic properties :
Rasa - Kaṭu, Tikta, Kaṣāya. Vīrya - Ūṣṇa.
Guṇa - Laghu, Rūkṣa Vipāka - Kaṭu.

भृङ्गारः कटुकस्तिक्तो रुक्षोष्णः कफवातनुत्।
केश्यस्त्वच्य कृमिश्वासकासशोथाम पाण्डुनुत्।।
(भा.प्र.)

Bhṛṅgārah kaṭukastiktorūkṣoṣṇaḥ kaphavātanut, Keśyastvacyaḥ kṛmiśvāsakāsaśosthama pāṇḍunut. (Bhāvaprakāṣa)

Actions/Uses : Kaphaghna, Vātaghna, Keśya, Cakṣuṡya, Viṣaghna, Rasāyana, Tvacya, Dantya, Amādiṣaghna, Pānḍughna, Tvagrogahara, Hṛdrogahara, Kṛmighna, Śvāsaghna, Netrarogahara.

Therapeutics :
Bhṛngaraja is acrid, bitter, hot and dry, reduces kapha and vaata and is a good rejuvenator. It is good for the hair and skin, expels intestinal worms, cures cough and asthma and strengthens body. It is a specific in nightblindedness, eye diseases, headache and diseases pertaining to hair and its growth. Herb is tonic and deobstruent in hepatic and spleen enlargements, emetic, used in skin diseases. Juice of plant in combination with aromatics administered for catarrhal jaundice. Leaves, in scorpion sting. Juice of leaves along with honey used as remedy for catarrh in infants. Decoction of leaf and root is liver tonic. Root is emetic and purgative; applied externally as antiseptic to ulcers and wounds in cattle. A preparation obtained from the juice of leaves boiled with sesamum or coconut oil is used for anointing the head to render the hair black and luxuriant. Fresh plant is considered anodyne and absorbent. It is rubbed on the gums in toothache and applied with a little oil for relieving headache. It is also applied with sesamum oil in elephamtiasis. Drug is extensively used against jaundice.

Elephantopus scaber Linn.

ENGLISH : Prickly-leaved elephantís foot.

INDIAN : Gojivhaa, Hastipata, Gobhi, Samudulan, Pathari.

SANSKṚTA : Gojihvā, Kharapatra, Karipadam, Dārvipatra, Gojī, Golī, Gobhī, Dārvika, Mūlaghna, Kṣīrinī, Dīrghapatra, Kharadala, Śubhā, Cakravat, Madhyādanḍasthira, Pīta, Godhūmika.

HABIT : An rigid annual erect herb with large leaves forming a rosette and numerous clusters of flower-heads.

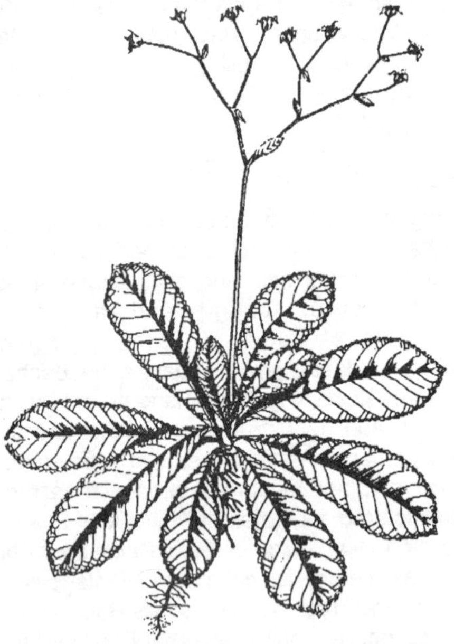

HABITAT : Throughout hotter parts of India.

CHEMICAL COMPOSITION : Epifriedelinol, lupeol, stigmasterol and a mixture of triacontan-1-ol and dotriacontan-1-ol. Sequiterpene dilactone. Alcoholic extract is antibiotic. Elephantopin.

Parts used : Roots and leaves.

Ayurvedic properties :
Rasa - Kaṭu, Madhura, Kaṣāya. Vīrya - Śīta.
Guṇa - Laghu, Snigdha. Vipāka - Madhura.

गोजिव्हा, कटुका, तिक्ता शीतला व्रणरोपणी।
पित्तं सर्व विषं हन्ति कासारुचि विनाशिनी।।
(निघण्टु रत्नाकर)

Gojivā kaṭukā tiktā śītalā vraṇaropaṇī
Pittam sarva viṣam hanti kāsāruci vināśinī.
(Nigaṇṭu ratnakara)

Actions/Uses : Balya, Grahi, Snehana, Śītala, Hṛdya, Arucighna, Mūtrajanana, Vraṇaropaṇī, Vraṇahara, Pramehaghna, Kāsaghna, Jvaraghna, Dantarogaghna, Viṣaghna, Raktapittahara, Śvāsaghna, Kaphapittaśāmak.

Therapeutics :

Gojihvā is mucilaginous, light, cooling, astringent, cardiac tonic, diuretic, alterative, febrifuge. Root is used to arrest vomiting and to cure filariasis. Decoction of roots and leaves is emollient; given in dysuria, diarrhoea and dysentery and for swellings or pain in stomach. Powder of root with pepper is applied to toothache. Paste of roots applied in rheumatism. Bruised leaves are boiled in coconut oil are applied to ulcers and eczema.

Eupatorium purpureum Linn.

SIMILAR : *Eupatprium ayapana* Vent.; *Eupatorium cannabinum* Linn.; *Eupatorium triplinerve* Vahl., *Eupaorium perfpliatum* Linn.

ENGLISH : Gravel root, Gravelweed, Queen of the medadow.

INDIAN : Aayaapaana.

SANSKṚTA : Āyāpāna.

HABIT : An erect annual herb with rhizome.

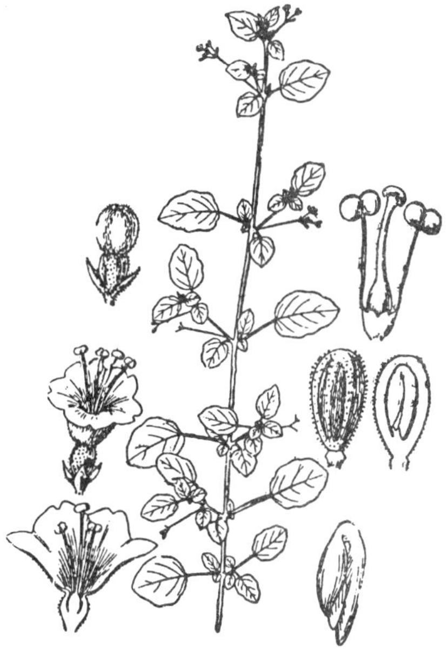

HABITAT : U.S.A.

CHEMICAL COMPOSITION : Plant contains a bitter principle eupatorin, inulin and a resin.

Parts used : Whole plant.

Ayurvedic properties :

Rasa - Tikta, Kaṣāya. Vīrya - Ūṣṇa.
Guṇa - Laghu, Rūkṣa. Vipāka - Kaṭu.

अजापर्णं तु तुवरं तिक्तं वीर्योष्णमेव च।
कफपित्तहरं हृद्यं ज्वरघ्नं रक्तरोधकम्।।

Ajāparṇa tu tuvaram tiktam viryoṣṇameva ca, Kaphapittaharam hṛdyam jvaraghnam raktarodhakam.

Actions/Uses : Kaphapittaśāmaka. Pauṣṭika, Dīpana, Pācana, Rocaka, Anulomana, Uttejaka, Cetanākāraka.

Therapeutics :

Ajaparṇa is diaphoretic, diuretic, antiscorbutic, emetic, and cathartic. An infusion of the plant is used as a stimulant, diaphoretic and tonic. A decoction of the plant and the juice of leaves are detergent, used in fevers, jaundice and applied to foul ulcers. Herb including its dried leaves, flowering tops and twigs is used in the form of infusion as a bitter tonic, expectorant, diaphoretic and antiperiodic. Fresh leaves bruised are applied to foul ulcers, sores and bites of reptiles.

HOMEOPATHIC USE : Eup. purp. is a Homeopathic remedy indicated in influenza and intermittent fever, but it has more trouble of kidneys and bladder. Fever or not, but it is a promissing remedy for some troubles of kidneys and bladder. Violent dysuria, chronic nephritis, smarting, burning in bladder and urethra on urinating are the chief symptoms indicating its use. It has cured impotency in males and sterility and uterine atony in females.

Inula helenium Linn.

SYNONYMS : *Corvisartia helenium Merat., Inula racemosa Hook. f.*

ENGLISH : Elecampane.

INDIAN : Raasnaa.

SANSKRTA : Rāsnā, Yuktarasā, Suvahā, Surasā, Rasanā, Sugandhā, Śreyasī, Atirasā, Rasāḍhyā, Gandhamūlā.

HABIT : A perennial erect herb with rhizome.

HABITAT : Indigenous to Iran, Afganistan and Baluchistan, Europe.

CHEMICAL COMPOSITION : Essential oil, bitter principle, benzoic acid. Roots contain essential oil and resin. Inulin, inulenin, pseudo-inulin, alantolactone, isoalantolactone and dihydroalantolactone from roots. Aerial parts, scopoletin and umbeliferone.

Ayurvedic properties :
Rasa - Tikta, Kaṭu. Vīrya - Ūṣṇa.
Guṇa - Laghu, Tīkṣṇa. Vipāka - Kaṭu.

Parts used : Roots and whole plant.

रास्ना तिक्ता गुरुश्चोष्णा पाचन्याम विनाशिनी।
वातरक्तं विषं श्वासं कासं च विषमज्वरम्।।
शोफं हिक्कां चामवातं कफं शूलं विनाशयेत्।
ज्वरं कम्पं चोदरं च सर्वान्वातांश्च नाशयेत्।।
(निघण्टु रत्नाकर)

Rāsnā tiktā guruścoṣṇā pācanyāmavināśinī,
Vātaraktam viṣam śvāṣam kāsam ca viṣamjvaram.
Śopham hikkām cāmavātam kapham śūlam vināśayet,
Jvaram kampam codaram ca sarvānvātānśca nāśayet.
(Nigaṇṭuratnākara)

Actions/Uses : Kaphaghna, Vātaghna, Kāsahara, Śvāsahara, Pācana, Arucihara, Uttejaka, Śopaghna, Pānduhara, Hikkāhara, Hṛtśulaghna, Vāyunāśī, Jvaraghna, Śothahara, Tvagroganaśana, Viṣahara.

Therapeutics :

Rāsnā is an aromatic tonic; and also used as diaphoretic, diuretic, expectorant and emmenagogue. Used in chronic bronchitis and rheumatism. Dried rhizomes and roots are used in lack of appetite, stomach trouble and as mild diuretic

Leontodon taraxacum Linn.

SYNONYMS : *Taraxacum dens-lenis Desf.*

Similar : Taraxacum officinale Weber ex Wigg.

ENGLISH : Taraxacum, Dandelion.

INDIAN : Bathur, Dudhalla, Dugdhafeni, Dulal, Udarjani, Barau, Kanphul.

SANSKṚTA : Dugdhaphenī, Payasvinī, Dudhala, Dudhalī, Lūtārī.

HABIT : A very variable, scapigerous, perennial erect herb.

HABITAT : A native of Europe and Asia Minor. Indigenus in India. Throughout Himalayas; Tibet and Nilagiries. Cultivated at Shaharanpur and Pune.

CHEMICAL COMPOSITION : A bitter principle taraxacin and taraxacerin; phytosterols taraxasterol and homotaraxasterol. Free amino acids; tryptophan found absent; taraxien and taraxanthin dipalmitate from flowers. Neoxanthin from petals.

Parts used : Root and stem.

Ayurvedic properties :
Rasa - Tikta, Kaṭu. Vīrya - Ūṣṇa.
Guṇa - Laghu, Rūkṣa, Tīkṣṇa. Vipāka - Kaṭu.

दुग्धफेनी कटुस्तिक्ताऽ शिशिरा विषनाशिनी।
व्रणापसारिणी रुच्या युक्त्या चैव रसायनी।।
(रा. नि.)

Dugdhaphenī kaṭustiktā aśiśirā viṣanāsinī,
Vraṇāpasāriṇī rucyā yuktyā caiva rasāyanī.
(Rājanighaṇṭu)

Actions/Uses : Kaphapittaśāmaka, Anulomika, Mūtrajanana, Dipana, Yakṛtottejaka, Krmighna, Pittaśaraka.

Therapeutics :
Rhizome and roots of Dugdhapheni are employed in medicine, is administered as liquid extract. It is a bitter tonic and a mild laxative and probably causes increased secretion of bile. It has an almost specific action on liver, by modifying and increasing its secretion. Hense it is extensively used in chronic diseases of digestive organs, especially hepatic affections, as jaundice, chronic inflammation or enlargement of liver, dropsies from hepatic obstruction and dyspepsia attended with deficient biliary secretion. It is also used as a diuretic, stomachic, hepatic stimulant. Root is diuretic, tonic and slightly aperient. Used in kidney and liver disorders. Leaves used for fomentation.

HOMEOPATHIC USE : Taraxacum is a Homeopathic remedy. The leading symptoms are : painless urging to urinate and frequent urging to urinate with copious discharge of urine. Taraxacum has caused pains in both splenic and liver region and has cured jaundice and enlarged indurated liver. The tongue is coated white and clears off in patches, leaving dark red very sensative spots. It is definitely a liver remedy, tonic and hepatic stimulant.

Sphaeranthus indicus Linn.

SYNONYMS : *Sphaeranthus mollis.; Sphaeranthus hirtus Willd.*

INDIAN : Gorakhamundi.

SANSKṚTA : Muṇḍī, Gorakhamuṇḍī, Muṇḍitika, Śrāvaṇī, Tapodhanā, Parivrajā, Pāvanī, Bhikṣu, Śrimatī, Śrāvaṇaśīrṣaka, Mahāmuṇḍī, Tiktā, Lobhanīyā, Chinnagranthinikā.

HABIT : An annual prostrate herb.

HABITAT : Throughout India.

CHEMICAL COMPOSITION : Essential oil containing methyl chavicol, α-ionone, d-cadinene, p-methoxycinnamaldehyde as major constituents and a-terpinene, citral, geraniol, geranyl acetate, β-ionone, sphaerene, indicusene and sphaeranthol as minor constituents. Leaves contain alkaloids.

Parts used : Whole plant.

Ayurvedic properties :

Rasa - Tikta, Kaṭu.	Vīrya - Ūṣṇa.
Guṇa - Laghu, Rūkṣa.	Vipāka - Kaṭu.

मुण्डिका कटुतिक्तास्यादनिलास्त्र विनाशिनी।
अपचीघ्नी ह्यपस्मारगण्डश्ली पदनाशिनी।।
(ध. नि.)

Muṇḍikā kaṭutiktāsyād anilāsravinaśinī,
Apacighni hyapasmāragaṇḍaśīīpadanaśinī.
(Dhanvantari nighaṇṭu)

Actions/Uses : Tridoṣaghna. Dīpana, Mūtrajanana, Anulomika.

Therapeutics :

Muṇḍī is laxative, tonic, deobstruent, alterative and aphrodisiac; used in insanity, bronchitis, diseases of spleen, pain in uterus and vagina, piles. Root and seed are anthelmintic. Flowers are alterative, cooling and tonic; are swallowed to cure conjuctivitis. Decoction of plant is used as a diuretic in urethral discharges. Rind of fruit as a fish poison.

Tagetus erecta Linn.

ENGLISH : Marigold.

INDIAN : Zenddu, Makhamala, Zergul.

SANSKṚTA : Jhaṇḍu, Sālapuṣpa, Gada, Galagoti, Vānti.

HABIT : An hairy erect annual herb 1-2 ft. high.

HABITAT : Native of France and Mexico, cultivated over the greater part of Europe. Grown in gardens in India.

CHEMICAL COMPOSITION : Salicylic acid, bitter substance calendulin and essential oil. Essential oil containing azulenogenic sesquiterpenes or sesquiterpene alcohols. Flowers contain calendulin, a tasteless substance analogous to bassorin, traces of essential oil, oleanonlic acid, a gum, a sterol, cholesterol, esters of lauric, myristic, palmitic, stearic and pentadecyclic acids, faradiol and arnidiol. Inulin in roots.

Parts used : Leaves and flowers.

Ayurvedic properties :
Rasa - Tikta, Kaṣāya. Vīrya - Śīta.
Guṇa - Laghu, Rūkṣa. Vipāka - Kaṭu.

झण्डुः कटुकषाया स्यातिक्ताशीता च वीर्यतः।
कफपित्तप्रशमनं रक्तसांग्रहिकं परम्।।

Jhaṇḍuḥ kaṭukaṣāyā syatiktā śīta ca vīryataḥ,
Kaphapittaprasamanam raktasāngrahikam param.

Actions/Uses : Kaphapittaśāmaka, Raktastambhana, Śothaghna.

Therapeutics :
Plant is vulnerary, astringent and styptic. A mild aromatic with diaphoretic, diuretic and stimulant properties. Leaves are resolvent and diaphoretic. Flowers, stimulant, antiseptic and emmenagogue. A tincture of dried flowers is administered in the treatment of amenorrhoea, and after dilution, as a lotion for sprains and bruises. Flowers and plant used to treat wounds and injuries.

Tricholepis glaberrima DC.

SYNONYMS : Lamprachenium microceohalum Benth.

INDIAN : Dhan, Kantaphalapatra, Motachor.

SANSKṚTA : Bramhadaṇḍī, Ajaladaṇḍi, Kanṭakapatraphala.

HABIT : A sub-erect annual herb.

HABITAT : Konkan and Western Ghats, Rajasthan, Gujarat.

CHEMICAL COMPOSITION : Sitosterol, stigmasterol campesterol; alkanes; betulin, spinasterol etc. isolated.

Parts used : Whole plant.

Ayurvedic properties :

Rasa - Kaṭu, Tikta. Vīrya - Ūṣṇa.
Guṇa - Laghu, Rūkṣa. Vipāka - Kaṭu.

ब्रह्मदण्डी कटुष्णा स्यात् कफशोफानिलापहा।।
(निघण्टु रत्नाकर)

Bramhadaṇḍī kaṭuṣṇa syāt kaphaśophānilāpahā.
(Nighaṇṭuratnākar)

Actions/Uses : Balya, Sāraka, Jvaraghna, Kāsaghna.

Therapeutics :

Bramhadaṇḍī is used in leucoderma and skin diseases and is considered nervine tonic and aphrodisiac; and is used in seminal debility. A decoction of plant is used in medicine.

Vernonia cinerea Less.

SYNONYMS : *Conyza cinerea Linn.*

ENGLISH : Ash coloured flea bane, Purple fleabane.

INDIAN : Sahadevi, Saayidevi, Sadodi, Daudotpala, Osari.

SANSKṚTA : Sahadevī, Jvarahara, Daudotpala, Ghaugavalli, Devasahra.

HABIT : An erect annual herb.

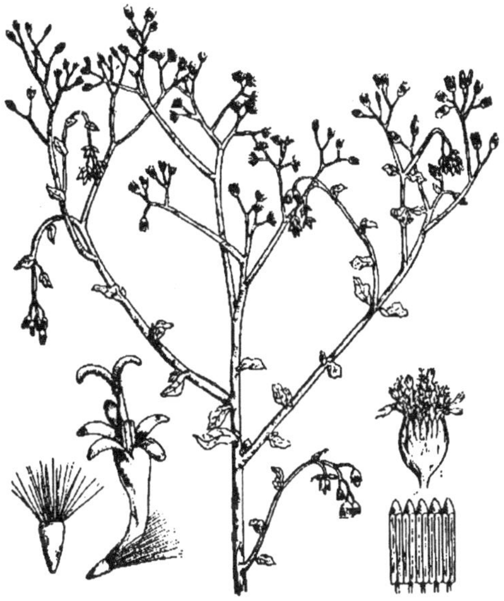

HABITAT : Throughout India.

CHEMICAL COMPOSITION : Seeds yield 38 per cent fatty oil. Plant contains β-Amyrin acetate; β-amyrin benzoate; lupeol and its acetate; β-sitosterol, stigmasterol, a-spinasterol and KCL.

Parts used : Root and leaves.

Ayurvedic properties :

Rasa - Tikta. Vīrya - Ūṣṇa.
Guṇa - Laghu, Rūkṣa. Vipāka - Kaṭu.

ज्वरं हन्ति शिरोबद्धा सहदेवीजटा यथा।
(च. सू. २५)

Jvaram hanti śirobaddha sahadevījaṭā yathā. (Caraka Sutra 25.)

Actions/Uses : Kaphavātaśāmaka.

Therapeutics :

Sahadevī is a febrifuge, diaphoretic, used in decoction to promote perspiration in febrile conditions. It is alterative and purifies blood, bile and semen. It is a specific for leucorrhoea and excessive bleeding. It is used in chronic skin diseases, dysuria, bladder stones, piles, worms and haematological disorders. Whole plant is a remedy for spasms of bladder and strangury. Juice of plant is given to children with incontinence of urine and in piles and is good for eyes. Root is bitter and used as an anthelmintic. A decoction of it is given in diarrhoea and stomach ache and its juice for cough and colic and also for dropsy. Flowers are administered for conjuctivitis and fever and for rheumatism. Seeds are used as anthelmintic, alexipharmic.

Xanthium strumarium Linn.

SYNONYMS : *Xanthium indicum Koenig, ex Roxb.*

ENGLISH : Cocklebur, Burweed.

INDIAN : Shankeshwara, Dutundi, Banokra, Chota-Gokhru.

SANSKṚTA : Śankesvara, Ariṣṭa, Artagala.

HABIT : A woody perennial.

HABITAT : Throughout India.

CHEMICAL COMPOSITION : Seeds contain Xanthostrumarin and oxalic acid. Strumaroside and stigmasterol from fruits. Xanthatin and xanthinin; xanthumin, xanthanol. Fruits are rich in vitamin C.

Parts used : Root, leaves, fruit.

Ayurvedic properties :

Actions/Uses : Svedajanana, Lālāsravavardhaka, Mūtrajanana, Śāmaka, Śothaghna.

Therapeutics :

Śankeśvara is diaphoretic, sedative, sudorific, sialogogue; useful in long-standing cases of malaria. Root is bitter, tonic, and is useful in strumous diseases and cancer. Fruit is cooling, demulcent, given in small-pox and for eye ailments as ointment.

Family : XVII - Averrhoaceae.

Averrhoa carambola Linn.

ENGLISH : Carambola apple, Coromandel gooseberry.

INDIAN : Karamalla, Kamraka, Kumrak.

SANSKṚTA : Karmaranga

HABIT : A small ornamental tree with close drooping branches.

HABITAT : Cultivated in gardens throughout hotter parts of India.

CHEMICAL COMPOSITION : Acid patassium oxalate; Vitamin A. Fruit contains proteins and amino acid compounds, vitamine C, oxalic acid. Flowers contain rutin, quercetin 3-glucoside.

Ayurvedic properties :
Rasa - Madhura, Āmla. Vīrya - Ūṣṇa, Śīta.
Guṇa - Laghu, Rūkṣa. Vipāka - Madhura, Āmla.

Parts used : Fruit, ripe, root and leaves.

कर्मरंग हिमं ग्राहि स्वाद्वम्लं कफवातहृत्।। (भा. प्र.)

Karmarangam himam grāhi svādvamlam kaphavātahṛt. (Bhāvaprakāṣa)

कर्मरकोऽम्ल उष्णश्च वातहृत् पित्तकारकः।
पक्वस्तु मधुराम्लः स्यात् बलपुष्टिरुचिप्रदः।। (रा. नि.)

Karmarako amla uṣṇaṣca vātahṛt pittakārakaḥ, Pakvastu madhurāmalaḥ syāt balapuṣṭi rucipradaḥ. (Rajanighaṇṭu)

Actions/Uses : Pittaghna, Raktapittaśāmaka, Balya, Brunhana, Śītala, Rocaka, Tṛṣṇāśāmaka, Jvaraghna.

Therapeutics :

Root is an antidote in poisoning. Leaves are antipruritic, antipyretic and anthelmintic. Crushed leaves or shoots are applied externally in chicken pox, ringworm, scabies and headache. Decoction of leaves used for aphthae and angina and to arrest vomiting. Flowers are vermicidal. Fruits are laxative, antiscorbutic, antidysentery, febrifuge and antiphlogistic. Fruit juice is used for bleeding piles and for relieving thirst and febrile excitement. Useful for hepatic colic. Ripe fruit is generally sour, antiscorbutic and highly cooling and a remedy for bleeding piles. Also useful in relieving thirst and febrile excitement. Seeds increase the flow of milk and in large doses act as an emmenagogue and cause abortion.

Family : XVIII - Basellaceae.

Basella alba Linn.

ENGLISH : Indian spinach, Malbar nightshade.

INDIAN : Maayaallu, Velbondi, Velgond.

SANSKṚTA : Upodakī, Apodikā, Niviti.

HABIT : A perennial creeping or climbing herb.

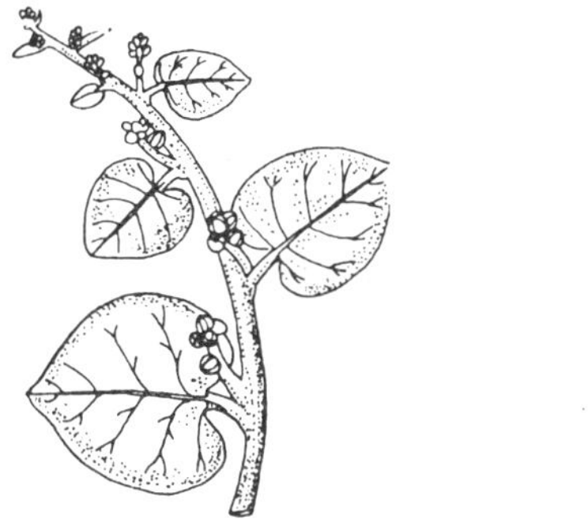

HABITAT : Occurs throughout India as a weed, wild or cultivated in gardens.

CHEMICAL COMPOSITION : Plant contains a good deal of mucilage and iron. Seeds contain protein, 23.1% and oil. Presence of l-norcarrenaline, dopamine. dopa, vitamins has been reprted.

Parts used : Root, leaves, stem and juice of plant.

Ayurvedic properties :
Rasa -Madhura, Kaṭu, Kaṣāya. Vīrya - Śīta.
Guṇa - Snigdha, Picchila. Vipāka - Madhura.

उपोदकी कषायोष्णा कटुका मधुरा च सा।
निद्रालस्यकरी रुच्या विष्टम्भश्लेष्मकृन्मतम्।।
श्लेष्मलं पौष्टिकं पथ्यं शीतलं चातिस्निग्णकम्।
स्निग्धं बल्यं कष्ठयं च वातपित्त मदापहम्।
रक्तपित्तं हरत्येवं मुनिभिर्भाषितं पुरा। (निघण्टु रत्नाकर)

Upodakī kasayosna katuka madhura ca sa

Nidralasya rucikaram malavastambha krnmatam.
Slesmalam paustikam pathyam sitalam catisnigdhakam,
Snigdham balyam kanthyam ca vatapittamadapaham.
Raktapittam haratyevam munibhirbhasitam pura.
(Nighanturatnakara)

Actions/Uses : Vātapittaśāman; Śītala, Snehana.

Therapeutics :

Upodakī is used as a cooling medicine in digestive disorders. Juice of plant if applied to body for Pitaamb. It reduces burning and itching of skin. It is beneficial in fistulae, pustules and inflammatory tumours. It is a specific for burns. Roots are rubefacient and are used as poultice to reduce swellings; sap is used in acne. A decoction of root is given to stop bilious vomiting and in intenstinal complaints. Leaves and stem are cooling, demulcent, laxative, appetiser, tonic, sedative and diuretic. They are useful in burning sensation, constipation, flatulence, anorexia, pruritus, leprosy, gonorrhoea, balanitis, strangury cattarhal affections and general debility. Used against syphilitic ulcers in nose. Vegetable of leaves is used for the same purpose. Leaf-juice mixed with butter is a soothing and cooling application for burns and scalds. It is popular application in urticaria caused by dyspepsia and in burns and scalds. Properties and uses are same as of *B. rubra.* and is considered as a variety of *B. rubra.*

Basella rubra Linn.

SYNONYMS : *Basella alba* Linn. *var. rubra* Stewart.

ENGLISH : Indian spinach, Malbar nightshade.

INDIAN : Velgond, Potaki, Lalbachlu.

SANSKṚTA : Potakī.

HABIT : A perennial climbing herb.

HABITAT : Throughout India, wild or cultivated in gardens.

CHEMICAL COMPOSITION : Plant contains protein, calcium, iron and Vitamins A and B. Loss of ascorbic acid content of leaves on heating.

Parts used : Roots, stems and leaves.

Ayurvedic properties :
Rasa - Madhura, Kaṭu, Kaṣāya. Vīrya - Śīta.
Guṇa - Picchila, Sthira, Sara. Vipāka - Madhura.

पोतकी शीतला स्निग्धा श्लेष्मला वातपित्तनुत्।
अकण्ठ्या पिच्छिला निद्राशुक्लदा रक्तपित्तजित्।।
बलदा रुचिकृत् पथ्या बृंहणी तृप्तिकारिणी।।
(भा. प्र.)

Potakī śītaḷā snigdhā ṣleśmalā vātapittanut,
Akanṭhyā picchilā nidrasukaḷadā raktapittajit,
Baladā rucikṛt pathyā bṛnhaṇī tṛptikāriṇī.
(Bhāvaprakāṣa)

Actions/ Uses : Agnidipana, Śital, Snehana.

Therapeutics :

Potakī is acrid, soporific, narcotic, laxative, aphrodisiac, fattening, improves appetite; useful in biliousness, dysentery, leprosy, ulcers. Root rubefacient and used as poultice to reduce local swellings; sap in acne. A decoction of root given to stop bilious vomiting and in intestinal complaints. Leaves demulcent, diuretic; useful in gonorrhoea and balanitis. Juice of leaves is used in urticaria to reduce itching and burning and in cases of constipation particularly in children and pregnant women. Fruit juice of red variety is applied to eyes in conjuctivitis.

Family : XIX - Berberidaceae.

Berberis aristata DC.

SYNONYMS : *Berberis tinctoria* Leschen.; *Berberis chitria* Ham.

ENGLISH : Ophthalmic barberry, Tree-turmeric, False calumba.

INDIAN : Daaruhallada, Jharki-halad, Zarishk, Chitra.

SANSKṚTA : Dāruharidrā, Dārvī, Kanṭakaṭerī, Kāminī, Kalīyaka, Pītikā, Hemakāntā, Parjanya, Pacampacā, Pītadru, Kaṣṭharanjanī, Dāruniśā, Sthirarāga, Kāmavatī, Hemavarṇavatī, Kusumbhaka.

HABIT : Genus commonly known in English as Berberry and in India as Kashmal or Kingor. An erect glabrous spinescent shrub or small tree with spines.

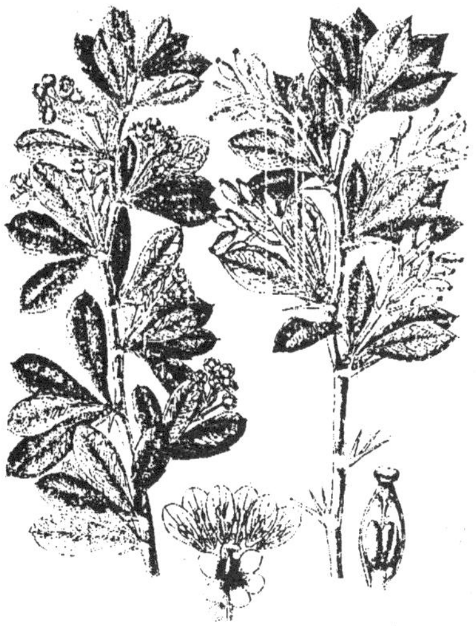

HABITAT : Himalayas, Nepal, Dhuna, Kunuwar.

CHEMICAL COMPOSITION : Root bark rich in alkaloid content, berberine principal alkaloid.

Parts used : Fruit, root-bark, stem and wood.

Ayurvedic properties :
Rasa - Kaṭu, Tikta, Kaṣāya. Vīrya - Ūṣṇa.
Guṇa - Rūksa, Laghu. Vipāka - Kaṭu.

तिक्ता दारुहरिद्रा स्याद् रुक्षोष्णा व्रणमेहनुत्।
कर्णनेत्रमुखोद्भूतां रुजं कण्डूं च नाशयेत्।। (ध. नि.)

Tiktā dāruharidrā syād rukṣoṣṇā vraṇamehanut, Karṇanetramukhodbhūtan rujam kaṇḍūm ca nāśayet.
(Dhanvantarinighaṇṭu)

Actions/Uses : Pittaghna, Kaphaghna, Lekhanī, Śodhanī, Vraṇahara, Meharogaghna, Raktadoṣahara, Tvagrogahara, Śothahara, Kṛimighna, Vārnakārī, Apacihara, Kaṇduhara.

Therapeutics :

Dāruharidrā is tonic, stomachic, astringent, anti-periodic, diaphoretic, antipyretic, emmenagogue and alterative, useful in the treatment of jaundice, enlargement of spleen etc. Root is purgative. Roots along with stem bark is a reputed drug in Ayurvedic medicine. Decoction of root bark is used in malarial fever. A compound "Rasaut" prepared from root bark, root chips, lower stem-wood is a colagogue, stomachic, laxative, diaphoretic, bitter tonic, antipyretic and antiseptic, used in skin diseases, menorrhagia, diarrhoea, jaundice and affections of the eyes; administered in indolent ulcers and haemorrhages. In Unani medicine used in leprosy. In Tibet, decoction used in piles, gastric disorders and other allied complaints.

Berberis vulgaris Linn.

SYNONYMS : *Berberis dumetorum Gouan.*

SIMILAR : Berberis lycium Royle.; Berberis aquifolium Pursh.

ENGLISH : Common Barberry; Pipperidge Bush.

INDIAN : Kashmal.

SANSKṚTA : Kaśmala.

HABIT : A bush with long, brittle branches bearing sharp thorns.

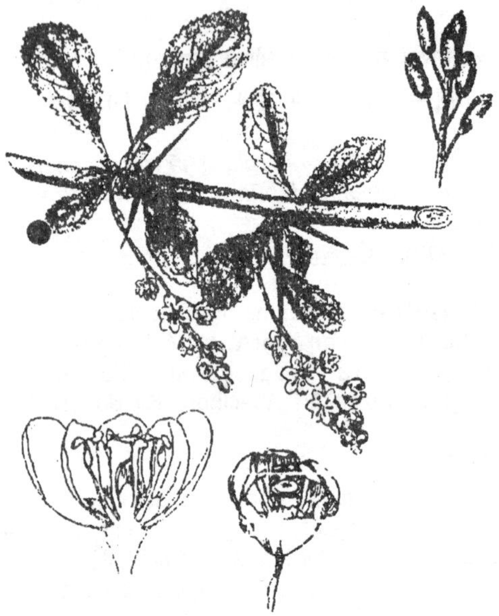

HABITAT : A common garden bush, growing wild in some parts of Europe and British Isles.

CHEMICAL COMPOSITION : Berries contain malic, tartaric and citric acid. It contains alkaloids berberine, oxyacanthine, berbamine. Carotenoids, tannin, carbohydrates, organic acids, Mn; and pectic substances and Vitamin C in berries.

Parts used : Root, root bark, stem and leaves.

Ayurvedic properties :
Rasa - Kaṭu, Tikta, Kaṣāya. Vīrya - Ūṣṇa.
Guṇa - Rūkṣa, Laghu. Vipāka - Kaṭu.

Actions/Uses : Pittaghna, Kaphaghna, Lekhani, Śodhanī, Vraṇahara, Meharogaghna, Raktadoṣahara, Twagrogahara, Śothahara, Kṛmighna, Varṇakārī, Apacihara, Kaṇḍūhara.

Therapeutics :

Kaśmala is tonic, purgative and antiseptic. Berberine is used in disorders of bile and urinary passages, especially in biliary and renal calculi, congestion of abdominal and pelvic cavities and rheumatism. Used in all cases of jaundice, liver complaints, general debility and biliousness. It acts more on the kidneys and bladder; then the liver and lastly on the mucous membranes.

HOMEOPATHIC USE : *B. vulgaris* is a Homeopathic remedy. The first indication is urinary trouble. All berberis pains radiate. Pains shoot out in every direction. Burning pain in bladder and kidney. Berberis when indicated throws out the stone. Renal colic, gall-stone colic or rheumatism, all are associated with some kind of urinary disturbance and Berberis is many times found useful.

Podophyllum emodi Wall. ex Hook. f. & Th.

SYNONYMS : *Podophyllum hexandrum Royle.*

ENGLISH : Indian Podophyllum.

INDIAN : Patvel, Paapraa, Bakracgimaka, Bhavanbakra.

SANSKṚTA : Hansapadi, Giriparpata, Vakrā, Vaiśākhaseva.

HABIT : A glabrous, 15-20 cm. tall, more or less succulent perennial herb growing in damp shady places, with a creeping 2-5 cm. long knotty rhizome bearing thick fibrous roots and small dark red fruits.

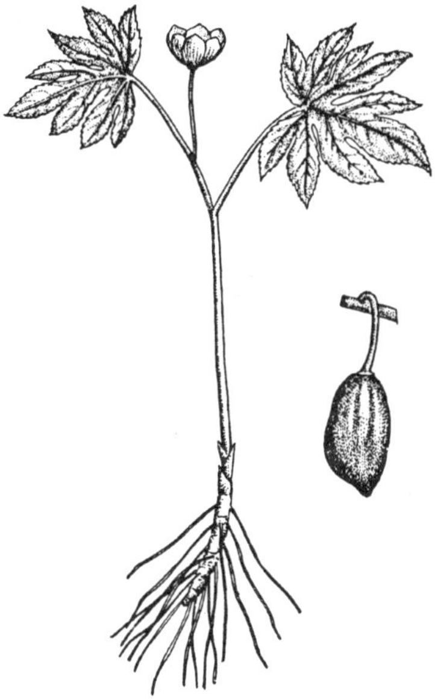

HABITAT : Interior ranges of Himalayas, Simla, Nepal.

CHEMICAL COMPOSITION : It contains a resin with essential oil, podophyllotoxin, picropodophyllin and quercetin. Dried rhizomes yield podophyllotoxin-B-D-glucoside. Roots and rhizomes yield a-lignan, a glycoside. Astragalin (kaempferol-3-glucoside) isolated. Indian podophyllum contains more podophyllin (up to 20 %) and no peltatin is present.

Parts used : Whole plant.

Ayurvedic properties :
Rasa - Tikta, Kaṭu. Vīrya - Ūṣṇa.
Guṇa - Laghu, Rūkṣa, Tīkṣṇa. Vipāka - Kaṭu.

हंसपादी गुरुः शीता हन्ति रक्तविषव्रणम्।
विसर्पदाहातिसारलूताभूताग्निरोहिनीः।।
(भावप्रकाश)

Hansapādī guruḥ śītā hanti raktaviṣavraṇam,
Visarpadāhātisāralūtābhūtāgnirohinīḥ.
(Bhāvaprakāṣa)

Actions/Uses : Kaphapittaghna, Pittaśodhana, Yakṛtotttejaka.

Therapeutics :

Hansapādī or Giriparpata is cholagogue, pugative, alterative, emetic and bitter tonic. It is a drastic but slowly acting purgative producing copious watery stools. It causes much griping and is therefore given in conjunction with belladonna and hyoscyamus. It is used for bilious complaints and constipation, also for inflammation of liver during typhoid fever. During recent years, podophyllin has acquired special importance for its possible use in controlling some forms of cancer.

Family : XX - Bignoniaceae.

Oroxylum indicum Vent.

SYNONYMS : *Bignonia indica* Linn.; *Calosanthes indica* Blume.

ENGLISH : Indian Trumpet flower.

INDIAN : Tentu, Ullu, Arlu, Saona, Kharasinga.

SANSKṚTA : Śyonāka, Bhallūka, Pṛthusimbi, Mayūrajangha, Aralū, Ṭeṭuka, Śoṣaṇa, Kunaṭa, Śūkanāsā, Kaṭambhara, Priyajīva, Ṭuṇṭuka, Naṭa, Kutannaṭa.

HABIT : A small to medium sized deciduous tree.

HABITAT : In Konkan on slopes of hills and semievergreen forests.

CHEMICAL COMPOSITION : Crystalline bitter oroxylin; alkaloids. Glucoside and bitter substances. Baicalein and its glucuronides.

Parts used : Root bark, stem bark, fruits and seeds.

Ayurvedic properties :

Rasa - Tikta, Kaṣāya.	Vīrya - Śīta, Ūṣṇa.
Guṇa - Laghu, Rūkṣa.	Vipāka - Kaṭu.

श्योनाको दीपनः पाके कटुकस्तुवरो हिमः।
ग्राही तिक्तोऽ निलश्लेष्मपित्तकास प्रणाशनः।।
(भा. प्र.)

Śyonāko dīpanaḥ pāke kaṭukastuvaro himaḥ,
Grāhi tikto anila śleṣmapittakāsapraṇāśanaḥ.
(Bhāvaprakāṣa)

श्योनाकः शोथहरः ....
(च. सू. ४)

Śyonakaḥ śothaharaḥ . . .
(Carka Sutra 4)

Actions/Uses : Kaphaghna, Pittaghna, Āmapācana, Tridoṣaśhāmaka. Śothahara, Śītapraśamana, Vraṇaropaṇa, Grāhi, Dīpana, Jvaraghna, Bastirogahara.

Therapeutics :

Śyonaka is an ingredient of Dasamula. It is bitter, diuretic, hot, astringent, carminative, stomachic, stregnth-giving and aphrodisiac. Root bark astringent, tonic, useful in diarrhoea and dysentery. It stimulates digestion, cures fever, cough, and other respiratory disorders. It is useful in diarrhoea, dysentery, abdominal pain, thirst, vomiting, anorexia, rheumatism, oedema and urinogenital disorders, leprosy and other skin diseases, worms. Powdered stem bark or its infusion is diaphoretic, useful in acute rheumatism, bitter tonic. Seeds purgative. Fruit is an expectorant; improves appetite. Clinical studies have shown that root bark is effective in the treatment of amoebic dysentery.

Stereospermum chelonoides DC.

SYNONYMS : *Bignonia chelonoides Linn. f., Dipterosperma personatum Hassk., Stereospermum personatum Hassk., Stereospermum tetragonum DC.*

ENGLISH : Trumpet flower, Yellow snake tree.

INDIAN : Paadalla, Padri, Paral, Kirsel, Koosga, Tuatuka.

SANSKṚTA : Pāṭalā-1, Kāṣṭhapāṭalā.

HABIT : A large deciduous tree.

HABITAT : Konkan on slopes of hills, in moist deciduous or ever green forests.

CHEMICAL COMPOSITION : Bark contains crystalline bitter substance. Antitumor agent lapachol, dinantin-7-glucuronide from leaves. β-sitosterol and n-triacontanol from root bark. Lapachol, dehydro-a-lapachone and dehydrotectol in root heartwood.

Parts used : Roots, leaves, bark and flowers.

Ayurvedic properties :

Rasa - Tikta, Kaṣāya. Vīrya - Śīta, Ūṣṇa.
Guṇa - Laghu, Rūkṣa. Vipāka - Kaṭu.

द्वितीया पाटला श्वेता निर्दिष्ठा काष्ठ पाटला।
सा चैव श्वेतकुम्भीका कुबेराक्षी फलेरुहा।।
पाटलायाः गुणैस्तद्वत्किंचिन्मारुतकृद्भवेत्।
(धन्वतरी निघण्टु)

Dvitīyā pāṭalā śvetā nirdiṣṭhā kāṣṭha pāṭalā,
Sā caiva śvetakumbhīkā kuberākṣī phaleruhā.
Pātalāyāḥ guṇaistadvātkincin mārutakṛdbhavet.
(Dhanvantari)

Actions/Uses : Śītala, Vātahara, Jvaraghna.

Therapeutics :

Root, leaves and flowers of Pātalā are cooling and used in decoction as a febrifuge. Juice of leaves mixed with lime juice is used in maniacal cases. Flower and fruit are used in scorpion sting. Properties and uses of Patala are the same as *S. suaveolens.*

Stereospermum suaveolens (Roxb.) DC.

SYNONYMS : *Stereospermum colais Mabber.*

INDIAN : Paatalaa, Paral, Kusgo, Kalagori.

SANSKṚTA : Pātalā-2, Kṛṣṇavṛntā, Madhudvatī, Amoghā, Sthalī, Alivallabhā, Kuberākṣī, Ambuvāsinī, Phaleruhā, Kāyasthā, Pātālī, Tāmarapuṣpī, Kumbhīpuṣpī, Kācasthālī, Madhudūtī, Viṭavallabha, Tālī, Kālasthālī, Raktapuṣpikā, Sthiragandhā, Kālavṛntī.

HABIT : Middle sized tree with red and fragrant flowers.

HABITAT : Dry deciduous forests throughout India, often planted.

CHEMICAL COMPOSITION : Root bark contains bitter substance. Lapachol isolated which showed highly significant activity against Walker 256 carcinosarcoma.

Parts used : Root, root bark, leaves, flowers, fruit and salts.

Ayurvedic properties :
Rasa - Tikta, Kaṣāya. Vīrya - Anuṣṇa.
Guṇa - Laghu, Rūkṣa. Vipaka - Kaṭu.

Flowers and fruit :
Rasa - Kaṣāya, Madhura. Vīrya - Śīta.
Guṇa - Laghu, Rūkṣa. Vipāka - Madhura.

पाटला तुवरा तिक्तानुष्णा दोषत्रयापहा।
अरुचिश्वासशोथास्त्रच्छर्दि हिका तृषाहरी।।
पुष्पं कषायं मधुरं हिमं हृद्यं कफास्त्रनुत्।
पित्तातिसारहृटकण्ठ्यं फलं हिक्कास्त्रपित्तहृत्।।
(भा. प्र.)

Pātalā tuvarā tiktānuṣṇā doṣatrayāpahā,
Aruciśvāsaśothāsracchardi hikā tṛṣāhārī.
Puṣpam kaṣāyam madhuram himam hṛdyam kaphasranut,
Pittātisārahṛt kaṇṭhyam phalam hikkāsrapittahṛt.
(Bhāvaprakāṣa)

Actions/Uses : Pittaghna, Vedanāsthāpana, Vraṇaropaṇa.

Therapeutics :

Pātalā is bitter, astringent, cardiotonic, cooling, tonic, and diuretic. It overcomes anorexia, difficult breathing, anasarca, piles, vomiting, hiccough and thirst. It is a constituent of Dashamoola. Root is bitter, heating, useful in kapha and vaata and for inflammations, eructations, vomiting, asthma, fevers, diseases of blood. Root bark is cooling, diuretic, tonic. Bark is Vaatakaphashaamaka. Fruit and flowers are Vaatapittashaamaka. Flowers are astringent, sweet, agrreable to heart, useful in bleeding diseases and diarrhoea and are rubbed up with honey to check hiccough. They are Vajikar, Sitala and Paustika, are taken in the form of a confection as an aphrodisiac. Fruit is useful in hiccough and blood diseases.

Family : XXI - Bixaceae.

Bixa orellana Linn.

SYNONYMS : *Bixa crallava.*

ENGLISH : Annatto, Safron tree.

INDIAN : Shendri, Sinduri, Kesri, Gowpurgee, Japhar, Latkan.

SANSKṚTA : Sindūrī, Raktabīja, Śoṇapuṣpī, Karacchada.

HABIT : A shrub or evergreen small handsome tree.

HABITAT : A South American plant related to Chaul-

moogra, found throughout the hotter parts of India. Cultivated to a small extent in Mysore and is grown in gardens as a hedge plant.

CHEMICAL COMPOSITION : Tetracyclic sesquiterpene-ishwarane from leaf oil. Tomentosic acid isolated. Roots, triterpenes, tomentosic acid and another unidentified one. Leaves, essential oil which contains sesquiterpene, bixaghanene. Seeds contain a yellow colouring matter bixin.

Parts used : Fruits, seeds, leaves and root-bark.

Ayurvedic properties :
Rasa - Kaṭu, Tikta, Kaṣāya. Vīrya - Śīta.
Guṇa - Śīta. Vipāka - Kaṭu.

सिन्दूरी विषपित्तास्त्रतृष्णावान्तिहरी हिमा।
(भा. प्र.)

Sindūrī viṣapittāsratṛṣṇāvāntiharī himā.
(Bhāvaprakāṣa)

सिन्दूरी कटुका तिक्ता कषाया श्लेष्मवातजित्।
शिरोर्त्तिशमनी भूतनाशा चण्डीप्रिया भवेत्।।
(रा. नि.)

Sindūrī kaṭukā tiktā kaṣāyā śleṣmavātajit,
Śirorttiśamanī bhūtanāśā caṇḍīpriyā bhavet.
(Rājanighaṇṭu)

Actions/Uses : Fruit : Grāhi, Sansrasana in large quantity.
Seeds, root : Rocaka, Jvaraghna, Grāhi.

Therapeutics :
Sindūrī is recommended for leprosy, eczema and elphantiasis. Root bark is antiperiodic and antipyretic. Leaves are used in jaundice and for snake bite. Decoction of leaves used as gargle for sore throat; infusion in dysentery, and is a popular febrifuge. Pounded leaves on maceration in water yield a gummy substance used in gonorrhoea and as diuretic. Twigs, in liver diseases and as emollient. Poultice applied to cuts and gashes as scar-preventive. Bixa paste is considered to be aphrodisiac. Fruit is astringent and purgative. Seeds are cordial, astringent, febrifuge, and a good remedy for gonorrhoea, antiperiodic and antipyretic. Seed pulp is astringent, haemostatic, antidysentary, diuretic, laxative febrifuge and helps digestion. Prescribed in epilepsy and skin diseases. Fresh pulp applied to bums to prevent blisters and scars. Seed fatty acid used in leprosy. Pulp useful in kidney diseases. Pulverised seeds are antisecretary, aqeous extract is antispasmodic and hypotensive. When used as a meat flavouring, impart a saffron-like taste and color.

Family : XXII - Bombacaceae.

Adansonia digitata Linn.

ENGLISH : Monkey bread tree, Baobab tree.

INDIAN : Gorakhachincha, Gorakshee, Sumpura, Gorak amli.

SANSKṚTA : Gorakṣī, Gorakṣakarkaṭī Gangerukīe, Gorakh-cinca.

HABIT : A curious-shaped, midium sized deciduous tree.

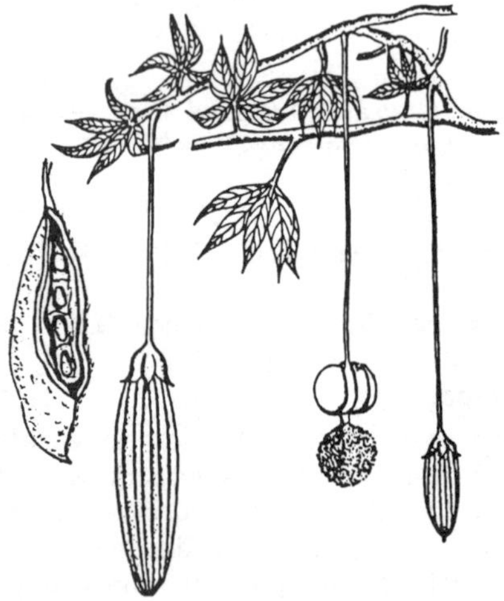

HABITAT : A native of tropical Africa, occasionaly cultivated in some parts of U.P; Bihar; Maharashtra and Tamilnadu.

CHEMICAL COMPOSITION : Pulp contains phlobaphenes, mucilage and gum, glucose, tartrate and acetate of potash and other salts. It is known as 'cream of tartar' and is an important source of Vitamin C. Leaves and bark contain wax, soluble and insoluble tannin, and a glucoside adansonin. Stem bark contains pectin, B-sitosterol. Seed oil, B-sitosterol, stigmasterol, cholesterol, etc.

Parts used : Fruit, bark and leaves.

Ayurvedic properties :
Rasa - Amla, Madhura. Vīrya - Śīta.
Guṇa - Laghu, Snigdha. Vipāka - Madhura.

गोरक्षी मधुराम्ला च शिशिरा दाहपित्तनुत्।
विस्फोटवान्त्यतीसार विषमज्वरनाशिनी।।
(रा. नि.)

Gorakṣī madhurāmlā ca śiśirā dāhapittanut,
Visphoṭavāntyatīsāra viṣamajvaranāśinī.
(Rājanighaṇṭu)

Actions/Uses : Pittaghna. Śītal, Snehana, Śransana, Ruchikara, Hṛdya.

Leaves : Grāhī, Snehana, Śītal, Dipan, Śangrahak.

Tender leaves : Vraṇaśothahara.

Bark : Snehana, Śīta Sangrāhī.

Therapeutics :

Tender leaves of Gorakṣī are applied over inflammations to reduce burning and pain of swellings. Leaves are used as diaphoretic and as a prophylactic against fever also used as fomentations and poultices for rheumatic affections of limbs and irriable inflammatory ulcers. Bark reduces pulse rate. Decoction of bark is useful for malaria. Fruit pulp is aperient, demulcent, and astringent, used in dysentery. A syrup with figs is cooling and refrigerant in fevers, diminishing the heat and quenching thirst. It relieves night-sweats and febrile flushes in consumption. It is useful in bilious dyspepsia and acid eructations.

Bombax ceiba Linn.

SYNONYMS : *Bombax malabaricum DC.; Salmalia malabarica Schott & Endl.*

ENGLISH : Silk cotton tree.

INDIAN : Shaalmali, Kaantisenbal, Huttian, Laala-saanwar, Kate-sawar, Nurma, Deokapaas, Shimal, Savari, Shembal.

SANSKṚTA : Śālmalī, Mocā, Raktapuṣpa, Sthirāyuhu, Kaṇtakādhya, Picchila, Tūlinī, Pūranī, Mahāvṛukṣa, Rocana, Rakta-śālmalī, Pancaparṇi.

HABIT : A lofty, deciduous tree buttressed at the base.

HABITAT : Throughout the hotter parts of India.

CHEMICAL COMPOSITION : Seeds contain hexacosanol, tocopherol, terenes etc. Bark contains mucilage which is used for haemoptysis in pulmonary T.B. and in influenza. Lupeol and β-sitosterol and a napththoquinone compound are present in stem bark and root bark. Roots yield triacontanol, β-sitosterol and a new glycoside. An essential oil and hentriacontane, hentriacotanol, quercetin, kaempferol, B-sitosterol and its glucosides are present in fresh petals of flowers.

Parts used : Root, stem, leaf, fruit, flowers, bark and gum.

Ayurvedic properties :
Rasa - Madhura, Kaṣāya. Vīrya - Śīta.
Guṇa - Snigdha, Laghu, Picchila. Vipāka - Madhura.

शाल्मलिः शीतला स्वाद्वी रसे पाके रसायनी।
श्लेष्मला पित्तवातास्त्रहारिणी रक्तपित्तजित्।।
शाल्मली पुष्पशाकं तु घृतसैन्धवसाधितम्।
प्रदरं नाशयत्येव दुःसाध्यञ्च न संशयः।।
रसे पाके च मधुरं कषायं शीतलं गुरु।
कफापित्तास्रजिद् ग्राहि वातळं च प्रकीर्तितम्।। (भा. प्र.)

Śālmaliḥ śītalā svādvī rase pāke rasāyanī,
Śḷeṣmaḷā pittavātāsrahāriṇī raktapittajit,
Śālmalī puṣpaśākantu ghrtasainthava sadhitam,
Pradaram nāśayatveva duḥsādhyñca na saṅsayaḥ.
Rase pāke ca madhuram kaṣāyam śitalam guru,
Kaphapittāsrajit grāhī vātalam ca prakīrtitam.
(Bhāvaprakāṣa)

शाल्मली पिच्छिलो वृष्यो बल्यो मधुरशीतलः।
कषायश्च लघुः स्निग्धः शुक्लश्लेष्मविवर्द्धनः।।
तद्रसस्तद्गुणो ग्राहि कषायः कफनाशनः।
पुष्पं तद्वच्च निर्दिष्टं फलं तस्य तथाविधम्।।
(राजनिघण्टु)

Śālmalī picchilo vṛṣyo balyo madhuraśītalaḥ,
Kaṣāyaśca lghuḥ snigdhaḥ
śukḷaśḷeṣmavivardhanaḥ,
Tadrasastadguṇo grāhī kaṣāyaḥ kaphanāśanaḥ,
Puṣpam tadvacca nirdiṣṭam phalam tasya tathāvidham.
(Rājanighaṇṭu)

Actions/Uses : Vātapittaśāmaka, Kaphakara, Grāhī, Vṛṣya, Balya, Raktastambhana, Vedanasthapana, Rasayani, Pramehaghna, Atisarasamana, Dahasamana, Sukravardhani, Sangrāhak but Snehan. Tubers are Sangrāhak, Pauṣṭik, Brunhan and Vāyasthāpan. Latex is hima, snigdha, vrsya and kaṣāya.

Therapeutics :

Various parts of Śālmalī are used in smallpox, bleeding gums, toothache and carries, sores in mouth, pain in leg, fever, enlarged spleen, atrophy, emaciation, rheumatism, spermatorrhoea, haematuria, cholera, pneumonia, pleurisy, intercosal neuralgia, leprosy and rinderpest. The plant is used in enteritis, dysentery menorrhagia, lymphoadenoma, hepatitis etc. Young tap roots are astringent and used in dysentery. Infusion of bark is used as demulcent and tonic. Externally it is used as styptic and for fomenting wounds. Paste of bark applied in skin eruptions, applied on boils, acne and pimples. Aq. extract with curd given in blood dysentery. Used in syphilis and gonorrhoea. Flowers are astringent and cooling. Paste of flowers as also of leaves applied in cutaneous troubles. Young fruits used as expectorant, stimulant and diuretic, beneficial in calculous affections, chronic inflammation of bladder and kidneys. Seeds are used in gonorrhoea, chronic cystitis and other catarrhal affections.

Family : XXIII - Boragianceae.

Cordia dichotama Forst.f.

SYNONYMS : *Cordia latifolia* Roxb.; *Cordia myxa* Roxb. *non* Linn.; *Cordia obliqua* Willd.

ENGLISH : Sebeston plum, Large sebestan.

INDIAN : Bhokara, Shelvant, Lasora.

SANSKṚTA : Śleṣmataka, Śleṣmaghātaka, Bahuvāra, Śailū, Śītala, Bhūtavṛukṣa, Uddāla, Śaila, Śīta, Gandhapuṣpa, Picchila, Śāpita, Lekhanātaka, Dvijakutsita, Śāka, Śelu, Karbudāraka, Uddālaka, Bahuvāraka.

HABIT : A perennial middle sized crooked tree.

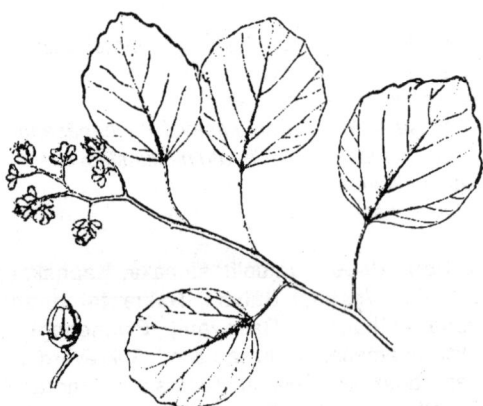

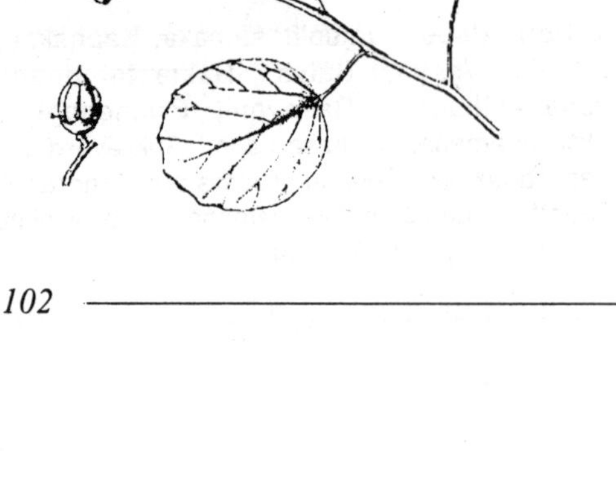

HABITAT : Common throughout India. In Konkan, on slopes of hills and in gardens.

CHEMICAL COMPOSITION : Mono- and polysacaccharides in fruit. Seed oil β-sitisterol. Seeds, flavonol glycosides. taxifolin 3-rhamnoside. Stem bark allantonin, β-sitisterol and a flavanone rhamnopyranoside. Roots β-sitisterol, flavon glycosides. Bark contains 20 % tannin.

Parts used : Leaves, bark, fruits and kernels.

Ayurvedic properties :
Rasa - Madhura, Kaṣāya, Tikta. Vīrya - Śīta.
Guṇa - Snigdha, Gurū, Picchila.
Vipāka - Madhura, Kaṭu.

श्लेष्मातकः कटु हिमो मधुरः कषायः।
स्वादुश्च पाचनकरः क्रिमिशूलहारी।।
आमास्त्रदोषमलरोधवहुव्रणार्ति।
विस्फोटशान्तिकरणः कफकारकश्च।।
(राजनिघण्टु)

Śleṣmātakaḥ kaṭuhimo madhuraḥ kaṣāyaḥ,
Svāduśca pācanakaraḥ krimiśūlahārī ca,
Āmāsradoṣamalarodhabahuvraṇārtti,
Visphotaśāntikaraṇaḥ kaphakārakasca.
(Rājanighaṇṭu)

Actions/Uses : Fruit : Vātapittaśāmaka.
Bark : Kaphapittaśāmaka, Vraṇaśodhan-ropaṇa, Kuṣṭhaghna, Gāhī, Viṣaghna, Kṛmighna.

Therapeutics :

Various plant parts of Śleșmātaka are used in dropsy, anasaraca, urticaria, cholera and dysentery. Eherial extract of aerial parts is hypothermic and diuretic. Leaves are applied to ulcers and in headache. Bark is mild astringent and tonic. Juice is used in gripes. Decoction used in dyspepsia and fevers. Fruit is astringent, antelmintic, diuretic, demulcent, expectorant and used in affections of urinary passages, diseases of lungs and spleen. Fruit mucilage is highly esteemed in coughs, in diseases of chest, uterus, urethra etc. and is a valuable anticapping agent for amidopyrine tablets. Powdered kernels mixed with oil is a remedy in ringworm. Seeds are anthelmintic.

Family : XXIV - Brassicaceae.

Brassica juncea (Linn.) Czern & Coss.

ENGLISH : Brown or Indian mustard.

INDIAN : Asal raaee, Laha, Sarson, Mohari.

SANSKṚTA : Rājikā, Āsuri, Rājakṣavaka, Siddhārthaka, Raktikā, Kṛṣṇavarṇaka.

HABIT : A tall much-branched annual herb.

HABITAT : First evolved in Middle East. Abundantly cultivated in U.P. also in low-lying hills in other parts of India.

CHEMICAL COMPOSITION : Seeds contain sinigrin, gluconapin, sinapine base and a volatile isothyocynate. Essential oil.

Parts used : Whole plant, seed and seed oil.

Ayurvedic properties :
Rasa - Kaṭu. Vīrya - Ūṣṇa.
Guṇa - Tīkṣṇa. Vipāka - Kaṭu.

राजिका कफवातघ्नी तीक्ष्णोष्णा रक्तपित्तकृत्।
किञ्चिद्रूक्षाग्निदा कण्डूकुष्ठकोष्ठकृमीहरेत्।।
अतितीक्ष्ण विशेषेण तद्वत्कृष्णाऽपि राजिका।
सरा हिमा गुरुर्ग्राही तत्पुष्पं प्रदरास्त्रजित्।।
(भा. प्र.)

Rājikā kaphavātaghnī tīkṣṇoṣṇā raktapittanut,
Kiñcidrūkṣagnidā kaṇḍukūṣṭhakoṣṭhakrimin haret.
Atitīkṣṇa viśeseṇa tadvat kṛṣnāpi rājikā,
Sarā himā gururgrāhī tatpuṣpam pradarāsrajit.
(Bhāvaprakāṣa)

Actions/Uses : Kaphavātaghna, Vedanāsthāpana, Visphotajanana, Śothahara, Lekhana, Vidāhī.

Therapeutics :

Rājikā is useful in inflammatory conditions of lungs, liver, bronchii etc. Whole plant possesses bitter, aperient and tonic properties. Seeds are acrid, bitter, carminative, anthelmintic, aperient, digestive, sudorific and tonic. They are useful in dengue fever, dyspepesia, abdominal colic, and worms. Given in small quantity it is stimulant and diaphoretic, In large doses it is emetic and cause burning sensation of body. External application on skin renders it red and part gets numb. A hot mustard bath is an emmenagogue. Oil is stimulant and counter irritant.

Brassica nigra (Linn.) Koch.

SYNONYMS : *Sinapis nigra* Linn.; *Brassica sinapioides* Roth.; *Brassica alba* Linn. *(White.)*; *Sinapis alba* Linn. *(White.)*.

ENGLISH : Black or true mustard.

INDIAN : Raaee, Saraso, Kaallee mohari, Banarasi rai.

SANSKṚTA : Sarṣapa, Kaṭuka, Tantubha, Kadambaka, Asuri, Sneha, Atitikṣva, Jvalanti, Krimika, Kṛṣṇasarṣapa, Kṣava.

HABIT : A much-branched annual herb.

HABITAT : Cultivated throughout India; widely distributed in Europe, Asia and Great Britain. Both the varieties Black and White are similar but the white variety is somewhat smaller. Pods of black mustard are smooth, whereas of the white are bristly.

CHEMICAL COMPOSITION : Seed contains glucoside Sinigrin, sinapine and a volatile isothiocyanate and essential oil. The black mustard and white mustard differ in containing a crystalline substance known as sulpho-sinapisin. The myrosine of white mustard yields with water a pungent oil of a different character from the volatile oil of black mustard.

Parts used : Leaves, seeds and seed oil.

Ayurvedic properties :
Rasa - Kaṭu-Tikta. Vīrya - Uṣṇa.
Guṇa - Tīkṣṇa, Rūkṣa. Vipāka - Kaṭu.

सर्षपस्तु रसे पाके कटुः स्निग्धः सतिक्तकः।
तीक्ष्णोष्णः कफवातघ्नो रक्तपित्ताभिवर्धनः।।
रक्षोहरो जयोकण्डूकुष्ठ कोष्ठकृमिग्रहान्।
यथा रक्तस्तथा गौरः किन्तु गौरो वरो मतः।।
(भा. प्र.)

Sarṣapastu rase pāke kaṭuḥ snigdhaḥ satiktakaḥ,
Tikṣṇoṣṇaḥ kaphavātaghno raktapittābhivardhanaḥ.
Rakṣoharo jayokaṇḍūm kuṣṭhakoṣṭhakṛmigrahān,
Yathā raktastathā gauraḥ kintu gauro varo mataḥ.
(Bhāvaprakāṣa)

Actions/Uses : Kaphavātaghna, Pittakara, Dīpana, Vidāhī, Kṛimighna.

Seeds : Ṡonitotkleṣakāraka, Lekhana, Varṇya, Kuṣṭhaghna.

Oil : Anulomik, Vātahar, Keṣya. Jantughna, Vedanāsthāpana, Snehana.

Therapeutics :

Both varieties of Sarṣhapa are irritant, stimulant, diuretic and emetic. Leaves are used as a vegetable which serves as a laxative. Seeds are a powerful stimulant and rarely employed in pure state. They are stomachic and used in neuralgic and rheumatic affections. Oil when applied to skin produces almost instant vesication. It is very useful application for chilblains, chronic rheumatism etc. Used chiefly as a poultice in acute local pains, pneumonia, bronchitis and other diseases of respiratory organs. The volatile oil is a powerful irritant, rubefacient and vesicant and combined with other remedies is an excellent application in rheumatic pains and colic etc. Oil is recommended to persons for use in their daily diet for chronic constipation. It is used to promote growth of hair and for maintaing its color.

HOMEOPATHIC USE : Sinapis alba and Sinapis nigra are two Homeopathicaly proved remedies. The well known emetic effects of mustard were produced by it and also "burning" sensations in many parts, notably anus, stomach and oesophagus. Flatulence, headache and salivation were other observed effects. The prover who had no sign of threadworms since thirteen years, passed many both living and dead; and had all the usual rectal symptoms which accompany their presence.

Lepidium sativum Linn.

ENGLISH : Garden cress.

INDIAN : Halleeva, Ahliva, Halim, Hurf, Chansaur, Akalam.

SANSKṚTA : Candraśūra, Vasapuspa, Carmahanti, Suvasara, Candrika, Pasumehanakarika, Nandini, Karavi, Bhadra, Aśalika, Darakṛṣṇa, Dīrghabīja.

HABIT : A small, herbaceous, glabrous annual herb.

HABITAT : Native of West Asia cultivated throughout India.

CHEMICAL COMPOSITION : Plant contains glucotropoeoline, sinapine, sinapic acid, β-sitosterol, benzylcyanide etc. A glycoside present in seeds, but absent in seed oil. Leaves contain protein, 5.8; fat 1.0, carbihydrate, 8.7 calcium, phosphate, iron, nickel, cobalt and iodine; vitamin A, thiamine, riboflavin, niacin and ascorbic acid. Essential oil. Seeds contain an alkaloid sinapic acid and its choline estere and a volatile oil. Seed muciliage, known as Cress Seed Mucilage allays irritation of mucous membrane of intestines in dysentery and diarrhoea.

Parts used : Whole plant, root, seeds and leaves.

Ayurvedic properties :
Rasa - Kaṭu, Tikta. Seeds : Snigdha, Picchila.
Vīrya - Ūṣṇa.
Guṇa - Laghu, Rūkṣa, Tīkṣṇa. Vipāka - Kaṭu.

चन्द्रशूरं हितं हिक्कावातश्लेष्मातिसारिणाम्।
असृग्वातगद्वेषि बलपुष्टिविवर्धनम्।।
(भा. प्र.)

Candraśūram hitam hikkāvātaśleṣmātisāriṇām,
Asṛg vātagadveṣi balapuṣṭivivardhanam.
(Bhāvaprakāṣa)

Actions/Uses : Kaphavātaghna, Vedanāsthāpana, Śūlaghna, Dīpana, Anulomika, Mūtrajanana, Jantughna, Tvagdoṣhara, Vātānulomana, Uttejaka, Malasangrāhī.

Therapeutics :

Candraśūra plant is used in cases of bleeding piles, asthma, and, cough with expectoration. Root is used in secondary syphilis and tenesmus. Seeds are aperient, diuretic, alterative, tonic, demulcent, aphrodisiac, carminative, galactagogue, emmenagogue and administered after being boiled with milk to cause abortion. Used during confinement as nutrition. Mucilage of seeds allays irritation of mucous coat of intestines and is used as aperant. Paste of seeds used as external application for lumber pain and rheumatism. Applied to pains and hurts as a poultice. Tea of seeds is useful in hiccough.

Raphanus sativus Linn.

SYNONYMS : *Raphanus caudatus* Linn.

ENGLISH : Radish.

INDIAN : Mulla, Muri.

SANSKṚTA : Mūlaka, Nīlakaṇṭha, Hastidanta, Śankhamūla, Mahākanda, Dīrghakandaka, Kunjarakṣāra, Rucya, Kṛdantābha, Mūlāhva, Sita, Haritaparṇa, Bhūkṣāra, Mṛttikākṣāra, Karidantabha, Mūlakapotikā, Śāleyaka, Miśra.

HABIT : An annual or biennial bristly erect herb.

HABITAT : Cultivated all over India.

CHEMICAL COMPOSITION : Seeds yield essential oil. Fleshy roots and seeds contain glucoside, enzyme, and methyl mercaptan. Sulforaphene from seeds.

Parts used : Root, leaves and seeds

Ayurvedic properties :
Rasa - Kaṭu. Vīrya - Ūṣṇa.
Guṇa - Tīkṣṇa. Vipāka - Kaṭu.

लघुमूलं कटुष्णं स्याद्रुच्यं च लघु पाचनम्।
दोषत्रयहरं स्वर्यं ज्वरश्वासविनाशनम्।।
नासिकाकण्ठरोगघ्नं नयनामयनाशनम्।
महत्तदेव रुक्षोष्णं गुरु दोषत्रयप्रदम्।।
(रा. व. नि.)

Laghumūlam kaṭuṣṇam syādrucyam ca laghu pācanam,
Doṣatrayaharam svaryam jvaraśvāsavināśanam.
Nāsikākaṇṭharogaghnam nayanāmayanāśanam,
Mahattadeva rukṣṇoṣṇam guru doṣatrayapradam
(Rājavallabha)

मूलकं तीक्ष्णमुष्णञ्च कटुष्णं ग्राहि दीपनम्।
दुर्नामगुल्म हृद्रोग वातघ्नम् रुचिदं गुरु ।।
(राजनिघण्टु)

Mūlakam tīkṣnamuṣṇañca kaṭuṣṇam grāhi dīpanam,
Durnāmagulma hṛdroga vātaghnam rucidam guru.
(Rājanighantu)

Actions/Uses : Tender root : Tridoṣahara.

Ripe root : Tridoṣakara. Dīpana, Pācana, Rucikara, Bījalekhana, Aśmārihara.

Therapeutics :

Mūlaka is refreshing and depurative. Its preparations are useful in liver and gall bladder troubles. Roots are used for urinary complaints, piles and gastrodynic pains. Juice of fresh roots is powerful antiscorbutic. Seeds and leaves are diuretic, laxative and lithontriptic. Juice of fresh leaves is diuretic, laxative. Seeds are expectorant, diuretic, laxative, carminative, peptic and emmenagogue, useful in gonorrhoea. Roots, leaves, flowers and pods are active against Gram-positive bacteria.

HOMEOPATHIC USE : Useful for neuralgic headaches, sleeplessness and chronic diarrhoea.

Family : XXV - Bromeliaceae.

Ananas comosus Merr.

SYNONYMS : *Ananas sativus Schult. f.; Bromelia ananas.*

ENGLISH : Pineapple.

INDIAN : Ananasa.

SANSKṚTA : Bahunetra, Anannāsa, Anāśapāzama, Kaṭahalasapharī, Ānārasa.

HABIT : An perennial herb.

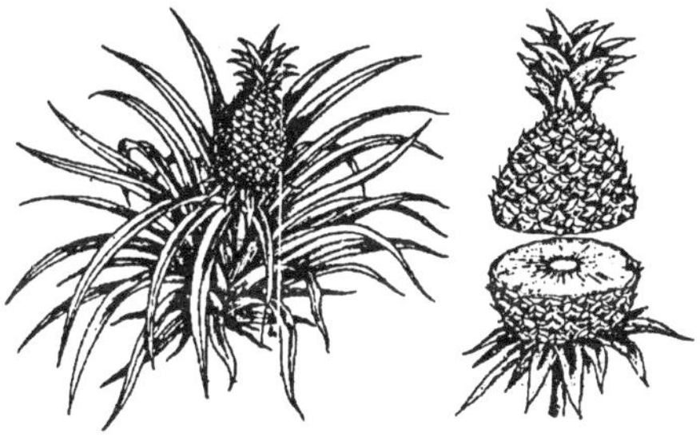

HABITAT : Native to tropical America and extensively cultivated throughout India.

CHEMICAL COMPOSITION : Stems contain enzyme bromelin and starch. Fresh juice also contains bromelin, an enzyme which aids digestion. Leaves, ergosterol peroxide, 5-stigmastene-3 B. 7a-diol, B-sitosterol, campesterol, stigmasterol and campestanol. Juice of unripe fruits showed marked anti-implantation and abortifacient activities in rats. Antioedema substance from rhizome and whole plant, fruits and leaves. Plant extract nematicidal. Ether extract spasmogenic. Alcoholic extract anthelmintic. Addition of 5% pineapple frower decoction and 2% ginseng root in cold creams can control eczema and wrinkles.

Parts used : Ripe and unripe fruit and leaves.

Ayurvedic properties :
Rasa - Ripe : Madhura, Unripe : Amla.
Vīrya - Śīta.
Guṇa - Gurū, Snigdha. Vipāka - Madhura.

बहुनेत्रफलं चाम्लं कृमिघ्नं मधुरं सरम्।
बल्य वातहरं रुच्यं श्लेष्मलं तर्पणं गुरु ॥ (राजवल्लभ)

Bahunetraphalam cāmlam kṛmighnam madhuram saram,
Balyam vātaharam rucyam śleṣmalam tarpaṇam guru.
(Rajavallabha)

Actions/Uses : Vātapittaśāmaka. Rucikara, Dīpana, Anulomana.

Fruit-juice : Vibandha, Hṛdya, Raktapittaśāmaka, Aśmaribhedana, Mūtrala, Balya.

Therapeutics :

Fresh juice of leaves or leaves themselves of Anannāsa are powerful purgative and anthelmintic and vermicide. Fresh juice of plant, is an excellent digest; blood purifier; applied to horny excrescences on skin, leprosy, elephantiasis, also found useful in diphtheria. Juice of ripe fruit is diuretic, antiscorbutic, diaphoretic, aperient and refrigerant and helps digestion. Juice of unripe fruit is acid, styptic, powerful diuretic and anthelmintic and emmenagogue; in large quantities it is abortifacient. Juice as also alcoholic extract of unripe fruit has antifertility activity. Rhizome also have antifertility activity.

Family : XXVI - Burseraceae.

Balsamodendron mukul Hook. ex Stocks.

SYNONYMS : *Commiphora mukul (Hook. ex Stocks) Engl.; Commiphora wightii Bhandari.; Balsamodendeon wightii Arnott.; Balsamodendeon roxburghii Stocks.*

ENGLISH : Gum gugul, Indian bedlelium tree.

INDIAN : Guggulla.

SANSKRṬ : Guggul̤a, Guggulu, Devadhūpa, Kauśika, Pūra, Marudviṣta, Palankaṣa, Pavanadviṣṭa, Bhavābhīṣṭha, Niśāṭaka, Jaṭāla, Rakṣīhā, Devestha, Śiva, Mahiṣākṣa, Kumbha, Jaṭāyu, Yātughna, Kālaniryāsa, Durga, Bhūtahara, Śāmbhava, Rukṣagandha, Niśāyāyī, Yāminīcara, Divya.

HABIT : A small tree or shrub with spines.

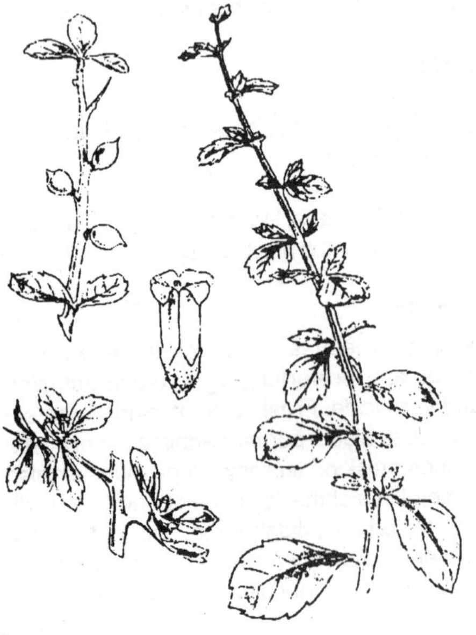

HABITAT : A native of Arabia and African coast, found in the arid and rocky zones in certain parts of India.

CHEMICAL COMPOSITION : Terpenoid and sterol fraction hypolipidaemic activity; Crude gum and its petroleaum ether extract caused decrease serum cholesterol level and lowered triglycerides. Monocyclic diterpines-a-camhorene and cembrene in resin. Allylcembrol in plant. Commercial product contains 1.45% essential oil besides gum and resin. Mericyl alcohol and β-sitosterol were isolated.

Parts used : Gum-resin.

Ayurvedic properties :
Rasa : Tikta, Kaṭu, Madhura, Kaṣāya. Vīrya - Ūṣṇa.
Guṇa - Laghu, Tīkṣṇa, Sūkṣma, Rūkṣa.
Vipāka - Kaṭu.

गुग्गुलविशदस्तिक्तो वीयोष्णः पित्तलः सरः।
कशायः कटुकः पाके कटुरुक्षो लघुः परः॥
भग्नसन्धानकृद् वृष्यः सूक्ष्मः स्वर्यो रसायनः।
दीपनः पिच्छिलो बल्यः कफवातव्रणापचीः॥
मेदोमेहाश्मस्त्रवातांश्च क्लेदकुष्ठांमारुतान्।
पिड़काग्रन्थीशोफार्शो गंडमालाकृमीन् जयेत्॥
स नवो बृहणो वृष्यः पुराणस्तवतिलेखनः।
(भा. प्र.)

Guggul̤ urviṣadastikto vīryoṣṇo pittalaḥ saraḥ,
Kaṣāyaḥ kaṭukaḥ pake kaṭurūkṣo laghuḥ paraḥ.
Bhagnasandhānakṛda vṛṣyaḥ sūkṣmaḥ svaryyo rasāyanaḥ,
Dīpanaḥ picchilo balyaḥ kaphavātavraṇapācih.
Medomehasmavātanscya kledakuṣṭhamamārutān,
Pidakagranthiśothārṣi gaṇḍamālā krimin jayet.
Sā navo bṛṇhano vṛṣyaḥ pūranāstvatilekhanaḥ.
(Bhāvaprakāṣa)

Actions/Uses : Fresh Guggula : Brihmana, Vṛṣya.

Old Guggula : Atilekhana, Hṛdya, Rasāyana,

Dīpani, Balya, Svarya, Vranaghna, Apacighna, Granthighna, Gandamālāhara, Medoghna, Mehaghna, Aśmaghna, Pidakaghna, Bhagnasandhānakara, Kṛmighna, Kledakuṣṭhapraśamanī.

Therapeutics :

Guggulu is highly rated in Ayurvedic medicine for treatment of rheumatic arthritis, obesity and several other disorders. Gum resin is astringent, demulcent, aperient, antiseptic, alterative, carminative, antispasmodic, emmenagogue, expectorant and aphrodisiac. It enriches blood, and is useful in indolent ulcers and as gargle in pyrrhoea alveolaris, chronic tonsilitis and pharyngitis. Resin smoke is mosquito repellent. Inhalation of fumes from burning gum is recommended in hay fever, chronic bronchitis, nasal catarrh, laringitis and phthisis. Guggulu is one of the constituents of several indigenous drugs. Bilarin is useful in infective hepatitis; Arogya-wardhini is used in acute viral hepatitis. Myostal useful in puerperal patients. Septilin prescribed in rhinosinal infection. Gugullipid useful in heart diseases, spondilitis and gout. It is a constituent of Chinese medicine used in extradural haematomas.

Boswellia serrata Roxb.

SYNONYMS : *Boswellia glabra* Roxb.; *Boswellia serrata var. glabra* (Roxb.) *Bennet., Boswellia serrata* Roxb. *ex Coleb., Boswellia thurifera* Roxb.

ENGLISH : Incense tree, Indian olibanum tree.

INDIAN : Saalaee, Shallaki, Kundur, Salphullie, Luban.

SANSKṚT : Sallakī, Suśravā, Bahusravā, Surabhiī, Suvaha, Piccha, Rasā, Maharuṇā, Gajabhakṣyā, Kundaru, Nāgavadhu, Pīta, Mahātaru, Nimbapatra, Kundurukī, Vallakī, Aśvanutrī.

HABIT : A medium to large-sized, deciduous balsamiferous tree.

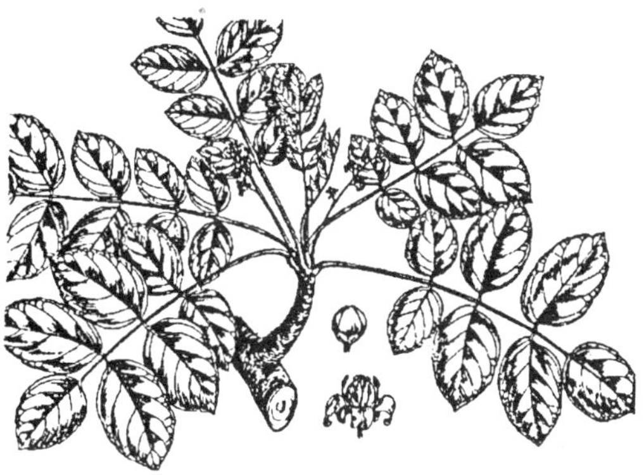

HABITAT : Commonly found in the dry forests from Punjab to West Bengal and in pensinsular India.

CHEMICAL COMPOSITION : Bark contains carbohydrates, glucosides and β-sitosterol. Gum resin contains a diterpine, alcohol, serratol, β-boswellic acid. Essential oil.

Parts used : Gum.

Ayurvedic properties :
Rasa - Madhura, Tikta, Kaṭu, Kaṣāya.
Vīrya - Śīta, Ūṣṇa.
Guṇa - Gurū, Snigdha, Laghu, Rūkṣa.
Vipāka - Madhura, Kaṭu.

सल्लकी तिक्ता मधुरा कषाया ग्राहिणी परा।
कुष्ठास्त्रकफवातार्शो व्रणदोषार्त्तिनाशिनी।।
(राजनिघण्टु)

Sallakī tiktā madhurā kaṣāyā grāhiṇī parā,
Kuṣṭhāsrakaphavātārśo vraṇadoṣārttināśinī.
(Rājanighaṇṭu)

Actions/Uses : Kaphapittaghna. Vraṇahārī, Puṣṭikārī, Arśoghna, Raktapittaśamana, Jvarāpaha, Svedāpaha, Pradaranāśaka, Grahi, Kuṣthaghni, Śothahara, Pācana, Vātānulomana, Vedanāsthāpana, Vraṇaśodhana-ropaṇa, Jantughna, Durgandhināśana.

Therapeutics :

Sallakī plant is cardiotonic, purifies vital spirit, dissolves wind, increases intellect and remedies forgetfulness. Bark is sweet, acrid, cooling, diuretic and tonic. It is useful for asthma, dysentery, piles, skin diseases, ulcers and cough. Gum is stimulant, antiseptic, diaphoretic, diuretic, astringent, emmenagogue, useful in diarrhoea, dysentery, piles, rheumatism, nervous and skin diseases. Gum s used as a fumigating agent to overpower unpleasant odours. It is used in subacute bronchitis, in chronic pulmonary affections, branchorrhoea and chronic laryngitis both internally and in the form of fumigation. Used in ointment for syphilis. Gum resin used as an ointment for sores. Non-phenolic fraction is antitumor. Grane used in plaster form for indurated as also ulcerated breasts, condylomata, cancer and scirrhous and uterine tumors.

anarium strictum Roxb.

SYNONYMS : *Canarium sikkimense King.*

ENGLISH : Black Damar

INDIAN : Dhupa, Kallaa-daamara, Raldhup.

SANSKRT : Mandadhūpa.

HABIT : A large handsome deciduous tree.

HABITAT : Forests of west coast of India.

CHEMICAL COMPOSITION : Plant contains canarone, taraxastanonol and taraxastanediol. Black dammar resin contains junenol, canarone, epikhustinal, a-amyrin, β-amyrin and its acetate. Essential oil from oleoresin.

Parts used : Gum

Ayurvedic properties :
Rasa : Madhura, Kaṣāya.   Vīrya - Ūṣṇa.
Guṇa - Gurū.   Vipāka - Madhura.

राल: सर्जरस: शाल: क्षण: कलकलोद्भव:।
ललन: शालनिर्यासो यक्षधूपोऽग्निवल्लभ:॥
राल: स्वादु: कषायोष्ण: स्तम्भनो व्रणरोपण:।
विपादिभूतहन्ता च भग्नसंधानकृत् यत:॥
(धन्वतरी निघण्टु)

*Rālaḥ svāduḥ kaṣāyoṣṇaḥ stambhano vraṇaropaṇaḥ,
Vipādibhūtahantā ca bhagnasandhānakṛt yataḥ.
(Dhanvantari)*

Actions/Uses : Vātahara, Grāhī, Vraṇaropana, Kīmighna.

Therapeutics :

Mandaphūpa gum is a stimulant to the skin and used as plaster and ointment. Gum used with gingili oil in rheumatic pains and chronic skin diseases such as psoriasis, pityriasis, etc.

Family : XXVII - Cactaceae.

Cereus grandiflorus Mill.

SYNONYMS : *Cactus grandiflorus Mill.*

ENGLISH : Night-blooming cereus, Sweet-scented Cactus.

INDIAN : Niwadunga-paanchakoni.

SANSKṚTA : Visarpin, Mahāpuṣpa, Uttamapuṣpa, Rātripraphulla.

HABIT : Sub-scandent shrub bearing large, attractive, sweet-scented flowers.

HABITAT : Indeginous to Mexico. Introduced into Indian gardens.

CHEMICAL COMPOSITION : Flowers contain flavonoids, glycosides, cacticin and narcissin.

Parts used : Flowers and tender shoots.

Ayurvedic properties :
Rasa - Madhura, Tikta, Kaṭu. Vīrya - Ūṣṇa.
Guṇa - Rūkṣa. Vipāka - Kaṭu.

रक्तं त्रिवृत्तु मधुरं रूक्षं वातकरं मतम्।
तुवरं च रसे तिक्तं कटु चोष्णं च रैचकम्।।
हितकृच्च मलस्तम्भं ग्रहणींच कफोदरम्।
शोथपाण्डुकृमीन्प्लीहां ज्वरं पित्तं कफं तथा।।
वातरक्तमुदावर्तं हृद्रोगं च विनाशयेत्।।
(निघण्टु रत्नाकर)

Raktam trivṛttu madhuram rukṣam vātakaram matam,
Tuvaram ca rase tiktam kaṭu coṣṇam ca recakam.
Hitakṛsca malastambham grahaṇīm ca kaphodaram,
Śothapāṇḍu kṛminplīhām jvaram pittam kapham tathā,
Vātaraktamudāvartam hṛdrogam ca vināśayet.
(Nighaṇṭuratnākara)

Actions/Uses : Vātakara, Recana, Hṛdayabalya, Mūtrajanana.

Therapeutics :
Visarpin is a cardiac stimulent and tonic, diuretic. It gives prompt relief in most cardiac diseases, such as palpitation, angina pectoris, cardiac neuralgia. Also useful in prostatic diseases, irritable bladder and congested kidneys. Fresh young shoots are used as cardiac stimulent and as partial substitute for digitalis. Liquid extract or tincture used in cases of dropsy and various cardic affections and goitre.

HOMEOPATHIC USE : Cactus is a Homeopathic remedy. It is particularly the heart remedy. The constictive sensation as from an iron band in various parts of body and especially in heart is an unerring indication for it. Heart and arteries at once respond to the influence of cactus. It should prove a very useful and valuable remedy for high blood pressure.

Opuntia dillenii Haw.

SYNONYMS : *Opuntia nigricans* Haw.; *Opuntia vulgaris* Mill.; *Opuntia elatior* Mill.

ENGLISH : Prickly pear, Slipper thorn.

INDIAN : Niwadunga-phadyaa, Vidara, Hathhathoria, Chapal, Naagaphannaa.

SANSKṚTA : Kanthārī, Kanṭaseva.

HABIT : An erect perennial under-shrub.

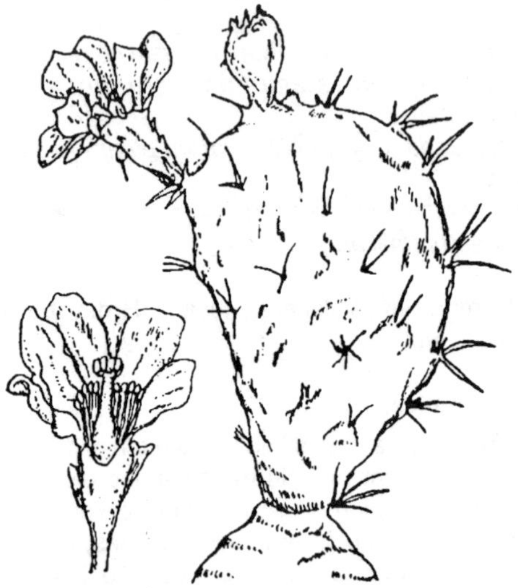

HABITAT : A native of South America. Introduced in India; probably indigenous to Mexico.

CHEMICAL COMPOSITION : Flowers contain gycoside of isorhamnetin and quercertain and glucoside narcissi. Isolation of isoquercitrin, β-sitosterol, opintiol. A polysaccharide contains galactose and arabinose from pods.

Parts used : Fruits, leaves and latex.

Ayurvedic properties :
Rasa - Kaṭu, Tikta. Vīrya - Ūṣṇa.
Guṇa - Tīkṣṇa. Vipāka - Kaṭu.

कन्थारी दीपनी रुच्या कद्वयुष्णा तिक्तकामता।
रक्तदोषं कफं वातं ग्रान्थिरोगं च नाशयेत्।।
स्नायुरोगं च शोफं च नाशयेदिती कीर्तिता।
(निघण्टु रत्नाकर)

Kanthārī dīpanī rucyā katvayuṣṇā tiktakāmatā,
Raktadoṣam kapham vātam granthirogam ca nāśayet.
Snāyurogam ca śopham ca nāśayeditī kīrtitā.
(Nighaṇṭuratnākara)

कन्थारी कटु तितोष्णा कफवातनिकृन्तनी।
शोफघ्नी दीपनी रुच्या रक्तग्रन्थीरुजापहा।।
(राजनिघण्टु)

Kanthārī kaṭu tiktoṣṇā kaphavātanikṛntanī,
Śophaghnī dīpanī rucyā raktagranthīrujāpahā.
(Rājanighaṇṭu)

Actions/Uses : Fruit : Kaphaghna, Sankocavikāsapratibandhaka, Dāhapraśamana. Hṛdayapoṣaka.

Therapeutics :

Kanthārī fruit is refrigerant, useful in gonorrhoea; baked and given in whooping cough; in form of syrup given to control spasmodic cough and expectoration. Milky juice is purgative. Ripe fruit is useful in gonorrhoea as a demulcent. Leaves mashed up and applied as poultice to allay inflammation and heat; made into a pulp applied to eyes in ophthalmia; heated and applied to boils to hasten suppuration.

Family : XXVIII-Caesalpiniaceae.

Bauhinia racemosa Lam.

SYNONYMS : *Piliostigma racemosa (Lamk.) Bentham.*

INDIAN : Ashmantak, Aapataa, Ashta, Jhinjeri, Kachnal, Sonan, Vanajaja.

SANSKṚTA : Aśmantaka, Ṡlakṣṇatvaga, Yugmapatra, Kuśalī, Aśmayoni, Amlapatraka, Mālukāpatra, Yamalapatraka, Anupuṣpaka, Pāpanāśana, Ṡvetakāncan, Kovidāra.

HABIT : A small, crooked, bushy tree with drooping branches.

HABITAT : Throughout India.

CHEMICAL COMPOSITION : Stem bark, octacosane, β-amyrin and β-sitosterol. Seeds contain protein, 10.5; pentosan, 7.2; mucilage, 4.0%. Oil, medicinal gum are other constituents.

Parts used : Whole plant, root, bark, leaves, young flowers, seeds and fruits.

Ayurvedic properties :

Rasa - Madhura. Virya - Ṡita.
Guṇa - Guru. Vipaka - Madhura.

कोविदारः कषायः स्यात् संग्राही व्रणरोपनः।
दीपनः कफवातघ्नो मूत्रकृच्छ्रनिवारणः।।
(राजनिघण्टु)

Kovidārah kaṣāyah syāt sangrāhī vraṇaropanaḥ,
Dīpanaḥ kaphavātaghno mūtrakṛcchranivāraṇaḥ.
(Rājanighaṇṭu)

कोविदारः कटुः पाके कषायो मधुरः परम।
कफपित्तहरो ग्राही किञ्चिन्मारुतकोपनम्।।
(शोधला निघण्टु)

Kovidāraḥ kaṭuḥ pāke kaṣāyo madhuraḥ param,
Kaphapittaharo grāhī kīncinmārutakopanam.
(Sodhala nighaṇṭu)

Actions/Uses : Kaphaghna, Pittaghna. Grāhī, Lekhana, Kṛmighna, Kuṣṭhaghna, Galagaṇḍaśamana, Viṣahara, Vraṇāpaha, Tṛṣṇaśamana, Raktavikāraśamana.

Therapeutics :

Root bark is astringent used in diarrhoea and dysentery. It is also used for inflammations, ulcers, skin diseases and has cholagogue activity; efficaceous remedy for goitre. Decoction of rootbark is useful in inflammation of liver and for gargle during aphthae. It also expels worms. Paste of bark is applied on wounds and swellings. Leaves are used for worms. Paste of young leaves is applied externally for headache during fever and decoction of leaves is used for headache and malaria. Bark and leaves are refrigerant, antipyretic, alexipharmic; cure biliousness, urinary discharges, thirst, headache, quartan fevers, anal fistula, and diseases of bloods. Young flowers or juice of leaves along with pepper or onion is used for diarrhoea and dysentery. They are anthelmintic. Flowers are useful in haemorrhage, piles, cough and as laxative.

Bauhinia variegata Linn.

ENGLISH : Buddhist bauhinia, Mountain ebony, Orchid tree.

INDIAN : Kovidaar, Kaanchana-laal, Raktakanchan. Kancanara has Red flowers while Kovidar has White.

SANSKṚTA : Kāncanāra, Kuhalī, Kovidāra, Gandārī, Dālaka, Pākārī, Camarika, Yugmapatra, Kāncana, Apsara, Śamya, Tarupungava, Kulī, Āsphotā, Girija, Svalpakesara, Kanakāraka, Kāntapuṣpa, Kāntāra, Yamalacchada, Mahāyamalaptraka.

HABIT : A medium-sized deciduous erect tree.

HABITAT : Dry forests in India.

CHEMICAL COMPOSITION : Quercitroside, isoquercitroside and rutoside from plant. Myricetol glycosides in seeds and kaempferol glycosisides in flowers. Tree yields gum; bark, tannin; seeds fatty oil.

Parts used : Bark, root, gum, leaves, seeds and flowers.

Ayurvedic properties :
Rasa - Kaṣāya. Vīrya - Śīta.
Guṇa - Gurū, Laghu, Rūkṣa. Vipāka - Kaṭu.

काञ्चनारो हिमो ग्राही तुवरः श्लेष्मपित्तनुत्।
कृमिकृष्ठगुदभ्रंशगण्डमालाव्रणापहः।।
कोविदारोऽपि तद्वस्यात् तयोः पुष्पं लघु स्मृतम्।
रुक्षं संग्राहि पित्तास्त्रप्रंद्ररक्षयकासनुत्।।
(भा. प्र.)

Kāñcanāro himo grāhī tuvaraḥ śleṣmapittanut,
Kṛmikuṣṭhagudabhranśagaṇḍamālā vraṇāpah.
Kovidāro api tadvatsyāt tayoḥ puṣpam laghusmrtam,
Rūkṣam sangrāhī pittāsrapradarkṣayakāsanut.
(Bhāvaprakāṣa)

Actions/Uses : Kaphaghna, Pittaghna, Kuṣṭhaghna, Dīpana, Grāhī, Mūtrakricchahara, Vraṇaropaṇa, Gandamālāśamana, Gudabhramṣaśamana.

Therapeutics :

Kāncānara is used in glandular diseases and as antidote to poison. Roots and bark light, cool, astringent, anthelmintic, acrid and styptic. They cure ulcers, swellings, leprosy, cough, menstrual disorders, glandular diseases and prolapse of rectum. Also useful in worms, dysentery, piles and diarrhoea. Root is carminative and is used in dyspepsia and flatulence. Decoction of root or rootbark is used for lessening fatness and against tumors. Bark is alterative, and tonic, useful in skin diseases, ulcers and in scrofula. Dried buds are used as laxative in dysentery, piles, diarrhoea and worms and against tumors. A decoction of buds is given in coughs, piles, haematuria and menorrhagia. Flwers are laxative.

Caesalpinia bonducella Fleming.

SYNONYMS : *Caesalpinia bonduc (Linn.)* Roxb.; *Caesalpinia crista* Linn.; *Caesalpinia nuga Ait. f.; Guilandina boncucella* Linn.

ENGLISH : Fever nut, Physic nut, Bonduc nut.

INDIAN : Saagaragotaa, Karanju, Kanja, Gajagaa.

SANSKṚTA : Latākaranja, Kuberākṣa, Kanṭakīkaranja, Kanṭaphala, Vīrāsya, Karanjī, Ghṛtakaranja, Prakīrya, Dusparśa, Pūtikaranja, Pūtika, Śakra, Viṭapakaranja, Vajravījaka, Dhanadākṣa, Gadaghna.

HABIT : A large scandent prickly climbing shrub.

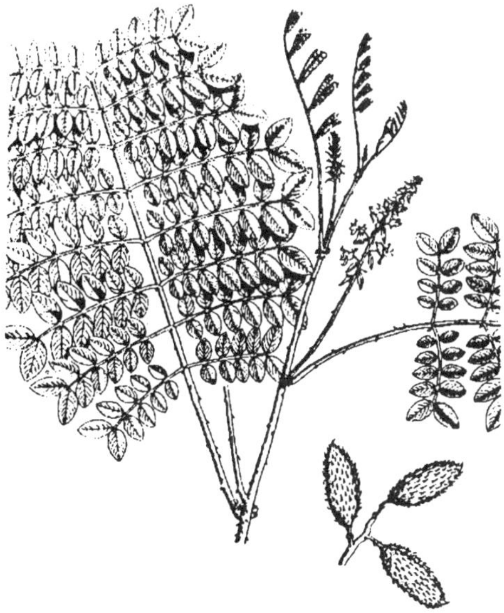

HABITAT : Throughout hotter parts of India grown as hedge plant.

CHEMICAL COMPOSITION : A bitter substance, bonducin. seeds contain bitter substance phytosterinin, bonducin, saponin, fatty oil, starch, sucrose, two phytosterols. Seeds contain three bitter compounds, α-, b-, y-, and d- caesalpin.
Note : C. bonducella and C. crita are synomymous with C. bonduc. Further C. bonduc and C. jayabo are two distinct species; however both are treated as one in Glossary.

Parts used : Seeds, leaves and bark.

Ayurvedic properties :
Rasa - Kaṭu, Tikta, Kaṣāya. Vīrya - Ūṣṇa.
Guṇa - Laghu, Rūkṣa, Tīkṣṇa. Vipāka - Kaṭu.

लताकंरञ्जपत्रं तु कटुष्णं कफवातनुत्।
तद्बीजं दीपनं पथ्यं शूलगुल्मव्यथापहम्।।
(रा. नि.)

Latakarañjapatram tu kaṭūṣṇam kaphavātanut,
Tad bījam dīpanam pathyam śūla gulma vyathāpaham.
(Rājanighaṇṭu)

कष्टकयुक्त करञ्जस्तु पाके च तुवरः कटुः।
ग्राहकश्चोष्णवीर्य स्यात्तिक्तः प्रोक्तश्च मेहहा।।
कुष्ठार्शोव्रणवातानां कृमीनां नाशनः परः।
पुष्पं चोष्णवीर्यं स्यात्तिक्तं वातकफापहम्।।
(शा. नि.)

Kanṭayukataḥ karañjastu pāke ca tuvaraḥ kaṭuḥ,
Grāhakaścoṣṇa vīryaḥ syātiktaḥ proktaśca mehahā.
Kuṣṭharśo vraṇavātānām kṛmīnām nāśanah paraḥ,
Puṣpam coṣṇavīryam syāttiktam vātakaphāpaham.
(Śaligramnighaṇṭu)

कुबेराक्षी यकृत्प्लींहवातघ्नी व्रणरोपणी।
(सो. नि.)

Kuberākṣī yakṛt plīhavātaghnī vṛaṇaropaṇī.
(Sodhalnighaṇṭu)

Actions/Uses : Kaphaghna, Vātaghna, Kaphavātaśāmaka, Bhedana, Sramsana, Stambhana, Śothahara, Vraṇaropaṇa, Raktaśodhaka, Kṛimighna, Virecanī, Vedanāsthāpana, Yakṛtarogahara, Plīhārogahara, Śvāsahara, Mūtrala, Niyatakālikajvarapratibandhaka, Raktapraśaman, Kaṭupausṭika.

Therapeutics :

Latākarañja or Kuberākṣa is bitter, germicidal, antipyretic, febrifuge and emmenagogue. Root bark is emmenagogue, febrifuge, stomachic, expectorant and anthelmintic and it is good for tumours and for removing the placenta. Decoction of root is useful in calculus and in honey to cure leucorrhoea. Leaves are emmenagogue, febrifuge and anthelmintic. Tender leaves are used in disorders of liver. Leaves and seeds are used in external applications for dispersing inflammatory swellings. Seeds are bitter, astringent, acrid, digestive, stomachic, liver tonic, antipyretic, antiperiodic, febrifuge, aphrodisiac and anthelmintic. They promote digestive power, heal ulcers, cure vomiting, hiccough, diabetes, leprosy, and piles. They are useful in asthma and snake bite. Bean meat is useful as laxative, blood purifier and in congestion of blood. Seed kernel is extensively used for intestinal worms, hydrocele and also against anasarca, liver and spleen diseases, malarial fever and mental disorders. Fruit is piscicidal. Oil from seeds is emollient and is used as embrocation to remove freckles from the face and for stopping discharge from ear. It is also useful in convulsions and paralysis.

Caesalpinia digyna Rottl.

SYNONYMS : *Caesalpinia oleosperma* Roxb.

ENGLISH : Teri pods.

INDIAN : Waakermulla, Waakeribhaate.

SANSKṚTA : Vākerī, Ghṛtakarañja.

HABIT : A large scandent prickly perennial climbing herb.

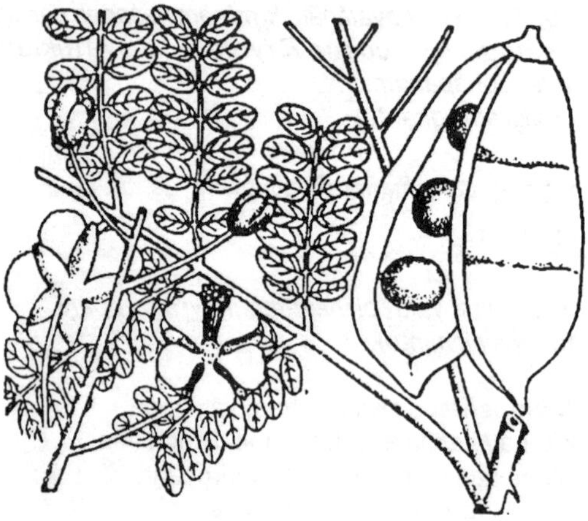

HABITAT : In Konkan, Bengal, Assam and Andamaan.

CHEMICAL COMPOSITION : Tannin in pod-cases but not in seeds; bitter substance bonducin, saponin.

Parts used : Leaves, roots and nodes on root

Ayurvedic properties :
Rasa - Kasāya, Katu. Vīrya - Ūsna.
Guna - Rūksa. Vipāka - Katu.

वाकेरी तु त्रिदोषघ्नी शोषघ्नी तुवरा परम्।
शस्यते गण्डमालायां प्रमेहे राजयक्ष्मणि।।
(द्रव्यगुण)

Vākerī tu tridoṣaghnī śoṣaghnī tuvarā param,
Śasyate gaṇḍamālāyām pramehe rājayakṣmaṇi.
(Dravyaguṇa)

Actions/Uses : Tridoṣaghna, Śodhana, Stambhana, Vraṇaropaṇa, Balya. Useful in Bhagandar, Nādivraṇa, Nāsura and Pathe.

Therapeutics :

Vākerī or Ghrtakaranja root is astringent, given in powder form mixed with milk, ghee, cumin and sugar, internally in phthisis and scrofula affections. Cures emaciation and is most efficacious in goitre. Powder is useful in diarrhoea and other chronic fluxes and also for diabetes. Known as Waakeribhate, used internally and applied externally.

Cassia alata Linn.

ENGLISH : Ringworm weed.

INDIAN : Daadamardan, Dadmari.

SANSKṚTA : Cakramarda, Dadrughna, Dvipagasti.

HABIT : An erect annual large shrub.

HABITAT : Everywhere in Konkan; Bengal.

CHEMICAL COMPOSITION : Plant contains chrysophanic acid; chrysophenol, emodin, rhein and aloe-emodi. Leaves, kaempferol and aloe-emodin and a volatile oil (sesquiterpene and phenolic compounds. Roots quinone pigments. Seeds yield galactomannan; emodin, aloe-emodin, B-sitosterol.

Parts used : Leaves and flowers.

Ayurvedic properties :
Rasa - Madhura, Kaṭu. Vīrya - Śīta.
Guṇa - Laghu, Rūkṣa, Ūṣṇa.
Vipāka - Madhura, Kaṭu.

चक्रमदर्दो लघुः स्वादु रुक्षः पित्तानिलापहः।
हृद्यो हिमः कफश्वासकुष्ठदद्रुक्रिमीन हरेत्।।
हन्त्युष्णं तत्फलं कुष्ठकण्डूदद्रुविषानिलान्।
गुल्मकासक्रिमींश्वासनाशनं कटुकं स्मृतम्।।
(द्रव्यगुण)

Cakramadardo laghuḥ svādu ruksaḥ pittānilāpahaḥ,
Hṛdyo himaḥ kaphaśvasakuṣṭhadadrukrimīn haret.
Hantyuṣṇam tatphalam kuṣṭhakaṇḍūdadruviṣānilan,
Gulmakāsakrimīśvāsanāśanam kaṭukam smṛtam.
(Dravyaguṇa)

Actions/Uses : Kaphaghna, Mūtrajanana, Kāsahara, Virecana.

Therapeutics :

Leaves of Dadrughna are sour; cure vaata; are antiparasitic. Decoction is astringent; tincture and extract are purgative. Leaves in decoction is considered as a cure for herpes and other skin diseases, even in venereal affections and all poisonous insect bites. Leaves are used in ringworm and in snake bite; also used among others as poultice and purgative. Extract of leaves antifungal. Leaf juice used against eczema. Decoction of leaves and flowers, used internaly in bronchitis and asthma and for washing eczematous patches. Bark used to treat skin diseases. Extract of aerial parts CNS depresant, diuretic and antiinflammatory.

Cassia angustifolia Vahl.

SYNONYMS : *Cassia senna L. var. senna Brenan.; Cassia acutifolia Del.; Cassia elongata Lem.-Lisanc.; Cassia obovata (L.) Collad.; Senna officinalis Roxb.*

ENGLISH : Senna, Tinnevellu senna.

INDIAN : Sonaamukhi, Bhuitarvad, Mulkacha.

SANSKṚTA : Kalyāṇi, Mārkanḍikā, Svarṇapatrī, Sanāya, Sonāmukhī, Bhūmiari, Bhupadma.

HABIT : A small perennial erect, undershrub, 0.6 to 0.75 m in height.

HABITAT : Indigenous to Somalia, Southern Arabia and Kutch area of Gujarat in India. Cultivated in different parts of India.

CHEMICAL COMPOSITION : Sennesides A,B,C and D, chrysophanol, emodin and physcion; aloe-emodin, rhein and rheum-emodin. Glucoside, kampferin, anthraquinone, essential oil, iso-rhamnetin, Ca-oxalate. Oxymethyl-anthraquinone in fruit.

Parts used : Leaves and fruits

Ayurvedic properties :
Rasa - Kaṭu, Tikta, Madhura, Kaṣāya. Vīrya - Ūṣṇa.
Guṇa - Laghu, Rūkṣa. Vipāka - Kaṭu.

विट्संग वह्णीमान्द्यञ्च यकृद्दात्युदरं तथा।
प्लीहोदरं बद्धगुदमजीर्णं विषमज्वरम्।।
कामलां पाण्डुरोगञ्च कल्याणी क्षपयेत् ध्रुवम्।।
(द्रव्यगुण)

Viṭsangam vaḥnimāndyam ca yakṛd dātyudaram tathā,
Plīhodaram badhagudamajirṇam viṣamajvaram,
Kāmalām panḍurogam ca kalyaṇi kṣapayet dhruvam.
(Dravyaguṇa)

Actions/Uses : Vātānulomana, Pittaśodhana, Lekhana, Tvagdoṣahara, Anulomana, Sransana, Yakṛtottejaka.

Therapeutics :
Kalyaṇi or Markanḍikā is useful in constipation, loss of appetite, liver complaints, splenic enlargements, dyspepsia, typhoid, jaundice, anaemia, leprosy. It is extensively used as a brisk and safe purgative. It is well adapted for children, elderly persons and delicate females. Leaves and fruits, laxative and purgative. In traditional medicine an infusion of leaves in the form of tea is used as laxative. It is also used in the form of calcium sennoside tablets as a laxative.

HOMEOPATHIC USE : Senna is also a Homeopathic remedy prepared from several species of Cassia; C. obovata, Alexandrian Senna, is the principal. Leaves are used to prepare tincture. Used as a laxative in ordinary practice, Senna has proved an excellent remedy in the colic of infants, with incarcerated flatulence and sleeplessness. Exhaustion is typical of Senna. "Repeated sneezing which caused heat, exhaustion and panting breathing."

Cassia auriculata Linn.

SYNONYMS : *Senna auriculata* Roxb.

ENGLISH : Tanner's cassia.

INDIAN : Tarawada, Tarod.

SANSKṚTA : Āvartakī, Pītapuṣpa, Pītakalika, Manojyna, Pītakalā, Carmaranga, Tindukinī, Vibhāndī, Nālī, Vāmaviṣānikā, Rangalatā, Mahadā, Vāmāvartā, Viṣānikā, Mahājālinī, Mahānālinikā.

HABIT : Woody perennial shrub with large bright yellow flowers.

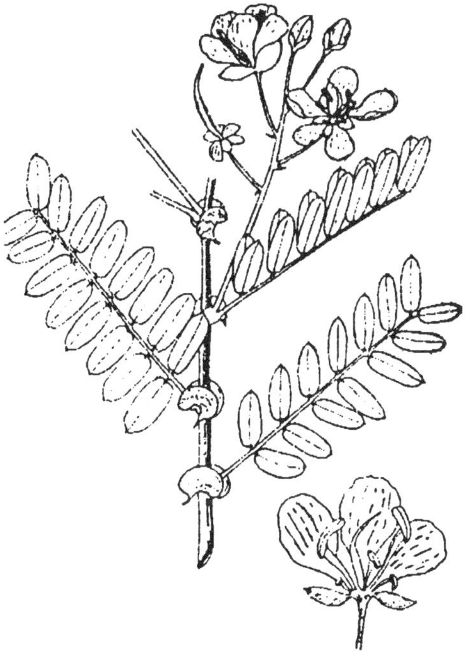

HABITAT : Western ghats and Konkan and other parts of India.

CHEMICAL COMPOSITION : Bark contains tannin. β-sitisterol and kaempferol from flowers. Three saturated higher fatty ketoalcohols and emodin from leaves. Goratensidine; Auriculacacidin.

Parts used : Bark, root, leaves, fruits and seeds.

Ayurvedic properties :
Rasa - Tikta, Kaṣāya. Vīrya - Śīta.
Guṇa - Laghu, Rūkṣa. Vipāka - Kaṭu.

आवर्तकी कषायातिस्तम्भनी तिक्तशीतला।
रक्तापित्तातिसारघ्नी कृमिकुष्ठविनाशिनी।
नेत्ररोगे प्रमेहे च तत्पुष्पं तु प्रयुज्यते।।

Āvartakī kaṣāyātistambhanī tiktaśītalā,
Raktapittātisaraghnī kṛmikuṣṭhavināśinī,
Netraroge pramehe ca tatpuṣpam tu prayujyate.

Actions/Uses : Kaphapittaśāmaka. Śulāpaha, Mukharukśamana, Kuṣḥtaghna, Kaṇḍūhara, Jantughna, Vraṇāpaha.

Flowers : Pramehaśamana.

Tender fruits : Tridoṣaghna, Vamihara, Kṛimighna, Akṣihita, Sarvapramehāpaha, Rucyam.

Seeds : Madhumehaghna, Viṣahara, Raktātisāraśamana.

Root : Raktapittaśamana, Tṛiṣṇāhaara, Pramehaghna, Śvāsaghna, Sukraksyayepujitum.

Therapeutics :

Bark and root of Āvartakī are astringent and tonic. Root is useful in urinary discharges; cures tumours, skin diseases and asthma. Tea of leaves is used in chronic fever. Leaves and fruits, anthelmintic. Powder of bark is used for fixing teeth and decoction for chronic dysentery. Twigs are used as tooth-brush. Decorticated seeds in fine powder or paste are valued local applications to purulent ophthalmia and conjuctivitis. Seeds as also tea of leaves or powder of all parts of plant or powder of seeds are taken internally in diabetes. It decreases thirst and frequency of urination. Used for turbid urine or chylous urine. Flowers are used for spermatorrhoea.

Cassia fistula Linn.

SYNONYMS : *Cassia rhombifolia* Roxb.; *Cassia excelsa H.B.& K.*

ENGLISH : Pudding Stick, Cassia pods, Golden shower. Purging cassia, Indian Laburnum.

INDIAN : Baahawaa, Amaltas, Girimalah.

SANSKṚTA : Āragvadha, Suvarṇaka, Rajavṛukṣa, Caturangula, Aruja, Sampaka, Ārevata, Kṛutamāla, Dīrghaphala, Svarṇabhūṣaṇa, Karṇa, Pragraha, Ārogyaśimbī, Karnikāra, Vyādhighāta, Hemapuṣpa, Rocana, Kaṇḍughna, Kṛutamālaka, Manthāna, Nṛupadruma, Rājataru, Mahādi, Kuṣṭhasūdana, Vyādhijita, Vyādhighna, Karṇābharaṇaka, Jvarāntaka, Adharecana.

HABIT : A middle-sized deciduous tree.

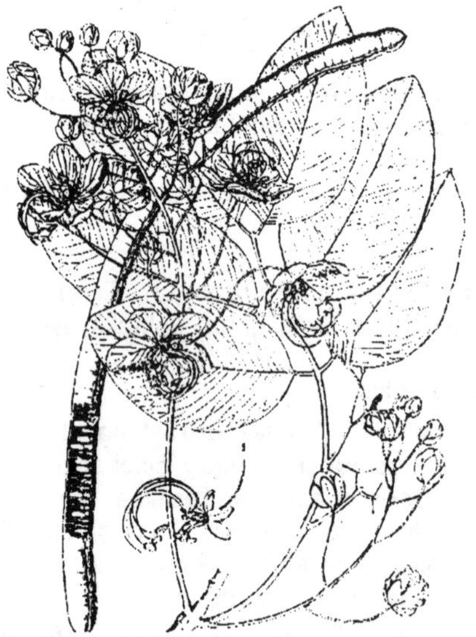

HABITAT : Throughout India.

CHEMICAL COMPOSITION : Leaves contain anthraquinone derivatives and very little tannin. Ceryl alcohol, fistulin, leucopelargonidin tetramer, kaempferol, rhein and glusoside in flowers. Root-bark, tannin, phlobaphenes and oxy-anthraquinone substance. Pulp contains rhein, glucose, sucrose and fructose volatile oil, three waxy substances and resinous substance. Fistulic acid from pods. Fistucacidin, from bark.

Parts used : Fruit pulp, root-bark, stem-bark, flowers, pods, leaves and root.

Ayurvedic properties :
Rasa - Madhura, Tikta. Vīrya - Śīta.
Guṇa - Gurū, Mṛudu, Snigdha. Vipāka - Madhura.

आरावधो गुरुः स्निग्धः शीतलः स्त्रसनो मृदुः।
ज्वरहृद्रोगपित्तास्त्रवातोदावर्तशूलनुत्।।
तत्फलं स्त्रसनं रुच्यं कुष्ठपित्तकफापहम्।
ज्वरे तु सततं पथ्यं कोष्ठशुद्धिकरं परम्।।
(भा. प्र.)

Aragvdho guruḥ snigdhaḥ śitalaḥ sransano mṛduḥ,
Jvarahṛdrogapittāsravātodāvartaśūlanut.
Tatphalam sransanam rucyam kuṣṭhapittakaphāpaham,
Jvare tu satatam pathyam koṣṭhaśudhikaram param.
(Bhāvaprakāṣa)

Actions/Uses : Vātaghna, Pittaghna. Kuṣṭhaghna, Kaṇḍūghna, Virecana, Śothahara, Vedanāsthāpana. Hṛdrogahara, Kaphodarahara, Pramehaghna, Samsrana, Jvaraghna, Raktapittahara, Gulmahara, Śulaghna, Viṣṭambhaghna, Udāvartaghna.

Therapeutics :
Āragvadhah is a reputed drug used in the treatment of skin diseases. It is found to be effective in the treatment of pyoderma. Root is astringent, tonic, febrifuge, and purgative and is useful in skin diseases, tuberculous glands, syphilis. It cures burning sensation. Root bark, seeds and leaves are laxative. Juice of leaves is useful in skin diseases. Pulp is used for inflammations. Flowers are cooling, astringent, cathartic, cure kapha and biliousness. Fruit cures leprosy, diseases of heart and is applied externally in rheumatism and snake bite. Seeds are emetic.

Cassia sophera Linn.

SYNONYMS : *Senna sophera (Linn.)* Roxb.

SIMILAR : *Cassia occidentalis* Linn.

ENGLISH : Senna sophera, Negro coffee.

INDIAN : Raantaakalla, Kaasamarda, Kasondi.

SANSKṚTA : Kāsamarda, Arimarda, Kāsāri, Karkaṣa, Kāla, Parimarda, Pītapuṣpaka, Kāsaghna, Kālamardaka, Kāsamardi, Anjana, Rājavṛukṣa, Karṣaka, Dīpana, Kanaka, Kartakāsaghna, Jaraṇa, Nāvara, Kola.

HABIT : A diffuse undershrub.

HABITAT : Throughout India.

CHEMICAL COMPOSITION : Emodin and chrysophanic acid. Seed contain tannic acid, mucilage, fatty oil emodin and a toxalbumin. Rhamnetin-3-O-B-D glucoside along with chrysophanol from flowers.

Parts used : Bark, leaves, seeds, root and root-bark.

Ayurvedic properties :
Rasa - Madhura, Tikta. Vīrya - Ūṣṇa.
Guṇa - Laghu. Vipāka -Kaṭu, Madhura.

कासमर्दः सुतिक्तः स्यान्मधुरः कफवातजित्।
विशेषतः पित्तहरः पाचनः कण्ठशोधनः।।
(ध. नि.)

Kāsamardaḥ sutiktaḥ syānmadhuraḥ kaphavātajit,
Viśeṣataḥ pittaharaḥ pācanaḥ kaṇṭhaśodhanaḥ.
(Dhanvantari)

Actions/Uses : Kaphavātaśāmaka, Pittaśāraka. Dīpana, Rucya, Mūtrala, Pācana, Ajīrṇahara, Vṛṣhya, Kāsahara, Kuṣṭhaghna, Vātānulomana Virecana.

Therapeutics :

Kāsamarda is recommended against cough, asthma and other respiratory ailments. It improves digestion, clears throat and purifies blood. Plant is used as febrifuge, purgative, diuretic and tonic. A decoction of whole plant is useful in diminishing urine and as an expectorant it gives relief in cases of acute bronchitis. Root is considered expectorant. It is used in snake bite. A paste made out of roots is a specific for ringmorm, eczema and other skin ailments. Leaves are anthelmintic and antiseptic. Infusion or decoction of leaves is given in asthma and hiccup. Infusion of fresh leaves is a useful injection in gonorrhoea in the sub-acute stage. Leaves, roots and seeds are purgative. Leaves and seeds are used externally in skin diseases, for ringworm and itch. Bark, leaves and seeds are cathartic. Bark in infusion or the powdered seeds with honey are given in diabetes. Properties of *C. sophera* and *C. occidentalis* are similar.

Cassia tora Linn.

SYNONYMS : *Cassia obtusifolia* Linn.; *Cassia toroides* Roxb.; *Cassia foetida* Pers.

ENGLISH : Foetid cassia.

INDIAN : Taakalla, Chakunda, Panevar.

SANSKṚTA : Cakramarda, Dadrughna, Meṣalocana, Cakrika, Eḍagaja, Prapuṣpa, Padmāṭa, Meṣakusuma, Prapunnāta, Vyāvartaka, Eḍahasti, Cakrila, Śakunāśana, Drudhabīja, Kuṣṭhakṛuntana, Mardaka, Meṣāhva, Kharjughna, Uraṇākṣa.

HABIT : A common herbacious fetid annual.

HABITAT : Throughout India as a weed.

CHEMICAL COMPOSITION : Obtusin, chrysoobtusin, aurantioobtusin from seeds. Content of rhein-like aglyconones in seeds; anthraquinone glucosides, glucoobtustifolin and glucoaurantioobstusin; glucosice cassiaside; robrofusarin glycoside; yellow pigment, torachrysone, polysaccharide; chrysophanol-B-gentiobioside; chrysophanic acid-9-anthrone. Seed oil, emodin, glucoside·chrysophanic acid. Seed torosachrysone, questin; naphthalenic lactones. Sennosides. Leaves and stems contain d-mannitol, myricyl alcohol and β-sitosterol.

Parts used : Leaves, seeds and roots.

Ayurvedic properties :

Rasa - Madhura, Kaṭu. Vīrya - Uṣṇa.
Guṇa - Laghu, Rūkṣa. Vipāka - Kaṭu.

चक्रमर्दः कटुष्णः स्यात् मेदोवातकफापहः।
दद्रू कण्डूहरः कान्तिसौकुमार्यकरो मतः।।
(ध. नि.)

Cakramardaḥ kaṭusnah syat medovata kaphapahah,
Dadrukanduharah kantisaukumaryakaro matah.
(D. Ni)

चक्रमर्दः कटुस्तीव्रः मेदोवातकफापहः।
व्रणकण्डूति कुष्ठार्तिदद्रुपामा दिदोषनुत्।।
(राजनिघण्टु)

Cakramardaḥ kaṭustīvraḥ medovātakaphāpahaḥ,
Vraṇakaṇḍūti kuṣṭharttidadrupāmādidoṣanut.
(Rajanighaṇṭu)

Actions/Uses : Kaphavātaśāmaka. Lekhana, Kuṣṭhaghna, Hṛdya, Viṣaghna, Medoghna.

Branch : Kuṣthapaha, Balya, Dadrupamacara.

Fruit : Viṣghna, Gulmahara, Kaṇḍūghna, Dadrughna, Śvāśahara, Kāsahara, Kṛmighna.

Therapeutics :

Cakramarda is a specific for all types of skin diseases. It is sweet, bitter, acrid, carminative, laxative and vermifuge and cures eczema, dermatosis, itching, ulcers, cough, dyspnoea and other respiratory diseases, anorexia, fever. It is used in constipation, gastro-intestinal disease, helminthic manifestation and obesity. It rehabilitates vitiated vaata, pitta and cough. Plant extract is antiviral, spasmolytic and diuretic, used against epilepsy, scabies and sores. Root rubbed into paste with lime juice is a specific for ringworm. Leaves are internally gentle aperient, externally germicide and antiparasitic. They have also aturant and anodyne action. Leaves pounded and applied on cuts, act like tincture iodine; applied against eczema. Tender, taken internally to prevent skin diseases; infusion, vermicidal. Seeds are used externally and internally for all sorts of eye diseases and are recommended for arthritis and hemicrania. Powder of seeds is given in abnormal delivery.

Saraca asoca (Roxb.) De Willde.

SYNONYMS : *Saraca indica* Linn.; *Jonosia asoka* Roxb.

NOTE : *In a revision of this genus, De Willd has treated Indian material as S. asoca. S. indica does not occur in India. Distinguishing characters are :*

S. asoca : Bracteoles erect, clasping the pedicel, persistent.

S. indica : Bracteoles spreading, persistent or fugacious.

HABIT : A middle-sized evergreen tree.

HABITAT : Cultivated in gardens throughout India.

ENGLISH : Ashoka tree.

INDIAN : Ashoka-laal, Ashoka-Sitechaa, Jasundi.

SANSKṚTA : Aśoka, Hemapuṣpa, Tāmrapallava, Rāgitaru, Doṣahārī, Naṭa, Smarādhivāsa, Pinḍipuṣpa, Apaśoka, Vicitra, Karṇapūraka, Kankelī, Subhaga, Raktapallava, Raktaka, Vītaśoka, Banjuldruma, Viśoka, Rāmā, Madhupuṣpa, Śokanāśa, Ṣatpadamanjarī, Varamanjarī, Gandhapuṣpa.

CHEMICAL COMPOSITION : 24-methylcholest-5-3n-3B-ol, -24-ethycholesta-5, 22-dien-3B-ol and 24-ethylcholest-5en-3B-ol isolated from bark; wax from bark contained n-alkanes; palmitic, steraric, linoleic and linolenic acid in fixed oil from flowers. β-sitosterol, quercetin, kaempferol quercetin etc. in flowers. Tannins.

Parts used : Bark, seeds and flowers.

Ayurvedic properties :
Rasa - Kaṣāya, Tikta. Vīrya - Śīta.
Guṇa - Laghu, Rūkṣa. Vipāka - Kaṭu.

अशोकः शीतलस्तिक्तो ग्राहिवर्ण्यः कषायकः।
दोषापचीं तृषादाहकृमिशोषविषास्त्रजित्।। (भा. प्र.)

Aśokaḥ śītalastikto grāhīvarṇyaḥ kaṣāyakaḥ,
Doṣāpacī tṛṣādāhakṛimiśoṣaviṣāsrajit.
(Bhāvaprakāṣa)

Actions/Uses : Kaphaghna, Pittaghna. Viṣaghna, Vedanāsthāpana, Stambhana, Raktaśodhana, Grāhī, Raktasangrāhaka.

Seeds : Mūtrala, Aśmarināsaka.

Therapeu``cs :

Aśoka is known as an uterine tonic. It imparts a

healthy tone and strength to uterus. In bloody leucorrhoea whether acute or chronic, it exerts an exhilarating and permanent curative action. It is an excellent remedy in suppressed menses for which colicky abdominal pains supervine, also useful for complaints of menopause and barrenness. Bark is refrigerant, astringent, alexiteric, demulcent, emollient, anthelmintic; cures dyspepsia, burning sensation, diseases of blood, biliousness, tumours, enlargement of abdomen, colic, piles, ulcers, bloody discharges from uterus, menorrhagia. Flowers pounded and mixed with water are used in haemorrhagic dysentery and for bleeding piles and retention of urine.

Tamarindus indica Linn.

SYNONYMS : *Tamarindus officinalis Hook; Tamarindus occidentalis Gaertn.*

ENGLISH : Tamarind.

INDIAN : Chincha, Aamali, Imli, Chinch.

SANSKRTA : Amlikā, Cincā, Cukrikā, Cukrā, Sutattiḍī, Śākacukrikā, Dantaśaṭhā, Amlā, Amlī, Tittiḍikā, Suktā, Suktikā, Cincikā, Yamadūtikā, Gurupatrā, Caritrā, Tintiṇī.

HABIT : Moderate-sized to large, evergreen tree.

HABITAT : Cultivated throughout India.

CHEMICAL COMPOSITION : Fruit contains tartaric acid, citric acid maleic acid and potassium bitartarate and traces of oxalic acid. Kernel, polysaccharides. Leaves contain flavonoid glycosides saponaretin, vitexin, orientin and homoorientin. Hordenine isolated from leaves, bark and flowers.

Parts used : Fruits, leaves, seeds, flowers and ash.

Ayurvedic properties :

Rasa - Amla. Vīrya - Ūṣṇa.
Guṇa - Gurū, Snigdha. Vipāka - Amla.

अम्लिकाम्ला गुरुर्वातहरी पित्तकफास्त्रकृत्।
पक्वा तु दीपनी रुक्षा सरोष्णा कफवातनुत्।।
(भा. प्र.)

Amlikāmlā gururvātaharī pittakaphāsrakṛt,
Pakvā tu dīpanī rukṣā saroṣṇā kaphavātanut.
(Bhāvaprakāṣa)

Actions/Uses : Leaves and seeds : Śothaghna.

Flowers : Netrābhiṣyanda, Raktasangrāhaka.

Fruit, unripe : Raktapittakāraka.

Fruit, ripe : Raktapittapraśaman. Tridoṣaghna, Dīpana, Pācana, Rucikara, Tṛṣṇānigrahana, Dāhaśāmak, Ānulomic.

Ash : Mūtrala.

Therapeutics :

Leaves of Amlikā reduce inflammatory swellings and are applied externally for inflammation of ulcers. Flowers are applied externally on eyes for Netraabhishyanda. Bark is astringent and tonic; heals ulcers. Fruit, refrigerant from the acids they contain, digestive, carminative, slightly laxative, antiscorbutic and antibilious; useful in diseases caused by deranged bile. An infusion of pulp forms a very grateful drink in febrile diseases. Pulp as well as a polutice of leaves are recommended for external applications to inflammatory swellings to relieve pain. Ash of shells of fruit used in menorrhagia and gonorrhoea.

Family : XXIX - Cannabaceae.

Cannabis sativa Linn.

SYNONYMS : *Cannabis indica* Lam.

ENGLISH : Indian Hemp (Indica), American Hemp (Sativa). Hashish, Marihuana.

INDIAN : Bhaanga, Gaanjaa, Charas, Siddhi, Jia.

SANSKṚTA : Bhangā, Ganjā, Ganjaka, Mātulanī, Siddhipatra, Jaya, Bhangikā, Vīrapatrā, Dīrgha, Vicitra, Varṣika, Mādanī, Uttejaka, Āanata, Vājikara, Sadahaka, Śivapriya, Vijayā, Harśanṇī, Sangikā, Ganjāī, Ānandā, Capalā.

HABIT : An tall annual herb.

HABITAT : Native of Western and Central Asia. Cultivated in India for narcotic drug.

CHEMICAL COMPOSITION : Cannabis yields 421 chemicals of various classes broadly divided as (i) cannabinoids, (ii) cannabispirans, and alkaloids. Cannabis drugs owe their narcotic and psychotomimetic properties to cannabinoids. Resins 2.5-15%. Tetrahydro cannabinol and cannabidiolic acid.

Parts used : Dried flowering or fruiting tops of the pistilate plant and resinous exudation of 3 varieties of cannabis viz. Ganjaa, Bhaanga and Charasa.

Ayurvedic properties :

Rasa - Tikta. Vīrya - Ūṣṇa.
Guṇa - Laghu, Tīkṣṇa. Vipāka - Kaṭu.

भंगा कफहरी तिक्ता ग्राहिणी पाचनी लघुः।
तीक्ष्णोष्णा पित्तला मोहमदवाग्वह्निवर्धिनी।।
(भावप्रकाश)

Bhangā kaphaharī tiktā grāhiṇī pācanī laghuḥ,
Tīksnoṣṇā pittalā mohamadavāgvahnivardhanī.
(Bhāvaprakāṣa)

Actions/Uses : Kaphaghna, Vātaghna, Pittakar. Dīpanī, Pācana, Vyāvayi, Grāhi, Agnidīpana, Āmapacana, Malasangrahaṇa, Vikāsi, Karṣaṇa, Rucya, Madakara.

Therapeutics :

Bhanga, Ganja and Charas are all prepared from C. Sativa. Bhanga is deep green in colour and consists of dried leaves. Bigger leaves used in Bhanga; smaller ones in Ganja. Marijuana consists of leaves and flowering parts of female plant only. Charas or Hashish is the resinous exudate from leaves and flowering parts of female plants. It has a peculiar odour. It is dark green or brown in colour and has a very strong odour. Bhanga and Ganja are prescribed as appetiser, nervous stmulant and in bowel complaints. Bhanga made into paste with water, mixed with milk and sugar is taken as an intoxicating liquor producing euphoria. Charas has a soothing action and is used in cases of mania and hysteria as well as asthma. It does not produce loss of appetite or constipation like opium. Leaves are used as snuff or for smoking. They are given internally to relieve pain and swelling in orchitis. Juice of leaves destroys worms and vermin. Seed and seed oil are diuretic and antidiarrhoeal. They are useful in chronic rheumatism and have action like ergot in delivery cases. The plant is used as a tonic, intoxicant, stomachic. It produces sleep without derangement of digestive organs. It is useful in gonorrhoea, menorrhagia, diarrhoea, cholera, hydrophobia, tetanus and rheumatism. Decoction of plant is efficacious in blood dysentery.

HOMEOPATHIC USE : Cannabis indica and Cannabis sativa are both proved Homeopathically and are two separate remedies. Botanically both the plants are identical; the difference in their properties is solely due to the difference of soil and climate in which they are grown. Both resemble closely in their Homeopathic symptoms but the mental symptoms and head symptoms are less pronounced in *C. sativa* than those in *C. indica*. Eye and genito-urinary symptoms are more prominent in *C.sativa*. Mind is the main seat of action of *C.indica*, has great soothing influence in many nerve disorders, like epilepsy, mania, dementia delirium tremens and irritable reflexes; violent shocks pass through the brain; feeling as if the top of head was shutting and opening. *C. sativa* has a reputation for gonorrhoea and more especially the chronic. Pus in kidney; pus and mucus in urine; urine does not pass even by the help of catheter; burning while urinating but especially immediately afterwards; swelling of prostate gland; urethral carbuncle and phimosis; aching pain in balls of eyes; cramp-like pulling in eyes; specks and opacity of cornea.

Family : XXX - Capparaceae.

Capparis zeylanica Linn.

SYNONYMS : *Capparis horrida* Linn. *f.*, *Capparis acuminata* Roxb.

HABIT : A many-branched thorny, sub-scandent climbing shrub.

HABITAT : Throughout the greater part of India as hedge plant.

INDIAN : Waaghaati, Gowindaphalla, Ardanda, Tarti.

SANSKṚTA : Gāndhārī, Ahimsra, Vyāghranakhī, Govindaphala, Kantakaphala, Karambha, Granthila, Kaṭukandari, Kinkani, Vartala, Tāpasapriya, Vyaghraghanti, Kṛṣṇāngi.

CHEMICAL COMPOSITION : An alkaloid, a phytosterol, a mucilaginous substance and water soluble acid. L-stachydrine, rutin and β-sitosterol. Seeds and leaves, thioglucoside glucocapparin, n-tricontane, a-amyrin and fixed oil.

Parts used : Root-bark, leaves and fruit.

Ayurvedic properties :
Rasa - Kaṭu, Tikta. Vīrya - Ūṣṇa.
Guṇa - Laghu, Rūkṣa. Vipāka - Kaṭu.

गान्धारी कटुतिक्तोष्णा कफवातनिकृन्तनी।
शोफघ्नी दीपनी रुच्या रक्तग्रन्थिरुजापहा।।
(रा. नि.)

Gāndhārī kaṭutiktoṣṇā kaphavātanikṛntanī,
Śophaghnī dīpanī rucyā raktagranthirujāpahā.
(Rājanighaṇṭu)

अहिस्त्रा चैव रास्ना च प्रलेपो वातशोफहृत्।।
(सुश्रूत)

Ahisrā caiva rāsnā ca pralepo vātaśophahṛt.
(Suśrūta)

Actions/Uses : Vatakaphaghna, Sothahara, Vedanasthapana, Rucya, Dipani, Raktasodhaka.

Therapeutics :

Various plant parts of Gandhari used in pain, swelling, rheumatism, hemiplegia, sores, colic, neuralgia, pleurisy, pleuritis, breast pain and dropsy. Paste of root bark applied to boils and swelling of testicles. Root bark is demulcent, sedative, stomachic, good appetizer and finds useful for internal application in colic. It is antihidrotic, bitter, cholagogue, and used in cholera. Leaves are counter irritant; ground into paste and used for external application in glandular swellings, piles and boils. Decoction of leaves is given in syphylis.

Crataeva nurvala Buch.-Ham.

SYNONYMS : *Crataeva religiosa Hook. f. & Thoms. non Frost. f.; Crataeva magna (Lour.) DC.; Capparis magna Lour.*

HABIT : A middle-sized deciduous tree.

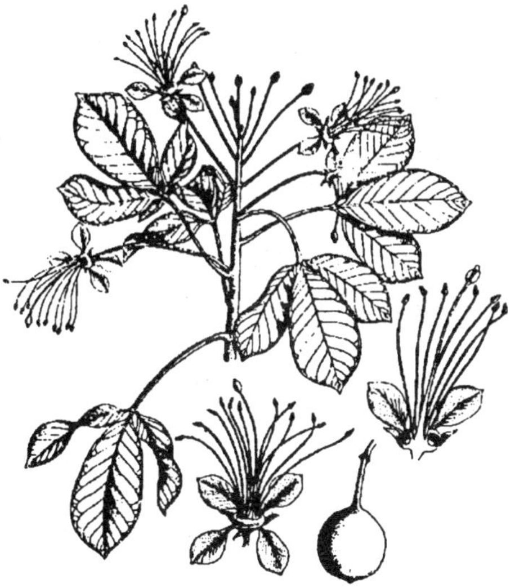

HABITAT : Almost all over India, wild or cultivated.

ENGLISH : Holy Garter Pear, Gartic Pear, Three-leaves caper.

INDIAN : Warunna, Waayawarnna, Barna, Bilasi, Tapia, Kawart, Bhatavarnna, Hadavarnna, Kumla.

SANSKṚTA : Varuṇa, Aśmaghna, Bilvapatra, Setuvṛukṣa, Aśmarīghna, Pādapa, Triparṇa, Bhramarapriya, Kumāraka, Vṛttaphala, Tikśāka, Tikta, Barhapuṣpa, Kaṣāyaka.

CHEMICAL COMPOSITION : Lauric, stearic, undecylic, oleic and linolenic acid from root bark. Ceryl alcohol, lupeol, friedelin, betulinic acid and diosgenin, saponin and tannin from bark. Fruits, cetyl alcohol, ceryl alcohol, triacontane, triacontanol, β-sitosterol and glucocapparin. Leaves, l-stachydrine. Root bark, rutin, quercetin and β-sitosterol.

Parts used : Bark, Leaves and root-bark.

Ayurvedic properties :
Rasa - Kaṣāya, Madhura, Tikta. Vīrya - Ūṣṇa.
Guṇa - Laghu, Rūkṣa, Snigdha. Vipāka - Kaṭu.

वरुणः कटुरुष्णश्च रक्तदोषहरः परः।
शीर्ष वातहरः स्निग्धो दीप्यो विद्रधिवातजित्।।
(राजनिघण्टु)

Varuṇaḥ kaṭurūṣṇaśca raktadoṣaharaḥ paraḥ,
Śīrṣavātaharaḥ snigdho dīpyovidradhivātajit.
(Rajanighantu)

वरुणः पित्तलोभेदि श्लेष्मकृच्छ्राममारुतान।
निहन्ति गुल्मवातास्त्र कृमीश्चोष्णोऽग्निदीपनः।।
कषायो मधुररस्तिक्तः कटुको रुक्षको लघुः।।
(भा. प्र.)

Varuṇaḥ pittalobhedi śleṣmakṛcchāmamārutān,
Nihanti gulmavātāsra kṛmīścoṣṇo agnidīpanaḥ,
Kaṣāyo madhurāstiktaḥ kaṭuko rūkṣako laghuḥ.
(Bhāvaprakāṣa)

Actions/Uses : Vātaghna, Kaphaghna, Pittakar, Pittasāraka, Vātaraktahara, Śothaghna, Aśmaribhedana, Bhedana, Dīpana, Anulomana, Kṛmighna, Raktotklista, Vidradhighna. Mūtrala, Mūtrakricchrahara.

Therapeutics :

Varuna is astringent, bitter, acrid, diuretic, anthelmintic, carminative, laxative and stomachic. Various plant parts used in baldness, sores, epidydimitis and hydrocele. Root and bark promote appetite, and increase biliary secretion. Extract of root bark mixed with honey is a valuable remedy for scrofulous enlargements of the glands of lower jaw. Leaves are stomachic and tonic. Fresh leaves and root bark are rubifacient. Juice of leaves mixed with coconut milk and butter is given internally in rheumatism. Bark is demulcent, alterative, tonic, stomachic, laxative, diuretic, antipyretic, useful in calculus affections and disorders of urinary organs. It is also used in snake bite. It is a contraceptive. Juice of bark is given to women after childbirth. Powdered bark is useful in urinary and renal troubles, gastro-intestinal and uterine affections. Decoction of bark, compounded with powder of root is efficacious in gravel.

Gynandropsis gynandra (Linn.) Briquet.

SYNONYMS : *Gynandropsis pentaphylla DC.; Cleome pentaphylla* Linn.; *Cleome gynandra* Linn.

HABIT : An annual, erect, glandular, pubescent herb.

HABITAT : A common weed abundant throughout warmer parts of India.

ENGLISH : Caravella.

INDIAN : Pandhari-tilwan, Tillawanna-paandhari, Mabli, Hurhur, Hulhul, Mabli, Kalvan, Kaaraliyaa, Kaanaphodi, Mothitilwan.

SANSKṚTA : Tilaparṇi, Sūryāvarta, Arkakānta, Varada, Suvarchala, Sūryabhakta, Ravivallī, Divyateja, Śītavṛukṣa, Mūlaparṇi, Pārthī, Sukhodbhava, Ravipriya, Arkapuṣpi, Pṛthvīka, Bramhasuvarcala.

CHEMICAL COMPOSITION : Seeds contain cleomin, glucocapparin and hexcosanol, β-sitosterol-B-D-glucoside and kaempferol. Seeds contain an unsaturated viscosic acid, a flavone, viscosin. Herb and seed yield an essential oil similar to garlic or mustard oil.

Parts used : Seeds, leaves and roots.

Ayurvedic properties :
Rasa - Kaṭu. Vīrya - Uṣṇa.
Guṇa - Tīkṣṇa. Vipāka - Kaṭu.

तिलपर्णी कटुष्णा स्यात्तीक्ष्णा विस्फोटकारिणी।
विदाहिकृमिशूलघ्नी स्वेदला कफवातनुत्।।

Tilaparnī kaṭuaṣṇā syāt tīkṣṇā visphotakāraṇī,
Vidāhi kṛmiśūlaghnī svedalā kaphavātanut.

अन्या तिक्ता कषायोष्णा सरा रुक्षा लघुः कटुः।
निहन्ति कफपितास्त्रश्वास कासा रुचि ज्वरम्।।
विस्फोट कुष्ठमेहास्त्र योनिरुक् क्रिमीपाण्डुता।।
(भा. प्र.)

Anyā tiktā kaṣāyoṣṇā sarā rūkṣā laghuḥ katuḥ,
Nihanti kapha pittāsraśvāsa kāsa ruci jvaram.
Visphota kuṣṭhamehāsra yoniruk krimipāndutaḥ.
(Bhāvaprakāṣa)

Actions/Uses : Kaphavataghna, Dipana, Pacana, Sulahara, Vidahi, Anulomana, Krimighna, Vedanasthapana, Vatahara, Uttejaka, Vistambhaghna, Vataraktahara, Kasaghna, Arucighna, Jvaraghna, Visphotahara, Kustaghna, Mehaghna, Pandughna.

Therapeutics :

Tilaparni or Suryavarta is used in scorpion sting and snake bite. Ether extract of plant is anticancer and spasmolytic. Decoction of roots is used in fever. Juice of leaves is used for otalgia. Juice of leaves and seed oil used to cure skin diseases. Bruised leaves are rubifacient and vesicant in rheumatism; remedy for muscular pain, headache and intestinal wounds. Seeds are antispasmodic, sudorific, anthelmintic and carminative. They are applied as poultice to sores having maggots and given as infusion for cough. Powder of seeds is a good remedy for piles.

Family : XXXI - Caricaceae.

Carica papaya Linn.

HABIT : A small erect tree with spongy trunk and milky sap.

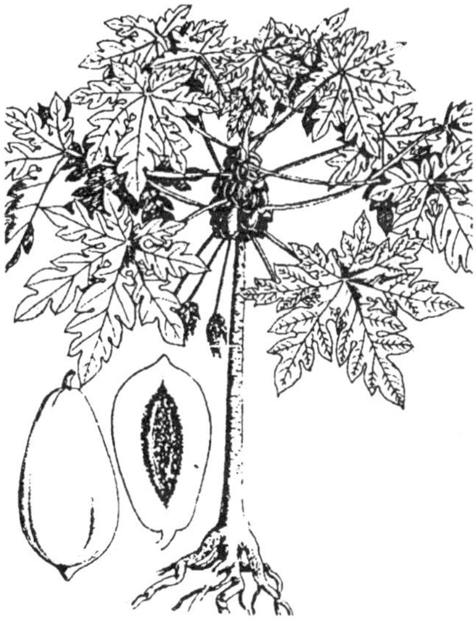

HABITAT : Cultivated in garden throughout India.

ENGLISH : Papaya, Papend tree.

INDIAN : Popaee, Papayaa, Arandkharbuza, Papita.

SANSKRTA : Pāriśa, Erandakarkaṭī, Madhukarkaṭa, Tālaptrī, Apakva, Harita, Erandacirbhaṭa, Chatrāpatrā, Saptāngulapatrā, Koṣaphala, Priya, Dugdhāpacanī, Pītaphala, Svāduphala, Mānsapācanī.

CHEMICAL COMPOSITION : Carpain, carposide and papain. Fresh fruit pulp contains sucrose, invert sugar, a resinous substance, papain malic acid and salts of tartaric and citric acid. Both ripe and unripe fruits are rich sources of pectins. Carotenoid pigments in fruits. Fruits rich source of vitamins. Seeds contain a sulphur containing basic substance carpasemine. Leaves contain a glycoside, carposide and an alkaloid carpaine. Carpain is also present in traces in bark, root and seeds. Leaves contain vitamins C and E. A sinigrin-like glycoside probably identical with carposide and a myrosin-like enzyme found in roots. Latex contains enzymes papain and chymopapain.

Parts used : Milky juice, fruits, seeds and leaves.

Ayurvedic properties :
Rasa - Madhura. Vīrya - Ūṣṇa.
Guṇa - Laghu, Rūkṣa, Tīkṣṇa. Vipāka - Madhura.

पारीशं शीतलं रुच्यम् दीपनं पाचनं सरम्।
मधुरं रक्तपित्तघ्नं विशेषादर्शसे हितम्।
पारशिक्षीर योगेन प्लीहागुल्मश्च नश्यति।।
(द्रव्यगुण)

Pārīśam śītalam rucyam dīpanam pācanam saram,
Madhuram raktapittaghnam viśeṣādarśase hitam,
Pārīśakṣīrayogena plīhāgulmaśca naśyati.
(Dravyaguna)

फलं कूश्माण्डवत् हृद्यं बृंहणं कृमिजित्परम्।
दीपनं श्वासकासघ्नं प्रीतिदं पित्तवर्द्धनम्।।
क्षीरं बीजं च कृमिहृत् त्वङ्मूलादिकमस्य तु।
त्वग्दोषवातरक्तादिरोगे ष्वप्युपयुज्यते।।
(स्व.)

Phalam kuśmāṇḍavat hṛdyam bṛnhaṇam kṛmijitparam,
Dīpanam śvāsakāsaghnam prītidam pittavardhanam,
Kṣīram bījam ca kṛmihṛt tvaṅ mūlādikamsya tu,
Tvagdoṣa vātaraktādi rogeṣvapyupayujyate.
(Svayamkrti)

Actions/Uses : Kaphavātaśāmaka, Mūtrala, Mandāgni, Śothahara, Kṛmighna, Arśghna, Stanyajanana, Agniuttejaka, Udararogaghna, Ārtavajanana.

Therapeutics :

Leaves of Erandakarkati tone the heart, are diaphoretic and diuretic. They are used to expel guinea worm. Ripe fruit is stomachic, carminative and diuretic, useful in digestive troubles. Fruit beverage is used for diarhhoea and dysentery. Unripe fruit is useful in excretory troubles of children. Milky juice or latex of unripe fruits is known as a very good digestive. It acts as lactagogue when taken internally or applied externally, anthelmintic, particularly effective in expelling lubrici, useful against ringworm and is used as cosmetic to remove freckles and other blemishes from the skin. Latex is antipruritic and is used for ulcer also in abortion. It is a constituent of "Gasex" useful in post operative period to relieve distension. Pulp of unripe fruit shows antifertility activity. Oral administration of powder of ripe fruit showed antifertility activity. Shampoos or detergents containing papaya extract, beneficial for skin. Seeds are vermifuge, emmenagogue used to quench thirst.

HOMEOPATHIC USE : The remedy has not yet been proved Homeopathicaly yet its homeopathic preparation is used with much efficacy in all the diseases and complications mentioned above. It is best adapted in fevers with enlarged spleen and liver, in dyspepsia and indigestion.

Family : XXXII - Celastraceae.

Celastrus paniculatus Willd.

SYNONYMS : *Celastrus montana* Roth.; *Celastrus multiflora* Roxb.

HABIT : An unarmed large perennial deciduous climbing or scrambling shrub.

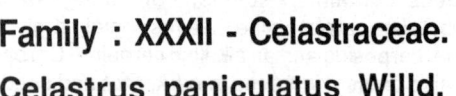

HABITAT : Found in tropical and subtropical Himalayas, Punjab, Assam, Bihar, South India and Konkan.

ENGLISH : Black oil tree, Staff tree.

INDIAN : Maalakaangonni, Kaanguni, Jyotishmati.

SANSKṚTA : Jyotiṣmati, Pārāvatapadī, Agnibhāsakā, Vanhiruci, Kākāndakī, Katavīkā, Pītatailā, Svarṇalatā, Kaṭabhī, Jyotirlatā, Kangunikā, Tanulaprabhā, Medhyā, Matidā, Vegā, Pingalā, Kṣriprā, Durbharā, Sarasvati, Amṛtā, Jyotiḥkāyā, Kangunī, Kundanī, Vṛṣā, Lavaṇī, Kukundanī.

CHEMICAL COMPOSITION : Acetic, benzoic, formic, linoleic, linolenic, palmatic and stearic acids. Celapagine, celapanigine, celapanine, celastrol, celastrine etc. from seeds. Polyhydric alcohol, malkanguniol; and four related alcohols. Seeds, triterpenes paniculatadiol, malkanguniol along with polyesters, β-amyrin and β-sitisterol. Malkangunin, a new sequiterpene and celapanine and celapanigine, two new sequiterpenoid tetra esters.

Parts used : Seeds, leaves, bark and oil.

Ayurvedic properties :

Rasa - Kaṭu, Tikta.	Vīrya - Ūṣṇa.
Guṇa - Tikṣṇa, Snigdha.	Vipāka -Kaṭu.

ज्योतिष्मती कटुस्तिक्ता सरा कफसमीरजित्।
अत्युष्णा वामनी तीक्ष्णा वन्हिबुद्धिस्मृतिप्रदा।।
(भा. प्र.)

Jyotiṣmatī kaṭustiktā sarā kaphasamīrajit,
Atyuṣṇā vāmani tīkṣṇā vahnibuddhismṛtipradā.
(Bhāvaprakāṣa)

Actions/Uses : Medhya. Kaphaghna, Vātaghna, Pittakar, Nibandha, Dīpana, Vātahara, Vedanāsthāpana, Vātānulomana, Gulmahara.

Therapeutics :

Various parts of Jyotismati are used in sore throat, anaemia, colic, syphilis, carbuncle. Leaves are emmenagogue. Leaves and roots used as poultice to cure headache. Bark is abortifacient. Decoction of bark used in bronchitis. Seeds are bitter, emetic, alterative, laxative, stimulant, nervine and aphrodisiac. Decoction of seeds is given in rheumatism, leprosy, gout, various fevers, paralysis. Seeds and oil stimulate intellect and sharpen memory. Oil from seeds is powerful stimulant, rubefacient and useful in beriberi. One part of oil and 8 parts of butter is known as Magzsudhi (Brain cleaner) is applied to head and believed to promote intelligence. Crude oil shows sedative and tranquillizing effect and drug combination containing *C. paniculata* is useful as depressent especially in hysteria.

Family : XXXIII-Chenopodiaceae.

Chenopodium album Linn.

ENGLISH : Goose Foot, Wild spinach.

INDIAN : Chaakawat, Bethu saag, Chandan betu.

SANSKṚTA : Vāstuka, Cillī, Yavaśāka.

HABIT : A small odourless annual erect herb.

HABITAT : Cultivated as a vegetable.

CHEMICAL COMPOSITION : Roots contain ecdysteroids; β-ecdysone and polypodine. Ascorbic acid, β-carotene, catechins, caffeic acid, p-coumaric acid, ferulic acid, B-sitosterol, stigmasterol etc. Essential oil contains carotine and vitamin C.

Parts used : Leaves, flowers, buds, seeds and oil.

Ayurvedic properties :
Rasa - Madhura. Vīrya - Śīta.
Guṇa - Rūkṣa, Gurū, Laghu. Vipāka -Kaṭu.

वास्तुकन्तु मधुरं सुशीतलं क्षारमीषदम्लत्रिदोषजित्।
रोचनं ज्वरहरं महाशिसां नाशनग्चमलमूत्रशुद्धिकृत्।।
(राजनिघण्टु)

Vāstukantu madhuram sūśitalam kṣāramīṣadamla tridoṣajit,
Rocanam jvaraharam mahārśasām nāsananca malamūtraśuddhikṛt.
(Rajanighantu)

वास्तुकं मधुरं शीतं क्षारं रुचिकरं सरम्।
पाचकं मलमूत्राणां शोधकं चाग्निदीपनम्।।
लघुशुक्रप्रदं बल्यं हृद्यं च ज्वरनाशनम्।
त्रिदोषार्शः कृमिप्लीहारक्तरुक्कुष्ठ पित्तनुत्।।
उदावर्तभवं वातं नाशयेदिति कीर्तितम्।
(निघण्टु रत्नाकर)

Vāstukam madhuram śītam kṣāram rucikaram saram,
Pācakam malamūtrāṇām śodhakam cāgnidīpanam.
Laghu śukrapradam balyam hṛdyam ca jvaranāśanam,
Tridoṣārsaḥ kṛmiplīhā raktarukkuṣṭhapittanut.
Udāvartabhavam vātam nāśayediti kīrtitam.
(Nighanturatnakara)

यवशाकं सरं शीतंमधुरं लवणं गुरु।
कफवातहरं रुक्षं रुच्यं विष्टम्भि पित्तलम्।।
चिल्ली तु मधुरा रुक्षा तुवरा शीतला सरा।
कटुपाका कटुः क्षारा रोचनी दीपनी लघुः।।
(कैयदेवनिघण्टु)

Yavaśākam saram śītam madhuram lavaṇam gurū,
Kaphavātaharam rūkṣam viṣṭambhi pittalam.
Cillī tu madhurā rūkṣā tuvarā śītalā sarā,
Kaṭupākā kaṭūḥ kṣārā rocanī dīpanī laghuḥ.
(Kaiyadevanighantu)

Actions/Uses : Tridoṣaśāmaka, Dāhaśāmaka, Pācana, Jvaraghna, Mūtrala, Virecana.

Therapeutics :

Vastuka is laxative, blood purifier, anthelmintic. Vegetable relieves indigestion. Infusion of leaves is laxative and anthelmintic and is used for intestinal ulcers and considered useful in piles and throat and eye trouble. Decoction is aperient, tonic, diuretic and aphrodisiac; given in billiousness, hepatic disorders and spleen enlargement. Powder of leaves is used externally as an antiseptic around genitalis of children. Flowers and buds are used in stomach troubles, weakness in children and for fattening. Seeds are prescribed for hepatic disorders and spleen enlargement and also for abortion. Oil kills worms.

Chenopodium anthelminticum A. Gray.

SYNONYMS : *Chenopodium anthelminticum* Linn.; *Chenopodium botrys* Linn.; *Chenopodium ambrosioides* Linn.

ENGLISH : Worm seed. Jerusalem oak.

INDIAN : Chandan-batawaa, Chukaa.

SANSKṚTA : Śvetacillī, Cāndila.

HABIT : An annual or perennial erect much branched aromatic herb.

HABITAT : A native of West Indies and South America. Now found in Southern India, Maharashtra, Kashmir and West Bengal.

CHEMICAL COMPOSITION : *C. ambrosioides* Linn. contains essential oil and ascaridole. Triterpene glycosides-chenopodosides A and B. Leaves contain kaempferol-7-rhamnoside and ambroside. *C. anthelminticum* contains similar amounts of essential oil and ascaridole. Twenty six compound and a new acid-chenopodic acid are identified in essential oil from *C. botrys.*

Parts used : Leaves, and oil.

Ayurvedic properties :
Rasa - Madhura. Vīrya - Śīta.
Guṇa - Śītala, Laghu. Vipāka - Kaṭu.

श्वेतचिल्ली सुमधुरा क्षारा च शिशिरा च सा।
त्रिदोषशमनी पथ्या ज्वरदोषविनाशनी।।
(राजनिघण्टु)

Śvetacillī sumadhurā kṣāra ca śiśirā ca sā,
Tridoṣa samanī pathyā jvaradoṣa vināśinī.
(Rajanighantu)

Actions/Uses : Tridoṣahara, Kṛmighna, Virecana, Agnidīpaka, Pācana, Plihādosahara, Raktapittasāmaka.

Therapeutics :
Caṇḍila is anthelmintic, antispasmodic. The main use is in oil for hook worm and round worm. Plant is employed in painful and profuse menstruation especially during climaxis. It is an antispasmodic useful in hysterias, chorea and other nervous affections. Chenopodium was used by the American Indians in infusions made from leaves and seeds as a household remedy against intestinal parasites.

HOMEOPATHIC USE : *C. anthelminticum* is a homeopathic remedy useful in aphasia, apoplexy, asthma, cerebral deafness, convulsions, dropsy, epilepsy, hemicrania, hemiplegia, leucorrhoea and suppressed menses. The most characteristic pain is dull pain below the angle of right scapula. *C. botrys* used in liver and stomach diseases, as diuretic and laxative.

Chenopodium olidum S. Wats.

SYNONYMS : *Chenopodium vulvaria* Linn.

ENGLISH : Goat's Arrach, Stinking Arrach.

INDIAN : Bathuwaa, Waastuka.

SANSKṚTA : Vāstuka-2.

HABIT : An annual erect herb growing mostly in waste places or on dunghills: It has an odor of decaying fish and it constantly gives out ammonia.

HABITAT : Europe and Great Britain also in North India.

CHEMICAL COMPOSITION : It contains Trimethylamene.

Parts used : Leaves

Ayurvedic properties :
Rasa - Madhura. Vīrya - Śīta.
Guna - Gurū, Rūkṣa. Vipāka - Madhura.

Actions/Uses : Sankochavikāasa-pratibandhaka, Āratavajanana.

Therapeutics :

Vastuka-2 is nervine and emmenagogue. Useful in hysteria of females and as a remedy for menstrual obstruction.

HOMEOPATHIC USE : *C. vulvaria* is a homeopathically proved remedy useful for constipation, enuresis, pain under scapula and spleen pain.

Family : XXXIV - Cleomaceae.

Cleome viscosa Linn.

SYNONYMS : *Cleome icosandra* Linn.; *Polanisia isocandra* (Linn.) *W. & A.*

ENGLISH : Wild mustard, Sticky Cleome.

INDIAN : Aryaval, Piwali-tilwan, Kaanaphodi, Hurhur.

SANSKṚTA : Ajagandhā, Ādityabhaktā, Tungī, Pūtimayūrikā, Kabarī, Kāvarī, Pūtikītā, Brāmhī, Karṇapotha, Tilaparṇī, Ugragandhā, Bastagandhā, Barbara, Kharapuṣpā, Pūtibarbara, Śakambharā, Śabarīgandhā, Paśugandhā, Śatapuṣpā, Arkakanṭa.

HABIT : An annual herb.

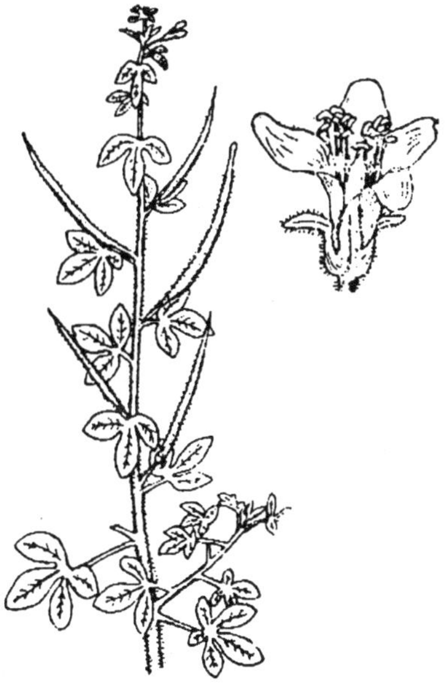

HABITAT : A common weed throughout India.

CHEMICAL COMPOSITION : Eriodicotyl-5-rhamnoside from whole plant. Kaempferide-3-glucuronide from roots. Seed oil rich in linoleic acid. Seeds contain 0.1% viscosic acid, 0.04% viscosin.

Parts used : Whole plant, seeds, leaves and root.

Ayurvedic properties :
Rasa - Kaṭu. Vīrya - Ūṣṇa.
Guṇa - Rūkṣa, Laghu, Tīkṣṇa. Vipāka - Kaṭu.

आदित्यभक्ता शिशिरा सतिक्ता कटुस्ताथोग्रा कफहारिणी च।
त्वग्दोषकण्डूव्रण कुष्ठभूतग्रहोग्र शीतज्वरनाशनी च।।
(राजनिघण्टु)

Ādityabhaktā śiśirā satiktā kaṭustāthogrāca kaphahariṇī
Tvagdoṣakaṇḍūvraṇa kuṣṭhabhūtagrahogra śītajvaranāśanī ca.
(Rajanighantu)

अजगन्था कटुष्णा स्याद्वातगुल्मोदरापहा।
कर्णव्रणार्तिशूलघ्नी कृमिघ्नी च ज्वरापहा।।
(रा. नि.)

Ajagandhā kaṭuṣṇā syādvāta gulmodarapahā,
Karṇavraṇārttiśūlaghni kṛmighnī ca jvarāpahā.
(Rajanighantu)

अजगन्धा कटुः पाके रसे रुक्षाग्निदीपनी।
हृद्या रुच्या लघुस्तीक्ष्णां दृकशुक्रकफवातहा।।
(कै. नि.)

Ajagabdhā kaṭuḥ pāke rase rūksāgnidīpanī,
Hṛdyā rucyā laghustīkṣṇā dṛk śukrakaphavātahā.
(Kaiyadevanighantu)

Actions/Uses : Agnidīpana, Udararogahara, Rucya, Hṛdya Gulmahara, Kṛmighna, Jvaraghna, Karṇārtihara, Karṇavraṇahara, Karṇaśūlaghna.

Therapeutics :
Leaves of Ajagandha are rubifacient, vesicant, sudorific and are used for external application for wounds and ulcers. Juice of leaves relieves earache. Bark is irritant and acrid. Externally it is rubefacient and vasicant. Seeds are carminative, vesicant, rubefacient, anthelmintic and antiseptic like mustard. They are a remedy for infantile convulsions. Polutice of paste of seeds is useful in chronic painful joints and in the form of extract used to kill maggots in unhealthy sores.

Family : XXXV - Clusiaceae.

Calophyllum inophyllum Linn.

ENGLISH : Alexandrian Laurel.

INDIAN : Undi, Undala, Unang, Surangi, Surpunka, Sultan champa.

SANSKṚTA : Punnāga, Puruṣa, Raktavṛukṣa, Devavallabha, Surapatī, Tunga, Suraparṇikā, Sugandhipuṣpa, Devaprasādajanaka, Nāgacampā, Sulatānacampaka, Rājacampaka.

HABIT : A middle sized evergreen sub-marine tree.

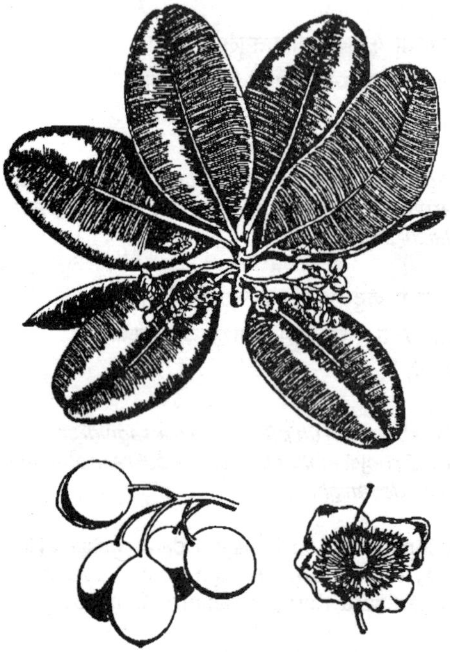

HABITAT : Coastal regions and littoral forest of India and grown as an ornamental tree.

CHEMICAL COMPOSITION : Friedelin, β-sitosterol, β-amyrin, canophyllol, 4-Phenylcoumarins-calophylloide, inophylloie and calophyllic acid from ripe seeds. Cinnamic acid, inophyllic and calophyssic acids from unripe seeds. Myricetin glucosid, myricetin, quercetin and leucocynanidin from flowers. Inophyssloidic acid from bark resin. Kernels yield 50-73 % of oil. Bark 11.9 % tannin; aromatic substance, resin obtained as exudation from the bark possesses emetic and purgative properties. Leaves contain saponin and hydrocyanic acid.

Parts used : Leaves, Bark, Seed, Gum and oil.

Ayurvedic properties :
Rasa - Madhura, Kaṣāya. Vīrya - Śīta.
Guṇa - Laghu, Rūkṣa. Vipāka - Madhura.

पुन्नागो मधुरः शीतः सुगन्धिः पित्तनाशकृत्।
(राजनिघण्टु)

Punnago madhuraḥ śītaḥ sugandgih pittanāśakṛt.
(Rajanighantu)

Actions/Uses : Pittaghna, Kaphaśāmaka, Raktapittakaphāhah.

Therapeutics :
Leaves of Punnaaga are useful in chicken pox, scabies, sunburn, skin inflammation, post natal lactation and sickness, eye catarrh and debility and as fish poison. Bark is astringent, diuretic, emmenagogue and given to babies with fever and running nose. It is used in decoction in internal haemorrhage and as a wash for indolent ulcers. Juice is purgative. Gum exuding from bark is a remedy for wounds and ulcers. It is emetic and purgative. Seeds are useful in the treatment of skin rash, rheumatism and as vermifuge. Oil expressed from seeds and known in Europe as Domba oil is rubefacient and irritant but on the mucous membrane of genito-urinary organs it is a specific. It is a highly esteemed external application in rheumatism; also in gonorrhoea and gleet; also applied to scabies.

Garcinia hanburyi Hook f.

SYNONYMS : *Garcinia morella Desr.; Garcinia morella (Gartn.) Desr.*

ENGLISH : Gambogia, Gamboge, Indian gamboge tree.

INDIAN : Tamal.

SANSKRTA : Kankuṣṭha, Tamāla, Kālakuṣṭha, Kolakākula, Recana, Ranga, Śobhana, Kinjavāluka, Vāraka, Viranga, Kākaphalama, Rhāsa, Pulaka, Rangadāyaka, Ranganātha, Nāmakuṣṭha, Mahākumbhī, Śodha, Phalaka, Kālapālaka, Jalavāluka, Kolavāluka,.

HABIT : A small or middle sized evergreen tree.

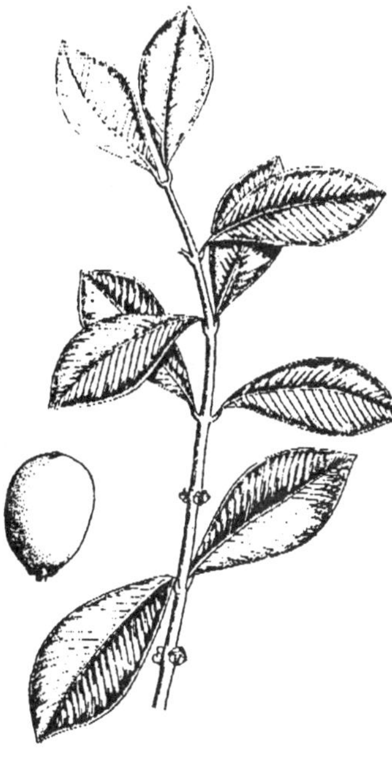

HABITAT : Siam ; Cochin; China. In the evergreen forests of Assam and Khasi hills.

CHEMICAL COMPOSITION : Garcinia species impart good dental health probably due to their possession of an antibiotic such as morellin, α- & β-guttiferins. Gum resin drastic cathartic, may produce nausea and vomiting. Seedcoat and gamboge are the cathartic principles.

Parts used : Gum resin.

Ayurvedic properties :
Rasa - Tikta, Kaṭu. Vīrya - Ūṣṇa.
Guṇa - Laghu. Vipāka - Kaṭu.

कंकुष्ठं कटुतिक्तोष्णं वीर्यंचोष्णं प्रकीर्तितम्।
गुल्मोदावर्तशूलघ्नं रसरञ्जं व्रणापहम्।। (ध. नि.)

Kankusṭham kaṭutiktoṣṇam vīryam coṣṇam prakīrtitam,
Gulmodāvartásūlaghnam rasarañjam vraṇāpaham. (Dhanvantari)

तमालः शालवद् वेद्यो दाहविस्फोटहृद् पुनः। (भावप्रकाश)

Tamālaḥ śālavad vedyo dāha visphotahṛd punaḥ. (Bhāvaprakāṣa)

Actions/Uses : Kaphapittahara, Vraṇaviśodhana, Vibandha, Recana, Gulmaghna, Udaraghna, Kṛmighna, Adhmanahara, Vraṇaghna, Śulahara, Śophaghna, Gudartihara, Anaha, Rucya.

Therapeutics :

Kankuṣṭha is a valuable drastic and hydragogue cathartic and also possesses anthelmintic and diuretic properties. It is used in dropsy when given in combination with acid tartarate of potash. In cerebral affections, such as apopelxy, when comined with calomel, it is held in some esteem. It is also useful in amenorrhoea and in ascites and externally applied to relieve pain and swelling. Morellin, an active constituent of the plant is widely used for promoting healing of wounds of different etiology.

HOMEOPATHIC USE : Gambogia is a homeopathic remedy reputed for its use in "profuse watery diarrhoea, particularly with old people. Burning in anus after stool; anus sore and excoriated. Frightful vomiting and purging with fainting. Pain in small of back as bruised or sprained. Gnawing in coccyx. These are other main symtoms.

Garcinia indica Chois.

SYNONYMS : *Garcinia purpurea* Roxb.

ENGLISH : Wild mangosteen, Kokam butter tree, Brindonia tallow tree.

INDIAN : Kokama, Amasula.

SANSKṚTA : Vṛkṣāmla, Śākāmla, Amlamahīrukha, Rasāmla, Amlaśākha, Cukrāmla, Amlapūra, Purāmla, Cudāmla, Bījāmla, Kaphāmla, Atyāmla, Śreṣṭhāmla, Raktapūraka, Tittidika, Phalacukra.

HABIT : A slender evergreen middle sized tree.

HABITAT : Often cultivated, in Konkan and Western Ghats.

CHEMICAL COMPOSITION : L-Leucine isolated from leaves; euxanthone and biflvonoids isolated fron heartwood. Heartwood, phenolic compounds. Fruit rind camboginol, benzophenone derivatives, garcinol and isogarcinol. Etherial extract of aerial parts semen coagulant and CNS depressant.

Parts used : Root-bark, fruit and oil.

Ayurvedic properties :
Rasa - Fruit-ripe: Madhurāmla
Fruit-unripe: Āmla Vīrya - Ūṣṇa.
Guṇa - Laghu, Rūkṣa. Vipāka - Āmla.

पक्वंतु गुरु संग्राहि मधुराम्लरसं तथा।
अल्पोष्णं रोचनं रुच्यं दीपनं वातनाशनम्।
तृष्णार्शोग्रहणी गुल्मशूलहृद्रोगजन्तुजित्।।
(भा. प्र.)

Pakvam tu gurū sangrāhi madhurāmlarasam tathā,
Alpoṣṇam rocanam rucyam dīpanam vātanāśanam,
Tṛṣṇārśograhaṇīgulmaśūlahṛdrogajantujit.
(Bhāvaprakāṣa)

Actions/Uses : Kaphavātaghna, Pittaśāmaka, Rucikara, Dīpana, Grāhī, Jvaraghna, Yakṛtottejaka, Tṛṣṇānigrahaṇa, Dāhapraśamana,

Fresh fruits : Hṛdya, Vātānulomana, Raktapittapraśaman, Grāhī. Dried fruits : Rocaka, Dīpnapācana, Grāhī, Raktapittasaman.

Rind : Stambhana.

Seed oil : Stambhan, Vraṇaropana.

Therapeutics :

Fruit of Vṛukṣāmla is antiscorbutic, cooling, cholegogue, emollient, demulcent. Bark astringent. Oil or kokam butter is a specific remedy in dysentery and mucous diarrhoea. It is soothing and used in skin diseases. Externally it has a healing property and is used as an application to ulcerations, fissures of lips, hands, chapped skin. It is used in the preparation of lintments, suppositories.

Mesua ferrea Linn.

SYNONYMS : *Calophyllum nagassarium Burm.f.; Mesua pedunculata Wight.; Mesua coromandelina Wight.*

ENGLISH : Ceylon iron wood, Mesua.

INDIAN : Naagachaafaa, Naagakeshara, Naagachampaa.

SANSKṚTA : Nāgapuṣpa, Nāgakeśara, Campeya, Keśara, Himābhaka, Kanaka, Kāncana, Nāga, Ahipuṣpa, Nāgareṇuka, Hemapuṣpaka, Duroha, Kinjalka, Nagakinjalka, Suvarṇa, Rukma, Pinjara, Phaṇipannaga, Vara, Gaja.

HABIT : A middle sized or large evergreen tree with short trunk.

HABITAT : Found in the eastern Himalayas, Assam, West Bengal, Eastern and Western Ghats.

CHEMICAL COMPOSITION : Mammeisin from seeds. Mesuagin from seed oil. Flowers contain essential oil, two bitter substances with bitter principle mesuol 1%. Antibiotic activity of mesuol, mesuone and kernelll and shell oils.

Parts used : Bark, leaves, buds, flowers, fruits and seeds.

Ayurvedic properties :
Rasa - Tikta, Kaṣāya. Vīrya - Iśata Uṣṇa.
Guṇa - Laghu, Rūkṣa. Vipāka - Kaṭu.

नागपुष्पं कषायोष्णं रुक्षं लघ्वामपाचनम्।
ज्वरकण्डुतृषा स्वेदच्छर्दिहृल्लासनाशनम्।
दौर्गन्ध्यकुष्ठवीसर्प कफपित्तविषापहम्।।
(भा. प्र.)

Nāgapuṣpam kaṣāyoṣṇam rūkṣam laghvāmapācanam,
Jvarakaṇḍutṛṣāsveadacchaṛdihṛllasanāśanam,
Daurgandgyakuṣṭhavīsarpa kaphapittaviṣāpaham.
(Bhāvaprakāṣa)

नागकेशरात्पोष्णं लघु तिक्तं कफापहम्।
वस्तिवातामयघ्नं च कण्ठशीर्षरुजापहम्।।
(राजनिघण्टु)

Nagakeśarālpoṣṇam laghu tiktam kaphāpaham,
Bastivātāmayaghnam ca kaṇṭhaśīrṣarujapaham.
(Rajanighantu)

Actions/Uses : Pittaghna, Kaphaghna, Vātakar, Śītala, Arśoghna, Bastirogahara, Pidāśāmaka, Raktasangrāhaka, Āmapācana, Tṛṣṇāhara, Kaṇḍuhara, Svedahara, Kuṣṭhaghna, Viṣaghna, Jvaraghna, Visarpahara.

Therapeutics :

Root bark is astringent and aromatic used in rheumatism. combined with ginger it is used as sudorific. Useful for burning in anus, blood dysentery and bleeding piles. Leaves are used as poultice for pustular errupations, applied on forehead in severe colds. Flowers are astringent, stomachic, used in cough attended with expectoration. Made into a paste with butter and sugar used in bleeding piles and burning of feet. Flower buds are used in dysentery. Unripe fruits are aromatic and sudorific. Fruits are astringent, useful in gastric troubles. Seed oil is applied during rheumatic pain and for scabies.

Family : XXXVI-Combretaceae.

Calycopteris floribunda Lam.

SYNONYMS : *Getonia floribunda Roxb.; Calyopteris floribunda (Roxb.) Poir.*

INDIAN : Ukshee, Baagullee, Kokaray, Kokoranj.

SANSKṚTA : Kāravallī, Toyavallī, Rājatvallī, Rajavallī, Suṣavī, Śveta-dhāṭakī, Palāsīkā.

HABIT : A diffuse or sub-scandent shrub.

HABITAT : Many parts of India in deciduous forest.

CHEMICAL COMPOSITION : n-Octacosanol, sitosterol, calycopterin ellagic acid, quercetin, proanthocynadinin and 7% tannin from leaves. Flowers contain gossypol calycopterin and quercetin.

Parts used : Leaves, roots and fruits.

Ayurvedic properties :
Rasa - Tikta. Vīrya - Śīta.
Guṇa - Laghu, Rūkṣa. Vipāka - Kaṭu.

कृमिपित्तहरा तिक्ता भेदिनी कफनाशिनी।
पाण्डुकुष्ठविकारघ्नी कारवल्ली ज्वरापहा।।
(म. नी.)

Kṛmipittaharā tiktā bhedinī kaphanāśinī,
Pāṇḍukuṣṭhavikāraghnī kāravallī jvarāpahā.
(Madanadinighantu)

जलजं कारवेल्लं स्यात् तिक्तं भेदकरं मतम्।
कफं कुष्ठं पाण्डुरोगं कृमीन् पित्तं च नाशयेत्।।
(नि. र.)

Jalajam Kāravalli syāt tiktam bhedakaram matam,
Kapham kuṣṭham pāṇḍurogam kṛmīn pittam ca nāśayet.
(Nighanturatnakara)

Actions/Uses : Kṛmipittahara, Pandukuṣṭhaghna, Bhedinī, Viṣaghna, Vātahara, Vāyunāsī, Sutikajvaraghna.

Therapeutics :

Kāravallī root is used in snake-bite. Leaves are bitter, astringent, anthelmintic, laxative, febrifuge and diaphoretic. They are useful in intestinal worms, colic, leprosy, malarial fever, dysentery, ulcers and vomiting. Leaves are chewed or their infusion given in colic and dyspepsia. They are used as a cure for fresh wounds and cuts; ground and administered with butter as cure for dysentery and malaria; used internally and applied externally to body during puerperal fever; and external application for ulcers. Juice from young twigs is used in diarrhoea and dysentery. Flowers are used as poultice for headache. Fruits are carminative and are used in jaundice.

Quisqualis indica Linn.

SYNONYMS : *Quisqualis densiflora Wall. ex. Miq.*

ENGLISH : Chinese-Shikunshi, Rangoon creeper.

INDIAN : Laala chameli, Rangunachi veli, Barmasi.

HABIT : A large, woody, scanent climbing shrub.

HABITAT : Indigenous in Africa and Indo-Malaysian region. Cutivated all over India.

CHEMICAL COMPOSITION : Rutin, trigonelline, L-proline, L-asparagine and quisqualic acid from leaves. Rutin and pelargonidin-3-glucoside from flowers Gum. Seeds yield 27% oil containing linoleic, oleic, palmitic, stearic and arachidic acids, and a sterol and a fatty acid, and an alkaloid used as anthelmintic; seeds contain an active principle resembling santonin.

Parts used : Leaves and seeds.

Actions/Uses : Krimighna, Jwaraghna.

Therapeutics :

Leaves are given in a compound decoction for flatulent distension of abdomen. Fruits and seeds are used as children's anthelmintic for ascaris. Roasted seeds used in diarrhea and fever. They are also used for rickets in children. Macerated in oil, seeds are used as an external application to parasitic skin diseases. Oil from seeds is purgative.

Terminalia arjuna W. & A.

SYNONYMS : *Pentaptera arjuna* Roxb.; *Pentaptera glabra.*

ENGLISH : Myrobalan, White marudha, Arjuna.

INDIAN : Arjuna, Arjunasaadaddaa, Sanmadat, Vellamarda, Sadaru, Kahu, Shardul.

SANSKṚTA : Arjuna, Kukubha, Dhavala, Gaura, Vīravṛkṣa, Pārtha, Indradru, Nadisarja, Mahīsaha, Karṇāri, Daṭhataru, Madhuratvak, Savyasācī, Śuklavṛukṣa, Pāndutaru, Śakaṭa, Nanditaru, Bharodvāha, Dhurandhara, Kaṣāya, Dhananjaya, Indrasūka, Pāndava, Gāndīvī, Kṛṣṇasārathī, Kaunteya, Pṛthāja, Kirītī, Vairāntaka, Mahīruha, Śvetavāha, Karṣaphala.

HABIT : A large evergreen tree.

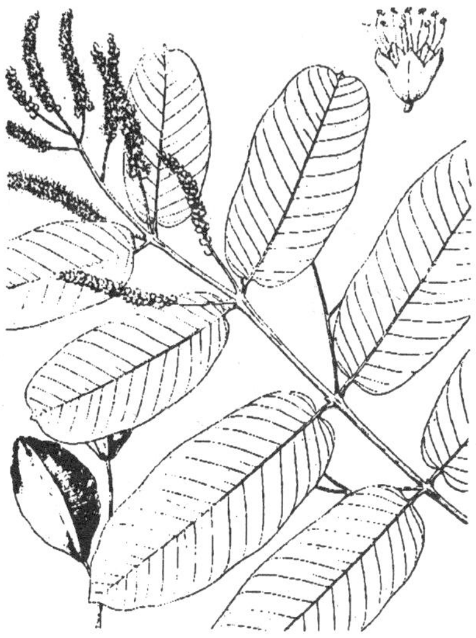

HABITAT : Common on the banks of rivers, streams and dry water-courses and throughout India in moist deciduous forests.

CHEMICAL COMPOSITION : Arjunolic acid, tomentosic acid, B-sitosterol, ellagic acid, saponin and (+) leucodelphinidin. Bark contains a crystalline compound arjunine, a laactone, arjunetin, essential oil, tannin (12 %) pyrocatachol, large quantities of calcium salts and traces of aluminium and magnesium salts, reducing sugars and colouring matter.

Parts used : Bark.

Ayurvedic properties :
Rasa - Kaṣāya.
Vīrya - Śīta.
Guṇa - Laghu, Rūkṣa. Vipāka - Kaṭu.

अर्जुनस्तु कषायोष्णः कफघ्नो व्रणनाशनः।
पित्तश्रमतृषार्तिघ्नो मरुतामयकोपनः।। (राजनिघण्टु)

Arjunastu kaṣāyoṣṇaḥ kaphaghno vraṇanāśanaḥ,
Pittaśramatṛsārtighno marutāmayakopanaḥ.
(Rajanighantu)

ककुभः शीतलो हृद्यः क्षतक्षयविषास्त्रजित्।
मेदोमेहव्रणान् हन्ति तुवरः कफपित्तकृत्।। (भा. प्र.)

Kakubhaḥ śitalo hṛdyaḥ kṣatakṣayaviṣāsrajit,
Medomehavraṇān hanti tuvaraḥ kaphapittakṛt.
(Bhāvaprakāṣa)

Actions/Uses : Hṛdya, Kaphaghna, Pittaghna, Vātavardhaka, Pūriṣa, Raktastambhana, Sandhānīya, Vraṇaropaṇa.

Therapeutics :

Leaves are used externally as a cover for sores and ulcers. Juice of fresh leaves is used for earache. Bark is astringent, antidysenteric, cardiotonic, styptic, febrifuge, cooling internally and wholesome for heart. It is used in heart diseases as a cardiac tonic, and in bilious affections. It cures wounds and urinary diseases. Powdered bark relieves hypertension, has a diuretic and a general tonic effect in case of cirrhosis of liver; is given internally with milk in bone fractrures and contusions with excessive ecchymosis. The decoction is used as a wash in ulcers and cancer. Ashes of bark prescribed in scorpion sting. Fruit is tonic and deobstruent.

Terminalia belerica Roxb.

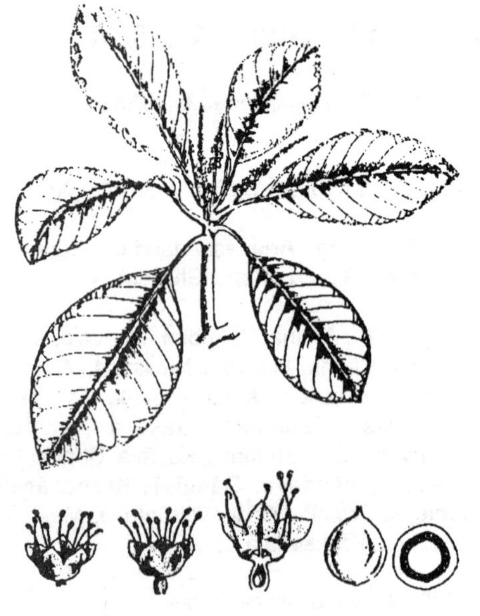

SYNONYMS : *Myrobalanus belerica Gaertn*

ENGLISH : Myrobalan, bastard. Belliric myrobalan.

INDIAN : Behaddaa, Bibheeta, Baheraa.

SANSKṚTA : Bibhītaka, Karṣaphala, Kalidruma, Samvaraka, Vasanta, Tilapuṣpaka, Bhutavasa, Dharmaghna, Tumula, Kalinda, Bahedaka, Kaliyugālaya, Vṛudhajata, Tailaphala, Anilaghnaka, Kāsaghna, Akṣa, Vidhyajātaka, Romaharṣa, Vanapriya, Hārya, Kalko.

HABIT : A large deciduous tree.

HABITAT : Found in the plains and lower hills, troughout the forests of India.

CHEMICAL COMPOSITION : Fruits contain about 17 % tannin and β-sitosterol, gallic acid, ellagic acid, ethyl galate, galloyl glucose and chebulagic acid. Heartwood and bark contain ellagic acid and the seed-coat of the fruit contains gallic acid. A new cardiac glycoside- bellericanin. Kernel oil had a purgative action, and fruit extract produced fall in blood pressure and significant increase of bile secretion in experimental animals.

Parts used : Fruits

Ayurvedic properties :
Rasa - Kasāya. Vīrya - Ūsna.
Guṇa - Gurū, Rūksa. Vipāka - Madhura.

विभीतकः कटुस्तिक्तः कशायोष्णः कफा पहः।
चक्षुष्यः पलितघ्नश्च विपाके मधुरो लघुः।।
(राजनिघण्टु)

Bibhītakaḥ kaṭustiktaḥ kasāyoṣṇaḥ kaphā pahaḥ, Cakṣuṣyaḥ palitaghnaśca vipāke madhuro laghuḥ. (Rajanighantu)

बिभीतकं स्वादुपाकं कषायं कफपित्तनुत्।
उष्णवीर्य हिमस्पर्श भेदनं कासनाशनम्।।
रुक्षं नेत्रहितं केश्यं कृमिवैस्वर्यनाशनम्।
बिभीतकमज्ञा तृट्छर्दिकफवातहरी लघुः।
कषाया मदकृच्चाथ धत्रीमज्ञापि तद्गुणा।।
(भा. प्र.)

Bibhītakam svādupākam kaṣāyam kaphapittanut, Ūṣṇavīryam himasprśaṁ bhedanam kāsanāśanam.
Rūksyam netrahitam keṣyam kṛmivaisvaryanāśanam,
Bibhītakamajjā tṛdcchardikaphavātaharī laghuḥ,
Kaṣāyā madakṛccātha dhatrīmajjāpi tad gunā.
(Bhāvaprakāṣa)

Actions/Uses : Kaphapittaghna, Dīpana, Pācana, Anulomana, Grāhī, Śothahara, Vedanāsthāpana, Kṛmighna, Kāsahara, Śvāsahara, Raktastambhana.

Seed-coat : Sangrāhak, Ślesmaghna, Pratiśyāya; used in Kās, Śvāsa and Svarabhanga by keeping it in mouth.

Kernel : Kaifi, Vedanāsthāpan, used externally to reduce swelling, burning and itching.

Therapeutics :

Fruit of Bibhitāka is bitter, astringent, tonic, laxative, antipyretic, used in piles, dropsy, diarrhoea, headache, leprosy, dyspepsia and billiousness; useful in coughs, hoarseness, eye diseases; Purgative when half ripe and astringent when ripe. Kernel is narcotic, useful in thirst, vomiting, bronchitis and corneal ulcers.

Terminalia chebula Retz.

SYNONYMS : *Myrobalanus chebula Gaertn.; Terminalia citrina* Roxb.

ENGLISH : Myrobalan, chebulic, Indian gall-nut.

INDIAN : Hiraddaa, Hareetakee, Harra, Baalhirade.

SANSKṚTA : Haritakī, Abhayā, Pathya, Cetanā, Rohiṇī, Pūtanā, Śivā, Amrtā, Prāṇadā, Kāyasthā, Vijayā, Nandinī, Medhyā, Pācanī, Haimamatī, Avyathā, Prapayyā, Jīvapriyā, Jīvanikā, Jīvantī, Jayā, Bhiṣagvarā, Jīvyā, Śreyasī, Cetakī, Jīvanīyā, Vayasthā, Divyā, Śakrasṛṣṭā, Devī, Rasāyanaphalā.

HABIT : A moderate sized or large deciduous tree.

HABITAT : Abundant in Northern India. Also occurs

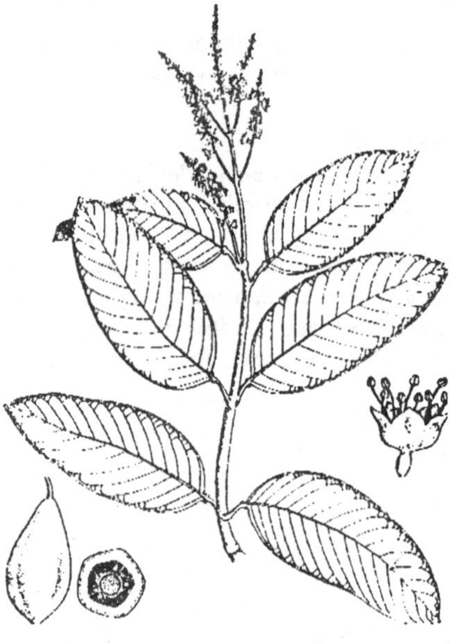

in forests of Bihar, WestBengal, Assam and Maharashtra especially in Konkan.

CHEMICAL COMPOSITION : Chebulin from flowers. Palmitic, stearic, oleic, linoleic, arachidic and behenic acids from fruit kernels. Fruits contain about 30 % of an astringent substance; astrigency is due to the charecteristic principle chebulnic acid. Also contain tannic acid 20-40 %, gallic acid, resin etc. and a purgative glycoside of anthraquinone derivative. Chebulin exhibited anti-spasmodic action on smooth muscle similar to papaverine.

Parts used : Fruits.

Ayurvedic properties :
Rasa - Kaṣāya.
Vīrya - Uṣṇa.
Guṇa - Laghu, Rūkṣa. Vipāka - Madhura.

हरीतकी पञ्चरसाऽ लवणा तुवरा परम्।
रुक्षोष्णा दीपनी मेध्या स्वादुपाका रसायनी।।
(भा. प्र.)

Harītakī pāncarasā alavaṇā tuvarā param,
Rūkṣoṣṇā dīpanī medhyā svādupākā rasāyanī.
(Bhāvaprakāṣa)

हरीतकी पञ्चरसा च रेचनी कोष्ठामयघ्नी लवणेन वर्जिता।
रसायनी नेत्रमयापहारिणी त्वगामयघ्नी किल योगवाहिनी।।
(राजनिघण्टु)

Harītakī pāncarasā ca recanī koṣṭhāmayaghnī lavaṇena varjitā,
Rasāyanī netramayāpahāriṇī tvagāmayaghnī kila yogavāhinī.
(Rajanighantu)

Actions/Uses : Tridoṣahara, Kaphaghna, Vātaghna, Śothahar, Arśoghna, Śothaghna, Śleshmāgna, Raktasanghrahaka, Balya, Pathya, Gulmahara, Vayasthāpan, Vedanāsthāpana, Vraṇasodhana-ropaṇa, Netrābhiṣyanda.

Therapeutics :
Haritaki fruit is astringent, light, digestive, antiseptic, alterative, laxative, diuretic and carminative. It is a safe and effective purgative. It promotes digestive power and heals wounds and ulcers; cures local swellings, skin and eye diseases, chronic and recurrent fever, anaemia, cardiac disorders, diarrhoea and dysentery, diabetes, cough and dyspnoea. It is useful in spleen enlargment, ascites, piles, hoarseness of voice, vomiting and blood pressure. Unripe fruits are more purgative and ripe

ones more astringent. They are used externally as a local application to chronic ulcers and wounds and as a gargle in stomatitis. Finely powdered, used as a dentrifrice and considered useful in cardiotonic. Bal-harade is highly useful in chronic diarrhoea and dysentery, flatulence, vomiting, hiccup, colic and enlarged spleen and liver.

Terminalia paniculata Roth.

SYNONYMS : *Pentaptera paniculata* Roxb.

ENGLISH : Flowering murdah.

INDIAN : Kinjalla, Kindal.

SANSKṚTA : Bījaka-1; Kinjala.

HABIT : A large deciduous tree.

HABITAT : Common in Konkan and Western Peninsula.

CHEMICAL COMPOSITION : β-sitosterol, triterpene carboxylic acid, a new glucoside, dl-methyl ellagic acid.

Parts used : Bark and flowers.

Ayurvedic properties :
Rasa - Kaṣāya.
Vīrya - Ūṣṇa.
Guṇa - Laghu, Rūkṣa. Vipāka - Madhura.

Actions/Uses : Sangrāhaka, Mūtrala, Hṛdayasamvardhaka, Asthisandha.

Therapeutics :

Bark of Bijaka-1 or Kinjala is diuretc and cardiotonic. Juice of fresh flowers rubbed with the root of Cocculus villosus used as a remedy in cholera and opiom poisoning.

Terminalia tomentosa W. & A.

SYNONYMS : *Terminalia crenulata* Roth.; *Terminalia tomentosa Cooke.; Terminalia tomentosa Bedd.; Terminalia glabra.; Terminalia alata Heyne ex* Roth.; *Pentaptera tomentosa.*

ENGLISH : Laurel.

INDIAN : Aina, Asan, Sain, Saj.

SANSKṚTA : Bijaka-2, Asana..

HABIT : A large, deciduous tree.

HABITAT : Common throughout India and Konkan except in Sind and Rajputana.

CHEMIICAL COMPOSITION : Bark gives waxy substance, crystalline substance characteristic of terpenoides. β-sitosterol, oleanolic acid, arjunolic acid and a triterpene acid.

Parts used : Bark.

Ayurvedic properties :
Rasa - Kaṣāya.
Vīrya - Uṣṇa.
Guṇa - Laghu, Rūkṣa Vipāka - Madhura.

बीजकः कुष्ठविसर्पश्वित्रमेहगुद् क्रिमीन्।
हन्ति श्लेष्मास्त्रपित्तश्च त्वच्यः केश्यो रसायनः।।
(भावप्रकाश)

Bījakaḥ kuṣṭhavisarpaśvitramehagudakṛīmin,
Hanti śleṣmāsrapittāsca tvacyaḥ keṣyo rasāyanaḥ.
(Bhāvaprakāṣa)

Actions/Uses : Kṛmighna, Sangrāhaka, Kesya, Rasāyana.

Therapeutics :
Bijaka-2 or Asana is astringent, antiseptic, bactericidal, demulcent, and detergent. It is efficacious in skin diseases, erysipelas, leucoderma, polyuria. It cures shleshma, haemorrhageic diseases and beneficial for skin and hair. It is considered to be a rejuvenator. Bark is bitter and styptic and has diuretic and cardiotonic properties. Juice of fresh bark is applied on swellings. Decoction of bark is astringent taken internally for atonic diarrhoea; applied locally to ulcers. It is Sangraahak and its action similar to Arjun.

Family : XXXVII-Convolvulaceae.

Evolvulvus alsinoides Linn.

SYNONYMS : *Evolvulvus hirsutus* Lamk.

INDIAN : Shankhapushpi, Shankhavali, Vishnukraantaa, Shuamakranta.

SANSKṚTA : Śankhapuṣpi, Viṣṇukrāntā, Nīlapuṣpi, Viṣṇugandhi.

HABIT : An very small perennial prostrate herb.

HABITAT : A common weed in open and grassy places in Konkan and Gujarat and also throughout India.

CHEMICAL COMPOSITION : Plant contains pentatriacontane, tricontane and β-sitosterol. Alkaloid evolvine is a powerful stimulent on respiration and B.P. Betain and a base isolated.

Parts used : Whole plant, leaves and oil.

Ayurvedic properties :
Rasa - Kaṭu, Kaṣāya, Tikta. Vīrya - Ūṣṇa.
Guṇa - Sara. Vipāka - Madhura.

शङ्खपुष्पी सरा मेध्या वृष्या मानसरोगहृत्।
रसायनी कषायोष्णा स्मृतिकान्तिवलाग्निदा।।
दोषापस्मारभूताश्रीकुष्ठक्रिमिविषप्रणुत्।।
(द्रव्यगुण)

Śaṅkhapuṣpī sarā medhyā vṛṣyā mānasarogahṛt,
Rasāyani kaṣāyoṣṇā smṛtikāntivalāgnidā,
Doṣāpasmarabhūtāśrīkuṣṭhakṛimiviṣapraṇut.
(Dravyaguna)

Actions/Uses : Dīpana, Anulomana, Jvaraghna, Garbhāśayauttejaka, Paustika.

Therapeutics :

Plant is a specific for all kinds of fevers. It is acrid, bitter tonic, alterative, anthelmintic, antiphlogistic, febrifuge and antisposmodic. It is a powerful brain stimulant and aphrodisiac. It is a sovereign remedy in bowel complaints esp. dysentery. It is used in leucoderma and as a vermifuge. With oil used for promoting growth of hair. Plant and leaves are used in insanity, epilepsy, nervine complaints, spermatorrhoea and bleeding. Roots are used in intermittent fever in children and for gastric and duodenal ulcers. They are helpful for uterus, brain and nerves. Leaf juice increases brain powerand is useful in internal haemorrhage and in conjuctivitis; given in bowel complaints, promotes constipation. Leaves made into cigarettes are smoked in chronic bronchitis and asthma. Flowers are good for uterine bleeding.

Ipomoea batatas Lam.

SYNONYMS : *Convolvulus batatas* Linn.

ENGLISH : Sweet potato.

INDIAN : Rataalle, Mitha alu, Shakarkand.

SANSKṚTA : Mukhāluka, Ratālu, Raktālu, Raktapiṇḍaka, Raktakanda.

HABIT : A slender, prostrate perennial herb.

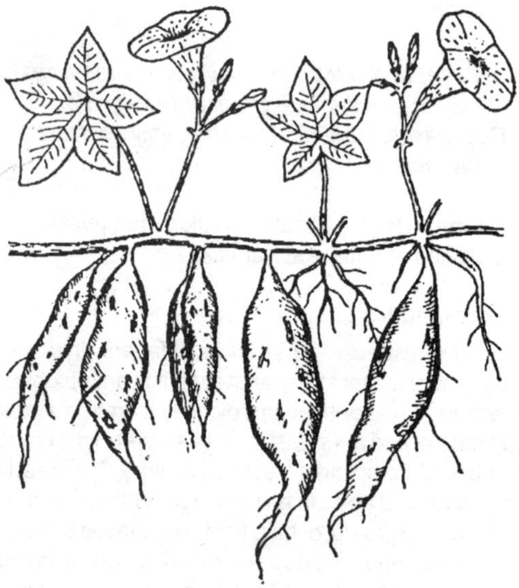

HABITAT : Cultivated in India, native of America.

CHEMICAL COMPOSITION : Tubers, hydrocyanic acid, 30% mg/100g, oxalic acid, 1.0% and phytic acid 8% . Good source of starch and vitamines A, B and C and enzyme. n-Pentacosane, n-heptacosane, n-nonacosane, β-sitosterol, oalmitic acid and NaCl isolated.

Parts used : Root, leaves and stem.

Ayurvedic properties :
Rasa : Madhura. Vīrya : Śīta.
Guṇa : Gurū. Vipāka : Madhura.

मुखालुकः स्यान्मधुरः शिशिरः पित्तनाशनः।
रुचिकृद्वातकृच्चैव दाहशोषतृषापहः।।
(राजनिघण्टु)

Mukhālukaḥ syānmadhuraḥ śiśiraḥ pittanāśanaḥ,
Rucikṛdvatakrccaiva dāhaśoṣatṛṣāpahaḥ.
(Rajanighantu)

Actions/Uses : Pittaghna, Kaphakara, Vātakara, Paustika, Grāhī, Anulomic, Mūtral, Kamoddīpaka.

Therapeutics :

Root is sweet, cooling, fattening and considered laxative. It is aphrodisiac; useful in stangury, burning sensation, thirst, urinary discharges; causes kapha and vata. In New Zealand whole plant or its infusion is used in cases of low fever and skin diseases. In Philippines it is used as antidiabetic. Leaves cooked in maize-soaked liquid to yield a lactagogue; ground with salt applied to whitlow. Bulb is used in vomiting, asthma and constipation.

Ipomoea digitata Linn.

SYNONYMS : *Ipomoea mauritiana* Jacq.; *Batatas paniculata* Choisy.; *Ipomoea paniculata* R.Br.; *Convolvulus paniculata* Linn.

ENGLISH : Giant potato.

INDIAN : Bhuee kohallaa, Bilaikanda, Vidaarikanda.

SANSKṚTA : Vidārī, Vidārīkanda, Bhūmikūṣmānda, Kṣīravidārī, Payasvinī, Kṣīravallī.

HABIT : An extensive perennial climber with tuberous roots.

HABITAT : Throughout tropical India in moist regions.

CHEMICAL COMPOSITION : Rhizome contains taraxcrol and sitosterol. Resin similar to jalap resin. Fixed oil from tubers contained palmitic, stearic, oleic, linoleic and linolenic acids. Alcoholic extract of tubers showed stimulent as well as depressant actions on different organ systems along with convulsant effect on central nervous system.

Parts used : Root and resin.

Ayurvedic properties :
Rasa - Madhura. Vīrya - Śīta.
Guṇa - Gurū, Snigdha. Vipāka - Madhura.

विदारी मधुरा स्निग्धा बृंहणी स्तन्यशुक्रदा।
शीता स्वर्या मूत्रला च जीवनी बलवर्णदा।
गुरुः पित्रास्त्रपवनदाहन् हन्ति रसायनम्।। (भा. प्र.)

Vidārī madhurā snigdhā bṛnhaṇī stanyaśukradā,
Śīta svaryā mūtralā ca jīvanī balavarṇadā,
Gurūḥ pitrasrapavanadāhān hanti rasāyanam.
(Bhāvaprakāṣa)

विदारी मधुरा शीता गुरुः स्निग्धास्त्रपित्तजित्।
ज्ञेया च कफकृत्पुष्टि वल्या वीर्य्यविवर्द्धिनी।।
(राजनिघण्टु)

Vidārī madhurā śītā gurūḥ snigdhāśrapittajit,
Jneyā ca kaphakṛtpuṣṭi balyā vīryavivarddhinī.
(Rajanighantu)

Actions/Uses : Recana, Brihmaṇī, Kaphaghna, Pittaghna, Jīvanī, Balavardhinī, Balya, Vṛṣya, Kaphakrit, Raktapittaghnī.

Therapeutics :
Vidari plant is sweet, cooling, restorative, tonic, diuretic, aphrodisiac, galactagogue, and demulcent and is found useful in fevers and bronchitis. It is used in many Ayurvedic formulations. Tuberous roots contain a resin similar to jalap resin and is considered tonic, alterative, aphrodisiac, demulcent, lactagogue, purgative, and cholagogue. It is used to increase weight reduced due to mental or physical fatigue, also in scorpion sting. Flour of raw rhizome is useful in enlargement of liver or spleen. Powdered root is given for emaciation in children and used in spleen and liver diseases, menorrrhagia, skin diseases. It is useful in liver complaints as a cholagogue. Powdered root-stalk is given with wine to women to increase secretion of milk. Juice of fresh root mixed with sugar and cumin is beneficial as lactogogue and given in spermatorrhoea in combination with coriander and fenugreck seeds. Ether extract of rhizome is antianphetaminic.

Ipomoea hederacea (Linn.) Jacq.

SYNONYMS : *Ipomoea nil (Linn.)* Roth.; *Convolvulus bilobatus* Roxb. *Concolvulus nil* Linn.

ENGLISH : Morning Glory, Pharbitis seeds.

INDIAN : Kaalla daanaa, Krushnavela.

SANSKṚTA : Śyāmabīja, Kṛṣṇabīja, Śankhanī, Kālādānā, Jhāramārīca.

HABIT : A perennial climbing shrub.

HABITAT : Throughout India both cultivated and wild.

CHEMICAL COMPOSITION : Seeds contain alkaloids lysergol, chanoclavine, penniclavine, isopeniclavine and elymoclavine. It also contains resin 14.2 % and glucosides. Seed oil contained palmitic, stearic, arachidic, oleic, linoleic and linolenic acids.

Parts used : Seeds and leaves.

Ayurvedic properties :

Rasa - Kaṭu, Madhura. Vīrya - Ūṣṇa.
Guṇa - Laghu, Rūkṣa, Tīkṣṇa. Vipāka - Kaṭu.

रेचनं श्यामवीजं स्यात् शोथोदरविनाशनम्।
ज्वरे पुरीषसङ् गे च दारुणे शिरसो गदे।।
उदावर्त्ते तयानाहे वुधैरेतत् प्रयुज्यते।।
(द्रव्यगुण)

Recanam Śyāmabījam syāt śothodaravināśanam,
Jvare purīṣasadge ca dāruṇe śiraso gade,
Udāvartte tathānāhe budhairetat prayujyate.
(Dravyaguna)

Actions/Uses : Kaphapittaśodhana, Kaphaghna, Vātaghna. Lekhana, Virecana, Jvaraghna, Śothahara, Raktaśodhana.

Therapeutics :

Śyamabīja seed is acrid, light, hot, anṭhelmintic, purgative and blood purifier. It is a substitute for jalap. It is laxative, carminative and cures inflammations, abdominal diseases, fevers, headache and bronchitis. It is a mild purgative useful to eliminate bile, cough and worms. Juice of leaves is also used for the same purpose. Beneficial in oedema, ascites, fever, constipation; relieves severe headache, flatulence and epistaxis.

Family : XXXVIII-Crassulaceae.

Kalanchoe pinnata Pers.

SYNONYMS : *Bryophyllum calycinum* Salisb.; *Bryophyllum pinnatum* Kurz.; *Cotyledon pinnata* Lamk.

SIMILAR : Kalanchoe laciniata DC.; Kalanchoe integra.

INDIAN : Paanfuti, Parnnabeeja, Ghaayamaari.

SANSKṚTA : Pāṣāṇabheda, Parṇabīja.

HABIT : A perennial succulent herb.

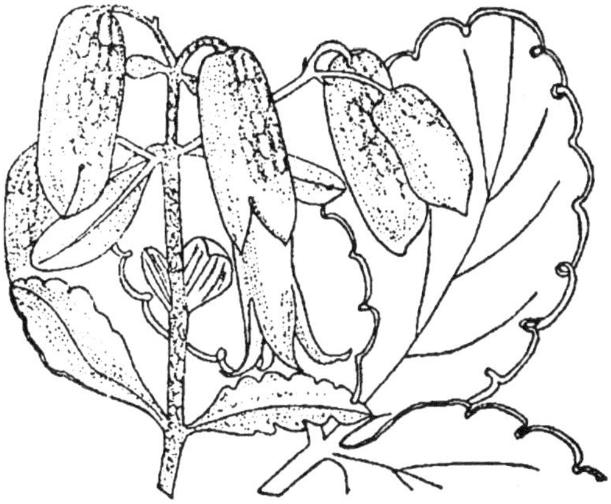

HABITAT : Indeginous to Africa, introduced in India, naturalized throughout hot and moist parts, cultivated in gardens, also found in forest.

CHEMICAL COMPOSITION : Plant contains n-alkane, n-alkanol; α- & β- amyrin and sitosterol. Leaves contain malic, isocitric and citric acids. Glycosides of quercetin and kaempferol; fumaric acid; phenolic components from leaves.

Parts used : Leaves.

Ayurvedic properties :
Rasa - Kaṣāya, Amla. Vīrya - Śīta.
Guṇa - Laghu, Rūkṣa. Vipāka - Madhura.

पर्णबीजं कषायाम्लं मधुरं शीतमेव च।
वातपित्तहरं रक्तस्तंभनं व्रणरोपणम्।।

Parṇabījam kaṣāyāmlam madhuram śītameva ca,
Vātapittaharam raktastambhanam vraṇaropaṇam.

पाषाणभेदो मधुरस्तिक्तो मेहविनाशनः।
तद्दामूत्रकृच्छ्रघ्नः शीतलश्चाश्मरहिरः।।
(राजनिघण्टु)

Pāṣāṇabhedo madhurastikto mehavināśanaḥ,
Trd dāmutrakṛcchghnaḥ śītalaścāśmarīharaḥ.
(Rajanighantu)

Actions/Uses : Vātapittaśāmaka, Raktaskandana, Raktastambhaka, Vraṇaropaṇa, Raktārśaghna, Raktapittaśāmaka, Raktapradaraghna.

Therapeutics :

Leaves of Parnabija are bitter, slightly toasted used as an application to bruises, wounds, boils and bites of insects. They are eaten to control diabetes. Juice of succulent leaves used in burns. Juice mixed with juice of leaves of Aegle marmelos, given in blood and amoebic dysentery.

Family : XXXIX-Cucurbitaceae.

Benincasa hispida (Thunb.) Cogn.

SYNONYMS : *Benincasa cerifera* Savi.; *Cucurbita hispida* Thunb.

ENGLISH : Ashgourd, Ash pumpkin, White pumpkin, White gourd melon.

INDIAN : Kohallaa, Bruhata phalla, Kumhra, Petha, Golkaddu.

SANSKṚTA : Kūśmāṇḍa, Puṣpaphala, Bṛhatphala, Somakā, Valliphala, Somasrṣṭā, Kuṁbhaphala, Kuśmaṇḍakī, Karkotikā, Suphalā, Amṛtā, Sthiraphalā, Kṣīraphalā, Nāgapuṣpaphalā, Rājakarkatī, Pītikā, Karkārū.

HABIT : A large climbing or trailing herb with stout hispid stems.

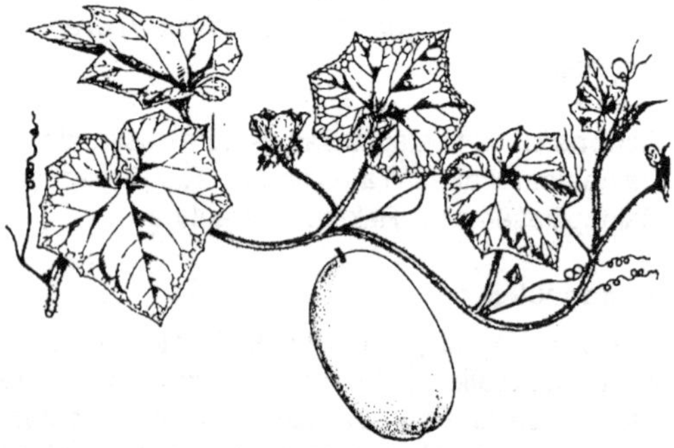

HABITAT : Native to Indo-Malaysian region. Cultivated in gardens throughout India.

CHEMICAL COMPOSITION : Fruits contain β-sitosterol, asparagine, luoeol, mannitol, proline, arginine, aspartic acid, glutamic acid, glutamine, hydroxyproline, isoleucine, cysteine, L-leucine, n-triacontanol, glucose and rhamnose. Source of vitamin B1.

Parts used : Leaves, fruit, seed, seed oil.

Ayurvedic properties :
Rasa - Madhura. Vīrya - Śīta.
Guṇa - Laghu, Snigdha. Vipāka - Madhura.

कूश्माण्डं बृंहणं गुरु पित्तास्त्रवातनुत्।
बालं पित्तापहं शीतं मध्यमं कफकारकम्।।
वृद्धं नातिहिमं स्वादु सक्षारं दपिनं लघुः।
वस्तिशुद्धिकरं चेतोरोगहृत् सर्वदोषजित्।।
(भा. प्र.)

Kuśmāṇḍam bṛnhaṇam vrstam guru pittāsravātanut,
Bālam pittāpaham śītam madhyamam kaphakārakam.
Vrddham nātihimam svādu saksāram dīpanam laghuḥ,
Bastiśuddhikaram cetorogahṛt sarvadoṣajit.
(Bhāvaprakāṣa)

Actions/Uses : Medhya, Pittaśāmaka, Vātaśāmaka, Sarvadoṣahara, Kṛmighna, Anulomana, Tṛṣṇānigrahaṇa.

Therapeutics :

The juice of leaves of plant is cooling and rubbed on bruices; poultice used in bruises. Fruit is laxative, diuretic, tonic, aphrodisiac, antiperiodic, specific for haemoptysis and other haemorrhages from internal organs. It is rejuvenative and highly nutritious, capable of improving intellect and physical strength. Fruit juice is used in insanity, epilepsy and other nervous diseases. It is an antidote for mercury and alcohol poisoning. A decoction of fruit is styptic, laxative, diuretic and nutritious and given to cure internal haemorrhages and diseases of resperatory organs. Syrup used in resperatory afflictions. Kushmanda lehyam or ghrit is recommended for epilepsy, constipation, piles, dyspepsia, syphilis and diabetes. Fruit as also seeds are used in tumors. Flesh is demulcent, thirst-quenching and cooling in fever. Raw flesh is a remedy for facial eruptions. Habitual consumption of seeds prevents hunger and prolongs life. Kernel used against skin eruption and seed ash as remedy for gonorrhoea. Seeds and oil from it is anthelmintic.

Citrullus colocynthis Schrad.

SYNONYMS : *Cucumis colocynthis* Linn.

ENGLISH : Bitter apple, Colocynth.

INDIAN : Indrayanna, Kadu-wrundaawana, Makal.

SANSKṚTA : Gavādanī, Indravāruṇī, Gavākṣī, Dhanuśreṇī, Citrā, Aindrī, Trapusī, Indravallī, Hemapuṣpī, Citraphala, Sūryāhva, Surendrāhva, Mahāphala, Sukhāruṇī, Galactirbhaṭī, Hastidantī, Ātmarakṣa, Sthāṇukarṇī, Mṛgabhakṣa, Vṛṣabhākṣī, Kṣudraphalā, Marusambhava, Citādevī, Mahendravāruṇī.

HABIT : An annual or perennial herb with a prostrate or climbing stem.

HABITAT : A native of warmer parts of Asia and Africa. Occurs wild throughout India, and sea shores of Gujrat, Maharashtra.

CHEMICAL COMPOSITION : Bitter oil 'citbittol', colocynthin, colocynthetin. Roots contain a-elaterin, hentriacontain, and saponins. Seeds contain fixed oil, a phytosterolin, 2 phytosterols, 2 hydro-carbons, a saponin, alkaloid, glycoside, tannin. Pulp contains a-elaterin, hentriacontane, citrullol, a phytosterol and a mixture of fatty acids. Pulp also contains an anticancer glycoside- a-elaterin-2-D-glucopyranoside. Fixed oil of seed contain myristic, palmitic, stearic, myristoleic, palmitoleic, oleic and linoleic acids.

Parts used : Fruit deprived of its rind, root, dried pulp.

Ayurvedic properties :
Rasa - Tikta, Kaṭu. Vīrya - Ūṣṇa.
Guṇa - Laghu, Tīkṣṇa, Rukṣa. Vipāka - Kaṭu.

गवादनीद्वयं तिक्तं पाके कटु सरं लघु।
वीर्योष्णं कामलापित्तकफप्लीहोदरापहम्।।
श्वासकासापहं कुष्ठगुल्म ग्रन्थिव्रणप्रणुत्।
प्रमेहमूढगर्भामगंडामयविषापहम्।।
(भा. प्र.)

Gavādanīdvayam tiktam pāke kaṭu saram laghu,
Vīryoṣṇam kāmalāpittakaphaplīhodarāpaham.
Śvāsakāsāpaham kuṣṭhagulma granthivraṇapraṇut,
Pramehamūḍhagarbhamagaṇḍāmayaviṣāpaham.
(Bhāvaprakāṣa)

Actions/Uses : Kaphaghna, Pittakar, Vātakar, Virecani, Vraṇahara, Kṛmighna, Svāsahara, Kāsahara, Apacighna, Kuṣṭhaghna, Gulmahara, Plihodarahara, Garaviṣaghna, Kamalahara, Pramehaghna, Granthighna, Gaṇḍamālāhara, Mūḍhagarbhaghna, Viṣaghna, Slipadaghna, Jvaraghna.

Therapeutics :
Gavādanī root is purgative, and is used in ascites, jaundice, urinary diseases and rheumatism. For inflammation of joints etc. awaleha or powder of root mixed with ginger and jaggery is used. Fruit is light, bitter, pungent, cooling, abortifacient, blood purifier, purgative, anthelmintic, antipyretic, cathartic and carminative; cures tumours, ascites, jaundice, enlargement of spleen. In moderate doses it is drastic hydrogogue, cathartic and diuretic. In large doses, emetic and gastro-intestinal irritant. Useful in abnormal presentation of foetus and in atrophy of foetus. Useful for inflammation of breasts, puerperal disorders, pain in joints. Externally used in ophthalmia and uterine pains. Fruit and root antidote to snake poison. Oil from seed, useful in hair growth and maintaining them black, sterility (maladu).

Coccinia grandis (Linn.) Voight.

SYNONYMS : *Coccinia cordifollia* Cogn.; *Coccinia indica* W. & A.; *Cephalandra indica* Naud.

INDIAN : Tonddali, Bimbi, Kunduri.

SANSKṚTA : Bimbī, Tuṇḍī, Tuṇḍikerī, Raktaphala, Vidruma, Titiḍī, Dantacchada, Vidrumaphala, Oṣṭhopamaphala, Tuṇḍikeraphala, Vimbaja, Pīluparṇikā, Vimboṣṭa.

HABIT : A slender dioecious perennial climber with tuberous roots.

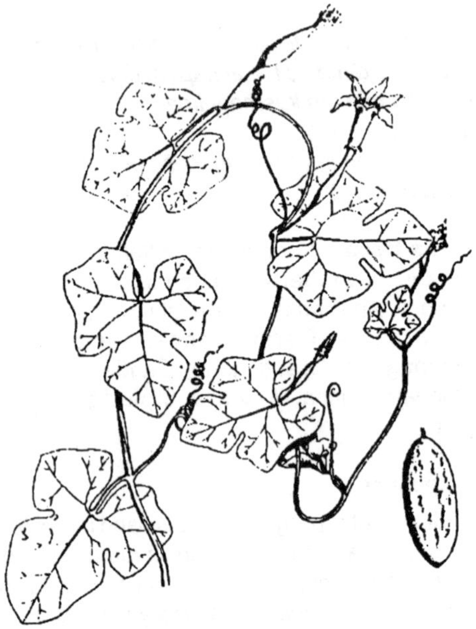

HABITAT : Cultivated throughout India.

CHEMICAL COMPOSITION : Aerial parts yield alcohol, cephalandrol, tritriacontane, β-sitosterol, cephalandrine A. & B; also heptacosane. Young green immature fruits of bitter variety, a glycoside of cucurbitacin B, β-amyrin and lupeol. Fruits, β-sitosterol and taraxerol. Ripe fruit carotenoids. Seed fat rich in palmitic, oleic and linoleic acids. Stigmast-7-en-3-one from roots. An orally effective hypoglycaemic principal, comparable to tolbutamide, isolated from roots.

Parts used : Leaves, root, fruit and bark.

Ayurvedic properties :
Rasa - Madhura, Tikta, Kaṣāya. Vīrya - Ūṣṇa.
Guṇa - Laghu, Rūkṣa, Tīkṣṇa.
Vipāka -Kaṭu, Madhura.

तिक्तबिम्बीफलं तिक्तं पित्तघ्नं वातकोपनम्।
विषघ्नं अतिरुच्यं स्यात् गुरु श्लेष्मकरं न च।
शोफास्त्रपाण्डून् जयति न मेध्यं छर्दिकृत् परम्।।
(कै. नि.)

Tiktabimbīphalam tiktam pittaghnam vātakopanam,
Visaghnam atirucyam syāt gurū śleṣmakaram na ca,
Śophāsrapāṇḍūn jayati na medhyam cchardikṛt param.
(Kaiyadevanighantu)

Actions/Uses : Kaphapittaśodhana, Yakṛtottejaka, Vraṇaropaṇa, Śothahara, Dīpanī, Svedakāri, Pūyamehahara, Vamanahara, Recanī, Mūtrasangrahaṇīya, Jvaraghna, Kuṣṭhaghna.

Therapeutics :
Plant is alterative and used internally in gonorrhoea. Juice of plant cures ear pain. It is highly valued for its antidiabetic potential. It is carrminative, antipyretic, viriligenic, galactagogue and roborant. Powder of root is taken with water to stop vomiting; paste applied to forehead in headache. Leaves applied externally in eruptions of skin. Juice of leaves is used to allay burning pains of boils and carbuncles and for bloody dysentery, billiousness and in diabetes. Leaves and stem are antispasmodic and expectorant. Dried bark is a good cathartic. Fruits, juice from leaves and roots used in diabetes. Various plant parts used in slow pulse, convulsions, sores, syphilis and gravel. In Ayurvedic practice it is used for fever, dropsy, haemorrhage of stomach, jaundice, flatulence. It cures cough and vomiting. It suppresses urine. In Siddha system it is used against infective hepatitis.

HOMEOPATHIC USE : Ghose has reported about Homeopathic provings of Cephalandra indica which confirms its usefulness in diabetes mellitus.

Cucumis callosus Cogn.

SYNONYMS : *Cucumis trigonus* Roxb.; *Cucumis pseudo-colocynthis* Royle.; *Bryonia callosa* Rottl.

INDIAN : Kaarette, Chitrawali, Wishaala.

SANSKRTA : Mṛgākṣi, Mṛgādanī, Śatapuṣpā, Citravallī, Viśāla.

HABIT : An annual climbing herb growing on hedges trees and homeyards.

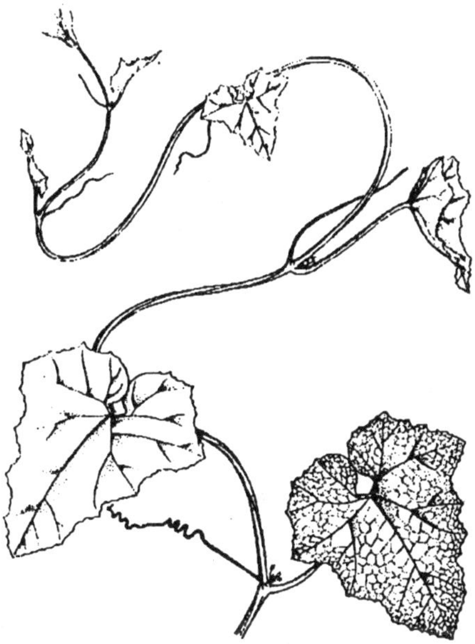

HABITAT : Wild throughout drier parts of India.

CHEMICAL COMPOSITION : Fruit contains steroid and triterpenoid compounds, cucurbitacin. Stigmast-7-en-3B-D-glucoside, alnusenone and alnusenol isolated. Fruit, proteolytic enzyme and fruit extract analgesic and antiinflammatory; diuretic in rats and stimulates uterus of guinea pigs.

Parts used : Fruit and root.

Ayurvedic properties :
Rasa - Kaṭu, Tikta. Vīrya - Śīta.
Guṇa - Rūkṣa. Vipāka - Amla.

मृगाक्षी कटुका तिक्ता पाकेऽम्ला वातनाशनी।
पित्तकृत् पीनसहरा दीपनी रुचिकृत्परा।।
(राजनिघण्टु)

Mṛgākṣi kaṭukā tiktā pāke amlā vātanāśanī,
Pittakṛt pīnasaharā dīpanī rucikṛtparā.
(Rajanighantu)

Actions/Uses : Smṛtipauṣṭika, Recana.

Therapeutics :

Pulp of fruit of Citravali is bitter and drastic purgative. Root and leaves are used in snake bite. Decoction of roots, milder prugative. Seeds, cooling, astringent, useful in billious disorders. Citrullus colocynthis and Cumimis callosus are used for the same purposes, without distinguishing.

Cucumis melo Linn.

ENGLISH : Sweet Melon, Muskmelon.

INDIAN : Kharabuja, Chibudda, Sakkar teti, Vaaluk.

SANSKṚTA : Kharbujā, Madhupāka, Kalinga, Kharvuja.

HABIT : An annual climbing herb.

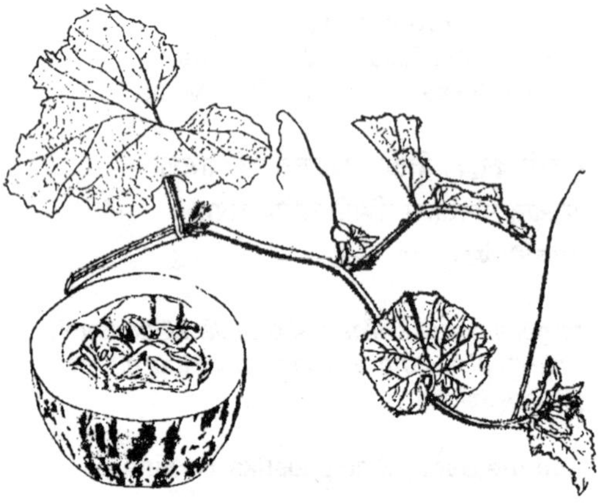

HABITAT : Extensively cultivated throughout India.

CHEMICAL COMPOSITION : Meloside A, meloside L and their caffeoyl ester from leaves. α-Carotene, β-carotene, C-carotene and three more carotenes isolated. Fruits contain ferulic, caffeic and chlorogenic acids. Fruit stalk contains cucurbitacin B & E. Seed kernel rich in oil.

Parts used : Fruit pulp, root, seeds and seed oil.

Ayurvedic properties :
Rasa - Madhura. Vīrya - Śīta.
Guṇa - Gurū, Snigdha. Vipāka -Madhura.

पक्वं तु खर्बुजं तृप्तिकारकं पौष्टिकं मतम्।
कफकृन्मूत्रलं बल्यं कोष्ठशुद्धिकरं गुरु।।
स्निग्धं सुस्वादु शीतं च वृष्यं दाहश्रमापहम्।
वातं पित्तं तथोन्मादं नाशयेदिति च स्मृतम्।।
कोमलं मधुरं तिक्तं किंचिदम्लं च तन्मतम्।
तत्तु वृद्धं च मधुरं रसे क्षारं तथाम्लकम्।।
रक्तपित्तं मूत्रकृच्छ्रं करोतीति बुधा जगुः।।
(निघण्टु रत्नाकर)

Pakvam tu kharbujam tṛptikārakam pauṣṭikam manam,
Kapha krnmūtralam balyam koṣṭhaśadhikaram gurū.
Snigdham susvadu śītam ca vṛsyam dāhaśrāmapaham,
Vātam pittam tathonmādam nāśayediti ca smṛtam.
Komalam madhuram tiktam kincidamlam ca tanmatam,
Tattu vṛddham ca madhuram rase kṣāram tathāmlakam.
Raktapittam mūtrakṛccham karotīti budha jaguḥ.
(Nighanturatnakara)

खर्बूजं मूत्रलं बल्यं कोष्ठशुद्धिकरं गुरु।
स्निग्धं स्वादुतरं शीतं वृष्यं पित्तानिलापहम्।।
(शालिग्रामनिघण्टु)

Kharbūjam mūtralam balyam kosthasuddhikaram gurū,
Snigdham svādutaram sītam vrsyam pittānilāpaham.
(Shaligramnighantu)

Actions/Uses : Seeds : Sītala, Mūtrajanana, Balya.

Fruit : Sītala, Mūtrajanana, Kuṣṭhtaghna.

Fruit (Old) : Raktapittavardhaka, Mūtrakṛcchakara.

Root : Recana, Vāmaka. Mūtrakṛcchaghna.

Therapeutics :

Kharbuja root is emetic and purgative. Fruit is eaten raw and cooked especially its pulp or juice forms a nutritive demulcent, diuretic and cooling drink. It is beneficial as a lotion in chronic and acute eczema as well as tan freckles and internally in cases of dyspepsia. The whole fruit is useful in chronic eczema. Alongwith other plant products it is used as antifertelity drug. Seeds are cooling, nutritive and diuretic. They yield a sweet edible oil which is nutritive and diuretic, useful in painful discharge, suppression of urine and urinary infection.

Cucumis momordica Roxb.

SYNONYMS : *Cucumis melo var. momordica Duthie & Fuller.*

INDIAN : Foota, Erwaaru.

SANSKṚTA : Ervāru.

HABIT : A climbing herb.

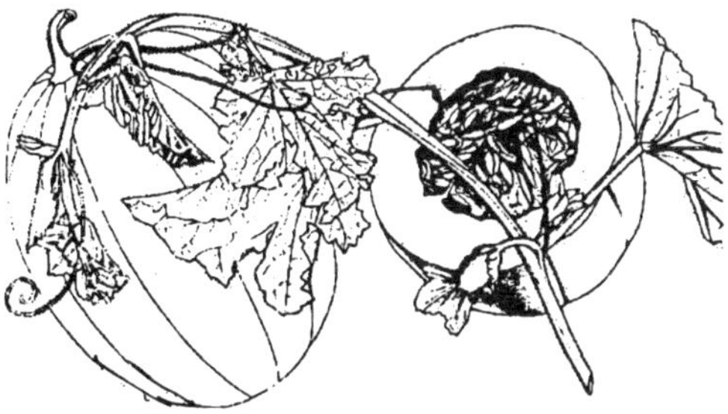

HABITAT : Cultivated in many parts of India.

CHEMICAL COMPOSITION : Seeds yield a sweet edible oil.

Parts used : Flowers, fruit and seed.

Ayurvedic properties :
Rasa - Madhura. Vīrya - Śīta.
Guṇa - Gurū, Rūkṣa. Vipāka -Madhura.

एर्वारुकं पित्तहरं सुशीतलं मूत्रामयघ्नं मधुरं रुचिप्रदम्।
सन्तापमूर्च्छापहरञ्च तृप्तिदं वातप्रकोपाय धनं तु सवितम्।!
(रा. नि.)

Ervārukam pittaharam śuśītalam,
Mūtrāmayaghnam madhuram rucipradam,
Santāpamūrcchapāharānca tṛptidam,
Vātaprakopāya ghanam tu sevitam.
(Rajanighantu)

Actions/Uses : Pittahara, Vātaprakopī in quantity.

Seeds : Dāhaśāmaka.

Therapeutics :

Flowers of Ervaru cause tridosha and dyspepsia. Unripe fruit is sweet, dry, astringent; cures kapha and biliousness; causes vaata. Ripe fruit is hot, causes biliousness. Seeds are used as a cooling medicine.

Cucumis sativus Linn.

ENGLISH : Cucumber.

INDIAN : Khiraa, Kaakadi.

SANSKṚTA : Trapusī, Pītapuṣpī, Kaṇṭālu, Bahuphalā, Tuṇḍiphala, Koṣaphalā, Muni, Trapusakarkaṭī, Sakusa, Sukasa.

HABIT : A trailing or climbing annual herb.

HABITAT : Cultivated in all parts of India.

CHEMICAL COMPOSITION : Fruits contain an enzyme erepsin, vitamin B1 and C; proteolytic enzymes, ascorbic acid, oxidase, succinic and malic dehydrogenases, rutin, palmitic, stearic and oleic acids, their glycerides, sterols and squalene. Seeds, Sitostero and stigmasta-7,22,25-trien-3B-ol, cucurbitaside. Leaves, free cucurbiasides B & C; ferredoxin.

Parts used : Fruit, seed and seed oil.

Ayurvedic properties :
Rasa - Madhura. Vīrya - Śīta.
Guṇa - Laghu, Rūkṣa. Vipāka - Madhura.

स्वादु पित्तापहं शीतं रक्तपित्तहरं परम्।
तद्बीजं मूत्रलं शीतं रुक्षं पित्तास्रकृच्छ्रजित्।।
(भा. प्र.)

Svādu pittāpaham śītam raktapittaharam param,
Tadbījam mūtralam śītam rūkṣam pittāsrakṛcchjit.
(Bhāvaprakāṣa)

Actions/Uses : Pittaśāmaka, Vātakaphavardhaka, Raktapittaśāmaka, Mūtrala, Tṛṣṇāhara, Mūtrakṛcchghna, Dāhaghna.

Therapeutics :

Medicinal properties of Trapusi are the same as those of C. melo var. utilissimuss. Fruit is nutritive and demulcent. Seeds are cooling, tonic and diuretic. Juice of fruit is a major constituent of anti-acne lotion. Pink eyes, sunburn and eye strain are relieved by application of cooling and refreshing slices to closed eyes. Plant is used in headache.

Cucumis utilissimus Roxb.

SYNONYMS : *Cucumis melo var. utilissimus Duthie & Fuller.*

INDIAN : Waalluka, Karkati.

SANSKṚTA : Karkaṭī, Vālungī.

HABIT : A climbing herb.

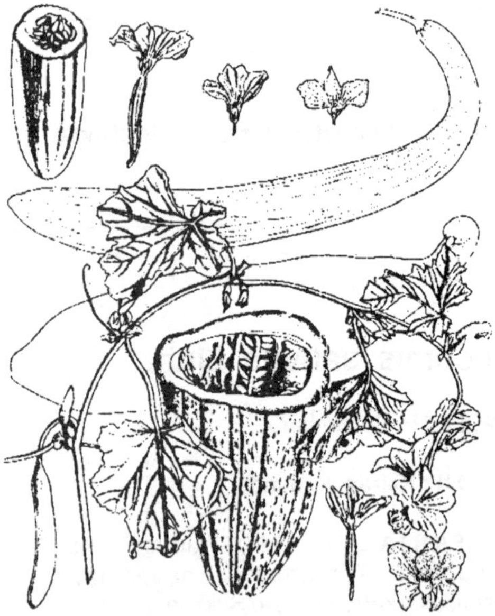

HABITAT : Cultivated in many parts of India.

CHEMICAL COMPOSITION : Seeds contain much farinaceous matter blended with a large proportion of a mild oil

Parts used : Fruit, seeds.

Ayurvedic properties :
Rasa - Madhura. Vīrya - Śīta.
Guṇa - Gurū, Rūkṣa. Vipāka - Madhura.

कर्कटी मधुरा शीता रुच्या लघ्वी च मूत्रला।
त्वचायां कटुका तिक्ता पाचकाग्नी प्रदीपनी।।
अवृष्या ग्राहिणी प्रोक्ता मूत्ररोधाश्मरी हरा।
मूत्रकृच्छ्रं वमिं दाहं श्रमं चैव विनाशयेत्।
सा पक्वी रक्तदोषस्य कारिण्युष्णा बलप्रदा।
(निघण्टु रत्नाकर)

Karkaṭī madhurā śītā rucyā laghvī ca mūtralā,
Tvacāyām kaṭukā tiktā pācakāgnipradīpanī.
Avṛṣyā grāhiṇī proktā mūtrarodhāśmarī harā.
Mūtrakṛccham vamim dāham śramam caiva vināśayeta,
Sā pakvā raktadoṣasya kāriṇyuṣṇā balapradā.
(Nighaṇturatnakara)

Actions/Uses : Śītala, Pācana, Pittanāṣana, Mūtrajanana.

Ash : Śleśmānissāraka.

Therapeutics :

Unripe fruit of Karkati causes vaata, kapha and flatulence. It is sweet, tasty, dry, cooling, diuretic and astringent. It cures biliousness. Ripe fruit is hot, tonic, stomachic; cures thirst, fatigue but causes biliousness and derangement of blood. Seeds are cooling, nutritive, diuretic; used in painful micturition and suppression of urine.

Cucurbita maxima Duch.

ENGLISH : Red quash gourd, Pumpkin.

INDIAN : Taambadda bhopallaa, Laal kumra, Mitha Kumra, Kadimah, Sitaphal, Daangar.

SANSKṚTA : Balābu, Pītakaṣmaṇḍū, Dangarī, Punyalatha, Dadhiphala.

HABIT : An trailing annual climbing herb with prickly or hairy stem.

HABITAT : Cultivated throughout India.

CHEMICAL COMPOSITION : Stigmasta-7,24(28)-dien-3B-ol and stigmast-7,25-dienol acetate and euglobunins from seeds. Seeds yield resin having vermicidal properties. Seed oil contains sterols and triterpenoids. Fruits, vitamin A.

Parts used : Fruit pulp, seeds and seed oil.

Ayurvedic properties :
Rasa - Madhura. Vīrya - Śīta.
Guṇa - Gurū. Vipāka -Madhura.

बलाबु शीतला रुच्या तृप्तिकृन्मधुरा स्मृता।
शोषं जाड्यं मूत्ररोधं दाहं रक्तरुजं तथा।।
(निघण्टु रत्नाकर)

Balābu śītalā rucyā tṛptikṛnmadhurā smṛtā,
Śoṣam jādyam mūtrarodham dāham raktarujam tathā.
(Nighanturatnakara)

Actions/Uses : Kaphakara, Grāhī, Raktapittahara, Kṛmighna.

Therapeutics :

Seeds of Pitakasmandu are anthelmintic, used as taenicide, diuretic and tonic. Fruit pulp is sedative, emollient and refrigerant and used as poultice, applied to burns, inflammations and boils. Useful in haemoptysis and pulmonary haemorrhage.

Plant is used in curing kapha, gonorrhoea and urinary diseases. Oil is nerve tonic.

Cucurbita pepo Linn.

ENGLISH : Vegetable marrow, Hungarian pumpkin.

INDIAN : Kaashiphalla, Pandharaa bhopalla, Kumra.

SANSKṚTA : Tunbī, Alābu, Dīrgha, Vartula, Kurlaru.

HABIT : An annual climbing herb.

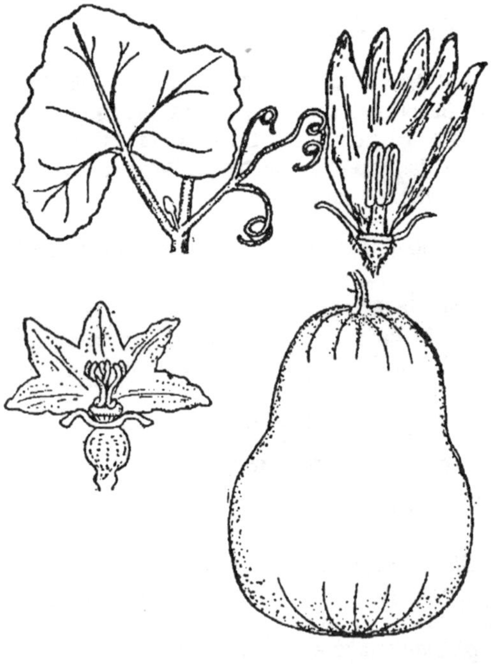

HABITAT : Native of America; naturalized and cultivated throughout India.

CHEMICAL COMPOSITION : Leaves and stem contain alkaloids. Flowers, flavonol glycosides. Vitamin C. Cucurrbitine, gibberellins and kaurenolide a resinous substance, phytosterarin and salicylic acid from seeds. Mannitol preparations. Analysis of fruit pulp for its nutritive value to suppliment South Indian diet.

Parts used : Seeds and leaves.

Ayurvedic properties :
Rasa - Madhura. Vīrya - Śīta.
Guṇa - Gurū, Snigdha, Sara. Vipāka - Madhura.

तुम्बी तु मधुरा स्निग्धा गर्भपोषणकारिणी।
वृष्या वातप्रदा बल्या पौष्टिकी रुच्य शीतला।।
मलस्तम्भकरी रुक्षा मेदका गुरु पित्तनुत्।
कफदाच फलं प्रोक्तं गुरु रुक्षंच शीतलम्।।
(निघण्टु रत्नाकर)

Tumbī tu madhurā snigdhā garbhapoṣanakāriṇī,
Vṛṣyā vātapradā balyā pauṣṭikī rucyaśītalā.
Malastambhakarī rūkṣā medakā gurū pittanut,
Kaphadāca phalam proktam gurū rūkṣamca śītalam.
(Nighanturatnakara)

Actions/Uses : Hṛdya, Balya, Rucya, Dhatupuṣṭivardhaka, Vṛṣya. Vatakara, Kaphakara, Garbhaposaka.

Therapeutics :

Fruit of Tumbi is cooling, astringent, increases appetite; cures leprosy, kapha and vaata, fatigue, purifies blood. Leaves are haematic, analgesic, remove biliousness. They are used as external application for burn. Seeds anthelmintic, useful as taenicide.

Diplocyclos palmatus Jeff.

SYNONYMS : *Bryonia laciniosa* Linn.; *Bryonopsis laciniosa (Linn.)* Naud.; *Bryonia palmata* Linn.

INDIAN : Shivalingee, Bajguriya.

SANSKṚTA : Linginī, Śivalingi, Baja, Bahupatra.

HABIT : An annual climbing herb.

HABITAT : Throughout India.

CHEMICAL COMPOSITION : Bryonin, the bitter principle. Seed oil a source of punicic acid.

Parts used : Whole plant, roots, leaves and seeds.

Ayurvedic properties :

Rasa - Kaṭu.	Vīrya - Uṣṇa.
Guṇa - Laghu, Rūkṣa.	Vipāka - Kaṭu.

लिंगिनी कटुष्णा च दुर्गन्धा च रसायनी।
सर्वासिद्धिकरा दिव्या वश्या रसनियामनी॥

Linginī kaṭuṣṇā ca durgandhā ca rasāyanī,
Sarva siddhikarā divyā vaśyā rasaniyāmanī.
(Nighantuadarsa)

Actions/Uses : Recana, Pittaśamana, Jvaraghna.

Therapeutics :

Lingini or Baja is bitter, aperient, tonic, used in billious attack, in fevers with flatulence. Leaves as external application for inflammation. Powdered seeds given to help conception in women; roots also for same purpose. Seed and plant serve as tonic and aphrodisiac. Various parts of plant used in headache, ague, colic pain, enlarged spleen, paralysis of tongue, delirium and convulsions, foaming at mouth, sores, adenitis, scrofulosa colli, syphilis, carbuncle, stomach swelling or tumor, constipation, phthisis and snakebite.

Lagenaria siceraria Standl.

SYNONYMS : *Lagenaria vulgaris* Seringe.; *Lagenaria leucantha* Rusby.

ENGLISH : Bitter bottle gourd, Calbash gourd.

INDIAN : Kadu bhopalla, Lauki tumri, Dudhyaa.

SANSKṚTA : Kaṭutumbi, Iśvāku, Piṇḍaphala, Dantabījā, Tiktālābū, Kati-tumbi, Tikta-tumbī, Alabu.

HABIT : A large, pubescent climbing or trailing herb.

HABITAT : African or Asian origin. Cultivated throughout India.

CHEMICAL COMPOSITION : A good source of vitamin B and ascorbic acid. The fruit is rich in pectin. Bitter fruit contains cucurbitacins B, D, G and H. Two triterenoids from fruits. Rhamnose, fructose, glucose, galactose, sucrose, raffinose and stachyose from seeds. Saponin and fatty oil. Palmitic, palmitoleic, oleic stearic and linoleic acids from seed oil.

Parts used : Root, leaves seeds, seed oil and fruit.

Ayurvedic properties :

Rasa - Tikta. Vīrya - Śīta.
Guṇa - Laghu, Rūkṣa. Vipāka - Kaṭu.

कटुतुम्बी कटुस्तीक्ष्णा वान्तिकृच्छ्वासवातजित्।
कासघ्नी शोधनी शोफव्रणशूलविषापहा।।
(रा. नि.)

Kaṭutumbī kaṭustīkṣṇā vāntikṛcchvāsavātajit,
Kāsaghnī śodhanī śophavraṇaśūlaviṣāpahā.
(Rajanighantu)

Actions/Uses : Kaphapittaśodhana, Tīvra-vāmaka, Kṛmighna, Bhedana, Śirovirecana, Kuṣṭhaghna.

Therapeutics :

Katitumbi fruit is bitter, hot, acrid, cooling, cardiotonic and emetic. It cures oedema, pain, ulcers, cough, asthma and other bronchial disorders. Pulp, emetic and purgative, applied to the soles in burning of feet. Decoction of leaves, mixed with sugar given in jaundice. Seeds are nutritive and diuretic, are used in dropsy and as anthelmintic; roots also in the treatment of dropsy. The seed oil is coolin and is applied in headache.

Luffa acutangula Roxb.

SYNONYMS : *Luffa amara* Clarke.; *Luffa amara* Roxb.

ENGLISH : Ridge Gourd, Ribbed gourd.

INDIAN : Kadu dodakaa, Shiraalle, Kali tori, Jhinga.

SANSKṚTA : Kośātakī, Kṛtachidrā, Jālinī, Kśevād, Jhongaka, Cali, Sutiktā, Ghanṭalī, Mṛdangaphalinī, Śvetadhāpā, Kośavati, Jvālā, Arkontī, Dhāmārgavā, Ghoṣakā, Tiktaghoṭālī, Kaṭudalinī, Jyotsnā, Kṛtavedanā, Mṛdangikā, Vedibhra.

HABIT : A large climber.

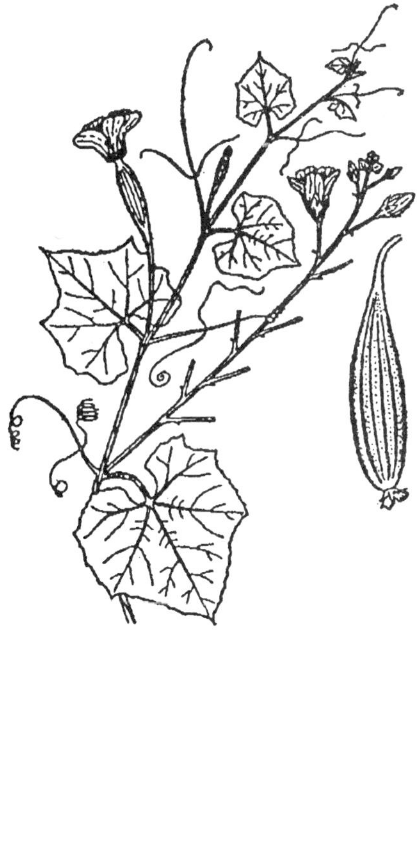

HABITAT : Throughout India, especially Western Peninsula.

CHEMICAL COMPOSITION : The fruit contains free aminoacids; arginine, glycine, threonine, glutamic acid, leucines. Ripe seeds contain bitter principles- cucurbitins B, D, G, and H. Amarin identified as Cucurbitacin B. Seeds contain 20 % of a saponin glycoside, enzyme and a fixed oil.

Parts used : Leaves and seeds

Ayurvedic properties :
Rasa - Kaṭu, Tikta. Vīrya - Śīta.
Guṇa - Laghu, Rūkṣa, Tīkṣṇa. Vipāka - Kaṭu.

अत्यर्थम् कटुतीक्ष्णोष्णं गाढेष्विष्ट गदेषु च।
कुष्ठपाण्ड्वामय प्लीहशोथ गुल्मगरादिषु।।
(च.क. ६)

Atyartham kaṭutīkṣṇoṣṇam gāḍheṣvisṭam gadeṣu ca,
Kuṣṭhapāṇḍvāmayaplīhaśothagulmagarādiṣu.
(Carakasanhita Kalpasthanam)

Actions/Uses : Kaphapittaśodana, Vamana, Virecana, Gulmahara, Yakṛtaplihā-vṛdhihara, Kāsahara, Śvāsahara.

Therapeutics :

Fruit of Kośātakī is demulcent, diuretic, bitter-tonic and nutritive. Fruit juice is bitter. Seeds, especially the ripe ones have emetic, expectorant and purgative properties. Juice of fresh leaves dropped into the eyes in granular conjuctivitis. Pounded leaves are applied to splenitis, haemorrhoeds and leprosy. Decoction of leaves is used in uraemia and amenorrhoea.

Melothria maderaspatana (Linn.) Cogn.

SYNONYMS : *Mukia maderaspatana (L.)* Roem.; *Mukia scabrella* Arn.; *Bryonia scabrella.; Cucumis maderaspatanus* Linn.

INDIAN : Ghugree, Chiraati, Agumaki, Bilari.

SANSKṚTA : Kośātakī, Ghanṭālī, Ahilekhana, Musimusikkayi.

HABIT : A prostrate or climbing annual herb.

HABITAT : Throughout India.

CHEMICAL COMPOSITION : Columbin isolated from roots.

Parts used : Leaves seeds and seed oil.

Ayurvedic properties :
Rasa - Kaṣāya. Vīrya - Śīta.
Guṇa - Laghu. Vipāka - Kaṭu.

कोषातकी तु शिशिरा कटुकाऽल्पकषायका।
पित्तवात कफघ्नी च मलाध्मान विशोधिनी।।
(राजनिघण्टु)

Kosātakī tu śiśirā kaṭukā alpa kaṣāyakā,
Pitta vāta kaphaghnī ca malādhmāna viśodhinī.
(Rajanighantu)

Actions/Uses : Vātapittakaphanāṣana, Mṛduvirecana, Mūtrajanana.

Therapeutics :

Kośātākī has expectorant properties and is prescribed against chronic diseases with cough as a predominant symptom. Decoction of root useful in flatulence and root is masticated for relief of toothache. Tender shoots and leaves are used as a gentle aperient and recommended in vertigo and billiousness. Decoction of seeds is sudorific; crushed and applied on aching bodies, especially on strained backs. The drug is an ingredient of some compound preparations for chronic diseases with cough as a predominant symptom.

Momordica charantia Linn.

ENGLISH : Carella fruit, Bitter gourd.

INDIAN : Kaaralle.

SANSKṚTA : Kāravelli, Angāravellī, Suṣavī, Pacāru, Bāhuvallī, Mandapī, Sukumārī, Kathillaka, Cīrīpatra, Karakā, Phalātmikā, Cīrītacchadā, Sukṣmavallī, Aphalā, Kantaphalā, Pītapuṣpā, Akhilī, Ambuvallikā, Toyavallī, Rājavalli.

HABIT : An annual slender climbing herb.

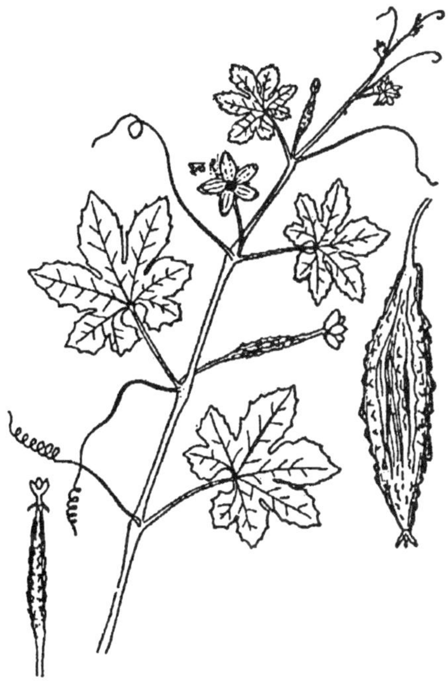

HABITAT : Cultivated throughout India.

CHEMICAL COMPOSITION : Contains alkaloid mormoridicine, saponin, carotene, glucosides and highly aromatic essential oil. Fruit contains a hypoglycaemic substance charantin and β-sitosterol-glucoside and stigmast-5,25-diene-3B-O-glucoside. Fruit juice orally administered showed activity in normal and diabetic rabbits and may have clinical usefulness in diabetes.

Parts used : Roots, leaves and fruit.

Ayurvedic properties :
Rasa - Tikta, Kaṭu. Vīrya - Ūṣṇa.
Guṇa - Laghu, Rūkṣa. Śīta. Vipāka - Kaṭu.

कारवेल्लं सकटुकं कटुपाकमवातलम्।
दीपनं भेदनं तिक्तं अवृष्यं अहिमं लघु।।
हन्त्यरोचक मित्तास्त्रकफपाण्डुव्रण क्रिमीन्।
श्वासकासप्रमेहार्शः कोठकुष्ठज्वरानपि।।
(कै. नि.)

Kāravellam sakaṭukam kaṭupākamadātalam,
Dīpanam bhedanam tiktam avṛṣyam ahimam laghu.
Hantyarocakamittāsra kaphāpaṇḍuvraṇakrimīn,
Śvāsa kāsa pramehārśaḥ koṭhakuṣṭhajvarāmapi.
(Kaiyadevanighantu)

Actions/Uses : Kaphapittaśamana, Kaphapittaghna, Rocana, Dīpana, Pācana, Bhedi, Śothahara, Yakṛtarogahara, Jvaraghna, Kuṣṭhaghna, Pānduhara, Pramehaghna, Kṛmihara, Vraṇahara, Kāsahara, Śvāsaghna.

Therapeutics :

Kāravellī is a reputed medicine for diabetes. Root is astringent and used in ophthalmia and prolapsus vagina; also useful in haemorrhoeds. Juice of leaves, emetic, purgative; given in bilious affections; rubbed in burning of soles of feet. Fruit and leaves, anthelmintic, useful in piles, leprosy, jaundice and as vermifuge. Fruit is very bitter and is eaten as a vegetable. It is carminative, tonic, stomachic, stimulant, emetic, anthelmintic, antibilious, laxative, cardiotonic, digestive stimulant and alterative. It is useful in cough, respiratory diseases, fever, intestinal worms, skin diseases, and poisonous affection; also for gout, rheumatism and subacute cases of spleen and liver. It improves digestion, calms down sexual urge, cures anaemia, anorexia, leprosy, ulcers, jaundice, flatulence and piles. Juice of fruit in snake bite. Powdered fruit is useful in healing wounds, psorous, intractable and malignant ulcers, a property atributed to proteolytic enzymes it contains.

Momordica dioica Roxb.

SYNONYMS : *Momordica subangulata Bl.*

INDIAN : Karattoli, Kaksa, Goljandra.

SANSKṚTA : Karkotakī, Vahisi.

HABIT : A perennial dioecious climber with tuberous roots.

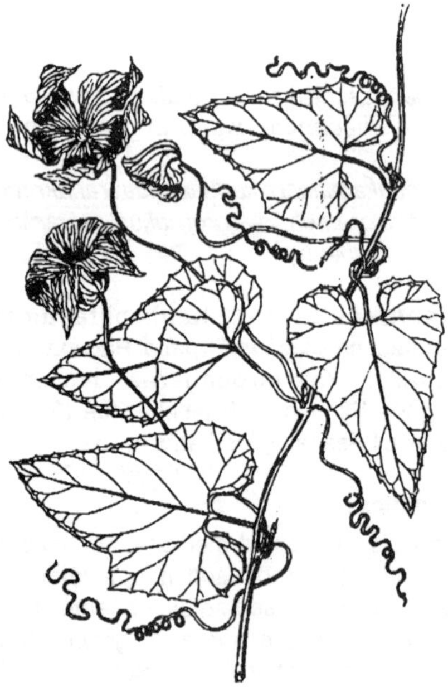

HABITAT : Throughout India.

CHEMICAL COMPOSITION : An alkaloid, a fragrant extractive matter and ash 3 to 4 per cent.

Parts used : Roots and ash.

Ayurvedic properties :
Rasa - Kaṭu, Tikta, Madhura. Vīrya - Ūṣṇa.
Guṇa - Laghu. Vipāka - Kaṭu.

कर्कोटकी रुचिकरा कट्वी चाग्निप्रदीपनी।
तिक्तोष्णा वातकफहृद्विषं पित्त विनाशयेत्।।
फलमस्यास्तु मधुरं लघु पाके कटु स्मृतम्।
कफकुष्ठ कासमेहश्वासज्वला किलासनुत्।
(निघण्टु रत्नाकर)

Karkoṭakī rucikarā kaṭvī cāgnipradīpanī,
Tiktoṣṇā vātakapha hṛdviṣam pitta vināśayet.
Phalamsyāstu madhuram laghu pāke kaṭu smṛtam.
(Nighanturatnakara)

Actions/Uses : Kapha-Vātanaśaka, Dīpana, Raktarśaghna, Vāmaka, Madhumehaghna.

Therapeutics :

Karkotakī is stimulant and astringent. Juice of root is antiseptic. Powder of root applied to skin renders it soft and supple and lessens perspiration. Roots toasted and used to stop bleeding from piles; used in urinary complaints; ground to paste smeared over the body as a sedative in high fever with delirium; used in snake bite and scorpion sting; juice is used as antiseptic. Powder or infusion of dried fruits, if introduced into nostril, produces a powerful errhine effect, and provokes a copious discharge from schneiderian mucous membrane.

Trichosanthes cucumerina Linn.

SYNONYMS : *Pentatrois microphylla W. & A.*

ENGLISH : Wild snake gourd.

INDIAN : Kaddu paddawalla, Raanpaddawalla, Patola.

SANSKṚTA : Paṭola-1, Tikta-paṭoli,

HABIT : An annual climbing herb.

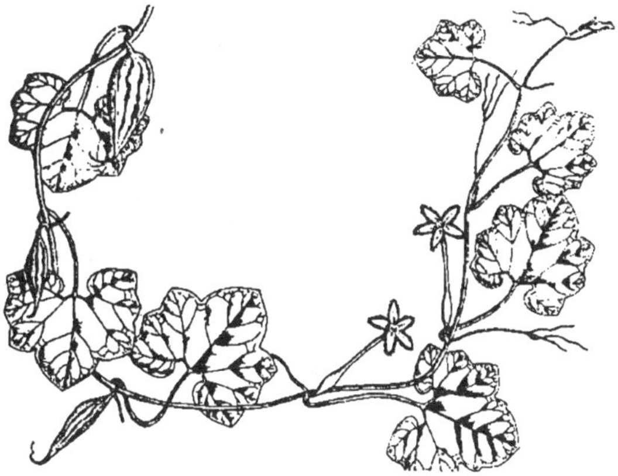

HABITAT : Throughout India.

CHEMICAL COMPOSITION : Fatty acids of seed fat.

Parts used : Whole plant, roots, leaves and fruits.

Ayurvedic properties :
Rasa - Tikta, Kaṭu, Madhura. Vīrya - Ūṣṇa.
Guṇa - Laghu, Snigdha. Vipāka - Madhura.

तिक्ता पटोली कटुका सारकोष्णा च तिक्तका।
भेदनी पाचनी चैव अग्निदीप्तिकरी परा।।
पित्तं कफं च कण्डूं च कुष्ठं चैवोपदंशकम्।
ज्वरं दाह तृषां कोठरोगं नाशयति क्रिमीन्।।
अस्याः फलं तिक्तकटु पाके स्वादु लघु स्मृतम्।
पित्तनाशकरं पर्णमूलं कफविनाशकम्।
कफनाशकरी वल्ली तैलं वातकफाकहम्।।
(नि. रत्नाकर)

Tiktā paṭolī kaṭukā sārakoṣṇā ca tiktakā,
Bhedanī pācani caiva agnidīptikarī parā.
Pittam kapham ca kaṇḍūm ca kuṣṭham caivopadanśakam,
Jvaram dāham tṛṣām koṭharogam nāśayati krimīn.
Asyāḥ phalam tiktakaṭu pāke svādu laghu smṛtam.
Pittanāśakaram parṇam mūlam kaphavināśakam,
Kaphanāśakarī vallī tailam vātakaphākaham.
(Nighanturatnakara)

पटोलं कटुकं तीष्णमुष्णं पित्ताविरोधीच।
कफासृक्कण्डूकुष्ठानि ज्वरदाही च नाशयेत्।।
पटोलं पाचनं हृद्यं वृष्यं लघ्वग्निदीपनम्।
स्निग्धोष्णं, हन्ति कासास्रज्वरदोषत्रयकृमीन्।।
पटोलस्य भवेन्मूलं विरेचनकरं सुरवात्।
नालं श्लेष्महरं पत्रं पित्तहारि फलं पुनः।।
(निघण्टु आदर्श)

Paṭolam kaṭukam tīkṣṇamuṣṇam pittavirodhī ca,
Kaphāsṛk kaṇḍūkuṣṭhāni jvaradāhī ca nāśayet.
Paṭolam pācanam hṛdyam vṛṣyam laghvagnidīpanam.
Snigdhoṣṇam hanti kāsāsrajvaradoṣatraya kṛmīn.
Paṭolasya bhavenmūlam virecanakara sukhāt,
Nālam śleṣmahara patram pittahāri phalam punaḥ.
(Nighantu Adarsya)

Actions/Uses : Pācana, Agnidīpana, Viryavardhaka.

Therapeutics :

Tikta-Paṭola is a general and cardioc tonic, alterative, febrifuge and antipyratic; useful for boils and intestinal worms. Root is purgative and tonic. Juice of roots is cathartic. Infusion of tender shoots and dried capsules is aperient. Juice of leaves emetic. Leaves and stems used in decoction for bilious disorders, skin diseases and emmenagogue. Fruit is bitter and laxative. Seeds are antifebile, anthelmintic; good in disorder of stomach.

Trichosanthes dioica Roxb.

ENGLISH : Pointed gourd.

INDIAN : Parvar, Potala.

SANSKṚTA : Paṭola-2, Putulika, Kulaka.

HABIT : A dioecious climber with a perennial rootstock.

HABITAT : Througout India especially in Gujarat.

CHEMICAL COMPOSITION : Roots contain an amorphous saponin, hentriacontane, a phytosterol a nonnitrogenous bitter principle, glucosidic in nature and resembling colocynth, small amount of essential oil, little fixed oil and traces of tannins.

Parts used : Whole plant, roots, leaves and fruit.

Ayurvedic properties :
Rasa - Madhura, Kaṭu, Tikata. Vīrya - Uṣṇa.
Guṇa - Laghu, Snigdha. Vipāka - Kaṭu.

पटोली बालकृत्स्वादुः पथ्या दीपन पाचनी।
रुच्या पुष्टिकरी ज्ञेया वातपित्तं ज्वरापहा।।
शोषत्रिदोषशमनी फलं वृष्यं रुचिप्रदम्।
मधुरं स्वादु पथ्यं च पाचकं लघु दीपकम्।।
हृद्यं स्निग्धं तथोष्णं ञ्च कफरक्तत्रिदोषनुत्।
कासज्वरक्रिमीन्हान्ति पर्णं वै पित्तनाशकम्।।
मूलं रेचकरं प्रोक्तं वल्ली चैव कफापहा।
(निघण्टु रत्नाकर)

Paṭolī balakṛtsvāduh pathyā dīpanapācanī,
Ṛucya puṣṭikarī jneyā vātapitta jvarāpahā.
Śoṣa tridoṣaśamanī phalam vṛṣyam rucipradam,
Madhuram svādu pathyam ca pācakam laghu dīpakam.
Hṛdyam snigdham tathoṣṇam ca kapharakta tridoṣanut,
Kāsa jvarakrimīnhanti parnam vai pittanāśakam.
Mulam recakaram proktam vallī caiva kaphāpahā.
(Nighanturatnakara)

Actions/Uses : Kapha-pittahara, Dipan, Pacana, Vṛṣya

Therapeutics :

Paṭola-2 is a good blood purifier and is beneficial in skin diseases. Plant is bitter, acrid in taste, alterative, tonic, hot in action, appetiser, digestive, germicidal, laxative and aphrodisiac. It aids digestionm. It cures haemetemesis, dermatosis, fever, cough bronchitis and ulcers. Root is tonic, hydragogue cathartic and febrifuge. Leaves are antipyretic, anthelminti, aphrodisiac; cure biliousness. Decoction of leaves with equal parts of coriander given in billious fever as a febrifuge and laxative. Juice of leaves and fruit is a cholagogue and aperient. Fruit is febrifuge, laxative and antibilious, and is used as a remedy for spermatorrhoea. Fresh juice of unripe fruit used as a colling and laxative, adjunct to alterative medicines.

Trichosanthes tricuspidata Lour.

SYNONYMS : *Trichosanthes bracteata (Lam.) Voigt.; Trichosanthes palmata* Roxb.; *Modecca bracteata* Lamk.

INDIAN : Kaundalla, Mahaakaala, Laal indraayan.

SANSKṚTA : Mahākāsa, Śvetapuṣpi.

HABIT : A large perennial, dioecious climber.

HABITAT : Throughout India in moist situations.

CHEMICAL COMPOSITION : Cyclotrichosantol and cycloeucalenol from leaves. Trichotetrol from roots.

Parts used : Fruit and seed oil.

Ayurvedic properties :
Rasa - Tikta. Vīrya - Uṣṇa.
Guṇa - Śīta. Vipāka - Kaṭu.

कण्ठरोगापचीश्वासकासप्लीह कफोदरान्।
मूढगर्भच हरती कुष्ठदुष्टव्रणाञ्जयेत्।।
(धन्वतरी)

Kaṇṭharogāpacī śvāsakāsa plīha kaphodarān,
Mūḍhagarbhaca haratī kuṣṭhaduṣṭa vraṇañjayet.
(Dhanvantari)

Actions/Uses : Vāmaka, Recaka, Sothaghna, Kaphahara, Kāsahara, Mūḍhagarbha-nivāraka.

Therapeutics :

Fruit of Mahākāsa is useful in asthma, earache, and ozoena. It is smoked in asthma and lung diseases. It is bitter, carminative, a violent hydrogogue cathartic, abortifacient; lessens inflammation; cures hemicrania, weakness of limbs, ophthalmia, leprosy; used in epilepsy, rheumatism. Oil obtained by boiling it in coconut is a remedy for earache, sore in ears and nostrils and ozoena in which it is instilled in drop; applied to scalp to cure hemicrania.

Family : XL - Cyperaceae.

Cyperus rotundus Linn.

SYNONYMS : *Cyperus hexastachys* Rottb.

ENGLISH : Nut-grass.

INDIAN : Naagarmotha, Mothaa, Mustaa.

SANSKṚTA : Mustaka, Vārida, Ghana, Cakrānkā, Nādeyī, Jalada, Ambuda, Pinḍamustā, Cūḍālā, Śiśirā, Bhadramustā, Kacchahvā, Kalāpinī, Anukesarocchātā.

HABIT : A pestiferous perennial weed with dark green glabrous culms.

HABITAT : Throughout India, common in waste grounds and gardens.

CHEMICAL COMPOSITION : Cyperene-1 and cyperene-2, cyperotundone, from tubers. Cyperene, β-selinene, cyperenone and α-cyperone from rhizomes. Essential oil contains mustakone. Root extract possessed tranquilising activity. It showed significant antipyretic and antiinflamatory acticities.

Parts used : Tuber or bulbous root.

Ayurvedic properties :

Rasa - Kaṭu, Tikta, Kaṣāya.	Vīrya - Śīta.
Guṇa - Laghu, Rūkṣa.	Vipāka - Kaṭu.

मुस्तं हिमं कटुग्राहि तिक्तं दीपनपाचनम्।
कषायं कफापित्तास्त्रतृड्ज्वरारुचिजन्तुजित्।।
(भा. प्र.)

Mustam himam kaṭu grāhi tiktam dīpanapācanam,
Kaṣāyam kaphapittāsra tṛd jvarārucijantujit.
(Bhāvaprakāṣa)

Actions/Uses : Pittaghna, Kaphaghna, Grāhī, Vātavardhaka, Dīpana, Pācana, Raktapittahara, Tṛṣṇāhara, Pittajvaragna, Arucihara, Jantughna, Atisaraghna, Krmighna.

Therapeutics :

Mustaka tuber is a home remedy for indigestion, sprue, diarrhoea and other intestinal problems of children. It is light, bitter, aromatic, astringent, demulcent, carminative, diuretic, anthelmintic, vermifuge, diaphoretic, galactagogue, emmenagogue and nervine tonic. Useful in disorders of stomach and irritation of bowels. An infusion or soup of tubers is useful in diarrhoea, dysentery, dyspepsia, vomiting, cholera, and fever. Root paste is applied to wounds, sores etc. and used in intestinal diseases.

Cyperus scariosus R. Br.

INDIAN : Lawala, Naagarmothaa (2).

SANSKṚTA : Bhadramustā, Nāgaramustaka, Musta, Amoda.

HABIT : A delicate slender sedge with deep brown tubers with aromatic odour.

HABITAT : Throughout India, especially in Gwalior State; damp places in Bengal, U.P.

CHEMICAL COMPOSITION : Essential oil, from tuber, sesquiterpenoid, cyperenone; cyperenol and patchoulenol, isopatchoulenone. Rotundene and rotundenol from oil; in addition to (-) β-selinene, isopathoula-3,5-diene. Leaves, glycoside of leptosidin.

Parts used : Tubers and oil.

Ayurvedic properties :
Rasa - Kaṭu, Tikta, Kaṣāya. Vīrya - Śīta.
Guṇa - Laghu. Vipāka - Kaṭu.

भद्रमुस्ता तु तुवरा शीता तिक्ता च पाचका।
कट्वग्निदीपनी ग्राही चाम्लपित्तकफापहा।।
अतिसारं रक्तदोषं ज्वरं चैव विनाशयेत्।
अरुचिं च तृषां चैव कृमीनपि विनाशयत्।।
(नि. रत्नाकर)

Bhadramustā tu tuvarā śīta tiktā ca pācakā,
Katvagnidīpanī grāhī cāmlapittakaphāpahā.
Atisāram raktadoṣam jvaram caiva vināśayet,
Arucim ca tṛṣām caiva krmīnapi vināśayet.
(Nighanturatnakara)

Actions/Uses : Śītal, Dīpan, Pācan, Vātānuloman, Grāhī, Mūtral, Svedahanan, Kaphagna, Medhya, Styanyajanan, Stanyaśodhan.

Therapeutics :

Tubers of Nāgaramustaka are aromatic, diuretic, astringent, cordial, stomachic and diaphoratic and are useful in diarrhoea and as a desiccant used for washing hair. Decoction is used in gonorrhoea and in syphilitic affections. Essential oil from tuber is hypotensive and CNS stimulant.

Family : XLI-Dipterocarpaceae.

Vateria indica Linn.

SYNONYMS : *Vateria malabarica Beddome.*

ENGLISH : Indian Copal, White dammar, Piṇey varnish.

INDIAN : Dhupa, Raalla, Chandrusa, Safetadamara, Kahruba.

SANSKṚTA : Sarja, Śāla, Aśvakarṇaka, Sasyasanvara, Ajakarṇa.

HABIT : A large, elegant, evergreen tree.

HABITAT : Western India, chiefly in evergreen forests, but also occasionally along rivers in deciduous forests.

CHEMICAL COMPOSITION : Damar resin. 1.06% essential oil in fruit. dl-Epicatechin, fistinidol and afzelechin from bark. Seeds yield a semi-solid fat, known as PINEY TALLOW or DHUPA FAT.

Parts used : Oil and Resin.

Ayurvedic properties :
Rasa - Tikta. Vīrya - Ūṣṇa.
Guṇa - Snigdha, Śīta. Vipāka - Kaṭu.

सर्जस्तु तुवरस्तिक्तः हिमः स्निग्धोऽतिसारजित्।
पित्तास्त्रदोषकुष्ठघ्नः कण्डूविस्फोटवातजित्।।
(रा. नि.)

Sarjastu tuvarastikaḥ himaḥsnigdho atisārajit,
Pittāsradoṣa kuṣṭhaghnaḥ kaṇḍūvisphoṭavātajit.
(Rajanighantu)

Actions/Uses : Atisārajita, Raktapittaśāmaka, Kuṣṭhagna, Snehana, Kaṇḍughna, Uttejaka, Vedanāsthapana.

Therapeutics :

Sarja Resin has tonic, carminative and expectorant properties and is used in several diseases such as throat troubles, chronic bronchitis, piles, diarrhoea, rheumatism, tubercular glands, boils, etc. Mixed with sesamum oil it is given in gonorrhoea, and with ghee, long-pepper for the treatment of syphilis and ulcers. Fatty oil from fruit used as a local application in chronic rheumatism.

Family : XLII - Ebenaceae.

Diospyros peregrina Gurke.

SYNONYMS : *Diospyros embryopteris Pers.; Diospyros malabarica (Desv.) Kostel ex Baker & Bakhuizen.f.; Carcinia malabarica Desv. Diospyros ebenum Koenig.*

ENGLISH : Riber eboney, Indian persimon.

INDIAN : Tembhurnnee, Tenddu, Timburi.

SANSKṚTA : Tinduka, Nīlasāra, Kālaskandha, Atimuktaka, Sphūrjaka, Tuṣta, Spandana, Rāmaṇa, Rava, Asitakāraka, Gābha, Tendū, Gāva, Ṭ imbarabī, Pānicikā, Tumika, Kṛṣṇasara, Virupaka.

HABIT : A middle sized evergreen tree.

HABITAT : Throughout India.

CHEMICAL COMPOSITION : Fruit pulp contains β-sitosterol and its glucoside, betulin, a monohydroxy triterpene ketone, betulinic acid, gallic acid, hexacosane, hexacosanol. Oleanolic acid and myricyl alcohol from leaves. Saponin from bark.

Parts used : Bark, fruit, seed and seed oil.

Ayurvedic properties :
Rasa - Kaṣāya.
Vīrya - Śīta.
Guṇa - Laghu, Rūkṣa. Vipāka - Kaṭu.

तिन्दकस्तु कषायः स्यात् संग्राहि वातकृत् परः।
पक्वस्तु मधुरः स्निग्धो दुर्जरः श्लेष्मलो गुरुः।।
(राजनिघण्टु)

Tindukastu kaṣāyaḥ syāt sangrāhi vātakṛt paraḥ.
Pakvastu madhuraḥ snighdo durjjaraḥ śleṣmalo guruḥ.
(Rajanighantu)

स्यादामं तिन्दुकं ग्राहिवातलं शीतलं लघु।
पक्वं पित्तमेहास्त्रश्लेष्मघ्नं मधुरं गुरु।।
(भा. प्र.)

Syādāmam tindukam grāhivātalam śītalam laghu,
Pakvam pittamehāsraśleṣmaghnam madhuram gurū.
(Bhāvaprakāṣa)

Actions/Uses : Kaphapittaśāmaka, Śothahara, Stambhana.

Therapeutics :
Tindduka plant or its parts are used in gravel, choleral menorrhagia and rinderpest. Bark is astringent and styptic and used in dysentery and intermittent fevers. A paste made from the bark is applied to boils and tumors. Unripe fruit is bitter, acrid, astringent, styptic and oleaginous. It is pounded and its extract used in diarrhoea. Chewing of fruit cures blisters in mouth. Infusion of fruits is used as gargle in aphthae and sore throat. Fruit juice is used as application for wounds and ulcers. An infusion of rind of fruit is useful in chronic dysentery and diarrhoea. Oil of seeds is given in diarrhoea and dysentery.

Family : XLIII - Euphorbiaceae.

Acalypha indica Linn.

SYNONYMS : *Acalypha spicata.; Acalypha ciliata.*

ENGLISH : Indian acalypha.

INDIAN : Khokalee, Kuppee, Khajoti, Diwali.

SANSKRTA : Haritamanjari, Rudra, Muktavarcas, Arittanunjaryie, Kokila, Kakali, Kutsyaparni, Vrsodari, Svetasunthi, Murakanta.

HABIT : An annual erect herb.

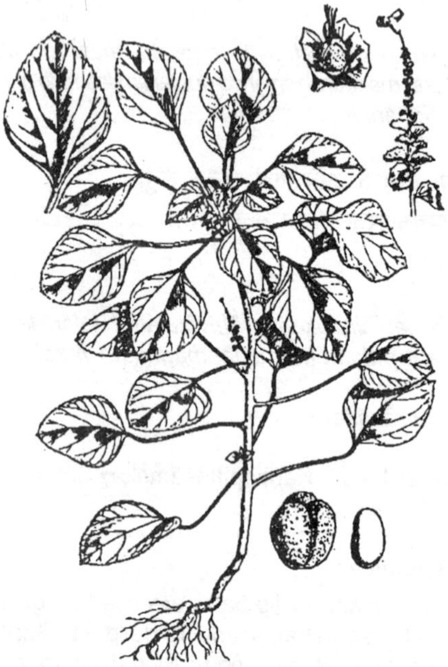

HABITAT : Throughout the plains of India as a weed.

CHEMICAL COMPOSITION : Plant contains kaempferol, sitosterol, triacetonamine. Leaves and twigs, acalyphine, acalyphamide and other amides, quinone, sterols and cyanogenic glycoside. Active principle HCN and an unknown subtance, extremely poisonous to rabbits; causes discoloration of blood and gastro-intestinal irritation.

Parts used : Herb, leaves and root.

Ayurvedic properties :
Rasa - Katu, Tikta. Virya - Usna.
Guna - Laghu, Ruksa. Vipaka - Katu.

कासश्वासहरी तिक्ता कफघ्नी कटुका तथा।
रेचनी मूत्रळा चैव ज्वरवातहरा स्मृता।।
तत्पत्रस्वरसं प्रोक्तं कर्णशूलहरं परम्।
पत्रकल्कस्य वर्तिस्तु बालानां मलबन्धाहृत्।।
(स्व.)

Kasasvasahari tikta kaphaghni katuka tatha,
Recani mutrala caiva jvaravatahara smrta.
Tatpatrasvarasam proktam karnasulaharam param,
Patrakalkasya varttistu balanam malabandhahrt.
(Sva.)

मुक्ता वर्चास्तथा रुदा वान्तिकृच्च विरेचनी।
कासश्वासगरघ्नी च ज्वरहृत् कफवातनुत्।।
एतस्याः स्वरसः पीतः कफोत्सारी च वामनः।
पायुलेपात मलोत्सारी कल्को वालेषु युज्यते।।
(द्रव्यगुण)

Muktavarcastatha rudra vantikrcca virechini,
Kasasvasagaraghni ca jvarahrt kaphavatanut.
Etasyah svarasah pitah kaphotsari ca vamanah,
Payulepat malotsari kalko balesu yujyate.
(Dra.)

Actions/Uses : Kasasvasahari, Kaphagna, Mutrala, Recani, Vamaka, Jvaravatahara, Virecani.

Therapeutics :

Haritamanjari plant is bitter, acrid and possesses diuretic, cathartic, expectorant, emetic, anthelmintic, anodyne and hypnotic properties. But it causes gastro-intestinal irritation. It is used as a substitute for ipecac and senega. A decoction of the herb is used as a cure for tooth and ear ache and is safe and speedy laxative. It is useful in bronchitis, pneumonia and asthma. Root is cathartic. Leaves are laxative, used in scabies and in snake bite. Fresh leaf juice useful in rheumatoid arthritis and skin affections. Juice with salt applied on eczema. Paste of leaves applied on burns; with juice of lime, useful in early cases of ringworm. Powder of leaves for bed-sores and maggot-infested wounds. Extract found useful in cardinal symptoms of wheezing cough; the raised eosinophil count also came down to normal.

HOMEOPATHIC USE : Herb is used as a remedy for severe cough associated with bleeding from lungs, haemoptysis and incipient phthisis. It has produced a number of symptoms in the alimentary tracts characterised by burning sense of weight in stomach, flatulence and spduttering diarrhoea.

Baliospermum montanum Muell.-Arg.

SYNONYMS : *Croton polyandrus* Roxb.; *Jatropha montana* Willd.; *Baliospermum polyandrum* Wight.; *Baliosoermum axillare* Blume.

HABIT : A stout, usually monoecious woody perennial shrub.

HABITAT : Distributed in tropical and subtropical Himalayas and into peninsular India in hills of Kerala.

INDIAN : Jamal Gota, Dantee.

SANSKṚTA : Dantī, Śīghrā, Śyenaghanṭā, Nāgasphotā, Nikumbhinī, Dantinī, Upacitā, Bhadrā, Rūkṣā, Rocanī, Niśalyā, Vakradantā, Madhupuṣpā, Viśalyā, Eranḍapatrikā, Eranḍaphalā, Udumbaradalā, Kumbhī, Viśodhinī, Anurevati, Taruṇī, Ghuṇapriyā, Makūlaka, Āsyā, Nikumbhā, Citrakūlā, Upacitrakā, Śāmalā, Drākṣāyanī, Ākhukarṇī, Madanakūlakā, Śabarī, Nyagrodhaka, Pratyakśreṇī, Sutaśreṇī, Jayā, Mūsikaṣambhavā, Vṛṣā, Nāgadantī, Ākhuparṇī, Phanjī, Vārāhāngī, Anukūlā, Citrakeśā.

CHEMICAL COMPOSITION : Roots yield 5 new phorbol esters belonging to diterpene hydrocarbo, tigliane skeleton viz. montain, balliospermin etc. Leaves contain β-sitosterol, B-D-glucoside and hexacosanol.

Parts used : Roots, leaves, seeds, latex and oil from seeds.

Ayurvedic properties :

Rasa - Katụ. Vīrya - Ūṣṇa.
Guṇa - Gurū, Tīkṣṇa. Vipāka - Kaṭu.

दन्तीद्वयं सरं पाके रसे च कटुदीपनम्।
गुदाङ्कुराश्मशूलार्शः कण्डूकुष्ठविदाहनुत्।।
तीक्ष्णोष्णं हन्ति पित्ताम्रकफशोफोदरकृमीन्।
(भा. प्र.)

Dantīdvayam saram pāke rase ca kaṭudīpanam,
Gudāṅkurāśmaśūlarśaḥ kaṇḍū kuṣthavīdahanut,
Tikṣṇosnam hanti pittāsrakaphasophodarakṛimin.
(Bhavaprakasa)

दन्ती कटूष्णा शुलामत्वग्दोषशमनी च सा।
अर्शोव्रणाश्मरीशल्यशोधनी दीपनी परा।।
अन्यादन्ती कटुष्णा च रेचनी क्रिमिहा परा।
शूलकुष्ठामदोषघ्नी त्वगामयविनाशिनी।।
(राजनिघण्टु)

Dantī kaṭuṣṇā śulamatvagdoṣaśamanī cs sā,
Arśovraṇāśmarisalyaśodhanī dīpanī parā.
Anyādantī kaṭusnā ca recanī krimihā parā,
Śūlakuṣṭhāmadoṣaghnī tvagāmayavinasinī.
(Rajanighantu)

Actions/Uses : Virecana, Āsukāri, Vikāsi.

Therapeutics :

Danti is used for treatment of abdominal tumours and cancer. Root is acrid, cathartic, digestive, diuretic, diaphoretic, febrifuge, anthelmintic, rubefacient and tonic, used in dropsy, anasarca flatulence, constipation, haemorrhoid, jaundice, skin diseases, leprosy, fever, vesical calculi, stangury and wounds. Leaves are purgative and also used for dropsy and for poulticing wounds. A preparation from leaves is given in abdominal tumors. Decoction or infusion of leaves is given in asthma. Seeds are drastic purgative, rubifacient, hydrogogue and stimulant are used in inflammation and flatulence. Oil from seeds is hydrogogue cathartic, external application in rheumatism. Latex is used in rheumatism.

Bridelia montana Willd.

INDIAN : Aasannaa, Hasan, Gondni, Kargandia, Kargnalia.

SANSKṚTA : Ekavīra.

HABIT : A large evergreen shrub or a small tree.

HABITAT : Various parts of India.

CHEMICAL COMPOSITION : Leaves and stem contain stigmasterol and friedelan-3-ol; leaves in addition, friedelin and β-sitosterol; its glucoside, hexacosanol and triterpenoid.

Parts used : Bark and leaves.

Ayurvedic properties :
Rasa - Kaṭu. Vīrya - Uṣṇa.
Guṇa - Uṣṇa. Vipāka - Kaṭu.

एकवीरा स्मृता तिक्ता चात्युष्णा वातहृत्स्मृता।
पक्षाघातं पृष्ठकटीशूलं चैव विनाशयेत्।।
(निघण्टु रत्नाकर)

Ekavīra smṛtā tiktā cātyuṣṇā vātahṛt smṛtā,
Pakṣāghātam pṛṣṭhakaṭīśūlam caiva vināśayet.
(Nighanturatnakara)

Actions/Uses : Vātaghna, Stambhan, Snehan, Aśmarighna, Śothaghna.

Therapeutics :

Ekavīra is anthelmintic and is used in treatment of bone fracture. Root and bark are astringent.

Bridelia retusa Spreng.

SYNONYMS : *Bridelia airy-shawii* Li.; *Bridelia spinosa* Willd.; *Cluytia retusa* Linn.

INDIAN : Fattarafodda, Asana-2, Ekdania, Gondai, Khaja, Kasai, Monja.

SANSKṚTA : Mahāvīra, Asana, Ekadivi.

HABIT : A shrub or a tree up to 18 m in height with strong conical spines 7 cm long.

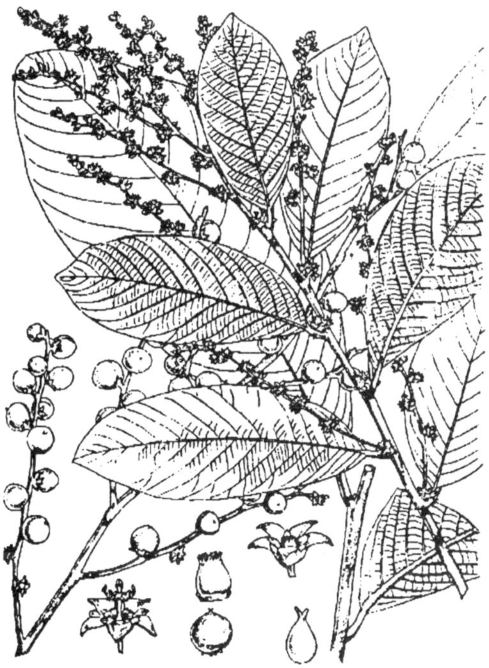

HABITAT : Throughout India.

CHEMICAL COMPOSITION : Fruit pulp, β-sitosterol, gallic and ellagic acids; seeds, ellagic acid. Bark contains a triterpenoid ketone and 16-40 % tannin.

Parts used : Root, stem and bark.

Ayurvedic properties :
Rasa - Kaṭu. Vīrya - Ūṣṇa.
Guṇa - Ūṣṇa. Vipāka - Kaṭu.

Actions/Uses : Vātaghna, Stambhan, Snehan, Aśmarighna, Śothaghna.

Therapeutics :

Root and bark of Mahavira is astringent. Bark with gingli oil is used as liniment in rheumatism. Extract of stem-bark is antiviral, hypotensive and anticancer. Decoction of stem bark with country liquor is used to prevent pregnancy; also for diarrhoea and earache. Paste of stem bark is applied to wounds and bark juice taken internally in snake bite.

Cicca disticha Linn.

SYNONYMS : *Cicca acida (Linn.) Merrill., Phyllanthus distichus Muell. -Arg.; Phyllanthus acidus Skeels.; Averrhoa acida* Linn.

ENGLISH : Country gooseberry, Otaheite.

INDIAN : Raayaaawallaa, Harfarauri, Narphal.

SANSKṚTA : Lavalī, Lavaliphala.

HABIT : A small deciduous tree.

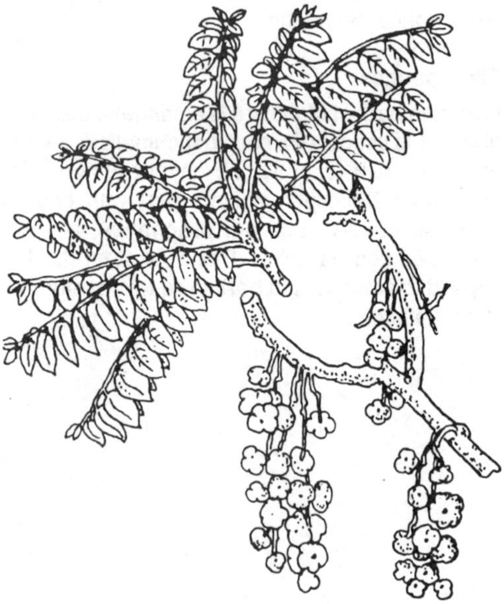

HABITAT : Cultivated in home yards and gardens in India.

CHEMICAL COMPOSITION : Root bark contains 18 % tannin, saponin, gallic acid and a crystalline substance. Bark contains β-amyrin and phyllanthol.

Parts used : Root, fruit and seeds.

Ayurvedic properties :
Rasa - Madhura, Amla, Kaṣāya. Vīrya - Śīta.
Guṇa - Gurū, Visāda, Rūksa. Vīpaka - Amla.

लवलीफलमस्त्रार्शः कफपित्तहरं गुरु।
विशदं रोचनं रुक्षं स्वाद्वम्लं तुवरं रसे।।
(भावप्रकाश)

Lavalīphalamasrārśaḥ kaphapittaharam gurū,
Viśadam rocanam rúkṣam śvādvamlam tuvaram rase.
(Bhavaprakasa)

Actions/Uses : Kaphapittahara.

Therapeutics :

Lavalī fruit is very sour, astringent, improves appetite, tonic to liver, useful in bronchitis, biliousness, vomiting, piles, constipation, urinary concretions, purifies and enriches blood. Root and seeds are cathartic.

Croton oblongifolius Roxb.

SYNONYMS : *Croton roxburghii* Balak.

INDIAN : Gannaasura, Naagadanti, Chuka.

SANSKṚTA : Nāgadantī, Śvetaghaṇṭā, Madhupuṣpī, Viśgodhinī, Nāgasphotā, Viśālākṣī, Nāgacchatrā, Vicakṣaṇā, Sarpapuṣpā, Śuklapuṣpī, Svādukā, Śītadantikā, Śītapuṣpī, Sarpadantī, Nāginī, Śvetadantā, Ekaparṇī, Garasphuṭā, Śasalyā, Kāṇḍinī, Śyenaghaṇṭā, Śūlinī, Rānadūtī, Viṣabhadrā, Keśaruhā, Bhutakusum, Bhuthala bhairi.

HABIT : A middle sized tree.

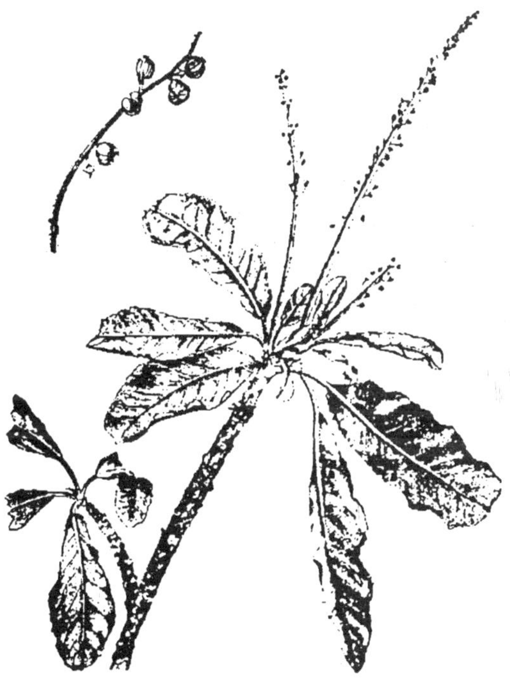

HABITAT : Common in the forests of Central and Eastern parts of Northern India, Bihar, West Bengal and Konkan.

CHEMICAL COMPOSITION : B-Sitosterol, a diterpene alcohol, oblongifoliol from seeds, roots and bark. Deoxyoblongifoliol. Ent-isopimara-7,15-diene, ent-isopimara-7,15-dien-19-ol and ent-isopimaradien-19-al; acetylaleuritolic acid; quercetin and procyanidin.

Parts used : Root, bark, leaves seeds and fruit.

Ayurvedic properties :

Rasa - Kaṭu. Vīrya - Ūṣṇa.
Guṇa - Laghu, Tīkṣṇa. Vipāka - Kaṭu.

नागदन्ती सरा चोष्णा कटुका क्रिमिनाशिनी।
शूलकुष्ठामदोषघ्नी मेदोरोगाश्मरीहरा।।
मुखरोगहरा चान्ये गुणास्तु लघुदन्तिवत्।
(नि. रत्नाकर)

Nāgadantī sarā coṣṇā kaṭukā kriminā́sinī,
Śūla kuṣṭhāmadoṣaghnī medorogāśmarī harā.
Mukharogaharā cānye guṇāstu laghudantivat.
(Nighanturatnakara)

Actions/Uses : Pittasodhan.

Therapeutics :

Bark and root of Nagadanti are alterative. Root made into paste is used against constipation. Decoction of rootbark with pepper given in diarhhoea and dysentery, also in inflammation and fever. Bark used in external applications for sprains, useful in liver diseases. Seeds and fruits are purgative. Various plant parts are used in headache, fever, icterus, scabies, spleen trouble, madness, epilepsy, convulsions, venereal sores, ulcer, syphilis, hydrocele, foul breath, cholera, neuralgia, pleurisy, pneumonia etc.

Croton tiglium Linn.

SYNONYMS : *Tiglium officinale Klotch.; Croton jamalgota f. Hamuion.*

ENGLISH : Tiglium, Purging Croton, Purging Nut.

INDIAN : Jamaalagottaa, Geyapal, Japal.

SANSKṚTA : Jayapāla, Dantibīja, Recaka, Sāraka, Tittirīphala, Maladrāvī, Bījarecanī, Kuntiñibīja, Kumbhībīja, Śodhanī, Nepaī, Ghanṭābīja, Cakradantī, Dantinībījaka, Nikumbhabīja, Nikumbhā, Jamālagoṭā, Kanakaphala.

HABIT : A woody perennial shrub.

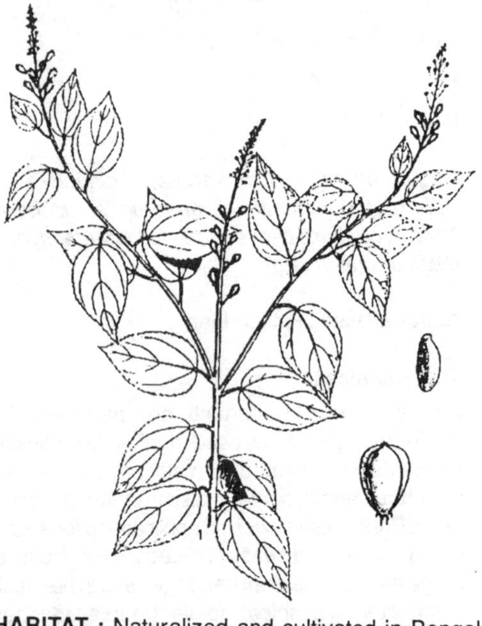

HABITAT : Naturalized and cultivated in Bengal, Assam and South India. Also cultivated in gardens in other parts of India.

CHEMICAL COMPOSITION : β-sitosterol from seeds. Active principle- phorbol-12-tiglate-13-decanoate from oil. Plant extract, used. in folk medicine for treating cancer, showed activity against P-388 lymphcytic leukaemia in mice. Seeds contain two tumor promoting principles. A new highly active tumor-enhancing compound C-3 from seeds. Seed kernel contain 35-57 % croton oil and two toxic proteins, croton globulin and croton albumin, sucrose and a glycoside crotonoside.

Parts used : Seed and fixed oil from seed.

Ayurvedic properties :
Rasa - Kaṭu. Vīrya - Ūṣṇa.
Guṇa - Gurū, Rūkṣa, Tīkṣṇa. Vipāka - Kaṭu.

Actions/Uses : Kaphapittahara.

जेपालः कटुरुष्णश्च कृमिहारी विरेचनः।
दीपनः कफवातघ्नो जलोदरविनाशनः।।
(रा. नि.)

Jepālaḥ kaṭuruṣṇaścya krmihārī virecanaḥ, dipanah kaphavataghno jalodaravinasanah. (Rajanighantu)

Therapeutics :

Jayapala seeds and oil are drastic purgative, diaphoretic, vasicant, vermifuge irritant, rubefacient and cathartic. Its action is prompt. Croton oil when rubbed on skin acts as a rubefacient and counter-irritant and vesicant. When administered internally it operates as a powerful hydrogue cathartic. It is found to be very useful in ascites, anasarca, cold, cough, fever, asthma, constipation, calculus, dropsy, and enlargement of abdominal viscera. It is given only when a drastic purgative is required as in dropsy and cerebral affections like convulsions, insanity and other fevers, attended with high blood pressure. Wood is diaphoretic in small doses and purgative and emetic in large doses.

HOMEOPATHIC USE : Croton tiglium is a Homeopathic remedy. The main indications are gurgling and rumbling in the intestines, as though only water were in them; stools coming like a shot followed by great prostration. Colic before stool; constant urging; aggravated from eating and drinking and from movement. Vesicular eruption on scrotum and penis. Frequent corrosive itching on glans and scrotum; eczema of scrotum, worse at night.

Emblica officinalis Gaertn

SYNONYMS : *Phyllanthus emblica* Linn.; *Cicca emblica* Kurz.

ENGLISH : Embelic, Indian gooseberry.

INDIAN : Aawallaa, Aamalaki, Aonla, Amla.

SANSKṚTA : Āmalakī, Śrīphalā, Dhātrikā, Sādhuphalā, Śivā, Amrutā, Śāntā, Āmalakā, Amrutaphalā, Jātiphalā, Bahuphalā, Śītā, Vṛṣyā, Dhātreyī, Vṛttaphalā, Rocanī, Tiśya, Dhārā, Divyā, Śukti, Amala, Amrutodbhavā, Akara, Pancarasā, Kāyasthā, Ādiphala, Dhātrī.

HABIT : A small or medium-sized deciduous tree with greenish-grey or red bark peeling off in scales and long strips.

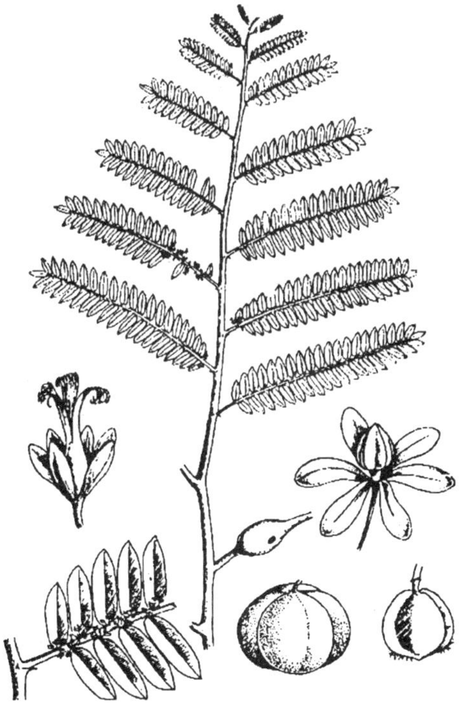

HABITAT : Common in mixed deciduous forests of India, often cultivated in gardens and homeyards.

CHEMICAL COMPOSITION : Fruits and leaves contain tannins; polyphenolic compound, 1,3,6-trigalloylglucose, terchebin, corilagin, ellagic and phyllenbic acids also alkaloids phyllantidine and phyllantine. Leaves and stem yield lupeol and β-sitosteerol. Seed fat contained linoleic acid and closely resembled linseed oil; ellagic acid and lupeol from roots. Fruit rich source of Vitamin C. Phyllembin from fruit pulp identified as ethyl gallate. It has mild depressant action on CNS and spamolytic action. It potentiates action of adrenalin. Anticholesterolaemic and antiatherogenic effects of the fruits have been studied by Thakur and Mandal (1984).

Parts used : Dried fruit, seed, leaves, root, bark and flowers.

Ayurvedic properties :
Rasa - Amla, Madhura, Kaṭu, Kaṣāya.
Vīrya - Ūṣṇa, Śīta.
Guṇa - Laghu, Rukṣa. Vipāka - Madhura.

आमलकं कषायाम्लं मधुरं शिशिरं लघु:।
दाहपित्तवमीमेह शोफघ्नं च रसायनम्।।
कटुमधुरकषायं किञ्चिदम्लं कफघ्नं।
रुचिकर मतिशीतं हन्ति पित्तास्त्रतापम्।।
श्रमवमनविबन्धाध्मान विष्टभदोष।
प्रशमनममृताभं चामलक्य: फलं स्यात्।।
(राजनिघण्टु)

Āmalakakam kaṣāyāmlam madhuram śiśiram laghu,
Dāhapittavamimeha śophaghnam ca rasāyanam.
Kaṭumadhurakaṣāyam kiñcidamlam kaphaghnam,
Rucikaramatiśītam hanti pittāsratāpam.
Śramavamanavibandhādhmanāviṣṭambhadosa,
Praśamanamamṛtābham cāmalakyaḥ phalam syāt.
(Rajanighantu)

Actions/Uses : Vātahara, Pittaśāmaka, Tridoṣahara.

Therapeutics :
Various plant parts of Aamalakee are used in toothache, sores, fever, anaemia, epilepsy, pimples, tubercular fistula, rinder pest, gonorrhoea,

and convulsions. Root and bark are astringent. Fresh roots as a remedy for jaundice. Leaves are cerebral and gastro-intestinal tonic, cardio-tonic, aphrodisiac, antipyretic, antidiabetic. Leaf extract is antibacterial. Decoction of leaves is useful for ulcers in mouth. Infusion of leaves mixed with Fenugreek seeds is useful in chronic dysentery. Flowers are cooling, refrigerant and aperient. Fruit is acrid, cooling, refrigerant, diuretic, and laxative, is a pronounced expectorant and has anticancerous properties. Fresh fruit is mild purgative, diuretic, improving liver function. Raw fruit is aperient. Dried fruit is cooling, and anti-haemorrhagic, useful in haemorrhage, diarrhoea and dysentery. It is especially good for abundant growth of hair. It has been found to be effective in the treatment of peptic ulcer and scurvy. Fruit, juice and its sediment and residue are antitoxidant due to gallic acid, carminative and stomachic. Fruit juice with lemon juice and sugar is taken for arresting bacillary dysentery. Juice with turmeric powder and honey used to cure diabetes insipidus. Fruit preparations are used in induration of liver, in collyrium and in warts of eyes. In combination with iron it is used for anaemia, jaundice and dyspepsia. Powder of fruit is useful for heamorrhoids, diarrhoea, menorhhagia. Fermented liquor prepared from the fruit is used in jaundice, dyspepsia and cough. Exudation from incisions on the fruit used as external application for inflammation of eye. Seeds used for asthma and bronchitis. A constituent of many Ayurvedic combinations such as Triphalaa used for constipation, Arogaywardhani used in viral hepatitis, SG-I Svitradilepa used for vitiligo; Chawanaprash used as a general tonic.

Euphorbia hirta Linn.

SYNONYMS : *Euphorbia pilulifera auct. non* Linn.; *Chamaesyce hirta* Millsp.

HABIT : An erect or ascending annual herb with hairy stems.

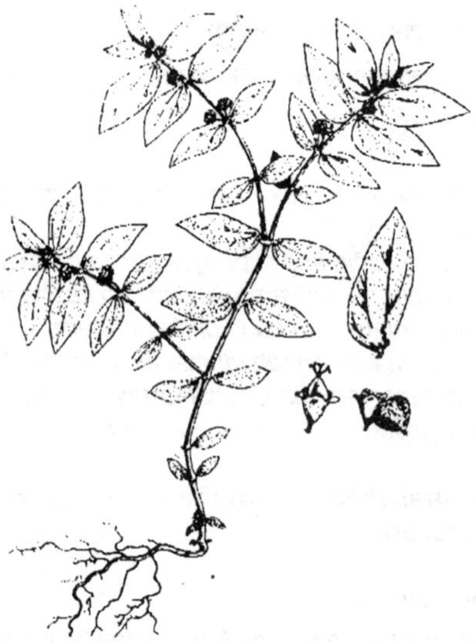

HABITAT : Common in waste places, road-sides, gardens, rice fields throughout hotter parts of India.

ENGLISH : Asthma-weed, Catshair, Euphorbia.

INDIAN : Naayatti, Mothidudhi, Dudhi, Dudali.

SANSKṚTA : Cāra, Pustoa, Amampachairarisi.

CHEMICAL COMPOSITION : Alkaloid, essential oil. l-inositol and an alkaloid xanthorhamnin Leucocyanidol, quercitol, camphol, quercitrin, quercitol derivatives containing rhamnose and chlorophenolic acid isolated. Aerial parts, choline, shikimic acid, l-inositol and sugars. L-hexacosanol, 24-Me-enacycloartenol, tenol, cycloartenol, β-sitosterol, euphorbol hexacosonate, B-amyrin-OAc, tinyatoxin, 2 derivatives of deoxyphorbol-OAc and ingenol-tri-OAc. Stem, hentriacontane, myricyl alcohol, taraxerol etc. Flowers, ellagic acid. Leaves and stems triterenoids and usual sterols. Roots 2 derivatives of deoxyphorbol-OAc, ingenol-tri-OAc and taraxerone.

Parts used : Herb.

Ayurvedic properties :
Rasa - Kaṭu, Tikta, Madhura. Vīrya - Ūṣṇa.
Guṇa - Guru, Rūkṣa, Tīkṣṇa. Vipāka - Kaṭu.

चारस्य च फलं पक्वं वृष्यं गौल्याम्लकं गुरु।
तद्बीजं मधुरं वृष्यं पित्तदाहार्तिनाशनम्।। (राजनिघण्टु)

Cārasya ca phalam pakvam vṛṣyam gaulyāmlakam gurū,
Tadbījam madhuram vṛṣyam pittadāhārttināśanam.
(Rajanighantu)

Actions/Uses : Kaphavātaśāmak.

Therapeutics :

Cara is demulcent, antispasmodic, anti-asthmatic pectoral, anthelmintic and local parasiticide. Plant is chiefly used in the affections of childhood, in worms, bowel complaints and cough, in postnatal complaints, failure of lactation, breast pain. Extract of plant has depressent action on cardiovascular system, a sedative effect on mucous membrane of respiratory and urino-genitory tract. Juice of plant is given in dysentery and colic, and milk applied to destroy warts. Plant alkaloid is effective in respiratory system and produces dilatation of bronchi. Decoction of plant is used in bronchial affections and asthma. Latex is vermifuge and used in diseases of urinogenitory tract, also used in application for warts. In Australia it is much esteemed as a remedy for cough, bronchial and pulmonary disorders, but more especially for the prompt relief it affords in paroxysmal asthma.

HOMEOPATHIC USE : Euphorbia pilulifera is a Homeopathic remedy. None of the irritant effects common to Euphorbias were produced, but a powerful action on the heart was noted. It has given excellent results in cases of acrid leucorrhoea, also in gonorrhoea, when there are intense pains at each micturition; humid asthma with prosration.

Euphorbia tirucalli Linn.

HABIT : An erect unarmed shrub or small tree.

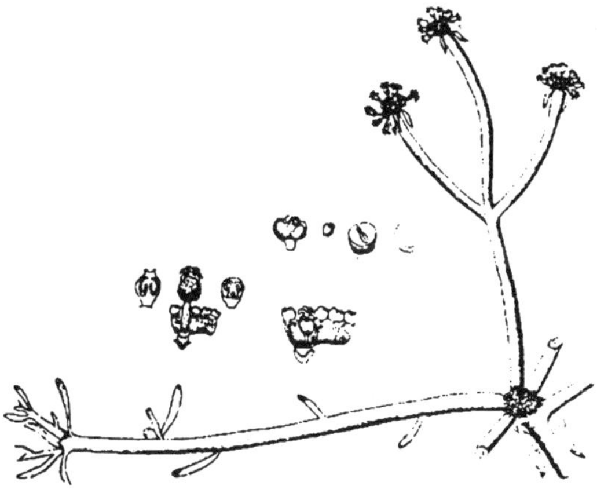

HABITAT : A native of America, naturalized in India, largely cultivated for hedges.

Similar : Euphorbia thymifolia Linn.; Chamaesyce thymifolia (Linn.) Millsp.

ENGLISH : Milk bush, Milk hedge, Indian tree spurge.

INDIAN : Shera, Konpal sehund, Vajraduhu.

SANSKṚTA : Śātalā, Trikantaka, Ganḍeri, Vajradruma, Dugdhikā.

CHEMICAL COMPOSITION : Hentriacontane, hentriacontanol, B-sitosterol, taraxerol, 3,3'-di-O-methylellagic acid and ellagic acid from stems. Taraxerone, tirucallol, sitosterol and 3,4,3'-tri-O-methylellagic acid isolated. Latex from Madagascar yielded 5 new highly irritant euphorbia factors. Latex from A.African plant gave 4 highly irritant diterpenoids. Fresh latex gave new triterpenoids, euphorbinol and cycloeuphomol.

Parts used : Tender branches, leaves stem, root, bark and latex.

Ayurv edic properties :

Rasa - Madhura.	Vīrya - Uṣṇa.
Guṇa - Guru, Rūkṣa, Tīksna.	Vipāka - Kaṭu.

शातला कटुका पाके वातला शीतला लघुः।
तिक्तशोथकफानाह पित्तोदावर्त्तरक्तजित्।। (भा. प्र.)

Sātalā katukā pāke vātalā śītalā laghḥ,
Tiktaśothakaphānāhapittodavarttaraktajit.
(Bhavaptakasa)

Actions/Uses : Kaphapittahara, Vātavardhaka.

Therapeutics :

Sātala is useful in biliousness, leprosy, leucorrhoea, whooping cough, asthma, dropsy, leprosy, enlargement of spleen, dyspepsia, jaundice, colic, tumours and stone in bladder. Milky juice, vesicant, rubifacient. In small doses a purgative but in large doses it is acrid, emetic and counter-irritant; application for warts, rheumatism, neuralgia, tooth-ache; used in asthma, cough and ear-ache; fish poison. Milky juice is applied to itch and scorpion bites. Decoction of tender branches as also that of roots is administered in colic and gastralgia.

Jatropha curcas Linn.

ENGLISH : Purging-nut tree, Curcas nut, Physic nut.

INDIAN : Mongali erandda, Bagbheranda, Safedaranda.

SANSKṚTA : Sthūlaeraṇḍa, Vyāghrairaṇḍa, Kānanaeraṇḍa, Parvataraṇḍa.

HABIT : A large glabrous shrub.

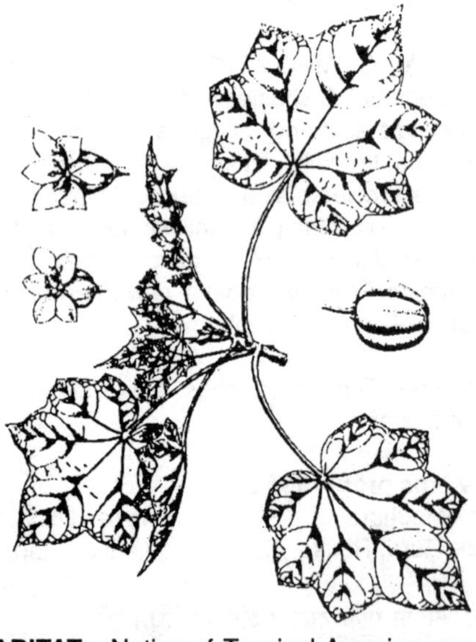

HABITAT : Native of Tropical America, grown in various parts of India as a field barrier. Common hedge-plant in Konkan.

CHEMICAL COMPOSITION : Leaves contain apigenin, vitexin and isovitexin. α-amyrin, stigmasterol, stigmastenes along with two new flaonoid glycosides found in leaves and twigs. Palmitic, oleic and linoleic acids in seed fat. Ash of seeds contained Ca., Mg., Na., K. and traces of P. Alcoholic extract of aerial parts is found to be CNS depressant and diuretic and showed activity against P-388 lymphocytic leukamia.

Parts used : Fruits, seeds, leaves, stem, root-bark and oil.

Ayurvedic properties :
Rasa - Madhura. Vīrya - Ūṣṇa.
Guṇa - Laghu. Vipāka - Madhura.

स्थूलैरण्डः गुणाढ्यः स्यात् रसवीर्यविपक्तिषु।
(राजनिघण्टु)

Sthūlairaṇḍaḥ guṇāḍhyaḥ syāt rasavīryavipaktiṣu. (Rajanighantu)

Actions/Uses : Most potent in respect to rasa, vīrya and vipāka. Vātaśāmaka.

Therapeutics :

Juice of Vyaghrairanda is a well known purgative and is useful in whitlow, convulsions, syphilis, neuralgia, dropsy, anasarca, pleurisy and pneumonia. Root bark is applied externally in rheumatism and is used in sores. Leaves are galactagogue, rubefacient, suppurative, incecticidal and are used in foul ulcers, tumours and scabies; given internally for jaundice. Leaves locally applied to breasts to increase secretion of milk. Leaves warmed and rubbed with castor oil and applied to boils and abscesses have suppurative effect. Decoction of leaves is antidiarhhoea useful in stomach-ache and cough also used for gargle to strengthen gums. Fresh stems are used as toothbrush. Fresh viscid luice flowing from stem is employed to arrest bleeding or haemorrhage from wounds. Stem bark is used for wound of animal bites. Fruit and seeds are anthelmintic; useful in chronic dysentery, urinary discharges, abdominal complaints, biliousness, anaemia, fistula, diseases of heart. Seeds are acro-narcotic, poisonous to human beings and cattle and used against warts and cancers, also to promote hair growth. Seed and oil are purgative, more drastic than castor oil. Wood causes dermatitis. Drug is bitter, acrid, astringent and anthelmintic. It serves to cleanse the entire system through its purgative property. It is useful in chronic dysentery, thirst, abdominal complaints, billiousness, anaemia, fistula, ulcer, diseases of the heart and skin.

HOMEOPATHIC USE : Jatropha is a Homeopathicaly proved remedy. It is one of the most important remedies for true Asiatic cholera, with great prostration and simultaneous vomiting and purging. Urgent gushing stools and irritation and pimples of the skin are other indications.

Jatropha multifida Linn.

HABIT : A large shrub or a small tree.

HABITAT : A native of S. America; cultivated in gardens everywhere.

ENGLISH : Coral plant, Physic nut.

VERNACULAR : Wirechani, Vishabhadra, Chini-yerandi.

SANSKṚIT : Viṣabhadra.

CHEMICAL COMPOSITION : Leaves of plant contain a saponin, a resin and tannin. Seeds contain a fatty oil and bitter substance.

Therapeutics :

Fruit is purgative; useful in piles, wounds, enlarged spleen, pains and skin diseases. Yellow fruits are poisonous particularly to children and cause vomiting and intense burning pain in stomach. A decoction of dried roots is given for indigestion and colic and also as a tonic. Seeds are powerful purgative and emetic. Leaves used in scabies. Latex is applied over wounds and ulcers. Oil from seeds is used both internally and externally as abortifacient. Sap used in cancer.

Macaranga peltata Muell.-Arg.

SYNONYMS : *Macaranga tomentosa* Wight.; *Mappa peltata* Wight.; *Osyris peltata* Roxb.; *Macaranga roxburghii* Wight.

HABIT : A small middle sized tree.

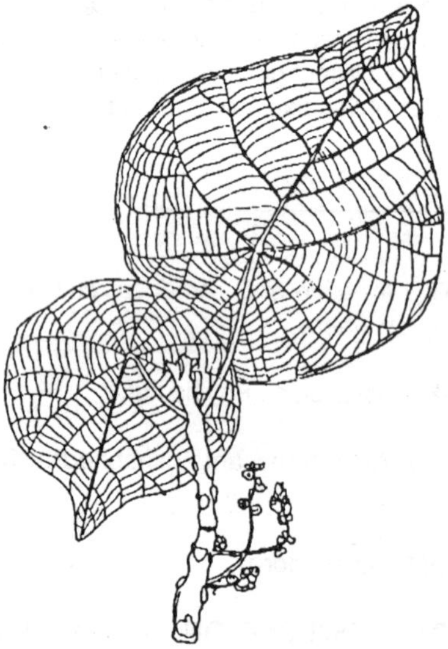

HABITAT : Hills of Orissa and Western Peninsula.

VERNACULAR : Chamdivado, Chanda.

SANSKṚIT : Candwar.

CHEMICAL COMPOSITION : Sitosterol from roots.

Therapeutics :

Country people use the bark of the tree for enlarged spleen. Gum, powdered and made into paste, a good application for venereal sores.

Mallotus philippinensis Muell.-Arg.

SYNONYMS : *Croton philippense* Lamk.; *Rottlera tinctoria* Roxb.

ENGLISH : Monkey face tree.

INDIAN : Kamal, Shendri, Kapilaa, Kunku, Rohini.

SANSKṚTA : Kampila, Kampillaka, Varṇaka, Karkaṣa, Raktānga, Recī, Recana, Recanaka, Ranjan, Lohitānga, Raktacurṇaka, Kamalānguḍi, Kapilī, Kahilā.

HABIT : A shrub or a small much-branched evergreen tree.

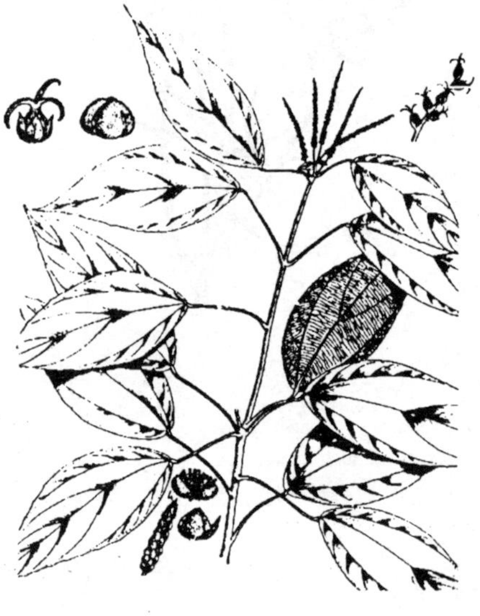

HABITAT : Widely distributed throughout tropical India.

CHEMICAL COMPOSITION : Rottlerin, isorottlerin, corotoxigenin and its rhamnoside, coroglaucigenin and its rhamnoside, from seeds. Resins, one of low m.p. and other of high m.p. plus wax. Kamal powder as oral contraceptive; the anti-fertility factor is rottlerin. Betulin-3-acetate, lupeol, lupeol acetate, sitosterol and bergenin from heartwood; acetylaleuritolic acid, a-amyrin, sitosterol and bergenin from bark.

Parts used : Glands and hairs from capsules or fruits.

Ayurvedic properties :
Rasa - Kaṭu. Vīrya - Ūṣṇa.
Guṇa - Laghu, Rūkṣa, Tīkṣṇa. Vipāka - Kaṭu.

काम्पिल्यः कफपितास्त्रक्रिमिगुल्मोदरव्रणान्।
हन्ति रेची कटूष्णश्च मेहानाहंविषाश्मनुत्।।
(भावप्रकाश)

Kāmpilyaḥ kaphapittāsrakirmigulmòdaravraṇān,
Hanṭīreci kaṭūṣṇaśca mehānāham viṣāśmanut.
(Bhavaprakasa)

Actions/Uses : Kaphavātaśāmaka, Kuṣṭhagna, Recana, Vraṇaropaṇa, Vraṇaśodhana, Kṛmighna, Gulmahara.

Therapeutics :
Glands and hair on the fruit of Kampila are bitter, styptic, anthelmintic and cathartic. Kamal powder has been known as an anthelmintic in India for very long time being especially adapted for the expulsion of tape-worm. It is well tolerated by children and debilitated individuals. With curd or milk, it expels worms from the intestine of children. An ointment of kamal powder is useful for ringworm, pityriasis and freckles. Kamal taken internally relieves leprous eruptions. Kamal powder alone is applied over syphilitic ulcers.

Phyllanthus niruri Linn.

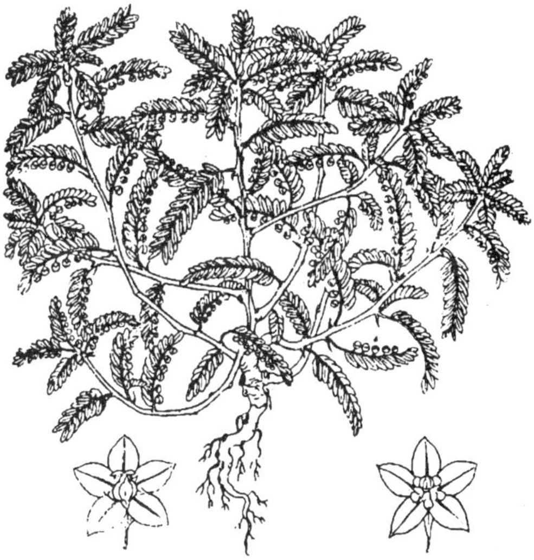

SYNONYMS : *Phyllanthus fratermus* Webster.

ENGLISH : Country gooseberry.

INDIAN : Bhuee aawalli, Jaramla, Bhuinanvalah, Bhonyabali.

SANSKṚTA : Bhūdhātrī, Bhūmyāmalak, Tāmalakī, Bahuphalā, Vitunnā, Drudhapādikā, Bahupatrā, Cāraṭī, Himālayā, Vīrā, Ajarā, Tāmalikā, Viṣvaparṇī, Bhayadā, Bindupatrī, Viṣaghnī, Tālī, Himā, Tamāla, Jatātāni, Tāmalā, Tālāmālā, Uttamā, Sukṣmaphalā.

HABIT : An erect annual herb.

HABITAT : Native to Western Pakistan and Western India. Common throuhgout the hotter parts of India in waste places and shady gardens.

CHEMICAL COMPOSITION : Leaves contain bitter substance phyllanthin, hypophyllanthin.

Three new lignans- niranthin, nirtetralin and phyltetralin from leaves. Kaempferol-4-rhamnopyranoside and eriodictyol-rhasmnopyranoside from roots. Stem contains saponin.

Parts used : All five parts.

Ayurvedic properties :
Rasa - Tikta, Kaṣāya, Madhura. Vīrya - Śīta.
Guṇa - Laghu, Rūkṣa. Vipāka - Madhura.

भूधात्री तु कषायाम्ला पित्तमेह विनाशिनी।
शिशिरा मूत्ररोगार्त्तिशमनी दाहनाशनी।।
(राजनिघण्टु)

Bhūdhātrī tu kaśayāmlā pittameha vināsanī,
Sisirā mūtrarogārtti samanī dāhanāsanī.
(Rajanghantu)

Actions/Uses : Kaphapittaśāmaka, Śothaghna, Vranaropaṇa, Kāsahara, Kuṣṭhaghna, Raktapittahara, Mūtrarogahara.

Therapeutics :

Bhudhatri or Bhumyamalaki is used as a diuretic in dropsical affections, gonorrhoea and other troubles of genito-urinary tract. Herb is bitter, astringent, deobstruent, diuretic, febrifuge and antiseptic. It is used in stomach troubles such as dyspepsia, colic, diarrhoea and dysentery and also for dropsy and diseases of urinogenital system. Fresh root is a remedy for jaundice. Leaves stomachic. Milky juice used as application to sore. Powdered leaves and roots, pulverized and made into polutice with rice-water used to lessen oedematous swellings and ulcers. Leaves are a popular remedy against fever. Infusion of young shoots given in dysentery.

Ricinus communis Linn.

SYNONYMS : *Ricinus europaeus* Nees.; *Ricinus laevis DC.; Ricinus viridis* Willd.

ENGLISH : Castor oil plant, Palma christi.

INDIAN : Erandi.

SANSKṚTA : Śvetaeranḍa, Eranḍa, Pancāngula, Vātārī, Citraka, Vardhamāna, Vyaḍambaka, Dīrghadanḍa, Vyāghrapuccha, Citra, Gandharvahasta.

HABIT : An annual or perennial bush or occasionally a soft-wooded small tree.

HABITAT : A native of tropical Africa. Found wild and also cultivated throughout hoter parts of India.

CHEMICAL COMPOSITION : Ricinine and ricin in plant. Castor oil contains palmitic, stearic, arachidic, hexadecenoic, oleic, linoleic, linolenic, ricinoleic and dihydroxystearic acids as methyl esters. Luoeol and 30-norupan-3B-ol-20-one in seed coat of castor bean.

Parts used : Oil, leaves, roots and seeds.

Ayurvedic properties :
Rasa - Madhura, Tikta, Kaṭu. Vīrya - Ūṣṇa.
Guṇa - Snigdha, Tīkṣṇa, Sūkṣma, Gurū.
Vipāka - Kaṭu.

श्वेतैरण्डः सकटुकरसातिक्त उष्णः कफार्ति।
ध्वंसं धत्ते ज्वरहरमरुत् कासहारी रसार्हः।।
(राजनिघण्टु)

Śvetairaṇḍaḥ sakaṭukarasatiktaūṣṇaḥkaphārti,
Dhvansam dhatte jvaraharamarut kāsahārī rasārhaḥ.
(Rajanighantu)

एरण्डयुग्मं मधुरमुष्णं गुरु विनाशयेत्।
शूलशोथकटीबस्तीशिरः पीडोदरज्वरान्।।
ब्रध्नश्वासकफानाहकास कुष्ठाममारुतान्।
तन्मज्ञा च विड्भेदी वातश्लेष्मोदरापहम्।।
(भावप्रकाश)

Eraṇḍayugmam madhuramuṣṇam guru vināśayet,
Śūlaśothakaṭībastīśīraḥ pīḍodarajvarān.
Braghnaśvāsakaphānahakāsakuṣṭhāmamārutān,
Tanmajjā ca viḍbhedī vātaśleṣmodarāpaham.
(Bhāvaprakāṣa)

Actions/Uses : Vātaghna, Āmapācana, Kaphaghna.

Therapeutics :

The name Erandah indicates the property of the plant to dispel diseases. Eranda is sweet, light, bitter, purgative and hot. From its mild action castor oil is especially adapted for young children and child-bearing women. It is a reputed remedy for all kinds of rheumatic affections. It cures dyspnoea, hydrocele, flatulence, dysentery, ascites, piles, cough, lumbago, headache, leprosy, arthritis, calculus and dysuria. It alleviates phantom tumour, splenic disorders, impurity of blood, dyspepsia and worm troubles. Root is used as an ingredient of various prescriptions for nervous diseases, pleurodynia and sciatica. Dried root is a febrifuge. Fresh leaves or leaves warmed over a fire when applied to the breasts of women act as a galactagogue. Leaves applied to the abdomen promote menstruation. They are applied to painful joints with much benefit. Tender leaves cure pain in bladder. Leaf applied to the head to relieve headache and as poultice for boils. Seed and oil from seed are purgative. Seeds are counterirritant, are used in scorpion sting, and as fish poison. Castoroil congeals to a gel-mass when the alcoholic solution is distilled in presence of sodium salts of higher fatty acids. This gel is useful in dermatosis and is a good protective in occupational eczemas and dermatitis.

HOMEOPATHIC USE : Its main action is on gastro-intestinal tract. Used in material doses for constipation. Five drops doses are used to increase quantity of milk in nursing women. 3rd potency is used for colic, incessant diarrhoea and purging even with fever; painless loose, incessant stool; stools green, slimy and bloody, rumbling in abdomen, nausea, profuse vomiting, contraction of rectal muscles; anus inflamed are other symptoms.

Family : XLIV - Fabaceae.

Abrus precatorius Linn.

SYNONYMS : *Glycine abrus* Linn.

ENGLISH : Indian Liquorice, Jequirity, Crab's Eyes.

INDIAN : Gunja. Ganchi, Gunchi, Rati.

SANSKṚTA : Gunjā, Gunjavallī, Cincapatri, Raktikā, Kākanautika, Bhilla, Angāravallī, Cūdāmaṇī, Raktaphalikā, Kṛṣṇaraktikā, Kākādanī, Kākanandikā.

HABIT : A perennial wiry, twining climber.

HABITAT : Throughout India.

CHEMICAL COMPOSITION : Roots and leaves contain glycyrrhizin. Precol, abrol, abrasine and precasine from roots. Gallic acid, abrine, hypaphorine, alanine, serine, valine, choline, trigonelline, precatorine and methyl ester, 5B-cholanic acid, abrin A and abrin B from seeds.

Parts used : Roots, seeds and leaves.

Ayurvedic properties :
Rasa - Tikta, Kaṣāya.
Guṇa - Rūkṣa, Laghu, Tīkṣṇa.
Vipāka - Kaṭu, Madhura.
Vīrya - Uṣṇa.

गुञ्जाद्वयं तु केश्यं स्यात् वातपित्तज्वरापहम्।
मुखशोषभ्रमश्वासतृष्णामदविनाशनं।।
नेत्रामयहरं वृष्यं बल्यं कण्डूव्रणापहम्।
कृमीन् प्रलुप्तकुष्ठनि रक्ता च धवळाऽपिच।।
(भावप्रकाश)

Guñjādvayam tu keṣyam syāt vātapittajvarāpaham,
Mukhaśoṣabhramaśvāsatṛṣṇāmadavināśanam.
Netrāmayaharam vṛṣyam balyam kaṇḍūvraṇāpaham,
Kṛmīn praluptakuṣṭhani raktā ca dhavaḷā api ca.
(Bhāvaprakāṣa)

Actions/Uses : Roots : Tridoṣaghna.

Leaves : Vātapittaghna, Śothaghna, Vraṇaropaṇa.

Seeds : Kaphavātaśāmak, Keṣya, Kuṣṭhaghna, Vraṇaropaṇa, Vedanāsthāpana.

Therapeutics :

Guñjā plant is bitter, emetic, tonic, vermifuge, demulcent, emollient, thermogenic, antihistaminic, antiseptic, aphrodisiac, and beneficial for hair. Root is emetic, alexiteric, diuretic, tonic, used in preparations prescribed for gonorrhoea, jaundice and haemoglobinuric bile. Leaves are sweetish in taste and are useful in billiousness and in leucoderma, itching and other skin diseases. Also used as a substitue for liquorice. Seeds are purgative, emetic, aphrodisiac used in nervous disorders; half boiled seeds taken as tonic. Paste of seeds applied locally in sciatica, stifness of shoulder joint and paralysis; used against skin diseases; useful in diarrhoea and dysentery. Poultice of seeds used as suppository to bring about abortion. Various plant parts are used in night blindness, inflammation of gums, muscular pain, convulsions, pain in loins, mucus in urine, gravel and diarhhoea. It produces a violent conjuctival inflammation and is likely to destroy corneal structures. Powdered seeds disturb uterine functions and prevent conception in women. Roots and leaves are useful in urinary troubles and dysuria and fever. Roots are abortifacient. Leaves with milk is aphrodisiac and antifertility.

Arachis hypogaea Linn.

SYNONYMS : *Arachis asiatica* Lour.

ENGLISH : Grounut, Peanut, Monkeynut.

INDIAN : Bhueemuga.

SANSKṚTA : Maṇḍapī, Tailamudga, Bhūmimudga, Bhūcaṇaka, Tribīja, Snehabījaka, Bhūmijā, Raktabīja, Bhūstha.

HABIT : A small, prostrate, diffuse or erect, branched, annual herb.

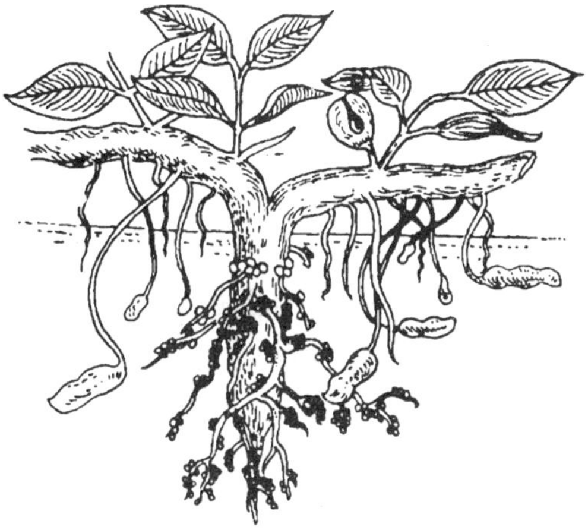

HABITAT : Native of Tropical Africa. Widely cultivated in India.

CHEMICAL COMPOSITION : Nut meal contains arachin, con-arachin; fat, protein, vitamins B1, B2, B6, nicotinic acid, vitamin E and lecithin. 5,7-Dihydroxychromone(1), eriodictyol and luteolin β-sitosterol, daucosterol and a saponin from shells. a-Resorcylic, p-hydroxybenzoic, cis- and trans-p-coumaric, phloretic and vanillic acids along with protocatechuic and chlorogenic acids from seed. Detection of aflatoxin in nuts. Average values on % dry basis of kernel are protein, 28.5; lipids, 47.5; reducing sugars, 0.2; disaccharide sugars, 4.5; starch, 4.0; pentosans, 2.5; and ash, 2.9.

Parts used : Roots, Seeds and oil.

Ayurvedic properties :
Rasa - Tikta, Kaṣāya. Vīrya - Uṣṇa.
Guṇa - Laghu, Rūkṣa, Tīkṣṇa. Vipāka - Kaṭu.

मण्डपी मधुरा स्निग्धा वातळा कफकारिका।
ग्राहिका बद्धवर्चश्य तत्तैलं तद्गुणं स्मृतम्।।
(शा. नि.)

Mandapī madhurā snigdhā vātaḷā kaphakārikā,
Grāhikā baddhavarccaśca tattailam tadguṇam smṛtam.
(Shaligrama nighantu)

Actions/Uses : Anulomika, Vraṇaropaṇa, Varṇaprasādana, Pauṣṭika.

Therapeutics :

Mandapī or Tailamudga fruit and oil are astringents. Unripe nuts are lactagogue. Oil is aperient, emollient and is used as a substitue of olive oil. It depressed whole blood-clotting time; coronary atherosclerosis by hydrogenated oil. A drug for haemophilia is prepared from oil. Shell is a folk medicine for hypertension.

Butea monosperma (Lam.) Kuntze.

SYNONYMS : *Butea frondosa* Koen. *ex* Roxb.; *Erythrina monosperma* Lam.

ENGLISH : Bastard tree.

INDIAN : Pallasha, Dhak, Khakara, Kakracha.

SANSKṚTA : Palāśa, Kinsuka, Parṇa, Raktapuṣpa, Kṣāraśreṣṭha, Kṛṣṇavṛinta, Tiktabījaka, Triparṇa, Vātapotha, Bramhavṛkṣa, Samidvāra, Yājnika, Vakrapuṣpa, Pūtadru, Kāṣṭhadru, Bījasneha, Samiduttama.

HABIT : A deciduous tree with a somewhat crooked trunk.

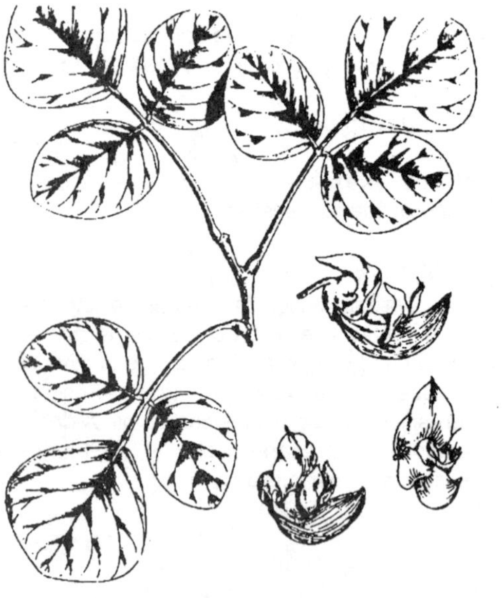

HABITAT : Common throughout India.

CHEMICAL COMPOSITION : α-amyrin, β-sitosterol, its glucoside and sucrose. Glycerides of palmitic, stearic, linoceric, oleic and linoleic acids from seed oil. A nitrogenous acidic compound along with palasonin, butrin, isobutrin, coreopsin, isocorepsin and sulfurein. Flowers contain butrin, butein and butin.

Parts used : Gum, seeds, flowers, bark, leaves, and salts.

Ayurvedic properties :
Rasa - Tikta, Kaṣāya, Kaṭu.
Vīrya - Uṣṇa. Flowers : Śita.
Guṇa - Snigdha, Sara.
Vipāka - Kaṭu. Flowers : Madhura.

पलाशो दीपनोवृष्यः सरोष्णो व्रणगुल्मजित्।
कषायः कटुकस्तिक्तः स्निग्धो गुदजरोगजित्।।
भग्न सन्धानकृद दोषग्रहण्यशः कृर्मान् हरेत्।
तत्पुष्पं स्वादुपाके तु कटुतिक्तं कषायकम्।।
(भावप्रकाश)

Palāśo dīpanovṛṣyaḥ saroṣṇo vraṇagulmajit,
Kaṣāyaḥ kaṭukastiktaḥ snigdho gudajarogajit,
Bhagna sandhānakṛd doṣagrahaṇyarśaḥ kṛmīn haret.
Tatpuṣpam svādupāke tu kaṭutiktam kaṣāyakam.
(Bhāvaprakāṣa)

फ्लाशस्तु कषायोष्णः क्रिमिदोषविनाशनः।
तद्बीजं पामा कण्डूतिददूत्वग्दोषनाशाकृत्।।
तस्य पुष्पं च सष्णिश्च कण्डूकुष्ठार्तिनाशनम्।।
(राजनिघण्टु)

Palāśastu kaṣāyoṣṇaḥ krimidoṣavināśanaḥ,
Tadbījam pāmā kaṇḍūtidadrutvagdoṣanāśakṛt.
Tasya puṣpam ca soṣnañca kaṇḍūkuṣṭhārtināśanam.
(Rajanighantu)

Actions/Uses : Kaphaghna, Vātaghna, Pittakar, Agnidīpanī, Śukravardhanī, Bhedana, Tṛiṣṇaśamani, Kṛimighna, Stambhana, Garbhādhananivāraṇa.

Flowers : Tṛṣṇasamani, Stambhana, Mūtrala.

Fruits : Pramehaghna, Arśaghna, Kṛmighna, Udararogaghna, Kuṣṭhaghna.

Seeds : Bhedana, Kṛimighna.

Kṣara : Bhedana, Anulomana.

Therapeutics :

Palāśa is efficaceous in the treatment of vaginal diseases, helminthic manifestations and haemorrhages. Root cures night blindness and other defects of sight and is useful in elephantiasis. Roots cause temporary sterility in women. Root bark is aphrodisiac and is used as analgesic and anthelmintic. Externally applied in sprue, piles, ulcers, tumors and dropsy. Leaves are astringent, tonic, diuretic and aphrodisiac and are used to cure boils, pimples, tumors and haemorrhages; given in flatulent colic, worms and piles. Juice of leaves mixed with curds and Curcuma aromatica is beneficial in heat eruption in children. Flowers are astringent, diuretic, depurative and aphrodisiac; are used as emmenagogue and as poultice in orchitis and to reduce swellings, in bruises and sprains. Lotion prepared from flowers is used for certain eye diseases. Decoction of flowers is given in diarrhoea and to puerperal women. Flowers alongwith Hygrophilia auriculata leaves and roots taken with milk to cure leucorrhoea. Juice given to women to induce sterility. Dried flowers are used for preventing sunstroke; effective in leprosy and gout. Seeds are acrid, bitter, aperient, sedative, rubefacient; used as a vermifuge and in snake bite. Decoction of seeds is given in gravel. Paste of powder with lemon is applied as a cure for ringworm and herpes for its cooling effect. Seeds are anthelmintic but not safe as they may cause nephrotoxicity. Extracts of seeds, flowers and leaves is reputed to have contraceptive properties. Bark is bitter, acrid, hot, oily, astringent, appetiser, anthelmintic, aphrodisiac and alterative. It is useful in abdominal tumours, colic, intestinal worms, bleeding piles, ulcers, haemorrhages, amenorrhoea and dysmenorrhoea. It lessens inflammation and biliousness and is useful in fracture of bones, diseases of anus, dysentery, bleeding piles and hydrocele. It cures ulcers and tumors. Decoction of bark is given in cold, cough, fever, haemorrhage and menstrual disorders. Gum is astringent and is used in diarrhoea and dysentery. Fresh gum from bark is applied to ulcers and sore throat and is given in phthisis, haemorrhages of stomach and bladder. Infusion is used as local application in leucorrhoea. Solution of gum is applied to bruises, erysipelatous inflammations and ringworm and also to check conception.

Cajanus cajan (Linn.) Millsp.

SYNONYMS : *Cajanus indicus* Spreng.

ENGLISH : Pigeon pea, Cadjan pea.

INDIAN : Toora, Arahara.

SANSKṚTA : Āḍhakī, Tuvari, Tuvarika.

HABIT : An annual or perennial shrub.

HABITAT : A native of Africa and grown in almost all the tropical countries; vastly cultivated throughout India as a pulse crop.

CHEMICAL COMPOSITION : Seeds contain two globulins, cajanin and concajanin. It also contains an enzyme, oxalic acid, calcium and phosphorus. Essential oil from plant contains α-, β-, & γ- selinenes, copaene and a mixture of eudesmols. Leaves, sterols, triterpenes and other lipidic constituents. Roots, isoflavon, cajanone. Rootbark, cajaflavanone.

Parts used : Leaves and seeds.

Ayurvedic properties :
Rasa - Madhura, Kaṣāya. Vīrya - Śīta.
Guṇa - Laghu, Rūkṣa. Vipāka - Kaṭu.

Parts used : Leaves and seeds

आढकी तुवरा रुक्षा मधुरा शीतला लघुः।
ग्राहिणी वातजननी वर्ण्या पित्तकफास्रजित्।।
(भा. प्र.)

Āḍhakī tuvarā rūksā madhurā sītala laghuh,
Grāhinī vātajanani arnyā pittakaphāsrajit.
(Bhāvaprakāṣa)

Actions/Uses : Kaphapittaghna, Vātakara, Sangrāhī, Varṇya, Mānsavardhaka.

Therapeutics :
Āḍhakī plant is used in convulsions, colic and leprosy. Leaves are astringent, diuretic, laxative, cooling and anodyne. They are useful in oral ulcers, strangury and inflammations. Juice of leaves is used in jaundice. Paste of leaves or flowers is applied in sores in mouth and tongue. Dried plant with dried flowers of Nymphaea alba are given for asthma. Seeds are cooling, acrid, anthelmintic, febrifuge and expectorant. Thay are useful for oral ulcers, tumours, bronchitis, cough, vomiting, fever and cardiac diseases. Seeds and leaves made into paste are warmed and applied over the mammae to check secretion of milk or induce lactation. It is also useful for reducing inflammation. Seeds with other vegetable ingredients are used as abortifacient.

Cicer arietinum Linn.

HABIT : An erect or spreading much-branched annual herb.

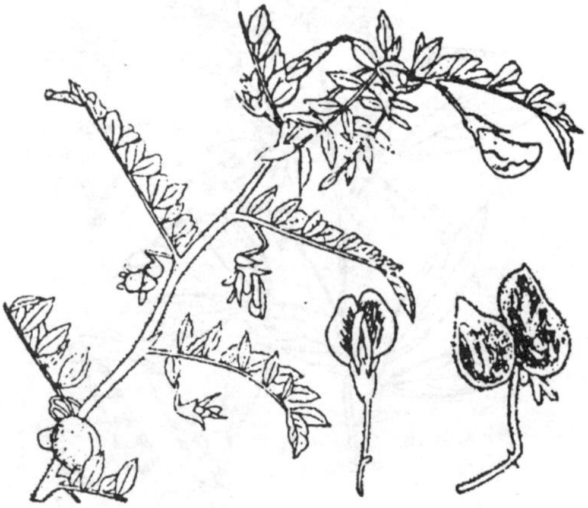

HABITAT : Largely cultivated as pulse crop in most parts of India.

ENGLISH : Bengal gram; Chick-pea.

INDIAN : Harabharaa, Channe.

SANSKṚTA : Caṇaka, Harimantha, Sakalapriya.

CHEMICAL COMPOSITION : Biochanins A, B, C from fresh sprouting germ. Vanillic and p-hydroxybenzoic acid from leaves. Oxalic, acetic, malic and another acid. Arginine, tyrosine, lycine, cystine, tryptophane; carotenoids and oil soluble vitamins A, D and E. Aerial parts, kaempgerol, quercetin, isorhamnetin and pratensein; phytolexin, cicerin. Seed and its lipid extract have hypocholesteroemic effect. Seeds reduced post-prandial plasma glucose.

Parts used : Seeds and leaves.

Ayurvedic properties :
Rasa - Kaṣāya. Vīrya - Śīta.
Guṇa - Laghu. Vipāka - Kaṭu.

चनको मधुरो रुक्षो मेहजित् वातपित्तकृत्।
दीप्रिवर्णकरो वल्यो रुच्यश्चाध्मानकारकः।। (राजनिघण्टु)

Canako madhuro rūkṣo mehajit vātapittakṛt,
Dīptivarṇakaro balyo rucyaścādhmānakārakaḥ.
(Rajanighantu)

Actions/Uses : Viṣṭambhī, Raktadoṣahara, Balya, Rucya, Jvaraghna.

Therapeutics :

Leaves of Cannaka are sour, improve taste and appetite. They are applied to sprains and dislocated joints. They cure bronchitis but cause flatulence. Juice of leaves is stomachic abortifacient. Unripe seed is stimulant, tonic and aphrodisiac; cures thirst and burning. Seed is sweet, refrigerant, nutritive, astringent, tonic and antibilious; used in leprosy, bronchitis. They are diuretic when fried. The acid exudation is astringent and usefull in dyspepsia, constipation and snake bite.

Clitoria ternatea Linn.

ENGLISH : Winged-leaved clitoria.

INDIAN : Gokarnna, Aparaajitaa, Kalina, Kajali.

SANSKRTA : Aparājitā, Viṣṇukrāntā, Śvetapuṣpī, Gokarṇikā, Abhedā, Asphotā, Supuṣpī, Aśvakhurā, Dadhipuṣpikā, Śvetanāmā, Mohanāśinī, Girikarṇikā, Bhūrdiśam, Nīlapuṣpa, Jayā, Kṣīrika, Nīladrikaraṇī, Ārdakaṇī, Kaṭabhī, Gardhabhī, Viṣahantrī, Nāgaparyāyakarṇī, Gajakaṇikā, Vājikhurā.

HABIT : A pretty perennial twiner with blue or white flowers.

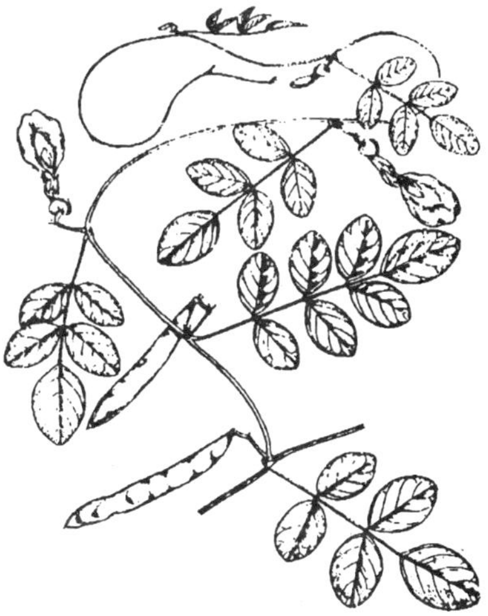

HABITAT : A common garden plant; also occurs among hedges all over India.

CHEMICAL COMPOSITION : Taraxerol and taraxerone from roots. Cinnamic acid, flavonol glucosides, hexacosanol, β-sitosterol and an anthoxanthin glucoside from seeds. Aparajitin, sitosterol, stearic, palmatic, oleic, linoleic and linolenic acid, and glycosides of kaempferol, stigmast-4-ene-3,6-dione from leaves.

Parts used : Root, bark, seeds and leaves.

Ayurvedic properties :
Rasa - Tikta-Kaṣāya, Kaṭu. Vīrya - Śīta.
Guṇa - Laghu. Vipāka - Kaṭu.

अपाराजिते कटु मेध्ये शीते कण्ठ्ये सुदृष्टिदे।
कुण्ठमूत्रत्रिदोषामशोथव्रणविषापहे।।
कषाये कटुके पाके तिक्ते च स्मृतिवुद्धि दे।
(भावप्रकाश)

Aparājite kaṭu medhyeśīte kaṇṭhye sudṛṣṭide,
Kuṣṭhamūtratridoṣāmasothavraṇaviṣāpahe.
Kaṣāye kaṭuke pāke tikte ca smṛtibuddhide.
(Bhāvaprakāṣa)

Actions/Uses : Tridoṣaghna, Dīpana, Pācana, Kaphaduṣṭighna, Āmapācana, Buddhipradā, Smṛtipradā, Medhakara, Cakṣuṣya, Rujāpahā, Mūtradoṣahara, Kuṣṭhaghna, Vraṇaghna, Jvaraghna.

Therapeutics :

Aparājitā root, bitter, emetic, cathartic, purgative and diuretic, useful in ascites and fevers; used by tribals to cause abortion. Root bark, diuretic and laxative. Infusion of root-bark is useful in irritation of bladder and urethra. In Konkan, root-juice is given in cold milk to remove phlegm in chronic bronchitis. Seeds are purgative and aperient. Seeds roasted and powdered are given in ascites and enlargement of abdominal viscera. Plant is used in snake bite.

Desmodium gangeticum DC.

SYNONYMS : *Hedysarum gangeticum* Linn.; *Desmodium collinum* Roxb.

INDIAN : Saalawanna, Shaalaparnni, Sarivan, Salpan, Darh.

SANSKṚTA : Sālaparṇī, Sthira, Vidārigandhā, Atiguhā, Śophaghnī, Saumyā, Dīrghamūlikā, Suputrā, Ansumatī, Mahāklītānika, Atiruhā, Sudalā, Kumudā, Tanwī, Pīvarā, Vātaghnī, Pittalī, Sarvānukāriṇī, Sudhā, Sumbhaptrikā, Niścalā.

HABIT : A woody perennial shrub.

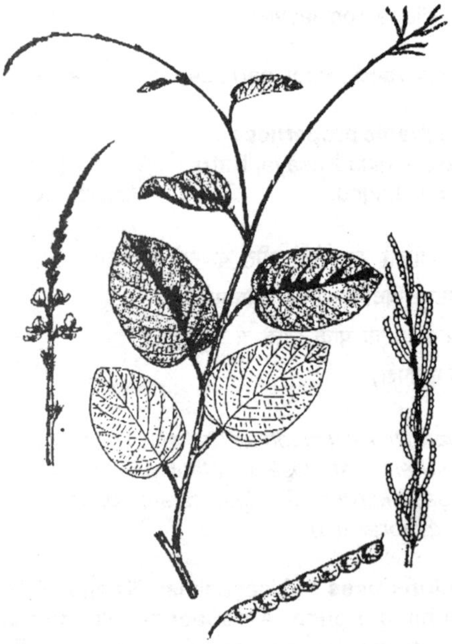

HABITAT : Throughout the plains of India and lower Himalayan regions.

CHEMICAL COMPOSITION : 5-Methoxy-N,N-dimethyltriptamine,N,N dimethyltetrahydroharman, 6-methoxy-2-methyl-β-carbolinium cation from aerial parts. Seeds contain β-carboline alkaloids. Gangetinin and desmodin.

Parts used : Whole plant, root and bark.

Ayurvedic properties :
Rasa - Madhura, Tikta. Vīrya - Ūṣṇa, Śīta.
Guṇa - Guru, Snigdha. Vipāka - Madhura.

शालिपर्णी रसे तिक्ता गुरुष्णा वातदोषनुत्।
विषमज्वरमेहार्शः शोफसन्तापनाशिनी।।
(राजनिघण्टु)

Śāliparṇī rase tiktā gurūṣṇā vātadoṣanut,
Viṣamajvaramehārśaḥ śophasantāpanāśinī.
(Rajanighantu)

Actions/Uses : Pittaghna, Vātaghna, Kaphaghna, Snehana, Hṛdya, Stambhana, Kṛmighna, Śothahara, Vṛṣya, Mūtrala, Mehaghna, Atisāraghna, Śvāsahara, Brihmana, Rasāyana, Tṛṣaghna.

Therapeutics :

Śāliparṇī root is a constituent of Dashamoola kwaatha used in post-natal care to avoid secondary complication. Plant is considered antipyretic and anticatarrhal and used for hazy vision and dysentery. It is a good cardiotonic. It is hot, sweet, diuretic, laxative and nervous tonic. It overcomes burning sensation, fever, thirst, cough, difficult breathing, dysentery and vomiting. Root is hot and bitter, astringent in diarrhoea, tonic, diuretic, alterative, aphrodisiac, anthelmintic; used in chronic fever, biliousness, cough, vomiting, asthma, snake-bite, scorpion sting.

Dolichos biflorus Linn.

SYNONYMS : *Dolichos uniflorus* Lamk.

ENGLISH : Horsegram.

INDIAN : Kulleetha, Gahat, Kulith.

SANSKṚTA : Kulittha, Tāmrabīja, Śvetabīja, Catrika, Aitucakra, Kulālī, Vanaja, Urvarā, Sthiramudrā, Aliskandha, Surāṣṭrajā, Kākavṛnta, Kuladhānakara, Khalakula, Kulastha.

HABIT : A branched sub-erect or trailing annual herb.

HABITAT : Cultivated in winter in many parts of India.

CHEMICAL COMPOSITION : Seeds rich source of urease. Seed extractive showed marked hypoglycaemic and hypocholesterolaemic effect.

Parts used : Seeds.

Ayurvedic properties :
Rasa - Kaṣāya.
Vīrya - Uṣṇa.
Guṇa - Laghu, Rūkṣa. Vipāka - Amla.

कुलत्थिका कटुस्तिक्ता स्यादर्शः शूलनाशनी।
विबन्धाध्मानशमनी चक्षुष्या व्रणरोपणी।।
(राजनिघण्टु)

Kulatthikā kaṭustiktā syādarśaḥ śūlanāśanī,
Vibandhādhmānaśamanīcakśyṣyā vraṇaropaṇī.
(Rajanighantu)

Actions/Uses : Kaphavātaśāmaka, Raktapittaprakopa.

Therapeutics :

Kulittha seeds are bitter, acrid, hot, astringent diuretic, antipyretic, anthelmintic and tonic. They cure kapha and vaata. Useful in abdominal complaints, asthma, bronchitis, urinary discharges, hiccough, ozoena, heart troubles, diseases of brain, intestinal colic, stangury, eye diseases, piles, leucoderma, inflammation, liver troubles; removes stones from kidney. Powder of seed is used as anti-diaphoretic. Decoction of seeds used in leucorrhoea and menstrual disorders. Soup prepared from pulses is an useful diet in sub-acute cases of enlarged liver and spleen.

Erythrina indica Lamk.

SYNONYMS : *Erythrina variegata* Linn. *var. orientalis (L.)* Merr.; *Erythrina orientalis* Linn.; *Erythrina mysorensis* Gamble.

ENGLISH : Indian Coral tree, Mochi wood.

INDIAN : Paangaaraa, Dadap, Mandar, Ferrud, Panara.

SANSKṚTA : Pāribhadra, Phalabhadra, Raktakusuma, Kanṭakimśuka, Kanṭakī, Nimbataru, Mandāra, Pārijātaka, Bahupuṣpa, Raktapuṣpa, Kṛmighna, Raktakesara, Kantakī pārijātaka, Mandalia, Prabhadra, Mimbataru, Saśimbu, Svalpakanṭi.

HABIT : A middle sized quick-growing tree.

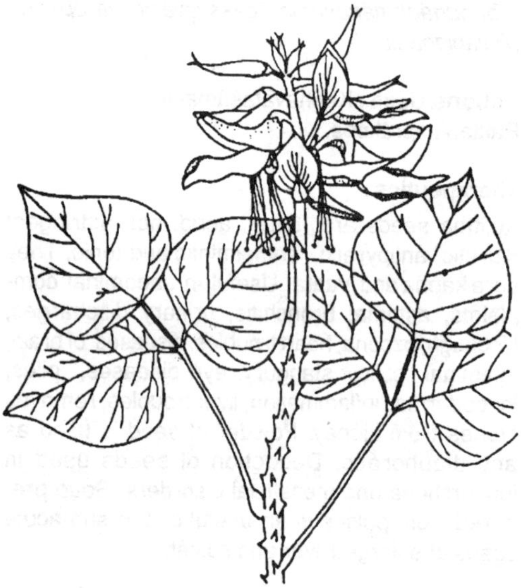

HABITAT : Wild in deciduous forests throughout India.

CHEMICAL COMPOSITION : Docosyl alcohol, β-sitosterol, y-sitosterol, d-sitosterol. Leaf alkaloids, erysotrine, erysodine, ethyralinehydrochloride and hypaphorine; erythrinine, erysodine and de-N-Me-orientaline. Trunk bark yields erysotrine, erysodine, erysovine, erysonine and hypaphorine, stachydrine.

Parts used : Bark, stem, leaves and their juice.

Ayurvedic properties :
Rasa - Kaṭu. Vīrya - Uṣṇa.
Guṇa - Laghu. Vipāka - Kaṭu.

परिभद्रः कटुष्णः स्यात् कफवातनिकृन्तनः।
अरोचकहरः पट्यो दीपनश्यापि कीर्तितः।।
(राजनिघण्टु)

Paribhadraḥ katuṣṇah syāt kaphavātanikṛntanaḥ,
Arocakaharaḥ pathyo dipanaśyapi kīrtitaḥ.
(Rajanighantu)

Actions/Uses : Kaphavātaśāmana.

Therapeutics :

Pāraibhadra is acrid, hot, anthelmintic, carminative, galactagogue, expectorant and febrifuge. Leaves are laxative, diuretic, anthelmintic, galactagogue, emmenagogue, applied externally for dispersing venereal buboes and for relieving pain in joints. Fresh juice of leaves, vermifuge and cathartic. It cures dysmenorrhoea, reduces excess fat and increases secretion of milk. Alcoholic extract of leaves and stem is spamogenic, used in convulsions, pimples, menorrhagia. Crushed leaves are applied to rheumatic joints to relieve pain. In Konkan, juice of young leaves is used to kill worms in sores. Bark is astringent and febrifuge, used in anorexia, liver troubles, helminthic manifestations, inflammation, intestinal worms and obesity. Also as a collyrium in opththalmia, as antidote to snake bite. It promotes appetite, destroys pathogenic parasites, arrests excessive micturition and cures edema, flatulence, colic, arthritis. Decoction of bark is used in dysentery, ophthalmia and other eye diseases. Juice of bark and young leaves is used to kill worms in sores.

Glycyrrhiza glabra Linn.

SYNONYMS : *Glycyrrhiza gladulifera W. & K.; Glycyrrhiza violacea* Boiss.

ENGLISH : Liquorice, Licorice.

INDIAN : Jeshthamadha, Mulhatti, Mithilakri.

SANSKṚTA : Madhuyaṣṭī, Yaṣṭī, Madhukā, Klītaka, Madhuparṇa, Madhurakāṣṭha, Madhurasrava, Valītaka, Yaṣṭika, Yaṣṭimadhu, Uprasiddha, Madhuramūlika, Yaṣṭhyāhvā, Jalayaṣṭi, Stalayaṣṭi.

HABIT : A hardy sub-erect perennial herb or undershrub.

HABITAT : Indigenous to Europe, Africa, Balucistan and Iran; cultivated in Jammu and Kashmir, Punjab and sub-Himalayan tracts.

CHEMICAL COMPOSITION : The main constituent of liquorice is a saponin-like glycoside Glycyrrhizin which ranges from 5 to 20% and is 50-60 times sweeter than sucrose. It also contains glycyrrhetinic acid, rhamnoglucoside, twenty seven flavonoids a triterpenoid-liquoric acid from roots. Root extract, estrogenic; contains β-sitosterol and stigmasterol. It contains herniarin and umbelliferone and flavones- liquiritin, liquiritigenin, isoliquiritin and isoliquiritigenin. Deglycyrrhised liquorice extract or liquorice extract as such is used for treatment of peptic and duodenal ulcers. Glycyrrhetic acid obtained from liquorice is used as antiinflammatory compound in the treatment of various skin diseases. Acid as well as its derivative, carbexonolone is also used as antiinflammatory in gastric ulcers.

Parts used : Peeled and unpeeled root.

Ayurvedic properties :
Rasa - Madhura. Vīrya - Śīta.
Guṇa - Gurū, Snigdha. Vipāka - Madhura.

मधुरं यष्टिमधुकं किञ्चित्तिक्तं च शीतलम्।
चक्षुष्यं पित्तहृद् रुच्यं शोषतृष्णाव्रणापहम्।।
(राजनिघण्टु)

Madhuram yaṣṭimadgukam kīncittiktam ca śītalam,
Cakṣuṣyam pittahṛd rucyam
śoṣatṛṣṇāvraṇāpaham.
(Rājanighaṇtu)

यष्टि हिमा गुरु स्वाद्वी चक्षुष्या बलवर्णकृत्।
सुस्निग्धा शुक्रला केश्यास्वर्या पित्तानिलास्तजित्।।
व्रणशोथविषच्छर्दितृष्णाग्ला निक्षयापहा।।
(भावप्रकाश)

Yaṣṭi himā guru svādvī cakṣuṣyā balavarṇakṛt,
Susnigdhā śukralā keśyā svaryā pittānilāsrajit.
Vraṇaśothaviṣacchardi tṛṣṇāglānikṣayāpahā.
(Bhāvaprakāṣa)

Actions/Uses : Pittaghna, Vātaghna, Kaphavardhaka, Rucya, Balya, Varṇakara, Śūkrala, Keṣya, Svarya, Śothahara, Vedanāsthāpana, Cakśuṣya, Ksayahara, Tṛṣṇapaha, Raktapittaśamana, Vraṇaghna, Gulmahara, Dāhaghna, Kāsaghna.

Therapeutics :
Madhuyaṣṭī is demulcent, expectorant, antitussive, laxative and sweetner. Root is sweet, slightly bitter; refrigerant, tonic, laxative demulcent, emolient, aphrodisiac, alexeteric, diuretic, alterative, galactagogue; good for the eyes; in incipent loss of sight, in diseases of eyelids; coughs and sore throat, removes biliousness, ear diseases due to biliousness, used in genito-urinary diseases, and in scorpion sting. Root is used in China

as a drug for strengthening muscle and bone, for increasing strenth and for curing wounds. Powdered root extracts and ammoniated extract are used as demulscent, expectorant and antiinflammatory components in cough mixtures. It is used as laxative and for treatment of Addison's disease. It is widely used to mask the bitter taste of galenials, tinctures and extracts like aloe, cascara, quinine etc. It is also used in medicinal teas as a taste corringent. Bulk of liquorice is used in food and flavour industries; for flavouring cigarettes, chewing and snuff tobacco; as a sweetner in low calorie foods, chocolates, chewing gums, beers, candies, baked goods and dietic foods. Constituent of Ayurvedic drugs "Madhuyasti"; "Geriforte"; and of Chinese drug for scleroderma and pneumonia.

Lens culinaris Medic.

SYNONYMS : *Lens esculenta Moench.; Ervum lens Linn.; Ervum himalayense A. Br.; Cicer lens Willd.*

INDIAN : Masoora, Masser.

SANSKṚTA : Masūra.

HABIT : A small annual sub-erect herb.

HABITAT : Native Asia Minor, Persia or possibly Hindu Kush. It is a cold weather crop. Cultivated throughout Iṇdia during cold season.

CHEMICAL COMPOSITION : Lentils are rich in proteins, and a good source of vitamin B. Tricetin, luteolin, a diglycosyldelphinidin and two proanthocyanidins from seed coat. Kaempfero glycoside and 3,4',7-trihydroxyflavone from cotyledons.

Parts used : Seeds.

Ayurvedic properties :
Rasa - Madhura. Vīrya - Śīta.
Guna - Laghu, Rūkṣa. Vipāka - Madhura.

मसुरो मधुरः शीतः संग्राही कफपित्तजित्।
वातामयकंरश्चैव मूत्रकृच्छ्हरो लघुः।।
(राजनिघण्टु)

Masuro madhurah śītaḥ sangrāhī kaphapittajit,
vātāmayakaraścaiva mūtrakrccha haro laghuḥ.
(Rājanighaṇṭu)

Actions/Uses : Śītalā, Pauṣṭika, Vātakāraka, Kaphapittajita, Sangrāhī, Mūtrakruchahara.

Therapeutics :
Masūra leaves are acrid and bitter. Seed is sweet and cooling, astringent, diuretic, improves appetite, removes kapha and biliousness; but causes pain and diseases due to vaata; cures strangury, tumours, dysentery, skin diseases; useful in diseases of eyes and of heart. Seeds, mucilaginous, laxative, enrich blood and are useful in diseases of chest, bronchitis, stomatitis, good for eye and inflammation of breast, also in cases of constipation and other intestinal affections; made into a paste, useful as cleaning application in foul and indolent ulcers.

Melilotus indica (Linn.) All.

SYNONYMS : *Melilotus parviflora Desf.; Desmodium triflorum DC.; Trifolium indica* Linn.

ENGLISH : Small melilot.

INDIAN : Raanmethee.

SANSKRTA : Methikā-1, Banamethikā, Bahupatrikā, Bahubījā.

HABIT : An erect annual herb.

HABITAT : Tropical zones of India.

CHEMICAL COMPOSITION : Benzo-1,2-pyrone from leaves and flowers.

Parts used : Leaves, and seeds.

Ayurvedic properties :
Rasa - Kaṭu. Vīrya - Ūṣṇa.
Guṇa - Laghu, Snigdha. Vipāka - Kaṭu.

मेथिका वातशमनी श्लेषघ्नीज्वरनाशिनी।
ततः स्वल्पगुणा वन्या वाजिनां सातु पूजिता।।
(भावप्रकाश निघण्टु)

Methikā vātaśamanī śleṣmaghnī jvaranāsinī.
Tataḥ svalpaguṇā vanyā vājinām sātu pūjitā.
(Bhāvaprakāṣa)

Actions/Uses : Grāhī, Stanyajanana, Pācana, Vraṇaropaṇa. Useful in Kaṣṭāvarta, Āmvāta, Atisāra, Udaraśūla.

Therapeutics :

Properties of Banamethikā are the same as those of Trignoella foenum-graecum. Plant is used as a discutient and emolient; externally as a fomentation, polutice or plaster for swellings. Seeds are useful in bowel complaints and infantile diarrhoea, given as a gruel.

Mucuna pruriens (Linn.) DC.

SYNONYMS : *Dolichos pruriens* Linn.; *Mucuna prurita* Hook.; *Mucuna monosperma* DC.; *Stizolobium pruriens (Linn.)* Medic.; *Carpopogon prupriens* Roxb.

ENGLISH : Cowage, Cowitch plant.

INDIAN : Khaajakuhili, Kiwanch, Kuhili.

SANSKṚTA : Kapikacchu, Ātmaguptā, Markaṭī, Kaṇḍūrā, Śūkaśimbī, Ruṣyaproktā, Svayamguptā, Romālu, Bṛmhaṇī, Prāvṛśāyanā, Lāngulī, Adhyandā, Duṣparśā, Kacchurā, Varāhitā, Tīkṣṇā, Vanasūrikā, Śītapittavṛanasughnī.

HABIT : A large half-woody perennial twiner.

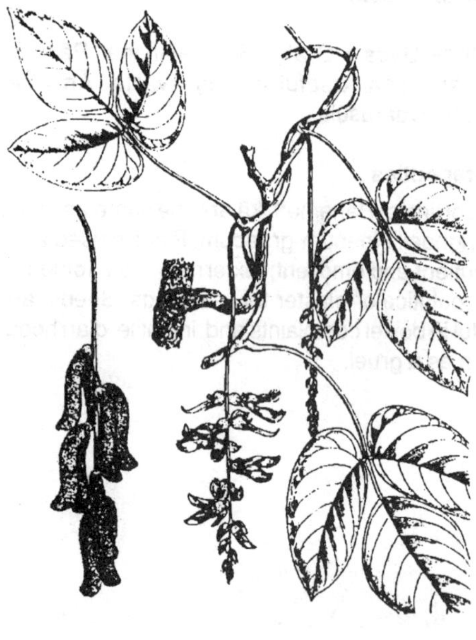

HABITAT : Native to the West Indies and tropical America. Almost throughout India and in Andaman and Nicobar Islands.

CHEMICAL COMPOSITION : Plant extract induced marked hypotensive and hupocholesterolaemic effect; the total indole alkylamines produced marked behavioural changes in rats. Seeds give mucuadine, mucuadinine, mucuadininine and pruriendine, and small amounts of nicotine; and reddish viscous oil.

Parts used : Roots, seeds and fruits.

Ayurvedic properties :
Rasa - Madhura, Tikta. Vīrya - Uṣṇa, Śīta.
Guṇa - Gurū, Snigdha. Vipāka - Madhura.

कपिकच्छुः स्वादुरसा वृष्या वातक्षयापहा।
शीतपित्तास्त्रहन्त्री च विकृता व्रणनाशिनी।। (राजनिघण्टु)

Kapikacchuḥ svādurasā vṛṣyā vātakṣayāpahā,
Śītapittāsrahantrī ca vikrtā vraṇanāśinī.
(Rajanighantu)

Actions/Uses : Vṛṣya, Brimhani, Balya, Vājikara, Vātahara.

Root : Yonisankocaka, Ārtavajanana, Mūtrala.

Seed : Mānsavaha, Vṛṣya.

Therapeutics :

Kapikacchu is a mechanical anthelmintic, it causes intestinal worms to writhe and become detached from intestines. An ointment prepared with the hairs acts externally as a local stimulant and mild vesicant. A decoction of plant or root is diuretic. Roots are purgative, given in dysentery and in uterine troubles. It acts as emmenagogue; also prescribed as remedy for delirium in fever. Powdered and made into paste applied to body in dropsy. Strong infusion mixed with honey given in cholera. Leaves are aphrodisiac, tonic, anthelmintic. Seeds are aphrodisiac, nervine tonic and alexipharmic, cure scorpion sting. Pods most active anthelmintic against Tenia cannina and paraphistonum.

HOMEOPATHIC USE : Dolichos is a Homeopathic remedy. The most characteristic symptom is a general intense itching without eruption. It is very useful during teething period for jaundice and constipation etc. but not for fever. In teething affections, if fever symptoms exist, always give a dose of Aconite before Dolichos. Where this precaution has been neglected, convulsions have followed the use of even high potencies.

Phaseolus trilobatus (Linn.) Schreb.

SYNONYMS : *Vigna trilobata* (Linn.) *Vendcourt.; Phaseolus trilobus sensu* Ait; *Dolichos trilobatus (Linn.)* Mant.

ENGLISH : Green gram.

INDIAN : Mooga, Raanamooga.

SANSKṚTA : Mudgaparṇī, Mudga, Kṣudrasahā, Kākamudga, Mada, Sūryapaparṇī, Mārjāragandhikā, Māsapatrikā.

HABIT : An annual or perennial herb.

HABITAT : Vastly cultivated in India.

CHEMICAL COMPOSITION : Methionine, tryptophan and tyrosine from beans. Strepogenin, uridine diphosphate-galacturonic acid. Seed protein of munga bean contained lysine, valine, leucine and phenylakanine. Vitexin and β-sitosterol from seed coat. Friedelin, epifriedelin, stigmasterol and tannins. (Asima)

Parts used : Whole plant and leaves.

Ayurvedic properties :

Rasa - Madhura, Tikta. Vīrya - Śīta.
Guṇa - Laghu, Rūksa. Vipāka - Madhura.

मुद्गपर्णी हिमा कासवातरक्तक्षयापहा।
पित्तदाहज्वरान् हन्ति चक्षुष्या शुक्रवृद्धिकृत्।।
(राजनिघण्टु)

Mudagaparṇī himā kāsavātaraktakṣayāpahā,
Pittadāhajvarān hanti cakṣuśyā śukravṛddhikṛt.
(Rajanighantu)

Actions/Uses : Tridoṣaśamaka especially Vātapittasāmaka, Rasāyanī, Balakārī, Caksusyā, Grāhi, Balya, Vrsya, Vīryavardhinī, Sothahara, Raktapittakara, Ksayaghna.

Therapeutics :

Mudagaparṇī is rejuvenating. It is bitter, cold, sweet on digestion, constipating, roborant, aphrodisiac, and cures cough, emaciation, pyrexia and haemetemesis. Leaves are sedative, cooling, antibilious and tonic. Seeds are cooling, astringent, used as diet in fever and to strengthen the eye and are given in fevers to relieve thirst. When applied externally they releave heat and burning of eyes. Paste is used as an external application in ophthalmia and haemorrhoids. Soup made of them is very nutritious and given after recovery from acute illness. A poultice is useful for checking secretion of milk and reducing distention of mammary glands.

Phaseolus mungo (Linn.) Mant.

SYNONYMS : *Vigna mungo* (Linn.) Hepper.; *Phaseolus radiatus* Linn. *Vigna radiata (L.) Wilczek var. radiata.; Phaseolus aureus* Roxb.; *Phaseolus radiatus L. var. aureus (Roxb.)* Prain.; *Phaseolus roxburghii.; Phaseolus mungo* Linn.

SIMILAR : Teramus labialis Spreng.

ENGLISH : Black gram.

INDIAN : Udeeda, Masha, Raanaudeeda.

SANSKṚTA : Māṣa, Māṣaparṇī, Sinhamukhī, Kṛṣṇavṛntā, Kāmbojī, Hayapucchikā, Svādumāṣa, Mahāsahā, Ārdamāṣā, Aśvapucchikā.

HABIT : An erect pubescent herb.

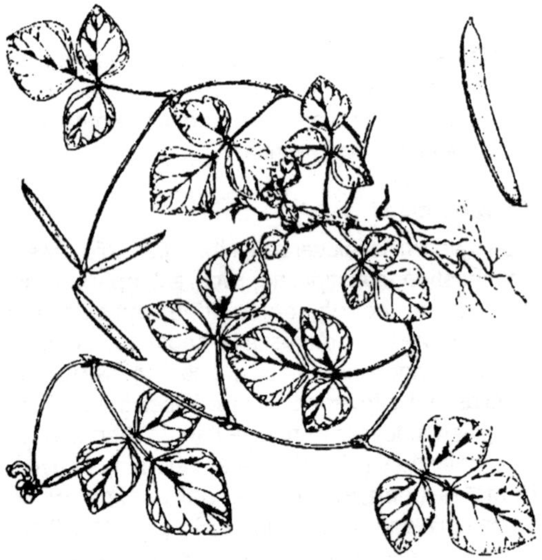

HABITAT : Extensively cultivated all over India.

CHEMICAL COMPOSITION : It contains albuminoids, starch and oil. Seeds contain saponin I, II, and III. It contains genistein, α-hydroxygenistein, dalbergioidin etc.

Parts used : Whole plant and roots.

Ayurvedic properties :
Rasa - Madhura, Tikta. Vīrya - Śīta.
Guṇa - Gurū, Laghu, Snigdha. Vipāka - Madhura.

माषो गुरुः स्वादुपाकः स्निग्धो रुच्योनिलापहः।
उष्णः सन्तर्पणो वल्यः शुक्रलो बृंदणः परः।।
माषपर्णी हिमा तिक्ता स्निग्धा शुक्रबलासकृत्।
मधुरा ग्राहिणी शोथवातपित्तज्वरास्त्रजित्।।
(भावप्रकाश)

Māṣo guruḥ svādupākaḥ snigdho rucyonilāpahaḥ,
Ūṣṇaḥ santarpaṇo balyaḥ śukralo bṛnhaṇaḥ paraḥ.
Māṣaparṇī himā tiktā snigdhā śukrabalāsakṛt,
Madhurā grāhaṇī śothavātapittajvarāsrajit.
(Bhāvaprakāṣa)

Actions/Uses : Vātapittaśāmaka, Kaphavardhaka, Snehana, Grāhī, Vātānulomana, Mānsavardhaka, Śothahara, Jvaraghna.

Therapeutics :

Māṣa is considered hot and tonic. Root is narcotic and is a remedy for aching bones. Seeds are sweet, oily,tonic, laxative, aphrodisiac, appetiser, diuteric, galactagogue; cures vaata, piles, asthma; good for the heart and in fatigue. Seeds are used both internally and externally. Internally, a clear decoction is useful in dyspepsia, in gastric catarrh, dysentery, diarrhoea, cough, cystitis, piles, paralysis, rheumatism, affections of liver and of the nervous system, and in fever.

Pongamia pinnata (Linn.) Merr.

SYNONYMS : *Pongamia glabra* Vent.; *Cytisus pinnatus* Linn.; *Derris indica (Lamk.)* Bennet.; *Pongamia pinnata (Linn.)* Pierre.; *Galedupa indica.*; *Pongamia pinnata* Benth.

ENGLISH : Indian beech, Pongamia oil tree.

INDIAN : Karanja, Kiramal, Kidamar.

SANSKṚTA : Karanja, Naktamāla, Gucchapuṣpaka, Prakīryā, Nakta, Snigdhapatra, Ghṛtapūra, Puṣpamanari, Rocana, Pītasāra.

HABIT : A medium-sized glabrous tree.

HABITAT : Native of Western Ghats. Found all over India on the banks of rivers and streams.

CHEMICAL COMPOSITION : Seeds contain 27 to 36.4% of a bitter fatty oil and traces of an essential oil. Seeds yield fixed oil and three crystalline substances karanjin, pongamol and glabrin. Roots contain four furoflavones, viz. keranjin, pongapin, pinnatin and gamatin. Seeds, pongapin; flowers, kaempferol and waxes; stem-bark, waxes. Tetra-O-methylfisetin and pongachromene from root and stem-bark. β-sitosterol from seeds.

Parts used : Fresh bark, roots, leaves, flowers and seeds.

Ayurvedic properties :
Rasa - Tikta, Kaṣāya, Kaṭu. Vīrya - Ūṣṇa.
Guṇa - Laghu, Tīkṣṇa. Vipāka - Kaṭu

करञ्जः कटुरुष्णश्च चक्षुष्यो वातनाशनः।
तस्य स्नेहोतिस्निग्धश्च वातघ्नः स्थिरदीप्तिदः।।
(राजनिघण्टु)

Karañjaḥ kaṭurūṣṇaśca cakṣuṣyo vātanāśanaḥ,
Tasya snehotisnigdhaśca vātaghnaḥ sthiradīptidaḥ.
(Rajanighantu)

करंजः कटुकस्तीक्ष्णो वीर्योष्णो योनिदोषहृत्।
कुष्ठोदावर्तगुल्मार्शोव्रणकृमिकफापहः।।
(भावप्रकाश)

Karañjaḥ kaṭukastīkṣṇo vīryoṣṇo yonidoṣahṛt,
Kuṣṭhodavartagulmarśovraṇakṛmikaphāpahaḥ.
(Bhāvaprakāṣa)

करंजः भेदनीयः लेखनीयः कण्डूघ्नश्च।।
(च. सू.)

Karañjaḥ bhedanīyaḥ lekhanīyaḥ kandūghnaśca.
(Caraka Sutra)

Actions/Uses : Kaphaghna, Vātaghna, Pittaprakopī, Dīpan, Pācana, Yakṛtottejaka, Virecaka, Kaṇḍughna, Kuṣṭhaghna, Yonidoṣahara, Gulmahara, Arśoghna, Vraṇahara, Pramehaghna, Raktadoṣahara, Udaraghna.

Therapeutics :

Karañja is an effective remedy for all skin diseases such as eczema, scabies, leprosy and ulcers. Root and bark are bitter, hot, acrid, anthelmintic, alexipharmic; useful in diseases of eye, vagina, skin; good for tumours, piles, wounds, ulcers, urinary discharges, itching, ascites and enlargement of spleen. They cure biliousness, vaata and kapha. Juice of roots, used for closing fistulous sores and cleaning foul ulcers; given internally with equal quantities of cocoanut milk and lime water for gonorrhoea. Leaves in form of polutice applied to ulcers infected with worms. Flowers are used as a remedy for diabetes. Fresh bark is used internally in bleeding piles. Seeds are used as external application in skin diseases. Powder of seeds is efficacious in whooping and irritating coughs of children. Expressed oil from seeds has antiseptic and stimulant healing properties, is useful in cutaneous affections, herpes and scabies; used in rheumatism. Internally the oil has sometimes been used as a stomachic and cholagogue in cases of dyspepsia with sliggish liver. Seeds and roots are used as fish poison.

Psoralea corylifolia Linn.

ENGLISH : Babachi.

INDIAN : Baawachi, Bavanchi, Bukchi.

SANSKṚTA : Bākucī, Avalguja, Kṛṣṇaphala, Pūtiphala, Kuṣṭhaghni, Somavallī, Kālameṣikā, Kuṣṭhahārī, Somarekhā, Sugandhakanṭaka, Sasankarekha, Sitavari, Suprabha, Putigandhika, Vakucibheda, Somakhanda, Kusthanasini.

HABIT : An erect annual herb.

HABITAT : Found on roadsides and waste places throughout India.

CHEMICAL COMPOSITION : Psoralone and isopsoralone, isopsoralidin, corylidin, triacontane and β-sitosterol-B-D-glucoside from seeds. A mixture of psoralen, isopsoralen and imperatorin caused hypertrophy of liver, kidney and spleen in experimental rats.

Parts used : Seeds and seed oil.

Ayurvedic properties :
Rasa - Tikta, Kaṭu. Vīrya - Uṣṇa.
Guṇa - Laghu, Rūkṣa. Vipāka - Kaṭu.

बाकुची मधुरातिक्ता कटुपाका रसायनी।
विष्टंभहृत् हिमा रुच्या सरा श्लेष्मास्त्रपित्तनुत्।।
रुक्षा हृद्या श्वासकुष्ठमेहज्वरकृमिप्रणुत्।
तत्फलं पित्तलं कुष्ठकफानिलहरं कटु।
केश्यं त्वच्यं वमिश्वासकासशोथामपाण्डुनुत्।।
(भावप्रकाश)

Bākucī madhurātiktā kaṭupākā rasāyanī,
Viṣṭabhahṛt himā rucyā sarā śleṣmāsrapittanut.
Rukṣā hṛdyā śvāsakuṣṭhamehajvarakṛmipraṇut,
Tatphalam pittalam kuṣṭhakaphānilaharam kaṭu,
Keśyam tvacyam
vamiśvāsakāsaśothāmapāṇḍunut.
(Bhāvaprakāṣa)

श्वित्रारिर्वाकुचीभेदः कुष्ठदोषत्रयास्त्रजित्।
वातरक्तहरो लेपात् सिध्मश्वित्रविनाशनः।।
(द्रव्यगुण)

Śvitrārirvākucībhedaḥ kuṣṭhadoṣatrayāsrajit,
Vātaraktaharo lepāt sidhmaśvitravināśanaḥ.
(Dravyaguṇa)

Actions/Uses : Kaphaghna, Vātaghna, Pittaprakopi, Dīpana, Pācana, Yakṛtottejana, Anulomana, Kṛimighna, Keṣya, Amaghna, Śothahara, Hṛdayottejaka, Tvacya, Kaṇḍūhara.

Therapeutics :
Bākucī root is useful in caries of teeth. Leaves are good to stay diarrhoea. Fruit is bitter, diuretic,

causes biliousness, cures leprosy, skin diseases, kapha, vaata, vomiting, asthma, difficulty in mucturation, piles, bronchitis, inflammations, anaemias, improves the hair and complexion. Seeds are sweet, bitter, astringent, stomachic, anthelmintic, diuretic, germicidal laxative, aphrodisiac, deobstruent, and diaphoretic in febrile condition. They are specially recommended in the treatment of leucoderma, leprosy, psoriasis and inflammatory diseases of the skin. They impart vigour and vitality, improve digestive power and receptive power of mind, improve texture and complextion of skin and help growth of hair. Oleo-resinous extract of seeds used as application to leucoderma. Essential oil from fruits used internally as tonic and aphrodisiac against impotency and externally for treatment of leucoderma, psoriasis and leprosy.

Pterocarpus marsupium Roxb.

ENGLISH : Indian Kino, Malbar Kino.

INDIAN : Pale-Aasan, Biballaa, Mahaakutaja, Bija. Dhorbenla

SANSKṚTA : Bījaka, Asana, Bījasar, Priyaka, Pītaśāla, Nilaka, Bandhukapuṣpa, Mahākuṭaja, Priyaka sauri.

HABIT : A large deciduous tree.

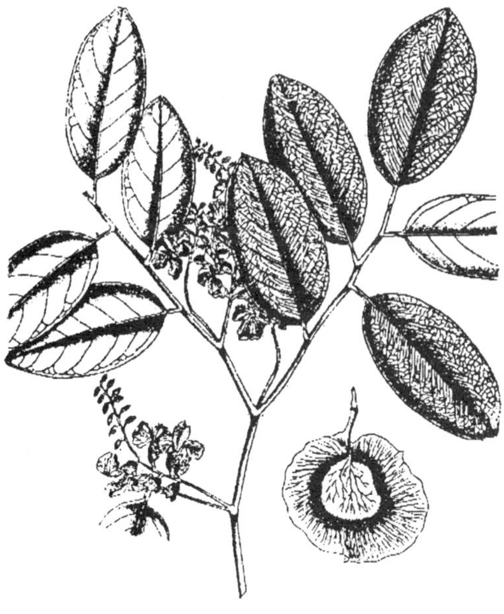

HABITAT : Common in hilly regions of throughout Western Peninsula and South India.

CHEMICAL COMPOSITION : Gum kino contains kino-tannic acid. Extractive of heartwood significantly lowered blood sugar and improved glucose tolerance in rabbits.

Parts used : Gum, leaves, bark, heart wood and flowers.

Ayurvedic properties :
Rasa - Tikta, Kaṣāya. Vīrya - Śīta.
Guṇa - Laghu, Rūkṣa. Vipāka - Kaṭu.

बीजकः कुष्ठवीसर्पश्वित्रमेहव्रणक्रिमीन्।
हन्ति श्लेष्मास्त्र पित्तं च त्वच्यः केश्यो रसायनः।।
(भावप्रकाश)

Bījakaḥ kuṣṭhavīsarpaśvitra mehavraṇakrimīn,
Hanti ślesmāsra pittam ca tvacyaḥ keśyo rasāyanaḥ.
(Bhāvaprakāṣa)

Actions/Uses : Kaphapittaśāmaka. Sothahara, Sandhāniya, Kuṣṭhaghna, Keṣya, Stambhana, Kṛmighna, Rasayana, Pramehaghna, Raktapittahara, Visarpahara, Śvitraghna, Vraṇahara.

Therapeutics :

Bījaka is a reputed drug for the treatment of leprosy and other skin diseases. It is an esteemed remedy for diabetes. Plant and gum are hot and bitter with a sharp taste; alterative, laxative, anthelmintic, cure vaata and kapha. Useful in diseases of blood, eruptions on body, leucoderma, erysipelas, leprosy, urinary discharges and also in eye troubles and elephantiasis; good astringent in diarrhoea and pyrosis. Bruised leaves are useful as external application to boils, sores and skin diseases. Bark is astringent and used in toothache. Heart wood is cool, bitter, astringent and acrid and cures morbid kapha and pitta, dysentery, diarrhoea and haemophilic disorders (raktapitta). Drug is a constituent of preparations like Asanavilvaadi tailam, Ayaskruti, Varaadi kashaayam, Khadirasaaraadi ghrutam. It is an important ingredient of "Asanaadi gana" of Vaagbhat which cures leucoderma, skin diseases, morbid kapha, diabetes, anaemia, krumi and disorders due to deranged fat. Gum is useful in toothache and pyorroea. It is locally applied in leucorrhoea and in passive haemorrhages.

Pterocarpus santalinus Linn. f.

ENGLISH : Red sanders, Red Sandlewood.

INDIAN : Raktachandana, Laal chandan.

SANSKṚTA : Raktacandana, Raktasāra, Arkacandana, Ārakta, Lohita, Kucandana, Soma, Ranjana, Raktānga, Tilaparṇī, Surakta, Tāmrābha, Patrānga, Bhāskarapriya, Pravālaphala, Haricandana.

HABIT : A tall deciduous tree.

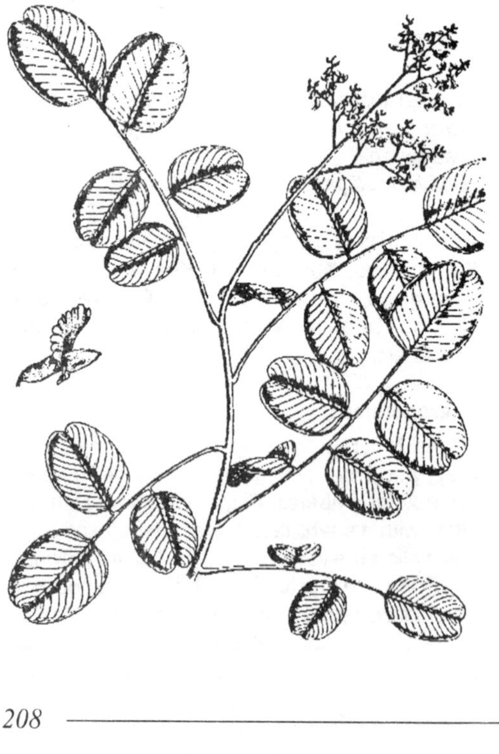

HABITAT : Mainly restricted to Andhra Pradesh, particularly Cuddapah district and neighbouring areas of Karnataka and Tamil Nadu, Maharashta and forests of South India.

CHEMICAL COMPOSITION : Fresh shoots yield glucosides and soloring matter. Pterocarpol from heartwood. Pterocarptriol, isopterocarpolone and pterocarpodiolone, together with β-eudesmol etc. from heartwood. Isolation of Santalin B.

Parts used : Wood.

Ayurvedic properties :
Rasa - Tikta, Madhura. Vīrya - Śīta.
Guṇa - Gurū, Rūkṣa. Vipāka - Kaṭu.

रक्तचन्दनमर्ताव शीतलं तिक्तमक्षिण गदास्त्रदोषनुत्।
भूतपित्तकफकासज्ञ्वरभ्रान्ति जन्तुवमिजित् तृषापहम्।।
(राजनिघण्टु)

Raktacandanamatīva śītalam
tiktamīkṣaṇagadāsradoṣanut,
Bhūtapittakaphakāsasajjvarabhrāntijantuvamijit
tṛṣāpaham. (Rajanighantu)

रक्तं शीतं गुरुः स्वादुच्छर्दितृष्णास्त्रपित्तजित्।
तिक्तं नेत्रहितं वृष्यं ज्वरव्रणविषापहम्।।
(भावप्रकाश)

Raktam śītam guruḥ svāducchardiṭṛṣṇāsrapittajit,
Tiktam netrahitam vṛṣyam jvaravraṇaviṣāpaham.
(Bhāvaprakāṣa)

Actions/Uses : Kaphapittaśāmaka, Stambhana, Kuṣṭhaghna, Śothahara, Dāhaśāmaka.

Therapeutics :

Raktacandan is bitter, cooling, antipyretic, antiperiodic, astringent, diaphretic and febrifuge and is used in abscesses, boils, eye troubles, headache, inflammation, thirst, vomiting and haemophilic disorders. It purifies blood and cures skin diseases and poisonous affections. Wood, astringent, tonic, diaphoretic, used as a cooling external application for inflammations and headache. It is a hot remedy useful in billious affections and skin diseases, in fever, boils and to strengthen the sight; also over swelling of eyelids for reducing swelling. Wood rubbed with water is employed as a wash in superficial extoriation of genital organs and in scorpion sting.

Sesbania bispinosa (Jacq.) Fawcett & Rendle.

SYNONYMS : *Sesbania aculeata* Pers.; *Aeschynomene bispinosa* Jacq.; *Coronilla aculeata* Willd.; *Sesbania egyptiana* Poir.; *Sesbania cannabina (Retz.)* Pers.; *Sesbania sesban (Linn.)* Merr.; *Aeschynomene sesban* Linn.

ENGLISH : Daincha, Prickly sesban.

INDIAN : Shewaree, Dhunchi jayat, Janjan.

SANSKṚTA : Jayantī, Balāmoṭā, Haritā, Jayā, Vijayā, Sūkṣmamūla, Vikrāntā, Aparājitā, Śītalā, Śataparṇī, Galagaṇḍahāri, Vātala, Itakata.

HABIT : A suffruticose, shrubby annual. A short-lived quick-growing soft-wooded tree.

HABITAT : Native to Australia. Cultivated throughout plains of India.

CHEMICAL COMPOSITION : Gum in endosperm of seeds consists of galatose and mannose.

Parts used : Roots, Bark, leaves, flowers and seeds.

Ayurvedic properties :
Rasa - Kaṭu, Tikta. Vīrya - Ūṣṇa.
Guṇa - Laghu, Rūkṣa. Vipāka - Kaṭu.

ज्ञेया जयन्ती गलगण्डहारी तिक्ताकटुष्णानिलनाशिनी च।
भूतपहा कण्ठविशोधनीच कृष्णा तुसातत्र रसायणीस्यात्।।
(राजनिघण्टु)

Jneyā jayantī galagaṇḍahārī
tiktākaṭuṣṇānilanāśinī ca,
Bhūtapahā kaṇṭhaviśodhanī ca
kṛṣṇa tu sā tatra rasāyaṇī syāt. *(Rajanighantu)*

बलामोटा कटुस्तिक्ता लघुः पित्तकफापहा।
मूत्रकृच्छ्रं विषंहन्ति विवादे कुरुते जयम्।। (कै. नि.)

Balāmoṭa kaṭustiktā laghuḥ pittakaphāpahā,
Mūtrakṛccham viṣam hanti vivāde kurute jayam.
(Kalyadevanighantu)

Actions/Uses : Kaphapittaghna, Dīpana, Grāhī, Kṛumighna, Vedanāsthapana, Vidradhi, Jvaraghna, Tridoṣahari.

Therapeutics :

Jayantī root is hot, bitter, alterative, anthelmintic, carminative, removes kapha, billiousness, inflammation, cures fevers, ulcers, diabetes, leucoderma, tuberculous glands, relieves throat troubles. Seeds are emmenagogue, stimulant, astringent; heal chronic ulcers and remove smallpox eruptions. They are useful in diseases of spleen, diarrhoea and excessive menstrual flow. Leaves are purgative, anthelmintic, maturant, demulcent; useful for hydrocele. Plant is considered a cure for wounds. Seeds mixed with flour applied to ringworm and skin diseases.

Sesbania grandiflora (Linn.) Pers.

SYNONYMS : *Agati grandiflora* Desv.; *Robinia grandiflora* Linn.

ENGLISH : Sesban, Agathi, Swamap Pea.

INDIAN : Agastaa, Hadagaa, Madga, Shevari, Agati, Anari.

SANSKṚTA : Agastya, Agastī, Śighrapuṣpa, Munidruma, Vraṇārī, Dīrghaphalaka, Vakrapuṣpa, Kumbhayoni, Surapriya, Panktipatra, Mahāruha, Madhuśingru, Vangasena, Vaka.

HABIT : A short-lived, quick-growing, soft-wooded tree.

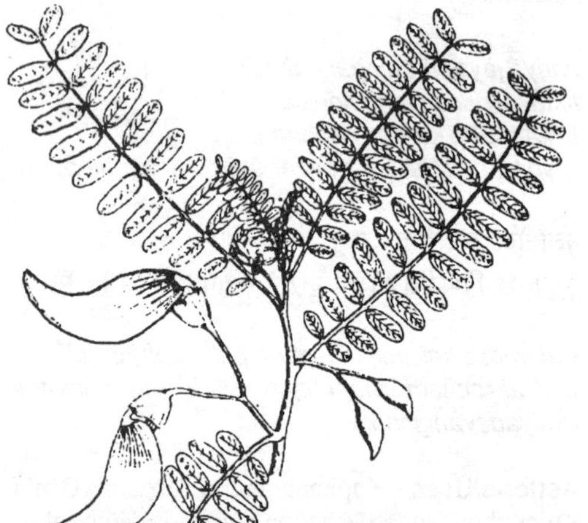

HABITAT : Native of Malaysia, grown throughout plains of India.

CHEMICAL COMPOSITION : A saponin from leaves yielded oleanolic acid, galactose, rhamnose and glucuroniic acid. Seed saponin on hydrolysis yielded oleanolic acid as major genin. Aqueous extract of flowers produces haemolysis of human and sheet erythrocytes even at low concentration. The haemolytic effect is due to methyl ester of oleanolic acid.

Parts used : Whole plant, flowers, leaves, bark and root.

Ayurvedic properties :
Rasa - Tikta. Vīrya - Śīta.
Guṇa - Laghu, Rūkṣa. Vipāka - Kaṭu.

अगस्त्यः पित्तकफजिच्चातुर्थिक हरो हिमः।
रुक्षो वातकरस्तिक्तः प्रतिश्यायनिवारणः।
तत्पुष्पं पीनसश्लेष्मपित्तनक्तान्ध्यनाशनम्।।
(भावप्रकाश)

Agastyaḥ pittakaphajiccāturthikaharo himaḥ,
Rūkṣo vātakarastiktaḥ pratiśyāyanivāraṇaḥ.
Tatpuṣpam pīnasaśleṣmapitta naktandhya nāśanam.
(Bhāvaprakāṣa)

Actions/Uses : Kaphapittaghna, Kaphanissāraka, Kāsahara, Dīpana, Grāhī, Anulomana, Vraṇaśodhana, Viresananasya, Kṛmighna, Śothahara, Śūlapraśamana, Viṣaghna.

Therapeutics :
Agastya root removes vaat, kapha and inflammation. Bark is tonic, astringent, in infusion given in small pox. Leaves are aperient, tonic, diuretic, laxative, antipyretic, are chewed to disinfect mouth and throat and are useful in sore mouth. They cure night blindness. A poultice made from leaf juice is applied to bruises. Flowers are used as a remedy in nasal catarrh and headache. Juice of flowers is put in eyes to cure dimness of vision. Seeds are used as emmenagogue.

Tephrosia purpurea (Linn.) Pers.

SYNONYMS : *Tephrosia villosa* Pers. *Cracca purpurea* Linn.; *Galega purpurea* Linn.; *Tephrosia maxima* Pers.; *Tephrosia lanceolata R.* Grah.; *Tephrosia hamiltonii Drumm ex* Gamble.

ENGLISH : Purple tephrosia, Wild indigo.

INDIAN : Unhaallee, Sharapunkhaa, Pleehaaree, Sarphonka, Dhamasia, Sirapankh, Jangli-kulthi.

SANSKRTA : Sarapunkha, Nīlavṛkṣakṛtī, Kālaśāka, Bānapunkhā, Sayakapunkha, Plihāśatrū, Śarpamukhī, Kriti, Plihāri, Indrapuṣpika.

HABIT : A polymorphic, much-branched sub-erect perennial herb.

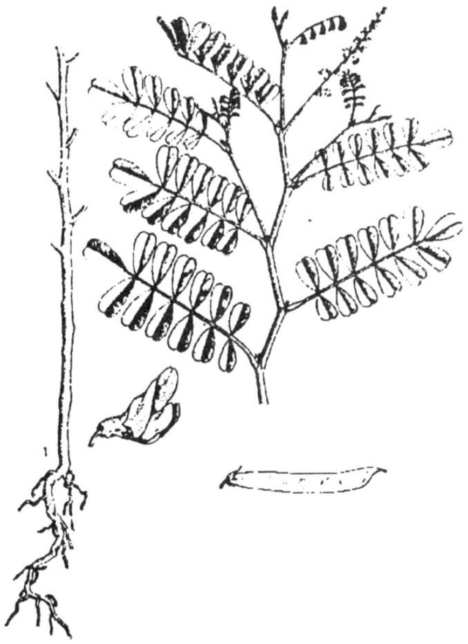

HABITAT : All over India.

CHEMICAL COMPOSITION : Caffeic acid from dormant seeds. Gucoside rutin and osyritin, β-sitosterol and lupeol from leaves. Delphinidin chloride and cyanidin chloride from flowers. Roots contain tephrosin, deguelin, isotephrosin, rotenone etc. Plant also contains tephrinose, villol, villinol, villosin, villosinol. Leaf extract exhibits dose dependent hypotensive activity and seed extract lowers blood glucose level considerably.

Parts used : Whole plant, roots, leaves, seeds and salts.

Ayurvedic properties :
Rasa - Tikta, Kaṣāya. Vīrya - Ūṣṇa.
Guṇa - Laghu, Rūkṣa, Tīkṣṇa. Vipāka - Katu.

शरपुंसा कटूष्णाच क्रिमिवातरुजापहा।
श्वेता त्वेषा गुणाद्यास्यात् प्रशस्ता च रसायने।।
(राजनिघण्टु)

Śarapunkhā katūṣṇāca kṛmivātarujāpahā.
Śvetā tveṣā guṇādya syāt praśastā ca rasāyaṇe.
(Rajanighantu)

Actions/Uses : Plīhaghna, Kaphaghna, Vātaghna, Śothahara, Dīpana, Kuṣṭhaghna, Jantughna, Vraṇasodhana-ropaṇa, Yakṛtottejaka, Pācana, Anulomana, Pittaśāraka.

Therapeutics :

Śarapunkha is highly beneficial in inflammation and enlargement of spleen and liver hence the names Plihasatru and Plihhari. Plant is bitter, astringent, acrid, diuretic, laxative, anthelmintic, and antipyretic; used internally as a purifier of blood and as a cordial. Dried herb is tonic, laxative, diuretic and deobstreunt. It is given for the treatment of bronchitis and bilious febrile attacks and also for the treatment of boils, pimples and bleeding piles. It is useful in coughs and in kidney disorders. Extracts of herb are useful in insufficiency of liver and improves its functioning. Decoction of the herb when given in Bright's disease with dropsy, showed mild diuretic effect. Herb is used as anthelmintic for children. Root is bitter and given in tympanitis, dyspepsia, chronic diarrhoea and obstinate colic. It is very efficaceous against hydrocele. Clinical studies have demostrated the roots are effective against inflammations of tonsils and adenoids. Fresh root bark, ground and made into a pill with a little pepper, is given in cases of obstinate colic. Leaves are useful in jaundice. Juice of leaves is beneficial in dropsy and diabetes. Seeds are employed as an anthelmintic for children.

Trigonella foenum-graecum Linn.

SYNONYMS : *Trifolium foenum graecum; Foenum-graecum officinale.*

HABIT : An aromatic, erect, annual herb.

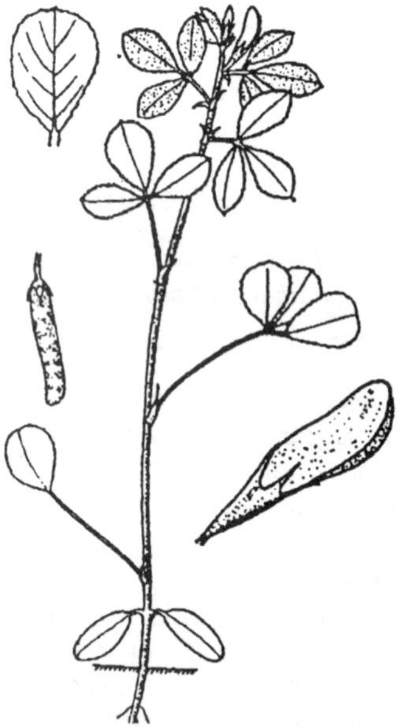

HABITAT : A native of the East. Cultivated in many parts of India.

ENGLISH : Fenugreek.

INDIAN : Methee, Muthi.

SANSKṚTA : Methikā-2, Methinī, Candrikā, Dīpanī, Jyotī, Pītabīja, Bahupatrikā, Gandhaphala, Vallarī, Kuncikā, Asumodhagam.

CHEMICAL COMPOSITION : Quercetin and luteolin; trigonelline and choline, nicotinic acid, fixed oil and an essential oil, mucilage, bitter extractive and coloring matter, two steroidal saponins from seeds. On hydrolysis, saponisn from stems yielded a sapogenin, while that from leaves gave diosgenin, tigogenin and gitogenin. Vitexin and isovitexin from seeds. Trigonelloside C from leaves.

Parts used : Whole plant and seeds.

Ayurvedic properties :

Rasa - Kaṭu.	Vīrya - Ūṣṇa.
Guṇa - Laghu, Snigdha.	Vipāka - Kaṭu.

मेथिका कटुरुष्णा च रक्तपित्त प्रकोपनी।
अरोचकहरा दीप्तिकरा वातघ्न दीपनी।।
(राजनिघण्टु)

Methikā kaṭuruṣṇā ca raktapitta prakopinī,
Arocakaharā dīptikarā vātaghnadīpanī.
(Rājanighaṇṭu)

Actions/Uses : Vātakaphaghna, Raktapittaprakopaka, Śothahara, Dīpana, Pācana, Anulomana, Śūlapraśamana.

Therapeutics :

Methikā plant and seeds are suppurative, aperient, diuretic, emanogogue, aphrodisiac; useful in dropsy, chronic cough, enlargerment of spleen and liver. They are recommended for use in dyspepsia with loss of appetite, in diarrhoea of puerperal women and in rheumatism. Leaves are used both internally and externally for their cooling properties. Seeds, carminative, tonic, antipyretic, anthelmintic; astringent; increase appetite; cure leprosy, vaata, vomiting, bronchitis, piles; an infusion given to small-pox patients as a colling drink; toasted and then infused, used for dysentery.

Uraria picta Desv.

SYNONYMS : *Hedysarum pictum* Jacq.

INDIAN : Raanagaanjaa, Prishnaparnnee, Chitraparnnee, Dabra, Deter, Pitvan.

SANSKRTA : Pṛṣṇiparṇī, Pruthakparṇī, Kalaśī, Guhā, Dhavani, Śṛugālavinnā, Citraparṇī, Lāngulī, Dhamanī, Andhrī, Mekhalā, Kroṣṭapucchī, Dīrghaparṇī, Jatilā, Sinhapucchī, Upacitra, Vedāvanhi.

HABIT : An erect, little-branched perennial weed.

HABITAT : In the dry grasslands, waste places and open forests and throughout the plains of India.

Parts used : Whole plant, pods and roots.

Ayurvedic properties :
Rasa - Madhura, Tikta. Vīrya - Uṣṇa.
Guṇa - Laghu, Snigdha. Vipāka - Madhura.

पृश्निपर्णी त्रिदोषघ्नीवृष्योष्णा मधुरा सरा।
हन्तिदाहज्वरश्वास रक्तातिसारतृड्वमीः।
(भावप्रकाश)

Pṛśniparṇī tridoṣaghnī vṛṣyoṣṇā madhurā sarā.
Hantidāhajvaraśvāsaraktātisāratuḍvamīḥ.
(Bhāvaprakāṣa)

Actions/Uses : Vātaghna, Pittaghna, Kaphaghna, Tridoṣghna.Hṛdya, Dīpana, Śonitasthāpana, Raktastambhana, Jvaraghna, Vātānulomana, Tṛṣṇāpraśāmaka.

Therapeutics :

Prusniparni is one of the components of Dasamoola. It has fracture healing porperties. It is employed for treating heart trouble. Root is aphrodisiac and its decoction is prescribed for cough, chills and fevers. Roots and pods are employed for treating prolapse of anus in infants. Fruit applied to the sore mouths of children. Plant considered useful in snake bite.

Family : XLV - Flacourtiaceae.

Flacourtia indica Merr.

SYNONYMS : *Flacourtia ramontchi L'Herit.; Flacourtia intermis auct (non Roxb.).; Gmelina indica Burm. f.; Flacourtia latifolia Cooke.; Flacourtia occidentalis Blatter., Flacourtia cataphracta.*

ENGLISH : Mauritius plum, Lovi-lovi, Tomi-tomi.

INDIAN : Taambat, Paan aawalla, Bhekal, Kaker.

SANSKṚTA : Vikantaka, Srvavrksa, Granthila, Kantaki, Svadukantaka, Yajnyavrksa, Vyaghrapada, Vrkhakantakari.

HABIT : Low shrub.

HABITAT : Punjab, Bihar, Maharashtra and Southern Peninsula.

CHEMICAL COMPOSITION : Alcoholic extract of aerial parts is diuretic and spasmogenic. A mixture of plant and gairika (red ochre) taken on 4th day of cycle makes a woman sterile.

Parts used : Fruit and bark. Root.

Ayurvedic properties :

Rasa - Madhura. Virya - Sita.
Guna - Laghu, Snigdha. Vipaka - Madhura.

विकंकतफलं पक्वं मधुरं सर्वदोषजित्।
(भावप्रकाश)

Vikankataphalam pakvam madhuram sarvadosajit. (Bha.)

Actions/Uses : Tridosaghna, Pittasaraka, Plihasothahari, Dipana, Pacana, Mutrala, Bhedana.

Therapeutics :

Vikantaka is used in burning sensation of chest, sores, fever, carbuncle, stab wound and cholera. Root is sweet, cooling, depurative, diuretic and is used in bilious affections, rheumatism, poisonous bites, skin diseases, urinary disorders and mental dieases. Bark is astringent and diuretic; applied to eczema; paste given internally only once in mad dog-bite. Fruits, digestive and stomachic, used in jaundice and enlarged spleen. Gum is given with other ingredients for cholera. Kerala physisians use it in preparations like "Nirgundyaadi gulika" and "Nirgundyaadi ghrutam".

Hydnocarpus laurifolia (Dennst.) Sleumer.

SYNONYMS : *Hydnocarpus pentandra* Oken; *Hydnocarpus wightiana* Blume.; *Hydnocarpus inebrians* Wall.; *Munnicksia laurifolia* Dennst., *Hydnocarpus venenata* Gaertn.

ENGLISH : Jangli almond.

INDIAN : Kadukavith, Kowti, Kastel, Chaulmoorga, Kantel.

SANSKṚTA : Tuvaraka, Katukapitha, Kusthavairi, Tuvari.

HABIT : A middle sized tree.

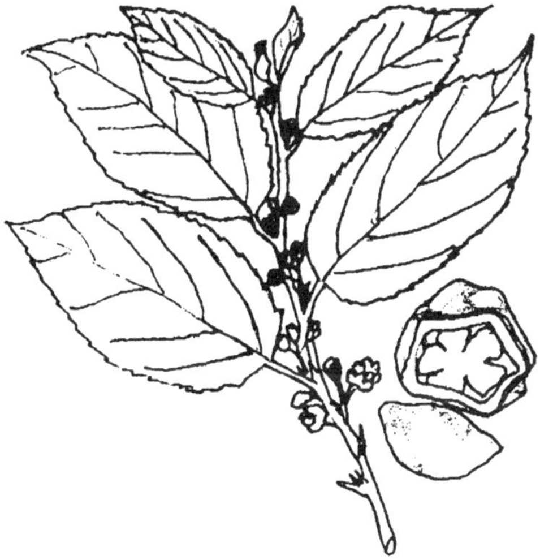

HABITAT : Grows over gardens and accessible places all over Western Peninsula, Konkan and South India.

CHEMICAL COMPOSITION : Pericarp contains leucopelargonidin. Seed hulls, flavonolignan, hydrocarpin, isohydrocarpin, methoxyhydrocarpin, apigenin, luteolin and chrysoeriol. Seed coat yields flavonolinans hydnowightin and neohydnocarpin. Hydrocarpin from seeds. Seeds yield fatty oil. Oil contains hydnocarpic, chaulmogric, gorlic, oleic, and palmitic acids.

Parts used : Seeds and seed oil.

Ayurvedic properties :
Rasa - Katu, Tikta, Kasaya. Virya - Usna.
Guna - Laghu, Tiksna, Snigdha. Vipaka - Katu.

तुवरस्तुवर चोष्णो रसे पाके च तिक्ततः।
कफव्रणकृर्मिमहे कुष्ठज्वर विनाशनः।
आनाहमर्शः शोफश्च नाशयेदिति ते जगुः।।
(नि. र.)

Tuvarastu varascosno rase pake ca tiktakah,
Kaphavranakrmimehah kusthajvaravinasanah,
Anahamarsah sophasca nasayediti te jaguh.
(Ni.rat.)

Actions/Uses : Kaphaghna, Vataghna, Kusthaghna, Kandughna, Jantughna, Vranasodhana-ropana, Vamaka, Virecana, Krmighna.

Therapeutics :
Tuvaraka fruits are piscididal. Seed and oil used in leprosy and skin diseases and as fish poison. Oil mixed with vaseline is efficacious in many skin diseases. Mixed with lime water it is used in sprains in rheumatic joints. Seeds in the form of paste and mixed with sulphur, camphor, castor oil find use as external application in wounds and ulcers.

Family : XLVI - Gentianaceae.

Exacum bicolor Roxb.

SYNONYMS : *Seboea carinata* Graham.

HABIT : An erect annual herb.

HABITAT : Maharashtra, Konkan and Western Ghats.

INDIAN : Udichirayat.

SANSKṚTA : Udicirayata.

CHEMICAL COMPOSITION : Plant extract antifungal against Helminthosporium sativum.

Parts used : Whole plant.

Ayurvedic properties :
Rasa - Tikta, Katu. Virya - Sita.
Guna - Laghu, Ruksa. Vipaka - Katu.

Actions/Uses : Dipana, Katupaustika, Jvaraghna, Saraka.

Therapeutics :

Udicirayata is tonic, stomachic and a substitute for gentian and chiretta.

Swertia chirata Buch-Ham.

SYNONYMS : *Gentiana chirayita* Roxb.; *Swertia chirata (Wall.)* Clarke.; *Swertia chirayita (Roxb. ex Flem.)* Kars.; *Ophelia chirata* Griseb.; *Swertia tongluensis* Burkill.

ENGLISH : Chiretta.

INDIAN : Chiraayata.

SANSKRTA : Kirata, Ciratatikta, Bhunimba, Kirataka, Naditikta, Naipala, Jvarantaka, Katutikta, Ramasevaka, Nidrari, Ramasenaka, Sannimataha, Kandatikta, Anaryatikta.

HABIT : An erect annual herb.

HABITAT : Semi-evergreen forests in India.

CHEMICAL COMPOSITION : Bitter principles (1.5 %), Ophelic acid, chiratin. Swertinin, swertianin, swerchirin, decussatin, isobellidifolin, friedelin, and B-sitosterol. Gentianine showed antipsychotic activity in rats and mice.

Parts used : Whole plant and ash.

Ayurvedic properties :
Rasa - Tikta. Virya - Sita.
Guna - Laghu. Vipaka - Katu.

किरातः सारको रुक्षः शीतलस्तिक्तको लघुः।
सन्निपातज्वरश्वास कफपितास्त्रदाहनुत्॥
कासशोथतृषाकुष्ठज्वर व्रणक्रिमिप्रणूत्॥
(भावप्रकाश)

Kiratah sarako ruksah sitalastiktko laghuh,
Sannipatajvarasvasa kaphapittasradahanut,
Kasasothatrsakustha jvaravranakrimipranut.
(Bha.)

Actions/Uses : Dipana, Pacana, Katupaustika, Jvaraghna, Dahasamana, Niyatakalika Jvarapratibandhaka, Anulomika.

Therapeutics :

Kirata is bitter and tonic without aroma or astringency. Unlike other bitters it does not constipate the bowels. Instead it tends to produce a regular action and causes a free discharge of bile. It is used as powder, infusion or as an extract. It is antimalarial, tonic, stomachic, febrifuge, laxative, anthelmintic and antidiarrhoeal. Plant extract is used as effective remedy for chronic fever and is used as a special remedy for bronchial asthma and liver disorders. If taken with sandal wood paste it stops internal haemorrhage of the stomach.

Swertia decussata Nimmo ex Grah.

SYNONYMS : *Swertia densiflora (Griesb.)* Kashyapa.

HABIT : An erect annual herb.

HABITAT : Semi evergreen forests in India.

INDIAN : Silaajita, Kaddu.

SANSKRTA : Silajita, Cirata.

CHEMICAL COMPOSITION : A monohydroxy triterpene and a dihydroxy triterpin from leaves and flowers.

Parts used : Whole plant.

Ayurvedic properties :
Rasa - Tikta. Virya - Sita.
Guna - Laghu. Vipaka - Katu.

शैलजं कटुकं तिक्तं मेहघ्नंच रसायनम्।
उष्णमुन्मादशोफघ्नं क्षयकुष्ठाश्मरीहरम्॥
शोफोदरापस्मारघ्नं वस्त्यशोरोगनाशनम्।
कण्डू च पाण्डुरोगं च छर्दिं वातं कफं जयेत्॥
वलीपलितकासघ्नं श्वासमूत्ररुजापहम्।
(निघण्टु रत्नाकर)

Sailajam katukam tiktam mehaghnam ca

rasayanam,
Usnamunmada sophaghnam ksaya kusthasmari haram.
Sophodarapasmaraghnam bastyasonroganasanam,
Kandum ca pandurogam ca cchardim vatam kapham jayet.
Valipalita kasaghnam svasamutrarujapaham.
(Ni.rat.)

Actions/Uses : Katupaustika, Sophaghna, Unmadanasaka, Vantighna.

Therapeutics :

Silajita is a good substitue for S. chirata and Gentiana kurroo. It cures urinary calculi, haemorhoids, asthma and cough.

Family : XLVII - Hypoxidaceae.

Curculigo orchioides Gaertn.

SYNONYMS : *Curculigo malabarica* Wight.; *Curculigo brevifolia* Dryand.; *Hypoxis orchioides* Kurz., *Carcinia malabarica* Desv.

INDIAN : Kaalli musalli.

SANSKṚTA : Musali, Talamuli, Krsna-musali, Talapatraka, Bhutali, Varsabhava, Dirghakandika, Hemapuspi.

HABIT : A sub-erect annual or biennial herb with rhizomes.

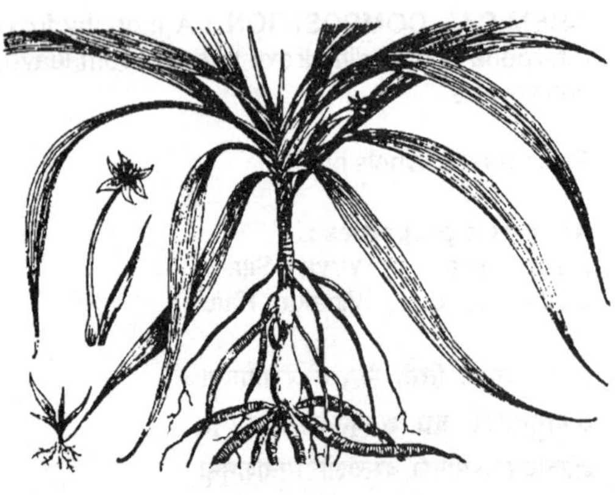

HABITAT : Sandy situations of hotter regions of India and Ceylon.

CHEMICAL COMPOSITION : Rhizome glycosides, also yuccagenin and alkaloid lycorine.

Parts used : Tuberous roots and bulbs.

Ayurvedic properties :
Rasa - Madhura, Tikta. Virya - Usna.
Guna - Guru, Snigdha, Picchila. Vipaka - Madhura.

मुशली मधुरा वृष्या वीर्योष्ण बृंहणी गुरुः।
तिक्ता रसायनी हन्ति गुदजन्यनिलं तथा।।
(भावप्रकाश)

Musali madhura vrsya viryosna brnhani guruh,
Tikta rasayani hanti gudajanyanilam tatha.
(Bha.)

Actions/Uses : Vatapittasamaka, Kaphavardhaka, Dipana, Anulomana, Vrsya, Sukrala, Brihmana, Rasayana, Mutrala, Mahakusthaghna.

Therapeutics :

Musali is a rejuvenative and aphrodisiac medicine. It is a bitter, aromatic tonic, viriligenic, roborant, easily digestible, diuretic and restorative. Roots are alterative and tonic. Rhizome is prescribed in piles, jaundice, asthma, dysuria, diarrhoea, gonorrhoea, menorrhagia, leucorrhoea and menstrual derangements. It is used as polutice for itch and skin diseases. It is pounded and applied on cuts and wounds; pounded with ajwain and its decoction given to unconscious children. Powder used externally to heal wounds. Leaves are used in whitlows.

Family : XLVIII - Iridaceae.

Crocus sativus Linn.

SYNONYMS : *Crocis officinalis var. A.* Hudson.; *Crocus orsinii* Parl.

ENGLISH : Saffron.

INDIAN : Keshara, Jafraana.

SANSKṚTA : Kumkuma, Ghusṛuṇa, Agniśekhara, Śoṇita, Vanhiśikha, Piṣuna, Vareṇya, Kāśmīrajanmā, Rakta, Asrugvara, Śaṭha, Pītaka, Rudhira, Kānta, Bāhlika, Kesara, Gaura, Kāleyaka, Cāru, Sankoca, Lohitacandana, Varayonika, Dipaka, Kusumātmaka, Bhavarakta, Saurabha, Agniśikhā.

HABIT : A small bulbous perennial herb with tuberous roots.

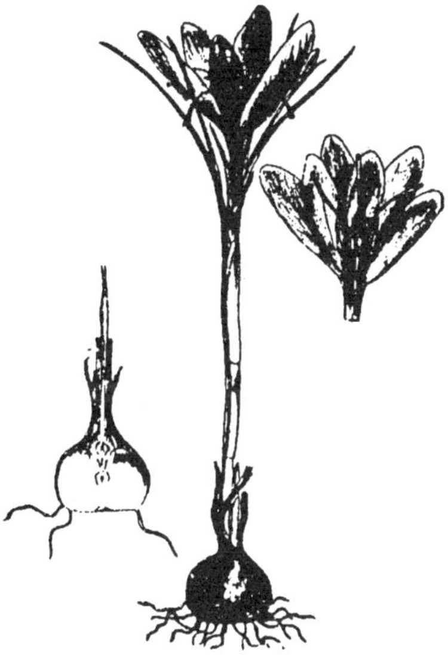

HABITAT : Greek and Roman origin, cultivated for generations in Spain, England and Kashmir in India.

CHEMICAL COMPOSITION : Saffron contains glucoside- crocin, crocetin, picrocrocin and an essential oil.

Parts used : Kesara.

Ayurvedic properties :
Rasa - Kaṭu, Tikta. Vīrya - Uṣṇa.
Guṇa - Snigdha. Vipāka - Kaṭu.

कुंकुमं कटुकं तिक्तमुष्णं श्लेष्मसमीरजित्।
व्रणदुष्टिशिरोरोगविषहृत् कायकान्तिदम्।। (ध. नि.)

Kunkumam kaṭukam tiktamuṣṇam ślasmasamīrajit,
Vraṇaduṣtiśirorogaviṣahṛt kāyakāntidam.
(Dhanvantari)

Actions/Uses : Tridoṣaśāmaka, Dīpana, Pācana, Rucikara, Grāhī, Cchardinigrahaṇa, Yakṛtottejana, Hṛdayottejaka, Mūtrajanana, Vājikaraṇa, Svedajanana, Garbhāśayasankocaka.

Therapeutics :

Kumkuma is carminative, stimulant, stomachic, diaphoretic, emmenagogue. It has been used with benefit in amenorrhoea, dysmenorrhoea, hysteria etc. It arrests chronic discharges of blood from uterus. Used as a diaphoretic, especially for children. It is given to children in exanthematous diseases from its reputed power in promoting eruption. Plant extract is uterine stimulant. Dried stigmas and tops of styles used in fevers, melancholia and enlargement of liver, catarrhal affections and snake bite. Different preparations have been used in various kinds of cancers. In Unani medicine, used as exhilarant, helps absorption of cardiotonics and strengthens inner part of body. An indigenous drug "Amber Mezhugu" useful in rheumatism. It is a constituent of preparations for preventing premature ejaculation.

HOMEOPATHIC USE : Crocus sativa is a Homeopathic remedy. The main action of the drug is on blood and nervous system; there are also a few but very important symptoms of eyes. The patient is obliged to wink or wipe the eyes; asthenopic. Extreme photophobia. Cannot read without gush of tears. In haemorrhages from different parts, the blood is black, viscid, clotted, and forming itself into long black strings. It is a specific for menorrhagia. It is very useful in hysterical conditions, particularly where there is great changeableness of mental symptoms.

Iris versicolor Linn.

ENGLISH : Blue flag, Flag Lily, Snake Lily, Liver Lily.

INDIAN : Waalavekhanda.

SANSKṚTA : Hemavatī, Vacā.

HABIT : A herb with rhizome.

HABITAT : A common British garden plant.

CHEMICAL COMPOSITION : Iridin.

Ayurvedic properties :
Rasa - Kaṣāya. Vīrya -Ūṣṇa.
Guṇa - Lekhanīya. Vipāka - Kaṭu.

Parts used : Roots.

Actions/Uses : Kaphapittasamana, Kāsahara, Āmavatahara.

Therapeutics :

Rhizome possesses cathartic and emetic properties. It is also alterative, diuretic and stimulant; chiefly used for its alterative properties. It is also valuable in liver complaints, fluor albus and dropsy. Eclectics prepare an oleoresin called IRIDIN which is believed to unite cholagogue and diuretic with aperient properties.

HOMEOPATHIC USE : Iris versicolor is a Homeopathic remedy. It has a very specific action on the digestive tract and mouth. It is a remedy for catarrh, congestion, heat and burning of all serous and mucous membranes in abdomen, throat and mouth. Its all secretions are acrid and corosive that every part is badly excoriated.

Family : XLVIII - Juglandaceae.

Juglans regia Linn.

ENGLISH : Walnut, Nux juglans, European walnut, Persian walnut.

INDIAN : Akroda, Akruta, Akhor, Krot.

SANSKṚTA : Akṣoṭaka, Pārvatīya, Phalasneha, Madhumajjā, Kīreṣṭa, Kandarāla, Bruhacchada, Karparāla, Rekhāphala, Śaīlabhava, Pīlu, Madanāhva, Svarāmaka, Guḍāśaya.

HABIT : A large deciduous monoecious tree.

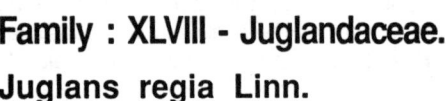

HABITAT : A native of Iran. Found throuhout Himalayas and hills of Assam, Kashmir and Khasia Hills.

CHEMICAL COMPOSITION : Cysteine, tryptophan, thiamin, ribiflavin, nicotinic acid, pantothenic acid, folic acid, vitamin B6, biotin, vitamin A, ascorbic acid are found in the leaves and fruits. Kernels yield 60-70 % of a drying oil known as Walnut Oil. Fresh leaves and unripe fruit are rich in ascorbic acid. Leaves yield an essential oil. Juglone, bernerine, cyclotrisjuglone, β-sitosterol from bark and root. Oxalic acid from fruit.

Parts used : Leaves, fruit, bark, seed and seed oil.

Ayurvedic properties :
Fruit :

Rasa - Madhura.	Vīrya - Ūṣṇa.
Guna - Gurū, Snigdha.	Vipāka - Madhura.

Bark :

Rasa - Kaṡaya.	Vīrya -Śīta.
Guṇa - Gurū, Snigdha.	Vipāka - Kaṭu.

अक्षोटकं गुणे स्निग्धं मधुरं रसपाकयोः।
गुरुष्णं बृंहणं वृष्यं बल्यं विष्टम्भीरोचनम्।
हृद्यं क्षयास्त्रपवनदाहघ्नं कफपित्तलम्।।
(नि. र.)

Akṡoṭakam guṇe snigdham madhuram rasapākayoḥ,
Guruṣṇam bṛnhanam vṛṣyam balyam viṣṭambhīrocanam,
Hṛdyam kṣayāsrapavanadāhagham kaphapittalam.
(Nighaṇṭuratnākar)

अक्षोटो मधुरो वल्यो स्निग्धोष्णो वातपित्तजित्।
रक्तदोषप्रशामनः शीतलः कफकोपनः।।
(राजनिघण्टु)

Akṣoṭo madhuro balyo snigdhoṣṇo vātapittajit,
Raktadoṣapraśamanaḥ sītalaḥ kaphakopanaḥ.
(Rajanighaṇṭu)

Actions/Uses : Vātasāmaka, Kaphapittavardhana, Dīpana, Snehana, Anulomana, Balya, Vājikaraṇa, Varṇya, Śothahara, Vedanā-śāmaka, Kuṣṭhaghna.

Therapeutics :
Leaves of Akśoṭaka are astringent, tonic, in decoction considered to be specific in strumous sores. Bark and leaves are anthelmintic, alterative, laxative and detergent. Used in herpes, eczema, syphilis and scrofula. Externally used as an application to skin eruptions, ulcers, etc. Bark is detergent, used against cancer; paste for toe sores. Fruits are tonic and useful in rheumatism. Shells of unripe fruits are vermifuge. Ripe fruit or kernel is palatable and possesses aphrodisiac properties. Seed oil is used for tape worms.

HOMEOPATHIC USE : Juglans regia is a Homeopathic remedy. It has similar symptoms as Juglans cinerea. Peevishness and mental indolence, exited as if intoxicated are the main mental symtoms. Lancinating pains in forehead, flatulence and bloating of abdomen, affections of spleen and liver, diarrhoea are some of the chief symptoms.

Family : XLIX - Lamiaceae.

Anisomeles malabarica (L.) R. Br.

SYNONYMS : *Anisomeles disticha.; Aninomeles fruticosa.*

ENGLISH : Malabar catimint.

INDIAN : Chaudhara, Kapurimadhuri, Vaikunttha.

SANSKṚTA : Devadroṇī, Sprukkā, Mahādroṇa, Alamūla, Butan-kuśa, Vaikuṇṭha.

HABIT : An aromatic, densely pubescent, perennial herb.

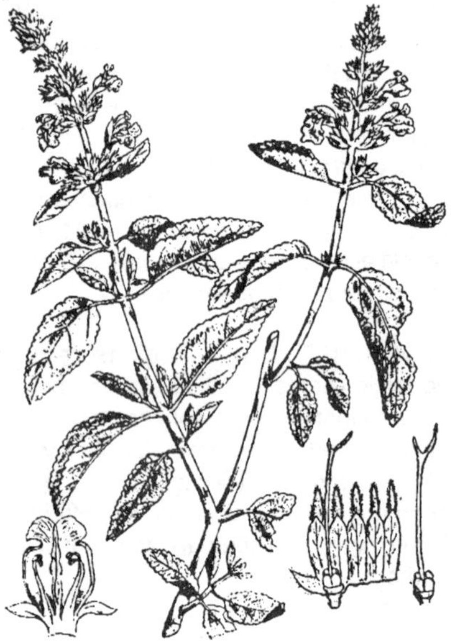

HABITAT : Commonly found in the western ghats from Maharashtra to Karnataka, Andhra Pradesh, and in Kerala and Tamil Nadu.

CHEMICAL COMPOSITION : Plant contains β-sitosterol; two diterpene lactones, ovatodiolide and anisomelic acid; also monocyclic diterpenes, anisomelolide, malbaric acid, anisomelyl-OAc and anisomelol. Aerial parts, anisomelin.

Parts used : Entire herb, leaves and essential oil.

Ayurvedic properties :
Rasa - Kaṭu, Tikta. Vīrya - Śīta.
Guṇa - Laghu. Vipāka - Kaṭu.

देवद्रोणी कटुस्तिक्ता मेध्या वातार्तिभूतनुत्।
कफमान्द्यापहा चैव युक्त्या पारदशोधने।।
(राजनिघण्टु)

Devadroṇī kaṭustiktā medhyā vātārttibhūtanut,
Kaphamāndyāpahā caivayuktyā pāradaśodhane.
(Rājanighaṇṭu)

Actions/Uses : Tridoṣaghna, Medhya, Svedajanana, Śitapraśamana, Uttejaka, Kṛmighna, Jvaraghna, Sutikāśulajita.

Therapeutics :
Devadroni is acrid, bitter, aromatic, intellect promoting, stomachic, anthelmintic, febrifuge, diaphoretic and sudorific. It has anti-spasmodic, anti-pyretic, emmenagogue, anti-periodic properties. It is useful in epilepsy, hysteria, amentia, colic,anorexia, dyspepsia, flatulence, worms. It cures itching, thirst, indigestion, fever, leprosy and haematological disorders. It is used in scorpion sting and snake bite. A decoction of the herb is beneficial in rheumatic joints. Leaves are used as stomachic, carminative, diaphoretic and astringent. The fresh juice or infusion of leaves is used in colic and dyspepsia; catarrhal affection and intermittent fever. It is administered to children in colic, dyspepsia and fever arising from teething. Essential oil is used externally in rheumatism.

Coleus amboinicus Lour.

SYNONYMS : *Coleus aromaticus* Benth.; *Plectranthus ambonicus (Lour.)* Spreng.

ENGLISH : Country borage, Indian borage.

INDIAN : Karmalo, Paatherachura, Paana owaa.

SANSKṚTA : Karpuravallī, Parṇayavānī, Pāṣāṇabhedī, Pātharacūra, Odhāpāna.

HABIT : A rather large succulent herb with aromatic leaves.

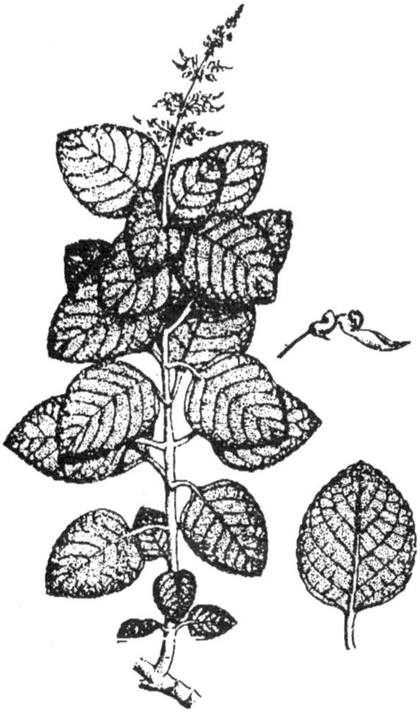

HABITAT : Cultivated in gardens throughout India.

CHEMICAL COMPOSITION : Carvacrol in essential oil. Cirsimaritin and β-sitosterol-B-D-glucoside, oleanolic, pomolic, euscaphic, tormentic and ursolic acids; salvigenin, quercetin, chrysoeriol, luteolin, apigenin, eriodictyol and taxifolin in leaves.

Parts used : Leaves.

Ayurvedic properties :
Rasa - Kaṭu, Tikta. Vīrya - Uṣṇa
Guṇa - Laghu, Rūkṣa. Vipāka - Kaṭu

तीक्ष्णा पर्णयवानुष्णा कटुस्तिक्ता रसे लघुः।
दीपनी पाचनी रुच्या मलसंग्राहिणी परम्।।
अग्निमांद्ये यकृद्रोगे ग्रहण्यामुदरौक्रिमी।
विषूचिकायामश्मर्यां मूत्रकृच्छ्रे च शस्यते।।

Tīkṣṇā parṇayavānyuṣṇā kaṭustiktā rase laghuḥ,
Dīpanī pācanī rucyā malasangrāhiṇī param.
Agnimāndye yakṛdroge grahaṇyāmudarau krimī,
Viṣūcikāyāmaśmaryām mūtrakṛcche ca śasyate.

Actions/Uses : Kaphavātaśāmaka, Vedanāsthāpana, Viṣaghna, Dīpana, Pācana, Grāhī, Rucikara, Vātānulomana, Yakṛtottejana, Kṛmighna.

Therapeutics :

Karpuravallī with its distinctive smelling leaves is a common home remedy for infantile cough, cold and fever. Plant is hot, digestive, carminative, diuretic, anthelmintic and constipating. It is antispasmodic, antilithic, cathartic, stimulant and stomachic. It is given in epilepsy and other convulsive affections, kidney and bladder stones. Leaves are used in urinary diseases and vaginal discharges. Juice of leaves mixed with sugar acts as a powerful aromatic carminative, given in colic and dyspepsia. In Samoa, weed is used for cold; herb against fractures; leaves applied as polutice for headache.

Leucas aspera Spreng.

SYNONYMS : *Phlomis aspera* Willd.; *Leuca aspera (Willd.)* Link.; *Leucas cephalotus.*

INDIAN : Ṫamba, Tumbo, Chota halkusa.

SANSKṚTA : Droṇapuṣpī, Droṇā, Phalepuṣpā, Citrak-supa.

HABIT : A herbaceous, much-branched, errect or diffuse annual.

HABITAT : Found throughout India and Ceylon.

CHEMICAL COMPOSITION : Two sterols, two alkaloids, galactose, oleanolic acid, ursolic acid and β-sitosterol. An alcoholic extract of leaves antibacterial against Micrococcus pyogenes and Escherichia coli.

Parts used : Whole plant, leaves and flowers.

Ayurvedic properties :
Rasa - Katu, Lavaṇa, Madhura. Vīrya - Ūṣṇa.
Guṇa - Gurū, Rūkṣa. Vipāka - Madhura, Kaṭu.

द्रोणपुष्पी गुरुः स्वादु रुक्षोष्णाकफपित्तनुत्।
सतीष्णा लवणां स्वादुपाका कट्वी च भेदिनी।।
कफामकामलाशोषतमकश्वासजन्तुजित् ।
पीतों मरिचचूर्णेन तुलसीपत्रजो रसः।
द्रोण पुष्पीरसो वाऽपि निहन्ति विषमज्वरम्।।
(भावप्रकाश)

Droṇapuṣpī guruḥ svādu rukṣoṣṇākaphapittanut,
Satīkṣṇā lavaṇā svādupākā kaṭvī ca bhedinī.
Kaphāmakamālaśoṣatamakaśvāsajantujit.
Pīto maricacurṇena tulasīpatrajo rasaḥ
Droṇapuṣpīraso vāpi nihanti viṣamajvaram.
(Bhāvaprakāṣa)

अंजने कामलातीतां द्रोणपुष्पीरसो हितः।
(वृंद्ध)

Anjane kāmalātītām droṇapuṣpīraso hitaḥ.
(Vṛnda.)

Actions/Uses : Kaphaghna, Vātaghna, Pittaśodhana, Agnidīpana, Anulomana, Pittasāraka, Recana, Kṛmighna.

Therapeutics :

Droṇapuṣpī is antipyretic and insecticide. It is a reputed home remedy for worms, fever and intestinal catarrh in children. It is antipyretic, antiseptic, carminative, febrifuge, wormifuge, antihistaminic. It is used in anorexia, cough, dyspepsia, fever, jaundice, psoriasis, respiratory and skin diseases. Leaves are considered useful in chronic rheumatism. Juice of leaves is applied in psoriasis, scabies and chronic skin eruptions. Flowers are used in cold.

Mentha arvensis Linn.

ENGLISH : Marchmint, Corn mint.

INDIAN : Pudinaa.

SANSKṚTA : Pudinā, Pūtinaśa, Vātaloma.

HABIT : A small annual herb.

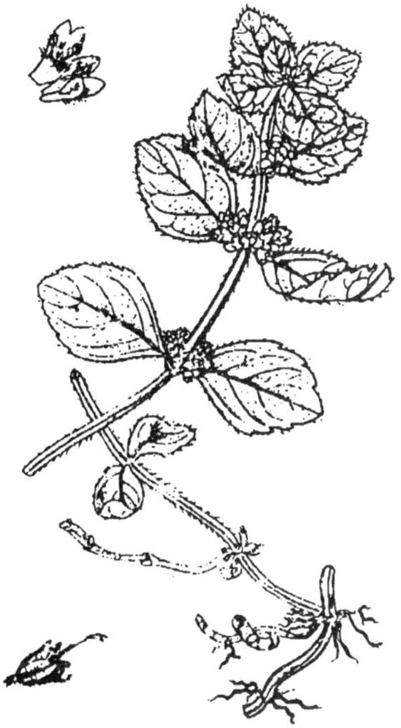

HABITAT : Western Himalayas, Kashmir, Punjab, Kumaon, Garhwal.

CHEMICAL COMPOSITION : Essential oil contains chiefly pulegone and d-isomenthone. Essential oil from leaves contains d-carvone , carene, d-sylvestrene and citronellol. Acacetin, apigenin, diosmetin, eriodictyol, hesperitin and luteolin from Polish herb. Alcoholic extract of leaves is anti-ovulatory and inhibitory of implantation in female experimental animals.

Parts used :

Ayurvedic properties :

Rasa - Madhura. Vīrya - Ūṣṇa.
Guṇa - Gurū, Rūkṣa. Vipāka - Madhura.

व्यञ्जनो वान्तिहारीच रुचिश्यः शाकशोभनः।
पुदिनस्तु गुरुः स्वादु रुच्यो हृद्यः सुखावहः।।
मलमूत्रस्तम्भकर कफकासमदापहः।।
अग्निमांद्य विषूचिघ्नः संगहण्यतिसारहा।।
जीर्णज्वरंकृमींश्चैव नाशयेदिति कीर्तितम्।
(भावनिघण्टु)

Vyanjano vāntihārī ca ruciścyaḥ śākaśobhanaḥ,
Pudinastu guruḥ svādu rucyo hṛdyaḥ sukhāvahaḥ.
Malamūtrastambhakara kaphakāsamadāpahaḥ,
Agnimāndya visūcighnaḥ sangrahaṇyatisārahā,
Jīrnajvaram kṛmīmscaiva nāsayediti kīrtitam.
(Bhāvanighaṇṭu)

Actions/Uses : Ūṣṇa, Rūkṣa. Vātaśamana, Dīpana, Ārtavajanana, Sankocavikāsapratibandhaka.

Therapeutics :

Infusion of leaves of Putināṣa is used in rheumatism. Dried plant is antiseptic, carminative, stomachic, stimulant, diuretic, refrigerant, antispasmodic and emmenagogue.

Mentha piperita Linn.

SYNONYMS : *Mentha officinalus; Mentha hircina* Hull.

HABIT : An annual or perennial herb.

HABITAT : Cutivated in Indian gardens.

ENGLISH : Peppermint.

INDIAN : Peppermittha.

CHEMICAL COMPOSITION : Methyl acetate, menthyl acetate, menthol, menthone and hydrocarbons. Viridiflorol, menthofuran, isomenthone, neomenthol, neoisomenthol, α-pinene, β-pinene, cineol and carvone.

Therapeutics :

Peppermint is the most agreeable and powerful of all the mints. It possesses aromatic, carminative, stimulant, mild antidiarrhoetic antispasmodic and stomachic properties. It is used for allaying nausea, flatulence, sickness, vomiting and as an infants' cordial. Essential oil from plant is antiseptic, stimulant and carminative.

Mentha spicata Linn.

SYNONYMS : *Menthà viridis* Linn.; *Mentha crispa.*

ENGLISH : Mackerel Mint, Spearmint.

INDIAN : Pahaadi pudinaa.

SANSKṚTA : Pūtihā, Phudino, Rocanī.

HABIT : An annual herb.

HABITAT : Cultivated in Indian gardens. Europe, North Africa, Great Britain.

CHEMICAL COMPOSITION : Diosmetin and

diosmin from leaves. Essential oil contains carvacrol, l-carvone, dihydrocarveol, limonene, dipentene and esters.

Parts used : Leaves.

Ayurvedic properties :
Rasa - Kaṭu. Vīrya - Ūṣṇa.
Guṇa - Laghu, Rūkṣa. Vipāka - Kaṭu.

पूतिहा कटुरुष्णश्च रोचनी दीपनो लघुः।
हन्ति वात कफाध्मानशूलच्छर्दिकृमीस्तथा ॥

Pūtihahā kaṭurusnasyca rocanī laghuh,
Hanti vātakaphādḥmāna sūlacchardikṛmimstathā.

Actions/Uses : Kaphavātaśāmaka. Rucikara, Dīpana, Vātānulomana, Cchardigrahaṇa, Kṛmighna.

Therapeutics :

Medicinal properties of Putihā are similar to peppermint, though a little feebler. Herb is considered to be stimulant, carminative and antispasmodic. Leaves are given in fever and bronchitis and in decoction used as a lotion in aphthae. Seeds are mucilaginous.

Ocimum americanum Linn.

SYNONYMS : *Ocimum canum* Sims.

ENGLISH : Sacred basil, Holy basil, Hoary bash.

INDIAN : Tullasa, Kaalli tullasa (2), Mamri.

SANSKṚTA : Arjaka, Ajabalā, Gambhīra, Kuṭheraka.

HABIT : An erect, sweet scented, pubescent herb.

HABITAT : Plains and lower hills of India. Cultivated in home gardens.

CHEMICAL COMPOSITION : Yields a volatile oil useful as a perfume for soaps and cosmetics. Three types of oil one containing methyl cinnamate as the chief constituent, second containing d-camphor and third citral. Polysaccharide containing xylose, arabinose, rhamnose, galactose; galacturonic and glucuronic acids.

Parts used : Leaves and whole plant.

Ayurvedic properties :
Rasa - Kaṭu. Vīrya - Ūṣṇa.
Guṇa - Laghu, Rūkṣa. Vipāka - Kaṭu.

अर्जकः कटुका रुक्षा कटुपाका विदह्रितः।
तीक्ष्णा रुचिकरा हृद्या दीपना लघवोऽ हिमाः।
पित्तलाः कफवातघ्नाध्नन्ति कण्डूविषकृर्मान्॥
(कैयदेव निघण्टु)

Arjakaḥ kaṭukā rukṣa kaṭupākā vidahinaḥ,
Tīkṣṇā rucikarā hṛdyā dīpanā laghavo himaḥ,
Pittalāḥ kaphavātaghnā dhnanti kaṇḍūviṣakṛmin.
(Kaiyadevanighaṇṭu)

Actions/Uses : Kaphanissāraka, Śītahara, Vātahara, Dīpana, Kṛmighna, Svedajanana, Kāsahara, Uttejaka.

Therapeutics :

Arjaka possesses aromatic, carminative, diaphoretic and stimulant properties. A decoction is taken for coughs, that of leaves for dysentery; it is also used as a mouth wash for relieving toothache. Juice of leaves is given to children for cold catarrh and bronchitis. Leaves made into paste used in parasitical skin diseases and applied to fingers and toe-nails during fever when the extremeties are cold.

HOMEOPATHIC USE : It is useful for diseases of kidneys, bladder and urethra. Uric acid diathesis. Red sand in urine is its chief symptom. Renal colic, especially right side, with micturition every fifteen minutes, pain causing patient to wring his hands, moaning and crying all the time. High acidity of urine with formation of spike crystals, symptoms of renal calculus. Cramps in kidneys.

Ocimum basilicum Linn.

SYNONYMS : *Ocimum caryophyllantum* Roxb.; *Ocimum pilosum* Willd.; *Ocimum anisatum.; Basilicum citratum.*

ENGLISH : Sweet basil, Garden basil, Common basil.

INDIAN : Sabzaa, Marvaa, Babui tulsi, Gulaal tulsi, Marua.

SANSKṚTA : Barbarī, Barbara, Munjariki, Surasa, Varvara.

HABIT : A perennial erect, almost glabrous herb.

HABITAT : Cultivated throughout India.

CHEMICAL COMPOSITION : Methyl cinnamate, methylchavicol, cineole, linalool, ocimene, borneol, sambulene and safrol from essential oil.

Parts used : Roots, leaves, flowers and seeds.

Ayurvedic properties :

Rasa - Tikta, Kaṭu.	Vīrya - Ūṣṇa.
Guṇa - Rūkṣa.	Vipāka - Kaṭu.

बर्बरीत्रितयं रुक्षं शीतं कटु विदाहिच।।
तीक्ष्णं रुचिकरं हृद्यं दोपनं लघुपाकिच।
पित्तलं कफवातास्त्तकण्डु क्रिमिविषापहम्।।
(भावप्रकाश निघण्टु)

Barbaī tritayam ruksam sītam katu vidahica,
Tīkṣṇam rucikaram hṛdyam dopanam laghupāki ca,
Pittalam kaphavātāsrakaṇḍu kṛmiviṣāpaham.
(Bhāvaprakāsa)

Actions/Uses : Uttejaka, Vātapraśamana, Svedajanana, Kṛmighna, Majjātantū-uttejaka.

Therapeutics :

Barbari is aromatic with carminative and cooling properties. It cures disorders due to kapha and vaata. dyspepsia, cough, constipation, bronchitis, intermittent fevers. Flowers are diuretic, carminative, stimulant and demulcent. Seeds are mucilaginous and given in infusion for gonorrhoea, dysentery and chronic diarrhoea. Roots are used in bowel complaints of children. Leaves are useful in treatment of croup, for which the warm juice with honey is given.

Ocimum gratssimum Linn.

SYNONYMS : *Ocimum frutescens.; Citratum zeylanicum.*

ENGLISH : Shrubby basil.

INDIAN : Raamtullasa, Ban tulsi, Ran tulasi.

SANSKṚTA : Vanabarbarika, Sumukha, Rāmatulasi.

HABIT : A tall, much-branched perennial shrub.

HABITAT : Cultivated throughout India.

CHEMICAL COMPOSITION : Essential oil contains thymol, eugenol, ocimene, cadinene, methyl chavicol, myrcene, monocyclic terpenes. Ocimol and gratissimin from leaves. Essential oil from leaves and soft stem contains citral, geraniol and citronellol.

Parts used : Leaves.

Ayurvedic propoerties :
Rasa - Tikta. Vīrya - Ūṣṇa.
Guṇa - Rūkṣa, Tīkṣṇa. Vipāka -Kaṭu, Madhura.

वनबर्वरिकाचोष्णा सुगन्धा कटुकाचसा।
पिशाचवान्तिभूतघ्नी घ्राणसन्तर्पणी परा।।
बीजं चास्या दाहशोषनाशकं परिकीर्तितम्।

Vanabarbarikā coṣṇā sugandhā kaṭukā ca sā,
Piśāca vānti bhūtaghnī ghrāṇa santarpaṇī parā,
Bījam cāsyā dāha śoṣa nāśakam parikīrtitam.
(Kayadevanighaṇṭu)

सुमुखो दाहनः स्वास्यो दोषोत्क्लेशी कटुच्छ्दः।
निद्राकरः सुप्रसन्नो गरघ्नः शोभनाननः।।
सुमुखो मधुरः पाके रुक्षोष्णो रोचनो लघुः।
कटुको दीपनो हृद्यो विदाही पित्तलो जयेत्।।
कफानिलविषश्वासकासहृत् पार्श्वजा रुजेः।।

Sumukho dāhanaḥ svāsyo doṣotkleṣī kaṭucchadaḥ,
Nidrākaraḥ suprasanno garaghnaḥ śobhanānanaḥ.
Sumukho madhuraḥ pāke rukṣoṣṇo rocano laghuḥ,
Katuko dīpano hṛdyo vidāhī pittalo jayet.
Kaphānila viṣa svāsa kāsahṛt pārśvajā rujaḥ.
(Kayadevanighaṇṭu)

Actions/Uses : Pūtihara, Vraṇaropaṇa, Vedanāsthāpana, Mūtrajanana.

Therapeutics :
Vanabarbarika or Ramatulasi is aromatic. A strong decoction is effectual in aphthae of children and baths of fumigations prepared with it is recommended in the treatment of rheumatism and paralysis. Decoction of leaves is useful in seminal weakness and is remedy for gonorrhoea. Seeds are given in headache and neuralgia.

Ocimum sanctum Linn.

SYNONYMS : *Ocimum hirsutam.; Ocimum tomentosum.; Ocimum virde.; Ocimum tenuiflorum* Linn.

ENGLISH : Sacred basil, Holy basil, Sacred bash.

INDIAN : Tullasa, Kaalli tullasa, Baranda, Chajadha.

SANSKṚTA : Tulasī, Surasa, Apetarākṣasī, Bhutaghni,Bahumanjari, Arjaka, Brinda, Manjari, Parnasa, Patrapuṣpaka, Suvāsatulasi.

HABIT : An erect herbaceous, much-branched, softly hairy annual.

HABITAT : Cultivated throughout India.

CHEMICAL COMPOSITION : Eugenol, its methyl ether, nerol, caryophyllene, terpeinen-4-ol, deccyladehyde, y-selinene, β-pinene, camphene and a-pinene from essential oil. Plant contains citric, tartaric and malic acids.

Parts used : Leaves, seeds and roots.

Ayurvedic properties :

Rasa - Kaṭu, Tikta. Vīrya - Ūṣṇa.
Guṇa - Laghu, Rūkṣa. Vipāka - Kaṭu.

तुलसी कटुकातिक्ता हृद्योष्णा दाहपित्तजित्।
दीपनी कुष्ठहृत्श्वासपार्श्वरुक्कफवातजित्।।
(भावप्रकाश)

Tulasī katukātiktā hṛdyoṣṇā dāhapittajit,
Dīpanī kuṣṭhahṛt śvāsaparśvaruk kaphavātajit.
(Bhāvaprakāṣa)

तुलसी श्वासहरा।च.
(अ. हृ. सू.)

Tulasi śvāsahara.
(Caraka)

Actions/Uses : Vātahara, Kaphaghna, Śītahara, Jvaraghna, Dīpana, Uttejaka, Vāyunāśi, Hikkāhara, Śvāsahara, Kāsahara, Viṣghna, Hṛdya, Pārśvaśūlahara, Kuṣṭhahara.

Therapeutics :

Tulsi is aromatic, carminative, antipyretic, diaphoretic and expectorant. It has been found to be very effective in treatment of viral encephalitis and tropical pulmonary eosinophilia in children. Plant is used in snake bite and scorpion sting. Fresh roots, stems and leaves are bruised and applied to the bites of mosquitoes. Decoction of roots is used as diaphoretic in malarial fevers. Leaves are expectorant. Juice of leaves is diaphoretic, antiperiodic and stimulating expectorant; used in catarrh and bronchitis; dropped into ear as remedy for earache. Infusion of leaves is given in malaria, used as stomachic, in gastric disorders of children and in hepatic affections. It improves appetite, afflictions of ear, destroys intestrinal worms and cures skin diseases. Dried leaves, powdered and used as snuff in ozaena. Seeds are deulcent, given in disorders of genito-urinary system.

Origanum majorana Linn.

SYNONYMS : *Majorana hortensis Moench.*

ENGLISH : Sweet Majoram.

INDIAN : Marawaa.

SANSKṚTA : Marubaka, Sukhatmaka, Maruvaka.

HABIT : A herb.

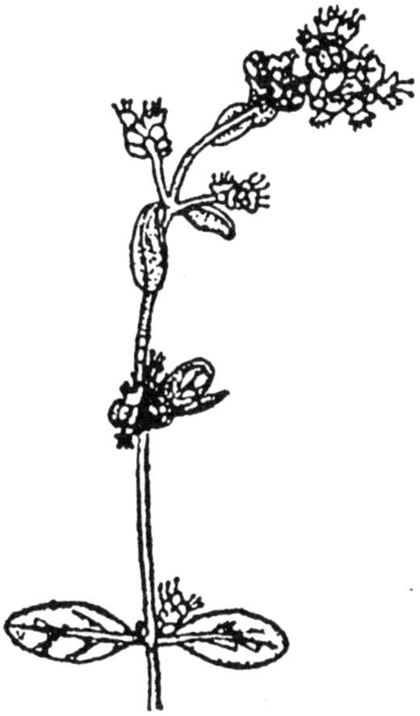

HABITAT : Europe and Great Britain. Extensively cultivated in India.

CHEMICAL COMPOSITION : Majoranin from leaves. Luteolin, apigenin and diosmetin.

Parts used : Whole plant, leaves, seeds and essential oil.

Ayurvedic properties :
Rasa - Kaṭu, Tikta. Vīrya Uṣṇa.
Guṇa - Laghu, Rūkṣa. Vipāka - Kaṭu.

सुखात्मकः कटुस्तिक्तो ग्रहजित पाचनो हिमः।
निहन्ति कफपित्तास्त्रकुष्ठकण्डूविषज्वरान्।
(कैय. नि.)

Sukhatmakah katustikto grahajit pacano himah,
Nihanti kaphapittasra kusthalanduvisajvaran.
(Kai.Ni.)

मरुबकः कटुष्णश्च दीपनस्तिक्ततीक्ष्णकः।
हृद्यः पित्त करो रुच्यो रुक्षो लघुसुगन्धिकः।
पाचकः पित्तकफहृद्ररक्तदोषविषज्वरान्।
कुष्ठ कण्डूरुचीवातश्वासशोफकृमीन्जयेत्।।
हृद्रोग वृश्चिकविषड्बन्धाध्मानशूलहा।
मान्धत्वग्दोषशमनः श्वेतकृष्णाविभेदतः।।
श्वेस्त्वौषधकृत्येषु योग्यः प्रोत्तुः पुरातनैः।
(निघण्टु रत्नाकर)

Marubakaḥ kaṭuṣṇasca dīpanastikṭatīkṣṇakaḥ,
Hṛdyaḥ pittakaro rucyo rukso laghusugandhikah.
Pacakah pittakaphahrd raktadosavisajvaran,
Kustha kandu sandhivata svasa sopha krmin jayet.
(Ni.rat.)

Actions/Uses : Pittakara, Kaphaghna, Pācana, Pittakaphadoṣahara, Viṣahara, Jvarahara, Hṛdya, Kuṣṭhahara, Kaṇḍūhara.

Therapeutics :
Marubaka is an emmenagogue and stimulant. Oleum majoranae is an excellent external application for sprains and bruises etc. Plant is carminative, expectorant, tonic to the liver. Leaves and seeds are astingent, remedy for colic. Essential oil from leaves used for hot fomentation in acute diarrhoea.

Pogostemon heyneanus Benth.

SYNONYMS : *Pogostemon patchouli* Pell.; *Origanum indicum* Roxb.; *Pogostemon patchouli* Hook. *f. non Pelletier.*

ENGLISH : Patchouli.

INDIAN : Paanch, Sugandhi paanadi, Peholi, Mali, Patchapan.

SANSKṚTA : Pācī.

HABIT : A highly aromatic herb.

HABITAT : Native of Malay Peninsula; Cultivated in Western Ghats, Kanara, also grows wild.

CHEMICAL COMPOSITION : Essential oil, sequiterpenes, ychellene and seychellene and glandular trichomes. Patchoulipyridine and epiguaipyridine.

Parts used : Leaves.

Ayurvedic properties :
Rasa - Kaṭu, Tikta. Vīrya - Śīta.
Guṇa - Laghu, Śīta. Vipāka - Kaṭu.

पाची कटुष्णा तिक्ता च सुगन्धिर्वात रोग हा।
ग्रहभूतव्रणघ्नीति त्वग्दोषापहरा नृपे।।
दाहतृष्णाविषघ्निति द्रव्यरत्नाकरे स्मृता।
(निघण्टु रत्नाकर)

Pācī kaṭuṣṇā tiktā ca sugandhir vātarogahā,
Grahabhūta vraṇaghniti tvagdoṣāpaharānṛpe.
Dāha tṛṣṇā viṣaghniti dravyaŕatnākare smṛta.
(Nighaṇṭuratnākara)

Actions/Uses : Raktastambhaka, Śonitasthāpana, Mūtrajanana, Vāyunāsi, Kaṇḍūghna, Viṣavraṇahari.

Therapeutics :

Pācī is diuretic, carminative and insecticidal. A lotion from roots is used for rheumatism. A poultice is applied to boils, and to relieve headache. A decoction of leaves is taken for cough and asthma and that of roots for dropsy.

Pogostemon parviflorus Benth.

SYNONYMS : *Pogostemon frutescens* Graham.; *Pogostemom pubescens* Benth.; *Pogostemom purpuricaulis* Dalzell., *Pogostemon benghalensis (Burm.f.)* Kuntze, *Pogostemon plectranthoides* Desf.

HABIT : A much-branched suffruticose herb.

HABITAT : Throughout India.

INDIAN : Paangli, Phangla.

SANSKṚTA : Phanijjaka.

CHEMICAL COMPOSITION : Alkaloids and essential oil.

Parts used : Whole plant.

Ayurvedic properties :
Rasa - Kaṣāya, Tikta. Vīrya - Śīta.
Guṇa - Śīta, Rūkṣa. Vipāka - Kaṭu.

फणिञ्जको लघुस्तिक्तो रुक्षोहृद्योऽग्निपित्तदः।।
कटुपाकरसो रुच्यस्तांक्ष्णोष्णः कफवातृजित्।
निहन्ति कृमिकृष्ठास्त्रं वृश्चिकाहिविषप्रणुत्।।
(कैयदेव निघण्टु)

Phanījjako laghustikto rukṣo hṛdyo agnipittadaḥ,
Katupākaraso rucyastikṣṇoṣṇaḥ kaphavātajit,
Nihanti kṛmikuṣṭhasram vṛśchikāhi viṣa pranut.
(Kaiyadevanighaṇṭu)

Actions/Uses : Raktasangrāhaka, Raktaskandana, Vraṇaropaṇa, Viṣahara, Uttejaka.

Therapeutics :

Root of Phannijjaka is a remedy for haemorrhages, useful in uterine haemorrhage; antidote to scorpion sting and snake bite. Fresh leaves are styptic, bruised and applied as a cataplasm to clean wounds and promote healthy granulation.

Family : LI - Lauraceae.

Cinnamomum cassia Blume.

SYNONYMS : *Cinnamomum aromaticum* Nees.; *Cassia lignea. Cinnamomum iners* Reinew., *Cinnamomum tanea.*

ENGLISH : Cassia bark, Cinnamum.

INDIAN : Daalachini, Tamaal saala.

SANSKṚTA : Tvak, Cinatvak, Guḍatvak, Virangam, Thracam.

HABIT : A middle sized tree.

HABITAT : Burma and China. Introduced into Ceylon. Cultivated in India.

CHEMICAL COMPOSITION : Volatile oil 1-2%. Cinnamic aldehyde and terpene aldehyde and esters.

Parts used : Dried inner bark of the shoots from truncated stalks and essential oil.

Ayurvedic properties :
Rasa - Kaṭu, Tikta, Madhura. Vīrya - Ūṣṇa.
Guṇa - Laghu, Rūkṣa, Tīkṣṇa, Picchila.
Vipāka - Kaṭu.

त्वक्कट्वी पित्तला स्वाद्वी कण्ठशुद्धिकरी लघुः
रुक्षा तिक्ता बस्तिशुद्धिकारिणी चोष्णदामता।।
कफहिक्कावातकासकण्डूहृद्रोगनाशिनी।
आमं च बस्तिरोगं चे पनिसंच विषतथा।।
शुक्रं चार्शं कृमीश्चैव नाशयेदिति कीर्तिता।
(नि. र.)

Tvak kaṭvī pittalā svādvī kaṇṭhaśudhikarī laghuḥ,
Rūkṣā tiktā bastiśudhikāriṇī coṣṇadā matā.
Kaphahikkā vātakāsa kaṇḍū hṛdroga nāśinī,
Āmam ca bastirogam ca pīnasam ca viṣa tathā.
Śukram cārśam kṛmīmścaiva nāśayediti kīrtitā.
(Nighaṇṭuratnākara)

Actions/Uses : Pittaśāmaka, Vātaśāmaka, Kaphanāśaka, Agnidīpana, Āmapācana, Yakṛtottejaka, Śukrala, Arucihara, Varṇya, Kaṇḍūghna, Āmadoṣaghna, Hṛdrogaghna, Bastirogaghna, Arśoghna, Kāsaghna, Kṛmighna, Pīnasaghna, Tridoṣaghna, Agnimandyahara, Sangrāhī.

Therapeutics :

Cinatvak is a mild astringent, a flavouring agent and oil, a powerful germicide. Properties of cassia are similar to those of cinnamon and uses are also the same as those of *C. zeylanicum.* A substitute for *C. zeylanticum.*

Cinnamomum zeylanicum Blume.

SYNONYMS : *Cinnamomum verum J.S.Presl.; Cinnamomum saigonicum.; Cinnamomum zeylanicum* Breyn.

ENGLISH : Cinnamum.

INDIAN : Tiki, Daalachini, Tamaal saala.

SANSKṚTA : Tvaca, Utkaṭa, Surasa, Sainhala, Bahugandha, Surabhi, Mukhaśodhana, Guḍatvak, Dārusitā, Tanutvak, Vanapriya, Valkala, Rāmavallabha, Coca, Hṛdya, Svarṇabhumika, Lataparṇa, Manaḥpriya, Gurutwak, Tamālpatra.

HABIT : A middle sized evergreen tree.

HABITAT : China, Ceylon, Cutivated in South India.

CHEMICAL COMPOSITION : Diterpenes-cinnzeylanin and cinnzeylanol. Cinnamon oil containin eugenol from bark. Cinnamaldehyde, camphor are other major constituents. Phlobatannin and mucilage. Leaves and stem extract is inhibitory against Ehrlich ascites tumor cells.

Parts used : Dried inner bark of the shoots from truncated stalks and essential oil.

Ayuevedic uses :
Rasa - Kaṭu, Tikta, Madhura. Vīrya - Uṣṇa.
Guṇa - Laghu, Rūksa, Tīksna, Picchila.
Vipāka - Kaṭu.

त्वचं लघूष्णं कटुकं स्वादु तिक्तं च रुक्षकम्।
पित्तलं कफवातघ्नं कण्ड्वामरुचिनाशनम्।।
हृद्बस्तिरोगवातार्शः कृमिपीनसकासजित्।
(राजनिघण्टु)

Tvacam laghūṣṇam kaṭukam svādu tiktam ca rūkṣam,
Pittalam kaphavātaghnam kaṇḍvāmarucināśanam.
Hṛd bastirogavātārśaḥ kṛmipīnasakāsajit.
(Rājanighaṇṭu)

त्वचन्तु कटुकंशीतं कफकासविनाशनम्।
शुक्रामशमनञ्चैव कण्ठशुद्धिकरं लघु।।
(राजनिघण्टु)

Tvacantu kaṭukam śītam kaphakāsavināśanam,
Śukrāmaśamanāncaiva kaṇṭhaśudhikaram laghu.
(Rājanighaṇṭu)

Actions/Uses : Pittaśāmaka, Vātaśāmaka, Kaphanāśaka, Agnidīpana, Āmapacana, Yakṛtottejaka, Śukrala, Arucihara, Varṇya, Kaṇḍūghna, Āmadoṣaghna, Kāsaghna, Hṛdrogaghna, Bastirogaghna, Arśarogaghna, Kṛmighna, Pīnasaghna, Tridoṣaghna, Agnimāndyahara, Sangrāhī.

Therapeutics :

Tvaca is a cordial, stimulant and tonic and is indicated in all cases characterised by feebleness and atony. As an astringent it is employed in diarrhoea, usually in combination with chalk, the vegetable infusion or opium. As a cordial and stimulant it is exhibited in latter stages of low fever. In flatulent and spasmodic affections of the alimentary canal, it often proves a very efficient carminative and antispasmosic. Bark is aromatic, astringent, stimulant and carminative; useful for checking nausea and vomiting. It is rarely prescribed alone but chiefly as an addition to other medicines to improve their flavour or to check their griping qualities. Bark is used in medical preparations as flu-preventive, for indigestion and flatulence control.

HOMEOPATHIC USE : Cinnamomum is a Homeopathic remedy the leading feature of which is haemorrhage; nose-bleed, haemoptysis, haemorrhage from bowels, post-partum haemorrhage. A strain in loins or false step brings on a profuse flow of bright blood. Heamorrhage is bright red and clear and increased by any physical exertion. It cures gastric disturbance and hysterical attacks; pains in bones and in muscles. Useful in cancer where pain and fetor are present; flatulency and diarrhoea; low blood pressure.

Laurus camphora Linn.

SYNONYMS : *Cinnamomum camphora T.* Nees & Eberm.*; Camphora officinarum* Nees.*; Cinnamomum camphora (L.)* Sieb.*; Cinnamomum camphora (Linn.)* Presl.

ENGLISH : Camphor.

INDIAN : Kaapura.

SANSKṚTA : Karpūra, Cinakarpūra, Twaka, Utkata, Surasa, Surabhi, Sainhala, Bahugandha, Mukhaśodhana, Guḍatvaka, Darusitā, Valkala, Tanutvak, Hṛdya, Coca, Rāmavallabha, Vanapriya, Svarṇabhumika, Gurutwak, Lataparṇa, Manaḥpriya.

HABIT : A large handsome evergreen tree.

HABITAT : Native of Central China, Japan and Formosa. Planted in gardens in India. Successfully cultivated.

CHEMICAL COMPOSITION : Essential oil sesquiterpenes camphorenone and campherenol. Roots, alkaloids laurolistsine and reticuliene. Leaves, palmitone. Heartwood, cyclopentenone.

Parts used : leaves and essential oil.

Ayurvedic properties :
Rasa - Tikta, Kaṭu, Madhura. Vīrya - Śīta.
Guṇa - Laghu, Tīkṣṇa. Vipāka - Kaṭu.

कर्पूरः शीतलो वृष्यश्चक्षुष्य लेखनो लघुः।
सुरभिर्मधुरास्तिक्तः कफपित्तविषापहः।।
दाहतुष्णास्यवैरस्यमेदोदौगन्ध्यिनाशनः।
आक्षेपशमनो निद्राजननो धर्म्मवर्द्धनः।।
वेदनाहारकः कामशान्तिकृच्छुक्रमेहहृत्। (द्रव्यगुण)

*Karpuraḥ śitalo vṛṣyaścakṣuṣyo lekhanam laghuḥ,
Surabhirmadhurastiktaḥ kaphapittaviṣāpahaḥ,
Dāhatṛṣṇāsyavairasyamedodaurgandhyanāśanaḥ.
Ākṣepasamano nidrājanano gharmmavarddhanaḥ.
Vedanāhārakaḥ kāmaśāntikṛcchukramehahṛt.
(Dravyaguṇa)*

Actions/Uses : Kaphaghna, Kāsahara, Jvaraghna, Dīpana, Pūtihara, Dāhaśāmaka, Vāyuhara, Svedajanana, Stanyanāśana, Vājikara, Sankocavikāsa-pratibandhaka.

Therapeutics :

Karpura is sedative, anodyne, antispasmodic, diaphoretic, anthelmintic, stimulant and carminative. Internally, camphor is used in colds, chills and in diarrhcea from colds. It is of great value in all inflammatory affections, fevers and hysterical complaints. It acts beneficialy in gout, rheumatic pains, and neuralgia, and is highly valued in all irritations of the sexual organs. It can be safely applied externally in all cases of inflammations, bruises and sprains. It is used in insecticidal preparations. Alcoholic extract of aerial parts is hypotensive and CNS depressant. Camphor useful in chronic pulmonary inflammation; also used against carbuncles.

HOMEOPATHIC USE : A gum obtained from the plant is used to prepare Homeopathic tincture of the remedy Camphora. Chill, cramp, convulsion with mental anguish are the chief symptoms of essential action of the remedy. Its main use is made generally for cholera only. But it is also very useful in epilepsy, hysteria, catalepsy, eclamsia, influenza, severe coryza. erysipelas, tetanus, small slow pulse, 60 beats per minute; body quite cold all over and low blood pressure are characteristic symptoms.

Family : LII - Lecythidaceae.

Barringtonia acutangula (Linn.) Gaertn.

SYNONYMS : *Eugenia acutangula* Linn., *Barringtonia spicata* Bl.

ENGLISH : Small Indian Oak.

INDIAN : Tiwar, Piwar, Samudraphalla, Hijjal, Dattephal.

SANSṚTA : Hijjala, Vidula, Vicula, Nadikānta, Jalaja, Nicula, Samudraphala, Adhiphala, Raktamanjara, Ambuvetasa, Toyajāta, Raktakārmuka, Nimnaga, Toyaja, Dhātriphala.

HABIT : A middle sized evergreen tree.

HABITAT : Throughout India.

CHEMICAL COMPOSITION : Heartwood contains triterpene dicarboxylic acid, barringtonic acid. Tanginol, barringtogenol E and barrinic acid. Leaves contain acutangulic and barringtogenic acids and stigmasterol; also tangulic and oleanolic acids, β-amyrin, β-sitosterol and stigmasterol. Fruits, barringtogenol D, C and B. Bark contains tannin 16 %. Bark and leaves contain small amount of a sapogenin.

Parts used : Roots, leaves, fruits, seeds and bark.

Ayurvedic properties :
Rasa - Tikta, Kaṭu, Madhura. Vīrya - Uṣṇa.
Guṇa - Laghu, Rūkṣa, Tīkṣṇa. Vipāka - Kaṭu.

हिज्जलः कटुरुष्णश्च पवित्रो भूतनाशनः।
वातामयहरी नानाग्रहसंचारदोषजित्॥
(रा. नि.)

Hijjala kaṭuruṣṇaścya pavitro bhūtnāśanaḥ,
Vātāmayaharī nānāgraha sancaradoṣajit.
(Rājanighaṇṭu)

जलवेतरावद् वेद्यो हिज्जलोऽयं विषापहः।
(भा. प्र.)

Jalavetarāvad veddyo hijjalo ayam viṣāpahaḥ.
(Bhāvaprakāṣa)

Actions/Uses : Vātaśamaka, Kaphapittaśodhana, Kṛmighna, Viṣaghna, Bījalekhana, Śirovirecana,

Vedanāsthāpana, Aśmari, Kāmala, Bhūtanāśanī, Cakṣuśrava, Galaganda, Kuṣṭhaghna.

Therapeutics :

Root of Hijjala is bitter, cooling, aperient, antipyretic, stimulant, and emetic. They are useful in catarrh, intermittent fever and constipation. Alcoholic extract of root is hypoglycemic. Leaves and roots are bitter tonic and possess properties similar to Cinchona. Tender leaves are edible. Juice of leaves is used in diarrhoea and dysentery. Bark is astringent and used in diarrhoea and blennorrhoea. Also given as a febrifuge and applied to relieve pain from bites and stings of insects. Alcoholic extract of stem bark antiprotozoal. Bark, root and seed are used as fish poison. Fruit is bitter, acrid, anthelmintic and vulnerary and is prescribed in gingivitis, as an astringent and tonic. Also used as an emetic and expectorant. They are useful for colic, intestinal worms, wounds, ulcers, leprosy, syphylis, cough, bronchitis, strangury, dysmenorrhoea and eye troubles. Powdered seed is aromatic, carminative, emetic, expectorant and used as snuff in headache. Plant used against cholera.

Barringtonia racemosa (Linn.) Roxb.

SYNONYMS : *Barringtonia racemosa* Blume.; *Eugenia racemosa* Linn.

ENGLISH : Small Indian Oak.

INDIAN : Samudraphalla-2, Sadphali, Nivar.

SANSKṚTA : Samudraphala, Nipa, Viṣaya, Samudrapad, Samstravadi.

HABIT : A moderate or small sized tree, with brown, somewhat fibrous bark.

HABITAT : Western sea coast of India.

CHEMICAL COMPOSITION : Gucoside- ṣaponin. Bulk of toxic principle of bark concentrated in resinous fraction. Bark contains eleagic acid and its mono, di-, and tri-OMe derivatives; flowers contain cyanidin- and delphenidin- 3-sambubiosides. Barringtonin composed of rhamnose and barringtogenoll from fruits.

Parts used : Root, fruit, seed and bark.

Ayurvedic properties :

Rasa - Tikta, Kaṭu, Madhura. Vīrya - Ūṣṇa.
Guṇa - Laghu, Rūkṣa, Tīkṣṇa. Vipāka - Kaṭu.

फलं समुद्रस्य कटूष्णकाश वातापहं भूतनिरोधकारी।
त्रिदोष दावानलदोषहारि कफामयभ्रान्ति विरोधकारी।।
(रा. नि.)

Phalam samudrasya kaṭuṣṇakari, vātāpaham bhūtanirodhakāri,
Tridoṣadāvānaladoṣahāri, kaphāmayabhrānti virodhakari.
(Rājanighanṭu)

Actions/Uses : Vātaśāmaka, Kaphapittaśhdhana. Bījalekhana, Śirovirecana, Vedanāsthāpana, Kṛmighna, Viṣaghna, Asmari, Bhūtanāśanī, Cakṣuśrava, Kāmala, Galaganda, Kuṣṭhaghna.

Therapeutics :

Root of Samudraphala is deobstruent and cooling. It is similar to cinchona in medicinal virtues. Fruit is used in cough, asthma and diarrhoea. Kernels of drupes with milk are given in jaundice and other bilious diseases. Seeds are aromatic and used in colic and opthalmia. Seeds and bark are tonic, vermifuge, fish poison and insecticidal.

Family : LIII - Liliaceae.

Aloe barbadensis Mill.

SYNONYMS : *Aloe indica* Royle.; *Aloe littoralis* Koening.; *Aloe vera Tourn. ex* Linn.; *Aloe vera (L.) Burm.f.*

ENGLISH : Indian Aloe, Barbodos aloe, Jafarabad aloe, Curacao aloe.

INDIAN : Koraphada, Kumaaree.

SANSKṚTA : Kumārī, Ajarā, Amarā, Virā, Bhrngeṣṭā, Vipulaśravā, Bramhaghnī, Taruṇī, Gṛhakanyā, Gṛhakumarī, Kanyā, Ghṛtakumarī, Dīrghapatrikā, Maṇḍalā, Sthaleruhā, Rāmā, Kapilā, Ambudhiśravā, Mātā, Akśakārakā, Bala, Atipittalā, Panktikandadala, Vistarī, Badakandinī, Viśāla, Picchasambhṛta, Dvajabhamadhyadaṇḍa, Mṛdu, Saruna, Rajīyuta, Pṛthu, Bahupatrā, Sukanṭakā.

HABIT : A rosettes herb with bulbs.

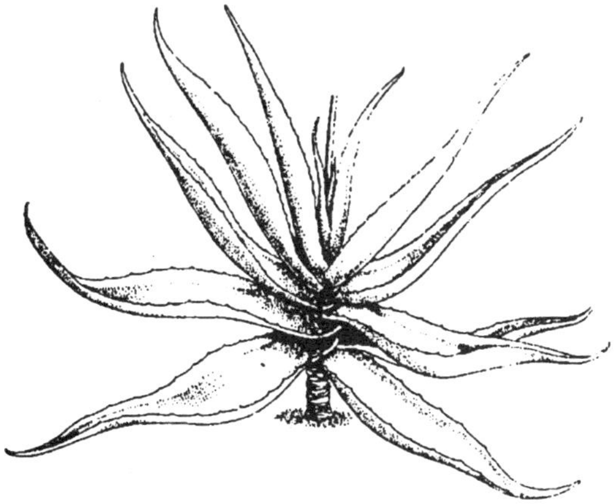

HABITAT : Planted in Indian gardens, found in a semi-wild state in many parts of India.

CHEMICAL COMPOSITION : Aloin, isobarbaloin, emodin, gum, resin, emodin chrysophanic acid, oxidase, catalase and sugars. Aloesin and aloesone; citric, malic and tartaric acids from leaves; jelly from leaves composed of four partially acetylated glucomannans. Contains glycoside barbaloin. Plant extract inhibited ovulation in rabbits.

Parts used : Whole plant, dried juice of leaves, pulp and root.

Ayurvedic properties :
Rasa - Tikta, Kaṭu, Madhura. Vīrya - Śīta.
Guṇa - Gurū, Snigdha, Picchila. Vipāka - Kaṭu.

कुमारी भेदिनी शीता तिक्ता नेत्र्या रसायनी।
मधुरा बृंहणी बल्या वृष्या वातविषप्रणुत्।
गुल्मप्लहियकृद्वृद्धि कफ ज्वर हरी हरेत्।
ग्रन्थ्याग्नि दग्ध विस्फोट पित्तरक्तत्वगामयान्।।
(भावप्रकाश)

Kumarī bhedanī śītā tiktā netryā rasāyanī,
Madhurā bṛnhaṇī balyā vṛṣyā vātaviṣapraṇut.
Gulmaplīhayakṛd vṛddhikaphajvaraharī bhavet,
Granthyāgnidagdhavisphoṭa
pittaraktatvagāmayān.
(Bhāvaprakāṣa)

Actions/Uses : Tridoṣahara, Bhedinī, Netrya, Rasāyani, Balya, Brihmana, Vṛṣya, Gulmagna, Viṣaghna, Pliharogaghna, Saruna, Yakṛtavṛdhihara, Kaphajvaraghna, Granthighna, Raktapittahara, Agnidagdhaśamanī, Kuṣthaghna, Śvaśaghna.

Therapeutics :

Kumārī is a reputed remedy for intestinal worms in children. It is bitter, cooling, aphrodisiac, hepatic stimulant, stomachic, anthelmintic, emmenagogue and purgative, acts particularly on the lower bowels. It is used in haemophilia, skin and uterine disorders, liver and spleen enlargement, chronic ulcers. Given to nursing mothers it causes purging in the suckling infant. Used in dyspepsia, constipation, rectal fissures, piles, suppression of menses. Dried juice is cathartic, given in constipation. Fresh juice is cathartic and cooling, useful in fevers. Pulp is used in menstrual suppression and root in colic. Aloe meat is eaten to alleviate colds; sap or meat with salt, keeps blood in good condition and relieves constipation. Leaves are used as hot polutice to relieve swellings. Aloe produces pelvic congestion and used in uterine disorders, also as tonic after pregnancy, but not prescribed to expectant women.

HOMEOPATHIC USE : Aloe socorotina is a Homeopathic remedy. The most pronounced action is the engorgement of pelvic vicera; increasing haemorrhoidal congestion. Very characteristic diarrhoea and a great deal of flatus in the abdomen with gugling are the chief symptoms.

Gloriosa superba Linn.

SYNONYMS : *Methonia superba* Lamk.

ENGLISH : Malabar glory lily.

INDIAN : Kallalaavi, Khadyaanaaga, Karihari, Laanguli, Indai, Kariannaag, Naagkaria, Bachhaaga.

SANSKṚTA : Kalihārī, Lāngalī, Viśalyā, Agnijivhā, Agniśikhā, Vanhiśikhā, Viṣā, Vanhimukha, Vanhijivhā, Agnimukhī, Vanhivaktrā, Garbhanut, Dīptā, Ananta, Śuklapuṣpī, Garbhāpatani, Vidyullatā, Halinī, Sīrī, Raktendupuṣpikā, Kalikārī, Lāngalinī, Halī, Vidyujvālā, Vraṇahṛta, Naktendu, Puṣpasaurabhā, Svarṇapuṣpā, Puspikā, Śakrapuṣpikā, Kalikālī, Puṣpeṇḍu, Indrapuṣpī, Prabhātā, Lāngulī, Ulaṭacandāla, Pradīpa, Dudhiyo, Ailnī, Garbhāghatini, Vilāngulī.

HABIT : An annual climbing herb with tuberous roots.

HABITAT : Throughout tropical India.

CHEMICAL COMPOSITION : Alkaloids, superbine and gloriosine. Colchicine and its derivatives from tubers. Sitosterol, its glucoside and β- & γ-lumicolchicines; β-sitosterol, its flucoside and 2-)H-6-MeO benzoic acid. Flowers contain luteolin, its glucoside, N-formyl-de-Me-colchicine its glucoside and 2-de-Me-colchicine. Colchicine increased basal release of growth hormone in experimental animal tissue. In primary tissue culture of malignant human gliomas, it induced mild and nonspecific reversible reduction in cell motility and did not change adhesion.

Parts used : Tuberous roots.

Ayurvedic properties :
Rasa - Tikta, Kaṭu. Vīrya - Ūṣṇa.
Guṇa - Laghu, Tīkṣṇa. Vipāka - Kaṭu.

कलिहारी सरा कुष्ठशोफार्शोव्रणशूलजित्।
सक्षारा श्लेष्मजित्तिका कटुका तुवरापिच।।
तीक्ष्णोष्णाकृमिहृल्लघ्वी पित्तला गर्भपातिती।
(भावप्रकाश)

Kalihārī sarā kuṣṭha śophārśo vṛaṇaśūlajit,
Sakṣārāśleṣmajitiktā kaṭukā tuvarapi ca,
Tīkṣṇoṣṇā kṛmihṛt laghvī pittalā garbhapātini.
(Bhāvaprakāṣa)

लांगली कटु रुष्णा च कफवातविनाशिनी।
तिक्ता सरा च श्वययूगर्भशल्यव्रणापहा।।
(सु. सां.)

Lāngalī kaṭuruṣṇā ca kaphavāta vināśanī,
Tiktā sarā ca śvayayūgarbhaśalyavraṇāpahā.
(Su.Sa.10)

Actions/Uses : Kaphavātaśāmaka, Dīpana, Balya, Vāmaka, Recana, Pittasāraka, Garbhanāśana, Vastiśodhaka, Balasghna, Arśoghna, Kṛmighna, Kuṣṭhaghna, Vraṇahara, Śophahara, Jantughna.

Therapeutics :
Lāngalī is a good abortifacient causing expulsion of foetus from the womb. It is bitter, acrid, astringent, anthelmintic and germicidal. It cures leprosy, swelling, piles, chronic ulcers, colic pain in bladder. Roots are antiperiodic, purgative, colagogue, anthelmintic; used in leprosy, parasitical affections of skin, piles, colic, in snake-bite and scorpion sting. Starch from root is given internally in gonorrhoea. Tubers are tonic, stomachic and anthelmintic when taken in doses of 5-10 grains. Tubers abortifacient; extract of root, ecbolic. Paste is antidote in snake bite. Powder of root is given in rheumatic fever. Various plant parts are used in spleen complaints, sores, tumors, erysupelas and syphilis. Extract of plant is spamolytic and CNS depressant. Juice of leaves is piscicidal.

Family : LIV - Linaceae.

Linum usitatissimum Linn.

ENGLISH : Linseed; Flax.

INDIAN : Jawasa, Aallashee, Tisi.

SANSKṚTA : Atasī, Devī, Umā, Ksumī, Hemavatī, Sunīlā, Picchalā, Madagandhā, Madotkaṭā, Nilapuṣpikā, Rudrapatnī, Rudranīlā, Celu, Pārvatī, Madhurā, Kṣaumī, Masruṇā.

HABIT : An erect annual herb, rarely shrubby.

HABITAT : Cultivated chiefly for oil throughout India.

CHEMICAL COMPOSITION : Linseed contains 16-31% protein, principally two globulins, e.g. linin and colinin. Linamarin and two other glucosides, crystalline and noncrystalline; phytin, lecithin and β-carotenoids. Cholesterol, campesterol, stigmasterol, sitosterol, 5-dehydro-avenasterol, cyclosterol and 24-methylenecycloartanol in seeds. Seeds contain about 30 to 40 % of fixed oil together with proteins, wax, resin, sugar, phosphates and a small quantity of linamarin.

Parts used : Seed, seed oil, leaves, bark and flowers.

Ayurvedic properties :
Rasa - Madhura, Tikta. Vīrya - Uṣṇa.
Guṇa - Gurū, Snigdha, Picchila. Vipāka - Kaṭu.

अतसी मधुरा तिक्ता स्निग्धा पाके कटुर्गुरुः।
उष्णा दृकशुक्रवातघ्नी कफपित्तप्रकोपिणी।।
(भावप्रकाश)

Atasī madhurā tiktā snigdhā pāke kaṭurguruḥ,
Uṣṇā dṛk śukravātaghnī kaphapittaprakopiṇī.
(Bhāvaprakāṣa)

अतसी मदगन्धा स्यात् मधुरा वलकारिका।
कफवातकरी चेषत् पित्तहृत् कुष्ठवातनुत्।।
(राजनिघण्टु)

Atasī madagandhā syāt madhurā balakārikā,
Kaphavātakarīceṣat pittahṛt kuṣṭhavātanut.
(Rājanighaṇṭu)

Actions/Uses : Vātaghna, Kaphapittavardhaka, Vraṇaśothahara, Snehana, Vātaśāmaka.

Therapeutics :

Dried ripe seeds of Atasī are used as demulcent and in form of polutices useful for rheumatic and gouty swellings and as laxative in cases of haemorrhoids. They are internally used for gonorrhoea and irritation of genito-urinary system. Crushed seeds are applied externally to accelerate suppuration of abscesses, boils etc. and also used in ulcerated and inflamed surfaces, gout and rheumatic swellings. Bark and leaves are used in gonorrhoea. Flowers are nervine and cardiac tonic. Oil mixed with limewater is used as an application to burns.

HOMEOPATHIC USE : Linum usitatissimum is a Homeopathic remedy. It is useful in hay-fever, asthma, trismus, urticaria and convulsions.

Family : LV - Lythraceae.

Ammannia baccifera Linn.

SYNONYMS : *Ammania vesicatoria* Roxb., *Ammania salicifolia* Hiern.

HABIT : An glabrous, erect, perennial, branching herb.

HABITAT : Very common all over India in rice fields and marshy localities.

ENGLISH : Blistering ammannia.

INDIAN : Aagiyaa, Dadmari, Jangali mendi, Aginbuti, Bharjambol.

SANSKṚTA : Kurandaka, Agnigarva, Agnivatī, Agnipatrī.

CHEMICAL COMPOSITION : Hentriacontine, dotriacontanol, betulinic acid, lupeol, ellagic acid, quercetin. Leaves contain lawsone.

Ayurvedic properties :
Rasa : Tikta, Kaṭu. Vīrya : Śīta
Guṇa : Sara, Gurū. Vipāka : Madhura/Kaṭu

Parts used : Whole plant, stem, inflorescence and leaves.

कुरण्डिका सरा रुच्या गुर्वी चाग्निप्रदीपनी।
नाशनी कफवातानां वैद्येस्तु परिकीर्तिता।।
(नि. र.)

Kuraṇḍikā sarā rucyā gurvī cāgnipradīpanī,
Nāśanī kaphavātānam vaidyaistu parikīrtitā.
(Nighaṇṭuratnākara)

Action/Uses : Kaphagna, Vātaghna.

Therapeutics :

The herb is a rich source of vitamin C and has anti-typhoid and anti-tubercular properties. It is bitter, acrid, appetiser, diuretic, aphrodisiac and lithontriptic. It is used in anorexia, colic, strangury, seminal weakness and renal and vesical calculi. Leaves are acrid and are used to raise blisters. They are laxative, stomachic, rubifacient and anti-pyretic and are prescribed in rheumatic pains, fevers and in ring worm and other skin diseases. Mixed with oil applied in herpatic eruptions. Extract of plant is antibacterial; extract of stem, leaf and inflorescence is more effective as compared to root and seed extract.

Lagerstroemia speciosa (Linn.) Pers.

SYNONYMS : *Munchausia speciosa* Linn., *Adambea glabra* Lamk. *Lagerstroemia reginae* Roxb., *Lagerstroemia flosreginae* Retz.

ENGLISH : Queen crape, Myrtle, Pride of India.

INDIAN : Taaman, Jarul, Mota-bondara.

SANSKṚTA : Tiniśa, Arjuna, Nebhi, Syandana, Tiriccha.

HABIT : A middle-sized to large deciduous tree.

HABITAT : Western and Eastern Ghats; Assam; Chittagong, Chota Nagpur.

CHEMICAL COMPOSITION : Leaves contain alanine, isoleucine, a-aminobutyric acid and menthoeonine. Lageracetal, amyl alcohol and ellagic acid; β-sitosterol, lagertannin etc. also from leaves. Leaves. fruits and bark contain tannin.

Parts used : Fruit, seeds, leaves, bark and root.

Ayurvedic properties :

Rasa - Tikta, Kaṣāya. Vīrya - Śīta.
Guṇa - Laghu, Rūkṣa. Vipāka - Kaṭu.

तिनिशः श्लेष्मपित्तास्त्रमेदः कुष्ठप्रमेहजित्।
तुवरः श्वित्रदाहघ्नो व्रणपाण्डुक्रिमिप्रणुत्।।

Tiniśaḥ śleṣmapittāstramedaḥ kuṣṭha pramehajit,
Tuvaraḥ śvitradāhaghno vraṇapāṇḍu krimipraṇut.
(Bhāvaprakāṣa)

Action/Uses : Vraṇaśothahara, Recana, Mūtral, Jvaraghna.

Therapeutics :

Root, astringent, stimulant and febrifuge. Bark and leaves, purgative. Aquious extract of leaves (called the drug "Banoba") has hypoglycemic activity. Decoction of bark is used in fever. Fruit is used as local application for aphthae of mouth. Seeds are narcotic.

Lawsonia inermis Linn.

SYNONYMS : *Lawsonia alba* Lam.

ENGLISH : Henna plant, Egyptian privet.

INDIAN : Mendee, Mehndi.

SANSKṚTA : Madayantikā, Nakharanjaka, Timira, Kokadanta, Mendikā, Dvivṛunta, Raktagarbha, Ragangi.

HABIT : A globrous, much branched shrub or small tree.

HABITAT : Cultivated and naturalized all over India, for its leaves.

CHEMICAL COMPOSITION : Leaves contain glucoside, coloring matter lawsome which is identical with 2-hydroxy-a-napthoquinone. Xanthones, laxanthones I, II and III. Seeds yield a non-drying oil. Essential oil contains β-ionone.

Parts used : Leaves, bark, flowers and seeds.

Ayurvedic properties :
Rasa - Tikta, Kaṣāya. Vīrya - Śīta.
Guṇa - Laghu, Rūkṣa. Vipāka - Kaṭu.

मादयन्ती लघु रुक्षा कषाया तिक्तशीतला।
कफपित्तप्रशमनी कुष्ठघ्नी सा प्रकीर्तिता।।
निहन्ति ज्वरकण्डति दाहासृक् पित्तकामलाः।
रक्तातिसार हृद्रोगमूत्रकृच्छ्रभ्रमव्रणान्।।
(द्रव्यगुण विज्ञान)

Mādayantī laghu rūkṣā kaṣāyā tiktaśītalā,
Kaphapittapraśamanī kuṣṭhaghnī sā prakīrttitā.
Nihanti jvarakaṇḍūti dāhāsṛk pittakāmalāḥ,
Raktātisārahṛdrogamūtrakṛcchabhramavraṇān.
(Dravyguṇa)

Actions/Uses : Kaphapittaśāmaka, Vedanāsthāpana, Varṇya, Keśya, Śothahara, Stambhana, Kuṣṭhaghna, Vraṇaśodhana-ropaṇa, Dāhapraśamana.

Therapeutics :

Madayantikā leaves are emetic, expectorant, allay burning sensation; cure leucoderma. They are used as external application in headache and rubbed over soles of feet in burning of feet. Decoction of leaves is used as astringent, gargle in relaxed sore throat. Juice of leaves mixed with water and sugar is given as a remedy for spermatorrhoea. Bark is alterative, sedative and astringent, is given in jaundice and enlargement of spleen; used in calculous affections; as alterative in skin diseases and leprosy. Oil and essence is rubbed over body to keep it cool. Flowers are refrigerant and soporofic.

Woodfordia fruticosa Kurz.

SYNONYMS : *Woodfordia floribunda* Salisb.; *Lythrum fruticosum* Linn.; *Grislea tomentosa* Roxb.

ENGLISH : Fire-flame, Shiranjitea.

INDIAN : Dhaayati, Dhavri, Dawi, Phulsatti, Santha.

SANSKṚTA : Dhātakī, Dhātuparṇī, Vanhijvālā, Madanī, Madanīyā, Pārvatīyā, Madyavāsinī, Dahani, Sadhupuspi, Kumuda, Kunjara, Bahupuspika, Gucchasangha, Gucchapuspaa, Bahujvala, Vanhiśikhā, Dhatukī, Subhakṣika, Pramadanī, Lodhrapuṣpiṇī, Adhahpuṣpī, Rakta-puṣpī, Ghṛtapuṣpikā.

HABIT : A woody perennial shrub.

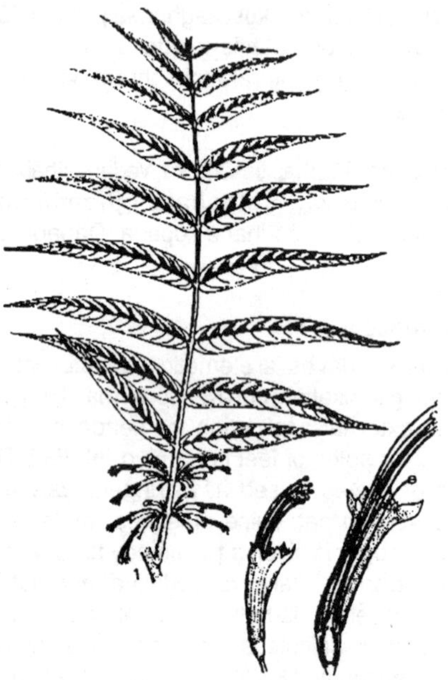

HABITAT : Throughout India and Baluchistan.

CHEMICAL COMPOSITION : Extract of flowers possessed significant abortifacient activity in mice. Flowers contain ellagic acid, β-sitosterol polystachoside, octacosanol, myricetin-3-galactoside, cyanidin-3,5-diglucoside, pelargonidin-3,5-diglucoside, chrysophanol-8-O-B-D-glucopyranoside.

Parts used : Bark, leaves and flowers

Ayurvedic properties :
Rasa - Kasāya, Kaṭu. Vīrya - Śīta.
Guṇa - Laghu, Rūkṣa. Vipāka - Kaṭu.

धातकी कटुरुष्णा च मदकृद विषनाशनी।
प्रवाहिकातिसारघ्नी विसर्पव्रणनाशनी।।
(राजनिघण्टु)

Dhātakī katurusna ca madakrd visanasani,
Pravāhikātisāraghnī visarpavraṇanāsanī.
(Rājanighaṇṭu)

Actions/Uses : Kaphaghna, Pittaghna, Vātavardhaka, Jantughna, Dāhapraśamana, Vraṇaropaṇa, Raktastambhana.

Therapeutics :

Dhātakī bark is pungent, acrid, cooling, anthelmintic, toxic, uterine sedative. In Konkan, leaves are used in bilious sickness. Dried flowers are used in the preparation of asavaas and aristaas as it helps fermentation. They are stimulant, astringent, used in dysentery, menorrhagia, in derangement of liver, disorders of mucous membrane and in haemorrhoids; it is considered a safe stimulant in pregnancy, it averts abortion. The drug is also used in headache and fever.

Family : LVI - Magnoliaceae.

Michelia champaca Linn.

ENGLISH : Yellow Champak, Golden Champak.

INDIAN : Sonachaafaa, Piwallaa chaafaa.

SANSKṚTA : Campaka, Svarṇapuṣpa, Cāmpeya, Śītala, Śītalacchada, Subhaga, Bhṛngamohī, Bhramarātithi, Cala, Surabhi, Divyapuṣpa, Sthiragandha, Atigandhaka, Sthirapuṣpa, Hemapuṣpa, Pītapuṣpa, Hemāhva, Sukumāra, Vanadīpa, Kāncana, Ṣatpadātithi, Ramya.

HABIT : A large or middle sized evergreen tree.

HABITAT : Found in Eastern Himalayas and Western Ghats. Cultivated, often found wild.

CHEMICAL COMPOSITION : Flowers and rinds of fruit contain essential oil. Mono- and sequiterpenes occur in essential oil. Fruit, fatty oil. Plant contains alkaloids.

Parts used : Bark, root and root bark, leaves, flowers, fruits and fruit rind.

Ayurvedic properties :
Rasa - Tikta, Kaṭu, Kaṣāya. Vīrya - Śīta.
Guṇa - Laghu, Rūkṣa. Vipāka - Katu.

चम्पकः कटुकस्तिक्तः शिशिरो दाहनाशनः।
कुष्ठकण्डूव्रणहरः॥
(राजनिघण्टु)

Campakaḥ kaṭukastiktaḥ śiśirodāhanāśanaḥ,
Kuṣṭhakaṇḍūvraṇaharaḥ.
(Rājanighaṇṭu)

Actions/Uses : Kaphapittaśāmaka, Dāhapraśamana, Tvagdoṣahara, Vraṇaśodhana, Ropaṇa, Rucikara, Dīpana, Āmapacana, Anulomana, Kṛmighna, Śothahara, Ārtavajanana, Garbhāśayottejaka.

Therapeutics :

Dried root and root bark of Campaka are purgative; in the form of infusion, useful as an emmenagogue; mixed with curdled milk useful as application to abscesses. Juice of leaves is given with honey in colic. Bark is bitter, tonic, febrifuge, stimulant, expectorant and astringent. Flowers and fruits are stimulant, antispasmodic, tonic, stomachic, carminative, bitter and cooling, used in dyspepsia, nausea and fever; also useful in renal diseases and in gonorrhoea; mixed with sesamum oil forms an external application in vertigo. Oil from flowers is useful application in cephalgia, ophthalmia and gout. Seeds and fruit are used for healing cracks in feet.

Family : LVII - Malpighiaceae.

Hiptage benghalensis (Linn.) Kurz.

SYNONYMS : *Hiptage madablota* Gaertn.; *Banisteria benghalensis* Linn.; *Hiptage parvifolia W. & A.*

INDIAN : Madhumaalati, Haladvel, Madhavilataa, Aneta, Kampti.

SANSKṚT : Mādhavī, Atimukta, Vāsantī.

HABIT : A large, handsome, evergreen climbing shrub.

HABITAT : Found throughout the warmer parts of Maharashtra, Konkan, Karnatak and other parts of India.

CHEMICAL COMPOSITION : Root contain glucoside hiptagin and root bark mangiferin. Alcoholic extract of aerial parts is CNS depressant and hypotensive.

Parts used : Leaves, vine and bark.

Ayurvedic properties :
Rasa - Kaṭu. Virya - Śita.
Guṇa - Laghu. Vipaka - Carapari.

माधवी कटुका तिक्ता कषाया मद्‌गन्धिका।
पित्तकासव्रणान् हन्ति दाहशोषविनाशिनी।।
(राजनिघण्टु)

Mādhavī kaṭukā tiktā kaṣāyā madagandhikā,
Pittakāsavraṇān hanti dāhaśoṣavināśinī.
(Rājanighaṇṭu)

Actions/Uses : Tridoṣahara, Kuṣṭhaghna.

Therapeutics :

Leaves of Madhavi are useful in chronic rheumatism, skin diseases and asthma. Juice of leaves is insecticide and used for application for scabies. Bark is useful in bronchial asthma and rheumatoid arthritis.

Family : LVIII - Malvaceae.

Abelmoschus esculentus Moench.

SYNONYMS : *Hibiscus esculentus* Linn.; *Abelmoschus esculentus* Linn.

ENGLISH : Lady's finger, Okra, Gumbo.

INDIAN : Bhenda, Gandhamulla, Bhindi, Ramturai.

SANSKṚTA : Ḍindiśa, Ṭinḍiśa, Gandhamūla.

HABIT : An annual erect herb.

HABITAT : Cultivated throughout India.

CHEMICAL COMPOSITION : Pods contain carotene, folic acid, thiamin, riboflavin, niacin, vitamin C, oxalic acid and essential amino acids. Thirteen flavonol glycosides and two anthocyanins from flower petals. Gossypetin and hibiscetin glucosides; oxalic acid and oxalates, volatile oil are other contents. A fair source of calcium, iron and vitamins. Calcium, 66.0 mg./100gm; iron, 1.5 mg./100gm. carotene 52.0; folic acid,105.ug; thiamine, 0.07; riboflvin,0.10; niacin o.6; vitamin 13.0. Ripe dry seeds contain 16-22 % edible oil. The stalks yield a blast fibre with potential for commercial explosion.

Parts used : Fruits and seeds.

Ayurvedic properties :
Rasa - Madhura. Vīrya - Śīta.
Guṇa - Gurū, Picchila. Vipāka - Madhur.

डिण्डिशो रुचिकृद् भेदी पित्तश्लेष्मापहः स्मृतः।
शिशिरो वातलो रुक्षो मूत्रलश्चाश्मरीहरः।।
(भावप्रकाश)

Ḍinḍiśo rucukṛd bhedī pittaśleṣmāpahaḥ smṛtaḥ,
Śiśiro vātalo rukṣo mūtralaścāśmarīharaḥ.
(Bhāvaprakāṣa)

Actions/Uses : Pittaśāmaka, Śleṣmāhara, Vṛṣya, Balya, Rucya, Grāhi.

Therapeutics :

Pods of Ṭinḍiśha are nutritious, appetizing and laxative. They are used as a vegetable and in the form of decoction used as emollient, demulcent, diuretic, in catarrhal affections, ardor urinae, dysuria and in gonorrhoea. Mucilage from fruits and seeds is emollient, demulcent and useful in gonorrhoea. Fruits are useful as aphrodisiac and in chronic dysentery. Tender pods are eaten in cases of spermatorrhoea and mucillage has aphrodisiac effect. Most serviceable in fevers, catarrhal attacks, irritable states of genito-urinary organs and in all cases attended with scalding, pain and difficulty in passing urine. Seeds are stimulant, antispasmodic. Infusion of roasted seeds is sudorific.

Abelmoschus moschatus Medic.

SYNONYMS : *Hibiscus abelmoschus* Linn.

ENGLISH : Ambrette plant, Musk mallow, Muskseed.

INDIAN : Kasturbhenda, Mushk bhendi bij, Musk mallow.

SANSKṚTA : Latākastūrī, Latākastūrikā, Gandhapūrva, Kastūrīlatikā.

HABIT : An erect annual or biennial, hirsute herb.

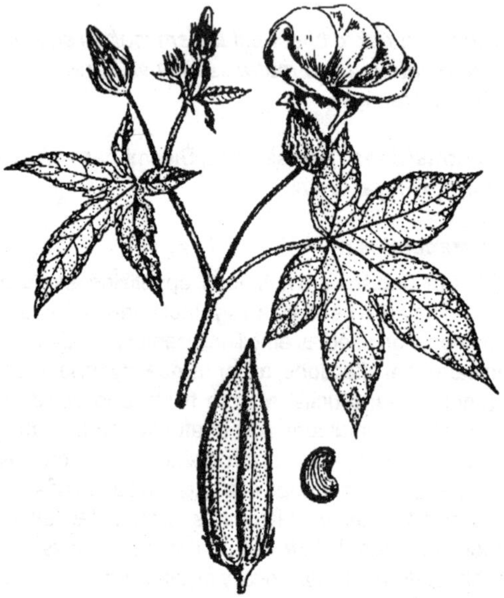

HABITAT : Cultivated in the hotter parts of India. Found wild in Deccan and Karnatak in the hilly regions and the foothills of the Himalayas.

CHEMICAL COMPOSITION : Leaves, flowers and seeds contain β-sitosterol and its B-D-glucoside, myricetin and myricetin-3'-D-glucoside, campesterol, stigmasterol. cholesterol, ergosterol, higher fatty acids, esters of acetic acid and ambretiolic acid. An aromatic absolute from seeds known as AMBRETTE SEED OIL.

Parts used : Whole plant, leaves and roots.

Ayurvedic properties :
Rasa - Tikta, Madhura, Kaṭu. Vīrya - Uṣṇa.
Guṇa - Laghu, Rūkṣa, Tīkṣṇa. Vipāka - Kaṭu.

लताकस्तुरिका तिक्ता स्वाद्वीवृष्या हिमा लघुः।
चक्षुष्या छेदिनी श्लेष्मतृष्णावस्त्यास्य - रोगहृत्॥
(भावप्रकाश)

Latākasturikā tiktā svādvī vṛṣyā himā laghuḥ,
Cakṣuṣyā chedanī śleṣmatṛṣṇāvastyāsyā rogahṛt.
(Bhāvaprakāṣa)

Actions/Uses : Kaphapittaghna, Mukhadurgandhināśanam Rucikara, Dīpana, Grāhī. Hṛdayottejaka, Mūtrala, Vṛṣya.

Therapeutics :
Mucilage made from root of Latakasturi and leaves are used in urinary discharges, painful micturition and as aphrodisiac. Seeds are diuretic, demulcent, tonic, stimulant, cooling, carminative, aphrodisiac, antispasmodic, and stomachic; rubbed to a paste with milk used to cure itch, in snakebite. A decoction, infusion or tincture of seeds is useful in nervous debility, hysteria, and other nervous disorders. Seeds are used in perfumery and as a substitute for musk. Juice of fresh plant is used as febrifuge and expectorant and a polutice made of whole plant is applied to chest in bronchitis.

Abutilon indicum (Linn.) Sweat.

SYNONYMS : *Abutilon indicum G. Don.; Abutilon asiaticum (Linn.) Sweet.; Sida guineensis Schumach.*

ENGLISH : Country mallow.

INDIAN : Petaaree, Mudraa, Jhampi, Chakrabenda, Kangori.

SANSKṚTA : Atibalā, Kakāntakā, Ṛṣyaproktā, Mudrikā, Balikā.

HABIT : A herbaceous woody perennial, softly tomentose shrub.

HABITAT : Found as a weed in sub-Himalayan tract and throughout the hotter parts of India.

CHEMICAL COMPOSITION : Gossypetin-8 and 7-glucosides and cyanidin-3-rutinoside isolated. Aerial parts contain alkanol, β-sitosterol etc. Leaves yield β-sitosterol and tocopherol oil.
Mucilage, asparagin. Plant contains fructose, galactose, glucose, n-alkane mixture, an alkanol fraction, B-sitosterol,vanillic acid, p-coumaric acid, p-hydroxybenzoic acid, caffeic acidm, fumaric acid, p-B-D-glycosyloxybenzoic acid, leucine, histidine, threonine, serine, glutamic acid, aspartic acid, and galacturonic acid. Leaves and stem contain vitamin C, 31.1 mg/100gm. Seeds yield a pale yellow semi-drying oil.

Parts used : Seeds, bark, root, leaves and flowers.

Ayurvedic properrties :
Rasa - Madhura. Vīrya - Śīta.
Guṇa - Laghu, Snigdha, Picchila. Vipāka - Madhura.

तिक्ता कटुश्चतिवला वातघ्नी क्रिमिनाशिनी।
दाह तृष्णा विषच्छदिर्द कलेदोपशमनी परा।।
(राजनिघण्टु)

Tiktā kaṭuścātibalā vātaghnī krimināśinī,
Dāha tṛṣṇā viṣacchadirda kaledopaśamanī parā.
(Rājanighaṇṭu)

बलाचतुष्टयं शीतं मधुरं बलकान्तिकृत्।
स्निग्धं ग्राहि समीरास्त्र पित्तास्त्रक्षतनाशनम्।।
(भावप्रकाश)

Balācatuṣṭayam śītam madhurm balakāntikṛt,
Snigdham grāhi samīrāsrapittāsraksatanāśanam.
(Bhāvaprakāṣa)

Actions/Uses : Balya, Vātaśamana, Mūtrajanana, Snehana, Samsranana, Kāsahara.

Therapeutics :
Atibalā is a drug in Ayurvedic and Unani medicine. It is useful as a febrifuge, anthelmintic, antiinflammatory, in urinary and uterine discharges, piles, lumbago. Juice is used as emollient to relieve soreness of nates in young children. Various plant parts are used in convulsions, cramps, colic, dysentery, consumption, bronchitis, menorrhagia, spermatorrhoea. Root is demulcent, diuretic and nervine tonic and is prescribed in fever, chest affections and urethritis. It is very useful in strangury, haematuria, stones in bladder and as wash in eye diseases. Infusion of roots is a good cooling remedy in fever and is also given in strangury and haematuria. Decoction of roots is given internally for stones in

bladder and applied as a wash in eye diseases. It is also used for piles. Powder of roots is used in cough and leprosy. Mixed with honey, chaulmoogra oil and fresh paste of sandlewood, it is an efficaceous treatment for leucoderma. Leaves are cooked and eaten in bleeding piles. Leaf extract is diuretic and as emollient; given in diarrhoea; decoction as mouth wash in toothache; also useful in gonorrhoea, inflammation of bladder; and as a wash for wound and ulcers and for enema and vaginal infection. Leaf paste applied in fever and headache. Flowers are applied to boils and ulcers. Their powder is eaten in ghee for blood vomiting and cough. Seeds are aphrodisiac, contain water soluble mucilage and are used as laxative in piles, demulcent and used in treating cough, are distinctly useful in gonorrhoea, gleet and chronic cystitis.

Gossypium arboreum Linn.

SYNONYMS : *Gossypium nanking* Meyen.; *Gossypium indicum* Tod.; *Gossypium neglectum* Tod.; *Gossypium sanguineum* Hassk.; *Gossypium intermedium* Tod.; *Gossypium cernuum* Tod.

HABIT : A woody perennial shrub.

HABITAT : Grown in gardens and about temples in India.

ENGLISH : Cotton plant.

INDIAN : Devakaapoosa, Rui, Kapaas, Tula.

SANSKṚTA : Udyānakarpāsa.

CHEMICAL COMPOSITION : Alcoholic extract of aerial parts is CNS depressant.

Parts used : Leaves, seeds and roots.

Ayurvedic properties :

Rasa - Madhura. Vīrya - Alpa Uṣṇa.
Guṇa - Laghu. Vipāka - Madhura.

Actions/Uses : Vātahara, Mūtrala, Dāhaśamana, Vraṇaropaṇa, Kaṇḍūghna, Netrābhiṣyanda.

Therapeutics :

Root of Udyānakarpāsa is used in fever; seeds in catarrh, consumption, gleet, gonorrhoea and chronic cystitis. Various plant parts are used in earache, madness, convulsions and labour complaints.

Gossypium herbaceum Linn.

SYNONYMS : *Gossypium otusifolium* Roxb.; *Gossypium wightianum* Tod.

ENGLISH : Cotton.

INDIAN : Kaapoosa.

SANSKRTA : Kārpāsa, Picu, Bādarā, Nagnajita, Tuṇḍī, Paṭampaṭī, Tuṇḍikerī, Nihanantā, Samudrāntā, Susumā, Sāṛinī, Cavyā, Guṇasū, Marudurvā, Ācchādanaphala, Pathyā, Anagnā, Paṭapradā, Bhadrā, Vandyaphala, Varadruma, Sthūlapicura, Somavallī, Cakrikā, Vāsanī, Viṣghnī, Vāmanī, Paṭatūla, Stanyadabīja.

HABIT : An annual woody shrub.

HABITAT : Greece, Turkey, Sicily, Malta, cultivated in India.

CHEMICAL COMPOSITION : Betaine, choline, salicylic acid etc. Gossypol, gossyfulvin, gossypurpurin, gossycaerulin, gossypin, gossipetin, gossipitrin herbacitrin, quercitin, carotenoids Vitamin E, methyl and higher alcohols, peptone, etc.

Parts used : Root. root bark, flowers and seeds.

Ayurvedic properties :

Rootbark :

Rasa - Kaṭu,Kaṣāya. Vīrya - Alpa Ūṣṇa.
Guṇa - Laghu, Tīkṣṇa. Vipāka - Kaṭu.

Seeds :

Rasa - Madhura. Vīrya - Alpa Usna.
Guṇa - Snigdha. Vipāka - Madhura.

कार्पासी मधुरा शीता स्तन्या पित्तकफापहा।
तृष्णादाहश्रमभ्रान्तिमूर्च्छाहृत् बलकारिणी।। (राजनिघण्टु)

Kārpāsī madhurā śītā stanyā pittakaphāpahā,
Tṛṣṇādāhaṣramabhrāntimūrcchāhṛt balakāriṇī.
(Rājanighaṇṭu)

कार्पासको लघुः कोष्णो मधुरो वातनाशनः।
तत्पलाशं समीरघ्नं रक्तकृत्मूत्रवर्धनम्।।
तत्कर्णापिडकानादपूयस्त्रावविनाशनम्।
तद्बीजं स्तन्यद वृष्यं स्निग्धं कफकरं गुरु।। (भावप्रकाश)

Kārpāsako laghuḥ koṣṇo madhuro vātanāśanaḥ,
Tatpalāśamsmīraghnam, raktakṛt mūtravardhanam.
Tat karṇapiḍakānādapūyasrāvavināśanam,
Tadbījam stanyadam vṛṣyam snigdham kaphakaram guru. (Bhāvaprakāṣa)

Actions/Uses : Root bark : Vātapittakara.
Seeds : Vātapittaghna, Vedanāsthāpana, Vraṇaropaṇa.

Therapeutics :

Root is diuretic and its paste mixed with rice-water is given in leucorrhoea. Bark of root is emmenagogue, galactagogue, parturient and oxytocic. Juice of leaves is antidysenteric and is used in scorpion sting and snakebite. Decoction of leaves finds special application in fever and diarrhoea. Flowers are used in uterine discharges and for safe and effective contraction of uterus. Useful in cases of difficult or obstructed menstruation. Seeds are demulcent, laxative, expectorant, galactagogue, aphrodisiac, abortifacient and employed to produce abortion. They are considered nervine tonic and given in headache.

Hibiscus cannabinus Linn.

ENGLISH : Bimli, Bimlipatam jute, Ambaari hemp, Kenaf, Mesta.

INDIAN : Ambaaddee, Patsan, Pitwa.

SANSKṚTA : Ambasṭhā, Mācikā, Nalita.

HABIT : An erect herbaceous annual with straight, slender, glabrous or prickly stem.

HABITAT : Apparently a native of Africa; generally cultivated.

CHEMICAL COMPOSITION : Leaves contain 5 flavon glycosides including rutin and isoquercetin. Kaempferol, sterols, triperpenes. Flowers, myricetin glucosides; root polyphenol. Seed hydrolytic enzymes and oil. Cannabinidin, cannabisccin, cannabiscitrin.

Parts used : Leaves, seeds and flowers.

Ayurvedic properties :

Rasa - Amla, Kaṣāya. Vīrya - Śīta.
Guṇa - Laghu. Vipāka -Amla.

अम्बष्ठा मधुरा तिक्ता लघ्वम्लारसपाकयोः।
कफपित्तहरा कण्ठ्या पक्वातीसारनाशिनी।।
(कैयदेवनिघण्टु)

Ambaṣṭhā madhurā tiktā laghvamlārasapakāyoḥ,
Kaphapittaharā kaṇṭhyā pakvātīsāra nāśinī.
(Kaiyadeva nighaṇṭu)

Actions/Uses : Virecaka, Rucikāraka, Hṛdya.

Therapeutics :

Ambaṣṭha is mainly cultivated as a fibre crop. Leaves are purgative. Juice of flowers with sugar and black pepper is a popular remedy for constipation and in billiousness with acidity. Seeds are aphrodisiac and fattening. They are used in puerperal fever, and as external application to pains and bruises.

Hibiscus rosa-sinensis Linn.

ENGLISH : Shoe Flower, Chinese hibiscus.

INDIAN : Jaasvanda, Jasut, Dasindache phula.

SANSKṚTA : Japā, Java, Rudrapuspa, Aundrapuspa, Trisandhyā.

HABIT : An evergreen woody, glabrous showy shrub.

HABITAT : Native of China. Grown as an ornamental plant in gardens throughout India.

CHEMICAL COMPOSITION : Leaves and stems yield taraxeryl-OAc and β-sitosterol. Flvones from flowers. Quercetin-3-diglucoside, 3,7-diglucoside, cyanidin-3,5-diglucoside and cyanidin-3-sophoroside-3-5-glucoside from deep yellow flowers; all above compounds and kaempferol-3-xylosylglucoside isolate from ivory white flowers. The glycoside material exhibited hypotensive action and antispasmodic action in experimental animal tissues.

Parts used : Roots, leaves and flowers.

Ayurvedic properties :

Rasa - Kaṣāya, Tikta. Vīrya - Śīta.
Guṇa - Laghu, Rūkṣa. Vipāka - Kaṭu.

जपा संग्राहिणी केश्या त्रिसन्ध्या कफवातजित्।
(भावप्रकाश)

Japā sangrāhiṇī keśyā trisandhyā kaphavātajit.
(Bhāvaprakāṣa)

Actions/Uses : Pittahara, Kaphavātaśāmaka, Arsávināsi, Keśya, Garbharodhi, Sangrāhī, Viṣahara, Hṛdya.

Therapeutics :

Root of Japā is demulcent and used for cough. A decoction of root is used for venereal diseases and fevers. Fresh root juice is given for gonorrhoea and powdered root for menorrhagia. Roots and other parts are used as remedies for gonorrhoea, vomiting of blood and stomachic troubles. Leaves are emollient, aperient, anodyne and laxative. Their juice is beneficial in gonorrhoea. A decoction is used as a lotion in fevers. Leaves and stem-bark are used for abortion. Staminal column is diuretic used in kidney trouble. Flowers are astringent, demulcent, emollient, refrigerant, constipating, hypoglycaemic, aphrodisiac, emmenagogue and are used in treating alopacia, burning sensation in the body, diabetes, menstrual disorders and piles. It is also used in fever, cough, menorrhagia, stranguary, cystitis and other irritable conditions of genito-urinary tract. They are made into a paste and applied to swellings and boils. A decoction of flowers is given in bronchial catarrh. They are fried in ghee and given in menorrhagia. Buds are used in treatment of vaginal and uterine discharges. Infusion of petals is given as a refrigerant drink in fevers. Leaves and flowers are good for healing ulcers and for promoting growth and color of hair. Flowers have been found to be effective in treating arterial hypertension and have significant antifertility effect. Japa is extensively used for blackening of hair.

Pavonia odorata Willd.

INDIAN : Kaallaa-waallaa, Sugandha-bala.

SANSKṚIT : Vālaka, Rhibera, Ambunamaka, Hrivela, Keśanāma.

HABIT : A pubescent herb with rhizomes having a musk-like odour.

HABITAT : Found in waste places and open woods in the Deccan, Madhya Pradesh, Rajasthan, U.P., Orissa and West Bengal.

CHEMICAL COMPOSITION : β-sitosterol, palmitic, stearic, oleic and linoleic acid from rhizomes; isovaleric acid, n-caproic acid, α-pinene and methyl heptenone from essential oil.

Parts used : Rhizomes.

Ayurvedic properties :
Rasa - Tikta. Vīrya - Śīta.
Guṇa - Laghu, Rūkṣa. Vipāka - Kaṭu.

वालकं शीतलं तिक्तं पित्तवान्तितृषापहम्।
ज्वरकुष्ठातिसारघ्नं केश्यं श्वित्रव्रणप्रणुत्।।
(राजनिघण्टु)

Vālakam śītalam tiktam pittavāntitṛṣāpaham,
Jvarakuṣṭhātisāraghnam keśyam
śvittravraṇapraṇut.
(Rājanighaṇṭu)

Actions/Uses : Snehana, Dīpana, Vātānulomana, Balya.

Therapeutics :

Roots of Valaka are aromatic and possess antipyretic, diaphoretic, diuretic, cooling, demulcent, carminative, tonic, stomachic, and astringent properties. They are prescribed in dysentery and inflammation and haemorrhage of intestines. Brised well with ghee they are applied to erysapelas as poultice. Their paste with rice water acts as an antiemetic. They enter into the composition of well known fever drink called "Snanga Paniya". Plant is cooling, light, carminative and tonic. It cures dyspepsia, indigestion, dysentery, vomiting, thirst, fever, skin diseases, ulcers and bleeding disorders. Is used as a cure for rheumatism.

Sida acuta Burm.

SYNONYMS : *Sida carpinifolia* Linn. *f.; Sida carpifnifolia sensu Masters (non L.f.).*

INDIAN : Naagabalaa, Jaglimethi, Bariara, Tupkaria, Tukati, Chikana, Kareta, Kharenta.

SANSKṚTA : Mahābalā, Jyeṣṭhabalā, Kaṭmbharā, Dhanyā, Keśarikā, Keśaruhā, Varṣapuṣpā, Mṛgādanī, Sāriṇī, Sahadevī, Prasāraṇī, Devabalā, Pītapuṣpī, Māngalyārhā, Cīrapuṣpī, Mahāgandhā, Maharṣi, Devadaṇḍā, Vipulā, Utpalasandnyakā.

HABIT : An erect annual or biennial undershrub or shrub.

HABITAT : Native to West Bengal but throughout hotter parts of India, particularly moist regions.

CHEMICAL COMPOSITION : Four alkaloids from aerial parts and three from roots. Ecdysterone isolated.

Parts used : Roots, Seeds and leaves.

Ayurvedic properties :
Rasa - Madhura. Vīrya - Śīta.
Guṇa - Laghu, Snigdha, Pichchila.
Vipāka - Madhura.

महाबला तु मधुरा धातु वृद्धि करी मता।
बल्या, वृष्या, त्रिदोषघ्नीज्वर हृद्रोगदाहनुत।।
वातार्शः शोष विषमज्वरान्मेह गणं तथा।
बहुसूत्रं नाशयेश्चेस्येचं वैद्यर्निस पितम्।।
(निघण्टु रत्नाकर)

Mahābalā tu madhurā dhātu vṛddhikarī matā,
Balyā vṛṣyā tridoṣaghnī jvarahṛdrogadāhanut.
Vātarśah śoṣa viṣamajvarānmeha gaṇam tathā,
Bahusūtram nāśayeścesyecam vaiddairnirupitam.
(Nighaṇṭuratnākara)

Actions/Uses : Vātagna, Pittaghna, Vedanāśāmaka, Śothahara, Vātānulomana.

Therapeutics :

Root of Mahabala is astringent, cooling, bitter tonic; useful in nervous and urinary diseases and in disorders of blood and bile; used as a febrifuge, stomachic, and for chronic bowel complaints and as aphrodisiac. Leaves are demulcent and diuretic and used in rheumatic affections as poultice. They are made warm and moistened with ginglili oil and used to hasten suppuration. They are also used as an abortifacient. Decoction of leaves and roots is emollient and tonic and is used in the treatment of haemorrhoieds and impotence. Juice of leaves is useful for relief in chest pain and as anthelminntic. Juice is boiled in oil and applied to reticular swellings and in elephantiasis.

Sida cordifolia Linn.

ENGLISH : Country-mallow.

INDIAN : Chikannaa, Balaa, Kungyi.

SANSKṚTA : Baja, Vaṭyālikā, Kharayaṣṭikā, Balāḍhyā, Odanikā, Vāṭyā, Bhadrā, Samangā, Bhadrodanī, Kalyānī, Varakāṣthikā, Moṭā, Bhadraphalā, Sanāsā, Varavāhinī, Śītapākī, Cirapuṣpī, Krurā, Pratāpī, Iṣyakā, Samā.

HABIT : An annual or biennial small shrub.

HABITAT : Grows wide along roadsides throughout India.

CHEMICAL COMPOSIṬION : Whole plant contains an alkaloid probably identical with ephedrin.

Parts used : Whole plant, roots, Seeds and leaves.

Ayurvedic properties :
Rasa - Madhura. Vīrya - Śīta.
Guṇa - Laghu, Snigdha, Picchila. Vipāka - Mạdhura.

वलातितिक्ता मधुरा पित्तातिसारनाशनी।
वलवीर्यप्रदा - पुष्टिकफरोगविशोधनी।।
(राजनिघण्टु)

Balātitiktā madhurā pittātisāranāsanī,
Balāvīryapradā puṣṭikapharogaviśodganī.
(Rājanighaṇṭu)

Actions/Uses : Vātaghna, Pittaghna, Vedanāśāmaka, Śothahara.

Therapeutics :

Juice of plant with water is given for spermatorrhoea. Roots are aromatic, bitter tonic, demulcent and diuretic. They are cooling, astringent, stomachic and aphrodisiac. They are supposed to cure vaatarakta and raktapitta, consumption, polyuria and ulcers and produce strength and impart beauty to the body. Juice of root is used for healing wounds. Infusion of roots is given in nervous and urinary diseases and disorders of blood and bile. It is also useful in bleeding piles, in strangury and haematuria, in gonorrhoea, cystitis, leucorrhoea, chronic dysentery, insanity, facial paralysis and in asthma as a cardiac tonic. Decoction of root with ginger is used as febrifuge. Root bark with seasmum oil and milk is effective in curing cases of facial paralysis and sciatica. Powder of root is given with milk and sugar for relief of frequent micturition and leụcorrhoea. Seeds are aphrodisiac, and are administered in gonorrhoea, given for colic and tenesmus. They are also useful in neurological disorders like hemiplegia, facial paralysis and sciatica. They are considered valuable in the treatment of rheumatism.

Thespesia populnea Soland ex Correa.

SYNONYMS : *Hibiscus populneus* Linn.

ENGLISH : Tulip tree, Portia, False rosewood, Umbrella tree.

INDIAN : Paarosaa pimpalla, Bhenda, Parsipu, Gajadanda.

SANSKṚTA : Pārīśa, Kapītana, Phaliśa, Kapicūḍha, Puspāṣvatha, Garbha-bhanḍa.

HABIT : A compact quick-growing evergreen middle sized tree.

HABITAT : Coast forests of India, largely grown as a roadside tree in tropical regions.

CHEMICAL COMPOSITION : Flowers contain populneol, gossypol, kaempferol, quercetin, rutin etc. Petals of flowers contain populin, populnetin and herbacetin. Thespesin from fruits. Seeds contain fatty oil. Calycopterin from heartwood.

Parts used : Root, leaves, flowers, fruits and bark.

Ayurvedic properties :
Rasa - Kaṣāya.
Vīrya - Śīta.
Guṇa - Laghu, Rūkṣa. Vipāka - Kaṭu.

पारीशो दुर्जरः स्निग्धः क्रिमिशुक्रकफप्रदः।
फलेम्लो मधूरो मूले कषायः स्वादुमज्ञकः।।
(द्रव्यगुण)

Pārīśo durjjaraḥ snigdhaḥ krimiśukrakaphapradaḥ,
Phalemlo madhūro mūle kaṣāyaḥ svadumajjakaḥ.
(Dravyaguṇa)

कपीतनो लघू रुक्षः कषयः शिशिरो हरेत्।
कफपित्तप्रमेहास्त्रकुष्णयोनिगदव्रणान्।।

Kapītano laghū ruksaḥ kaṣayaḥ śiśiro haret,
Kaphapittapramehāsra kuṣṭhayonigadavraṇān.

Actions/Uses : Kaphapittaghna, Sandhānīya, Śothahara, Kusthaghna, Stambhana, Raktaprasādana, Dāhaśāmaka.

Therapeutics :
Pārīsá is a reputed remedy for skin diseases. It is light, acrid, cooling and astringent. It is useful in dysentery, piles, diabetes, haemorrhoides. It cures ulcers, itching, scabies and other skin diseases and urinary disorders. Root is tonic. Bark is astringent, given internally as an alterative. In the form of hot polutice leaves are beneficial in painful joints. Fruits, leaves and root are applied externally to scabies, psoriasis and other skin diseases.

Family : LIX - Meliaceae.

Aphanamixis polystachya (Wall.) Parker.

SYNONYMS : *Amoora rohituka W. & A.; Aglaia polystachya* Wall.

Similar : Tecomella undulata (G. Don.) Seem.

INDIAN : Rohitaka, Harinhara.

SANSKṚTA : Rohītaka, Plīhāri, Janavallabha, Saptavha, Sitavhaya.

HABIT : A large handsome evergreen tree, with a dense spreading crown and a straight cylindrical bole.

HABITAT : Usually found in forests of Assam, Western Peninsula, South India, Maharashtra, Konkan and Andaman Islands.

CHEMICAL COMPOSITION : Fixed oil. Amoorin and a sterol from essential oil of seeds. Aphanamixin, a triterpene closely related to flindissol and melianone from fruit shell. Rohitukin from seeds and eight other limonoids isolated. Seed shows antibacterial and antifungal activity and alcoholic extract of stem anticancer.

Parts used : Bark and seeds.

Ayurvedic properties :
Rasa - Kaṭu, Tikta, Kṣāya. Vīrya - Śīta.
Guṇa - Laghu, Rūkṣa. Vipāka - Kaṭu.

रोहितकौ कटुस्निग्धौ कषायौ च शुशीतलौ।
क्रिमिदोष व्रणप्लीहरक्तनेत्रामयापहौ।।
(राजनिघण्टु)

Rohitakau kaṭusnigdhau kaṣāyau ca śuśītalau,
Krimidoṣa vraṇaplīharaktanetrāmayāpahaū.
(Rajanighantu)

Actions/Uses : Plihasankocaka, Kaphapittaghna, Dīpana, Anulomana, Kṛmighna, Raktaśudhikara, Stambhana, Lekhana, Viṣaghna.

Therapeutics :

Rohitaka is used in enlarged glands, corpulence, liver and spleen diseases. Plant used in eye diseases by tribals of Orissa. Seeds are anthelmintic, laxative and refrigerant. Bark is astringent, beneficial in abdominal complaints and is used in spleen and liver diseases and tumours. Externally applied as poultice in rheumatsm. Seed oil is used as liniment in muscular pains and rheumatism.

Azadirachta indica A. Juss.

SYNONYMS : *Melia azadirachta* Linn.; *Melia indica* Brandis.

ENGLISH : Nim, Margosa, Indian Lilac.

INDIAN : Kadunimba, Nimb, Bakayan, Balanrtnimba.

SANSKṚTA : Nimba, Picumarda, Arista, Pāribhadra, Hinguniryāsa, Prabhadra, Kākaphala, Kiresta, Dhamana, Sarvatobhadra, Varatikta, Pitasaraka, Chardana, Pavanesta, Neta, Agnidhamana, Śukrapriya, Viśīrnaparnaka, Sutiktaka, Subhadra, Śīta, Kṛmighna, Sutikta.

HABIT : A large, evergreen tree.

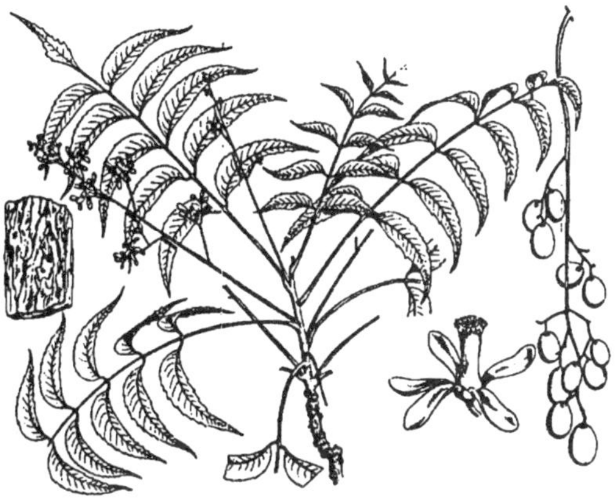

HABITAT : Wild in forest of Maharashtra, often planted all over India.

CHEMICAL COMPOSITION : Nimbin, nimbidin and nimidol from margosa oil; a paraffin alcohol-sugiol and oxyphenol nimbiol, nimbosterol nimbolin A and nimbolin B from trunk bark. Epoxyazadiradione, azadiradione and azadirone also meliantriol and meldenin from seed oil. Nimbolide, quercetin and B-sitosterol from leaves. Deaetylnimbin from seed and bark. Nimbidinin from bitter principle.

Parts used : Flowers, leaves, bark, seeds and oil.

Ayurvedic properties :

Rasa - Tikta, Kṣāya.	Vīryạ - Śīta.
Guṇa - Laghu, Rūkṣa.	Vipāka - Kaṭu.

निम्व:शीतो लघुर्ग्राहि कटुपाकोग्निवातनुत्।
अहृघ: श्रमतृढ्कास ज्वरारुचिक्रिमिप्रणुत्।।
व्रणपित्तकफच्छर्दिकुष्ठहृल्लासमेहनुत्।
निम्वपत्रं स्मृतं नेत्र्यं क्रिमिपित्तविषप्रणुत्।
वातलं कटुपाकज्व सर्वारोचक कुष्ठनुत्।
निम्बफलं रसे तिक्तं पाके तु कटु भेदनम्।
स्निग्धं लघूष्णं कुष्ठघ्नं गुल्मार्श: क्रिमिमेहनुत्।।
(भावप्रकाश)

Nimbaḥ śīto laghurgrāhi kaṭupākognivātanut,
Ahṛdyah śramatṛtkāsa jvrārucikrimipraṇut,
Vraṇapittakaphacchrddikuṣṭhahṛllāsamehaṇut,
Nimbpatram smṛtam netryam krimipittaviṣaprnit,
Vātalam kaṭupākancya sarvārocakakuṣṭhanut,
Nimbphalam rase tiktam pāke tu kaṭu bhedanam,
Snigdham laghūṣṇam kuṣṭhaghnam gulmārśaḥ krimimehanut. *(Bhavaprakasa)*

Actions/Uses : Kaphaghna, Pittaghna, Vātakar, Rocana, Grāhī, Kṛmighna, Yakṛtottejaka, Stambhana, Raktaja.

Therapeutics :

Nimba is used in Ayurvedic medicine for leprosy and skin diseases, fever; for purification of blood. Leaves are applied as polutice to boils. Decoction of leaves is antiseptic, used in ulcers and eczema. Bark, root bark and young fruit are bitter tonic, alterative, astringent, anthelmintic and antiperiodic. Gum is demulcent, tonic, in catarrh affections. Dry flowers are tonic and stomachic. Oil is stimulant, antiseptic, alterative, useful in rheumatism and skin diseases. Bark, gum, leaf and seed are used in snakebite and scorpion sting. Flowers and berries are purgative, emollient and anthelmintic. Alcoholic extract of bark is anticancer, antiviral and spasmogenic.

HOMEOPATHIC USE : Azadirachta indica is a Homeopathic remedy. It is a grand remedy in chronic fever. Fever generally comes in the afternoon at about 3 to 4.30 p.m, which commences with a very slight chill or none at all and abates about 7.30 p.m. Glowing heat and burning, especially in face, eyes, palms of hands and soles of feet in open air; sweat copious, commencing on forehead, gradually extending towards trunk; no sweat on lower part of body. It is useful in cases previously maltreated with quinine.

Naregamia alata W. & A.

ENGLISH : Goanese ipecacuanha.

INDIAN : Pitpaapadaa, Teen paani, Pitvel, Kapurbhendi.

SANSKṚTA : Triparṇī, Kaṇḍalu, Bhūmināgaranga.

HABIT : A small branched undershrub.

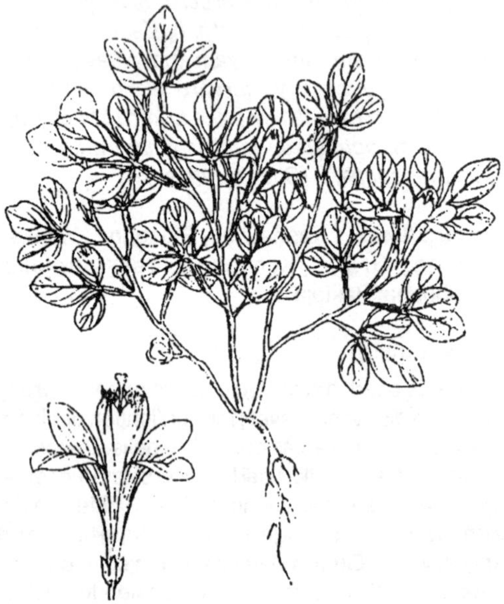

HABITAT : Western Ghats and Konkan.

CHEMICAL COMPOSITION : Root bark contains alkaloid naregamin. Heneicosane, β-sitosterol, stearic and palmitic acids from stems and roots.

Parts used : Root and leaves.

Ayurvedic properties :
Rasa - Madhura. Vīrya - Śīta.
Guṇa - Śīta. Vipāka - Madhura.

त्रिपर्णी मधुरा शीता श्वासकासविनाशनी।
पित्तप्रकोपशमनी विषव्रणहरा परा।।
(राजनिघण्टु)

Triparṇī madhurā śīta śvāsakāsa vināśanī,
Pittaprakopa śamanī viṣavraṇaharā parā.
(Rajanighantu)

Actions/Uses : Kaphanissāraka, Pittaśāmaka, Viṣahara, Kaṇḍughna.

Therapeutics :

Root of Triparni is emetic, cholagogue, expectorant, useful in acute dysentery. Decoction of leaves and stem is given with bitters and aromatics for billiousness. Plant is used in rheumatism and itch.

Soymida febrifuga A. Juss.

SYNONYMS : *Swietenia febrifuga* Roxb.

ENGLISH : Bastarol cedar, Indian redwood.

INDIAN : Rohan, Ruhin, Rakat rohan.

SANSKṚTA : Mānsarohiṇī, Atiruhā, Vrttā, Carmakaśā, Mansaruhā, Praharavallī, Vikasā, Kaśā, Carmakasā, Mānsaraktā, Pinśitarohinī, Ruhā, Agniruhā, Mansaruhā.

HABIT : A tall glabrous deciduous tree.

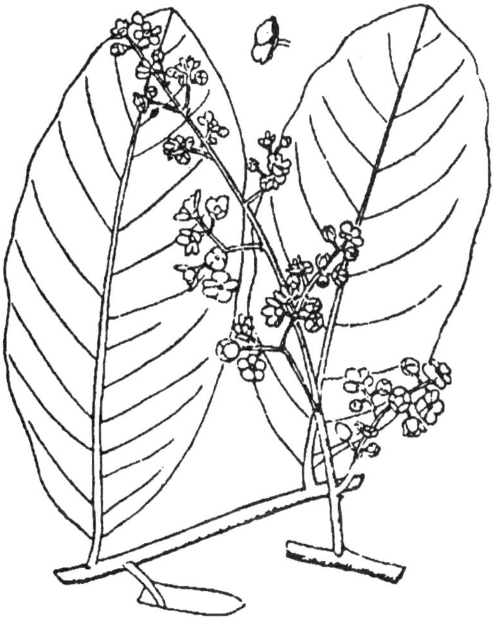

HABITAT : Common in the forests of Central India, Bihar and other hilly areas of Konkan and Khandesh on rocky soil.

CHEMICAL COMPOSITION : Bark contain bitter substance. Lupeol, sitosterol, methyl angolensate, deoxyandirobin from wood bark. Quercetin-3-O-L-rhamnoside and 3-O-rutinoside from leaves.

Parts used : Bark.

Ayurvedic properrties :
Rasa - Kaṣāya, Kaṭu. Vīrya - Śīta.
Guṇa - Laghu, Rūkṣa. Vipāka - Kaṭu.

स्यान्मांसरोहिणी वृष्यासरादोषत्रयापहा।
रसे पाके तु कटुका तुवराशीतला च सा।।
शीता कषाया रुच्या च क्रिमिघ्नी कण्ठशोधिनी।।
(भावप्रकाश)

Syānmansarohiṇī vṛṣyā sarā doṣatrayāpahā,
Rase pāke tu kaṭukā tuvarāśītalā ca sā,
Śītā kaṣāyā rucyā ca krimighnī kaṇṭhaśodhinī.
(Bhavaprakasa)

Actions/Uses : Stambhana, Vraṇaropaṇa, Grāhī, Pauṣṭika.

Therapeutics :

Mansarohini bark is astringent, bitter tonic, febrifuge, particularly prescribed in malaria, used in general debility, intermittent fevers, diarrhoea and dysentery. It is applied to rheumatic swelling and used as a gargle in stomatitis. Decoction is used as vaginal douch in leucorrhoea.

Toona hexandra (Wall. ex Roxb.) Roemer.

SYNONYMS : *Cedrela toona* Roxb.; *Cedrela hexandra* Wall. *ex* Roxb.; *Toona ciliata Roemer, syn.* Hesper.

ENGLISH : Red cedar, Toon.

INDIAN : Toon, Nandi-wruksha, Deodari, Kuruk.

SANSKṚTA : Nandivṛkṣa, Tūnī.

HABIT : A tall tree with coloured wood.

HABITAT : Common in hotter parts of India, Konkan and Western Ghats.

CHEMICAL COMPOSITION : Bitter substance, red coloring matter nyctanhin. Cedrelone, 1,2-dihydrocedrelone, bergapten and β-sitosterol from heartwood. Leucocyanidins from toon wood. Geranylgeraniol. Wood yields essential oil.

Parts used : Bark and flowers.

Ayurvedic properties :
Rasa - Madhura, Kaṣāya, Tikta. Vīrya - Śīta.
Guṇa - Gurū, Snigdha. Vipāka - Kaṭu.

नन्दीवृक्षः कटुस्तिक्तः शीतस्तिक्तारस्त्रदाहजित्।
शिरोर्त्ति श्वेतकुष्ठघ्नः सुगन्धिः पुष्टिवीर्यदः।।
(राजनिघण्टु)

Nandīvṛkṣaḥ kaṭustiktaḥ śītastikttārsadāhajit,
Śirortti śvetakuṣṭhaghnaḥ sugandhiḥ
puṣṭivīryadaḥ.
(Rajanighantu)

Actions/Uses : Grāhī, Balya, Vedanāsthāpana, Kaṭu-pauṣṭika, Vīryavardhaka, Garbhāśaya-sankocaka.

Therapeutics :

Bark of Nandivrksa is astringent, tonic, antiperiodic, used in chronic infantile dysentey and as external application for ulcers. Flowers are emmenagogue; used in menstrual diorders.

Family : LX - Menispermaceae.

Anamirta cocculus (Linn.) W. & A.

SYNONYMS : *Anamirta paniculta* Colebr., *Cocculus suberosus* DC. *Menispermum cocculus* Linn., *Cocculus indicus*.

ENGLISH : Levant berries, Cocculus Indicus, Crow killer.

INDIAN : Kadu-phal, Kaakamaari, Garudphul, Kaakaphala.

SANSKṚTA : Kākamārī, Kākaghnī, Garalạphala.

HABIT : An extensive climber.

HABITAT : Moist deciduous forests in Deccan, Orissa and Bengal.

CHEMICAL COMPOSITION : Picrotoxin, cocculin, anamirtin. Pericarp contains alkaloids menispermine and paramenispermine. Alkaloids berberine, palmatine, magnoflorine and columbamine in stem and roots. Major tertiary alkaloid 1-8-oxotetrahydropalmatine.

Ayurvedic properties :
Rasa : Tikta, Kaṭu. Vīrya : Ūṣṇa
Guṇa : Tīkṣṇa, Kaṣāya. Vipāka : Kaṭu

Parts used : Leaves, fruits and seeds.

काकमारीफलं तीक्ष्णं कषायोष्णं कटु स्मृतम्।
रक्तशुद्धिकरं तिक्तं कृमिजन्तुविनाशकम्।।
शोफव्रणश्वासकासकण्डूत्वग्दोषनाशनम्।
कफवातहरं प्रोक्तं दुष्टव्रणहरं परम्।।
(स्व)

Kākamārīphalam tīkṣṇam kaṣāyoṣṇam kaṭu smṛtam,
Raktasudhikaram tiktam krmijantuvināśakam.
Śophavraṇa śvāsakāsa kaṇḍū tvagdoṣa nāśanam,
Kaphavātaharam proktam duṣṭavraṇaharam param.
(Svayamkrti)

काकमारी काकवैरी तिक्ता कीटाविनाशिनी।
विचर्चिकरक्षतदोष जरात्वग्दोषनाशनी।।
(आ. चंद्रिका)

Kākamarī kākavairī tiktā kīṭavināśanī,
Vicarcikarakṣatadosa jāratvagdosanāsanī.
(Ayurvedartha Candrika)

Action/Uses : Kaphahāraka, Vātahāraka

Therapeutics :
Fruits of Kākāmarī are bitter, astringent, expectorant, antifungal, anthelmintic, thermogenic. They are used in bronchitis, foul ulcers, chronic skin diseases and ring worm. Tender leaves are used by tribles for the contraction of uterus immediately after delivery. Snuff powder prepared from dried leaves is a remedy in malarial fever. The dried fruits constitute the drug "COCCULUS". The powdered drug is employed in external applications for pedicular and other obstinate and chronic affections of skin. Stimulant and parasiticide. Seeds and berries used as piscicide and for night sweats of phthisis. Plant is used in the treatment of skeletal fractures.

Alcoholic extract of aerial parts is anticancer and diuretic. Ointment prepared from drupes is used as insecticide, to destroy pediculi and chronic skin diseases.

HOMEOPATHIC USE : Cocculus is a Homeopathic remedy. Spasm, paralytic weakness, extreme irritability of nervous system, sea and car sickness and empty hollow feeling are the most notable features.

Cissampelos pareira Linn.

SYNONYMS : *Chondodendron tomentosum* R. & P.; *Cissampelos hexandra* Roxb.; *Cissampelos convovulacea* Willd.; *Cissampelos hernandifolia* Wall.

ENGLISH : Pareira brava; Virgin Vine; Velvet Leaf.

INDIAN : Pahaadawela, Venivel, Akauadi, Harjeuri, Pari.

SANSKṚIT : Laghupathā, Ambaṣṭhā, Pāṭha, Garudi, Avihakarni, Brihattikta. Cchinnaveśika, Devi, Ekaṣthila, Kuceli, Mahanjasi, Mālavi, Pāpacelī, Tikta, Prācina, Pratanini, Rucisýa, Rasā, Siśira, Sthapini, Susthira, Tiktapuṣpa, Triśira, Venivalli.

HABIT : A climbing shrub.

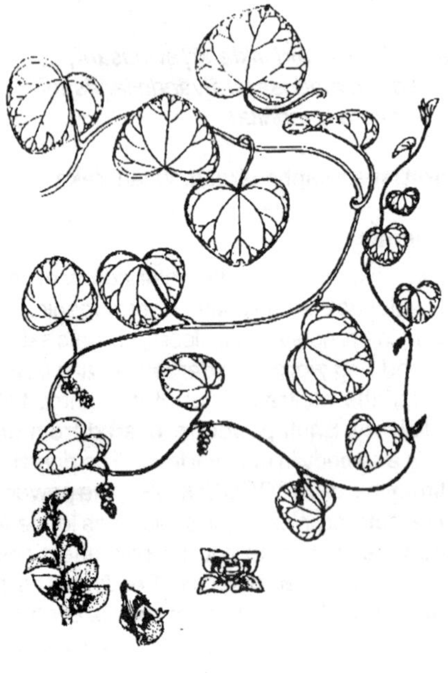

HABITAT : Native of West Indies, Central America and India.

CHEMICAL COMPOSITION : Hayatin, hayatinin, L-curine and isochondrodendrine from roots and vines, all having curare-like activity. Cycleanine, (-) bebeerine, from leaves. Hayatine salts preparations, hayatine methiodide are muscular relaxtant.

Ayurvedic properties :

Rasa : Tikta, Kaṭu.	Vīrya : Ūṣṇa
Guṇa : Tīkṣṇa, Kaṣāya.	Vipāka : Kaṭu

Parts used : Root, leaves and bark.

लघुपाठा तिक्तरसा विषघ्नी कुष्ठकन्डूनुत्।
छर्दिहृद्रोगगरजित् त्रिदोषशमनी मता।।
(राजनिघण्टु)

Laghupaṭhā tiktarasā visaghṇī kuṣṭhakaṇḍūnut,
Chardihṛdrogagarajit tridoṣasamanī matā.
(Rajanighantu)

Action/Uses : Kaphahāraka, Vātaharaka, Tridoṣahāra.

Therapeutics :

Plant is mild stomachic, bitter tonic, diuretic and antilithic. It exerts an astringent and sedative action on mucous membrane of genito-urinary organs. Root is bitter tonic, antiperiodic, diuretic, purgative, stomachic, used in dyspepsia, diarrhoea, dropsy, cough and urinary troubles like cystitis. Used as a stimulant to kidneys, relieving urinary irritation and in chronic inflammation of bladder and various urinary diseases. Pills prepared from root, pepper, assafoetida and ginger and mixed with honey are found useful in indigestion and colic. Leaves are applied externally in scabies and septic ulcers and leaf extract for external application for itch, sores and sinuses.

HOMEOPATHIC USE : Pareira ia a Homeopathic remedy. Its characteristic symptoms, clinically confirmed are violent pains with stangury; must go on all fours and press head against floor to pass water at all; urethritis with severe pain when passing water, and discharge of mucus from urethra. Pain down thighs and even to feet when attempting to pass water. Violent pain in glans penis with the straining.

Cocculus hirsutus (Linn.) Diels.

SYNONYMS : *Cocculus villosus DC.; Menispermum hirsutum* Linn.

ENGLISH : Broom creeper, Ink berry.

INDIAN : Vaasanawela, Tan, Jamtikibel, Parvel, Chireta, Hunder.

SANSKṚTA : Chilihinta, Pāṭhā, Pātālagāruḍī, Ambaṣṭhā, Viddhakarṇī, Jaliamni, Vasati-tikta, Sthāpanī, Vanatiktakā, Rasā, Ekāsthīlā, Pāpavelī, Prācīnā, Śreyasī.

HABIT : A scandent tomentose climbing shrub.

HABITAT : Tropical and subtropical India.

CHEMICAL COMPOSITION : Trilobine, isotrilobine, coclaurine and magnoflorine from stems and roots. D-Trilobine and DL-coclaurine from roots. β-sitosterol, ginnol and monomethyl ether of inositol isolated. Essential oil. Water soluble extract of root is sedative, hypotensive, bradycardiac, cardiotonic and spasmokytic.

Parts used : Roots and leaves.

Ayurvedic properties :
Rasa - Tikta. Vīrya - Ūṣṇa.
Guṇa - Laghu, Snigdha, Picchila. Vipāka - Kaṭu.

पाठोष्णा कटुकातीक्ष्णा वातश्लेष्महरी लघु।
हन्तिशूलज्वरच्छर्दिकुष्ठातिसारहृद्रजः।।
दाहकंडूविषश्वासकृमिगुल्मगर व्रणान्।।
(भावप्रकाश)

Pāṭhoṣṇā kaṭukātīkṣṇā vātaśleṣmaharī lghuh,
Hanti śūla jvara chhrdi kuṣṭhātisāra hṛdrujaḥ.
Dāhakaṇḍu viṣaśvāsa kṛmugulmagara vraṇān.
(Bhavaprakasa)

छिलहिण्टः परं वृष्यः कफघ्नः पवना पहः।।
(भावप्रकाश)

Chilihiṇṭaḥ param vṛṣyaḥ kaphaghnaḥ pavanāpahaḥ.
(Bhavaprakasa)

Actions/Uses : Tridoṣaśāmaka, Dīpana, Pācana, Śāmaka, Kaphaghna, Snehana, Jvaraghna, Tvagrogahara, Anulomana, Raktaśudhikara, Viṣaghna, Kuṣṭhaghna.

Therapeutics :

Root of Pātālagāruḍi is refrigerant, alterative, laxative, sudorfic; used to allay irritation, fever; useful in chronic rheumatism and venereal diseases. Decoction of leaves is laxative. Powder of leaves mixed with water applied to eyes, giving cooling effect. Juice of leaves when mixed with water forms a jelly which is taken as a cooling medicine for gonorrhoea and used externally for eczema, prurigo and impetigo.

Tinospora cordifolia (Willd.) Miers ex Hook.f & Thoms.

SYNONYMS : *Cocculus cordifolius* DC.; *Menispermum cordifolium* Willd.; *Tinospora glabra (N. Burm.)* Merr.

HABIT : A large, glabrous, deciduous climbing shrub.

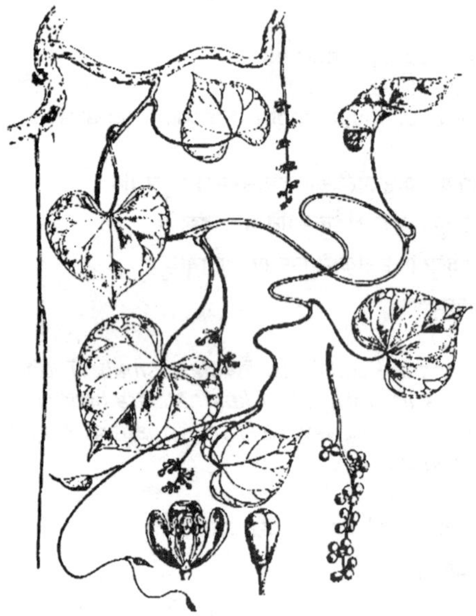

HABITAT : Throughout tropical India and Andamans.

ENGLISH : Heart leaved moon seed, Gulancha tinospora.

INDIAN : Gullawela, Amrutvel, Jiwantika, Gurch, Gulancha, Ambarvel, Giroli.

SANSKṚTA : Guḍuci, Amrutā, Madhuparṇī, Cinnaruha, Vatsadani, Kundalinī, Cakralakṣaṇā, Kandodbhava, Amṛtavallī Jvarāri, Chinnā, Amṛtalata, Rasāyanī, Vayasthā, Kundali, Chinnāngā, Mandalī, Yamamvara, Surakṛtā, Viśalyā, Bhiṣakpriya, Tikta, Jivantikā, Candrahāsā, Nāgakumārikā, Ādhārā, Tandrikā, Nirjara, Somavalli.

CHEMICAL COMPOSITION : Furanoid bitter principle tinosporine and a furanoid diterpene tinosporide and tinosporidine, β-sitosterol from stems. Cordifol, heptacosanol and octacosanol from leaves.

Parts used : Stem and leaves.

Ayurvedic properties :
Rasa - Tikta, Kaṭu, Kaṣāya. Vīrya - Uṣṇa.
Guṇa - Snigdha, Mṛdu, Rūkṣa, Laghu.
Vipāka - Madhura.

गुडूची कटुकातिक्ता स्वादुपाका रसायनी।
संग्राहिणी कषायोष्णा लघ्वी बल्याग्निदीपनी।
दोषत्रयाम तृड्दाहमेहकासांश्च पाण्डुताम्।
कामला कुष्ठवातास्त्रज्वरकृमिवमी हरेत्।।
(भावप्रकाश)

Guḍucī kaṭukatiktā svādupakā rasāyanī,
Sangrāhiṇī kaṣāyoṣṇā lghvī balyāgnidīpanī,
Doṣatrayām tṛḍdāhamehakāsānsca pāṇḍutām,
Kāmalā kuṣṭhavātāsrajvarakṛmivamīrharet.
(Bhavaprakasa)

Actions/Uses : Tridoṣaghna, Vedanāsthāpana, Kuṣṭhaghna.

Therapeutics :

Guḍucī is tonic, antiperiodic and diuretic properties. It is valuable in general debility and fevers and other exhausting diseases. It is a remedy in secondary syphilitic affections, chronic rheumatism and mild forms of intermittents. Starch from roots and stems of Amrutaa is nutrient, used in chronic diarrhoea and chronic dysentery. Mature stem is acrid, bitter, hot, restorative and aphrodisiac. It alleviates all the three doshaas and is a digestive tonic. It cures fever, jaundice, thirst, burning sensation, diabetes, piles, skin ailments, resperatory disorders, neurological diseases and improves intellect. Juice of fresh plant is diuretic and useful in gonorrhea. Plant is a constituent of several Ayurvedic preparations used in general debility, dyspepsia, fevers and urinary diseases. Powdered and made into an infusion, used as altereatived and aphrodisiac. Externally the drug is used against rheumatic complaints.

Tinospora malabarica (Lam.) Miers

SYNONYMS : *Tinospora tomentosa (Colebr.)* Miers.; *Menispermum malbaricum* Lamk.; *Cocculus malabaricus DC.; Tinospora sinesis (Lour.)* Merr.; *Campylus sinensis* Lour.; *Tinospora malabarica* (Lam.) *Hook.f. & Thoms.*

HABIT : A giant deciduous climber.

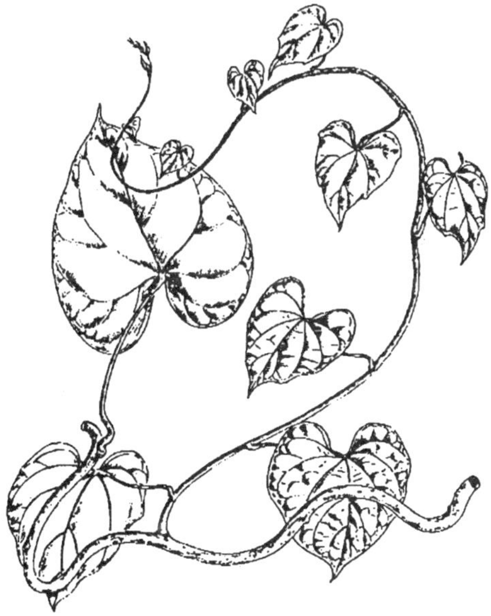

HABITAT : Konkan, Karnataka, Tamilnadu, Bengal, Orissa.

INDIAN : Sudarshana, Gullawela-2, Vhadli-amrutvel, Giloe, Gulancha, Nimkathia bour gurcha.

SANSKRTA : Kandodbhava-guruci, Sudarśana.

CHEMICAL COMPOSITION : Root and stem contain starch, a bitter principle and a tace of berberine.

Parts used : Leaves and stem.

Ayurvedic properties :
Rasa - Tikta, Kaṭu, Kaṣāya. Vīrya - Ūṣṇa.
Guṇa - Snigdha, Mṛdu, Rūkṣa, Laghu.
Vipāka - Madhura.

कन्दोद्भवा गुरुचि च कदुष्णा सन्निपातहा।
विषघ्नी ज्वरभूतघ्नि वलीपलितनाशिनी।।
(राजनिघण्टु)

Kandodbhava guruci ca kadusnā sannipātapahā,
Viṣaghnī jvarabhūtaghni balīpatitanāśinī.
(Rajanighantu)

Actions/Uses : Tridoṣaghna, Vedanāsthāpana, Kuṣṭhaghna.

Therapeutics :

Sudarśana is tonic, used almost in the same way as T. cordifolia. Leaf-juice mixed with that of Coleos amboinicus and honey is employed in gonorrhoea. Fresh leaves and stems are used in chronic rheumatism in China and Tongking.

Family : LXI - Mimosaceae.

Acacia arabica Willd. var indica (Benth.) Brenan.

SYNONYMS : *Acacia nilotica (L.)* Willd. *ex Delile ssp. indica (Benth.)* Brenan.

ENGLISH : Acacia, Babul.

INDIAN : Baabhulla (Vedibabul).

SANSKṚTA : Babbūla, Barbura, Yugalānkṣa, Panktībīja, Dṛuḍhabīja, Tīkṣṇakanṭaka, Kaphāntaka, Ajabhakṣa, Malaphala, Sūkṣmapatra, Kaṣāya, Gośṛnga, Yugmakāṣṭa.

HABIT : A moderate-sized almost evergreen tree with a short trunk, a spreading crown and feathery floiage.

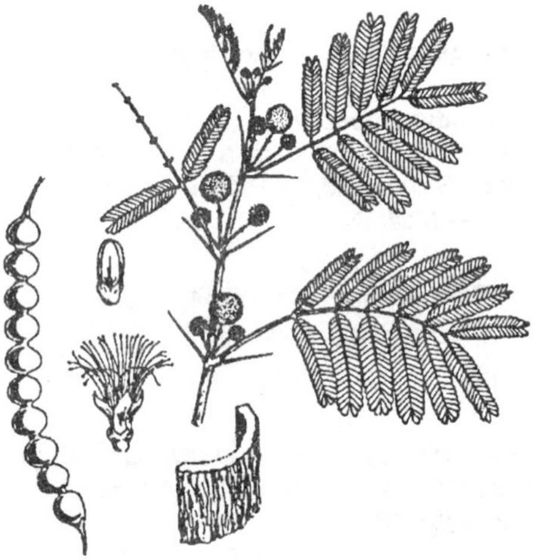

HABITAT : North Africa; naturalized in all drier parts of India.

CHEMICAL COMPOSITION : Bark contains 7-20% tannnin; pods contin 12-19 % tannin. Several polyphenols are present in pods. Quercetin, gallic acid, catechin, epicatechin, dicatechin, leucocyanidin, epigallocatechin, polyphenolic phlobaphenes consisting mainly of catechol and pyrogallol, from bark. Gum contains 2-O-B-L-arabinofuranosyl-L-arabinose and 3-O-B-L-arabinopyranosyl-L-arabinose. Leaves and fruits contain 32 % tannin.

Parts used : Bark, pods, gum and stem.

Ayurvedic properties :
Rasa - Madhura, Kaṣāya. Vīrya - Śīta.
Guṇa - Gurū, Rūkṣa. Vipāka - Kaṭu.

बब्बलस्तिक्तमधुरः स्निग्धः शीतोष्णतुवरः।
आमरक्तातिसाराणां नाशनो ग्राहको मतः।।
कफं कासं च पित्तं च दाहं रक्तातिसारकम्।
वातं प्रमेहं शमयेत् पर्णन्तु ग्राहकं मतम्।।
रुच्यं कटुष्ण कासघ्नं वातपुंस्त्वकफार्शनुत्।
बाबूलस्य फलं रुक्षं विशदं स्तंभनं गुरु।।
कषायं मधुरं शीतं लेखनं कफपित्तहृत्।
(नि. र.)

Babbulastiktamadhuraḥ snigdhaḥ śītoṣṇatuvaraḥ,
Āmaraktātisārañām nāśano grāhako mataḥ.
Kapham kāsam ca pittam ca dāham raktātisārakam,
Vātam prameham śamayet parṇantu grāhakam matam.
Rucyam kaṭuṣṇakasaghnam vātapunstvakapharśanut,
Babbulasya phalam ṙukṣam viśadam stambhanam gurū.
Kaṣāyam madhuram śītam lekhanam kaphapittahṛt.
(Nighantu ratnakara)

Actions/Uses : Kaphapittaśāmaka.

Gum : Vātapittaśāmaka, Raktastambhana, Vraṇaropaṇa, Stambhana, Samkocaka, Kṛmighna.

Therapeutics :

Babbula is demulcent, and is used as an emulsifying, binding and suspending agent. Bark is astringent, alexipharmic, cooling and anthelmintic; cures skin diseases, bleeding piles; used in asthma and bronchitis; decoction in leucorrhoea, diarrhoea, dysentery, biliousness, leucoderma, urinary discharges, good is ascites. Gum is a substitute for true gum arabic; used in sore throat, diarrhoea and dysentery stops bleeding and urinary and vaginal discharges; also useful in diabetes. Along with latex of Caletropis procera given to cure asthma. Pods used in impotency and effective in urinogenital disorders. Flowers, pods and gum-resin used as tonic in diarrhoea and dysentery. Bruised leaves are applied to sore eyes in children; eaten in throat infection and polutice used in sore eyes. Paste of burnt leaves effective ointment in itch. Various parts are used in hair-fall, earache, symphysis, dysentery, cholera, leprosy and rinderpest.

Acacia catechu (Linn.f.) Willd.

SYNONYMS : *Mimosa catechu* Linn. *f.; Mimosa catechuoides* Roxb.

ENGLISH : Cutch tree.

INDIAN : Khaira, Khadira.

SANSKṚTA : Khadira.

HABIT : A moderate-sized, deciduous tree.

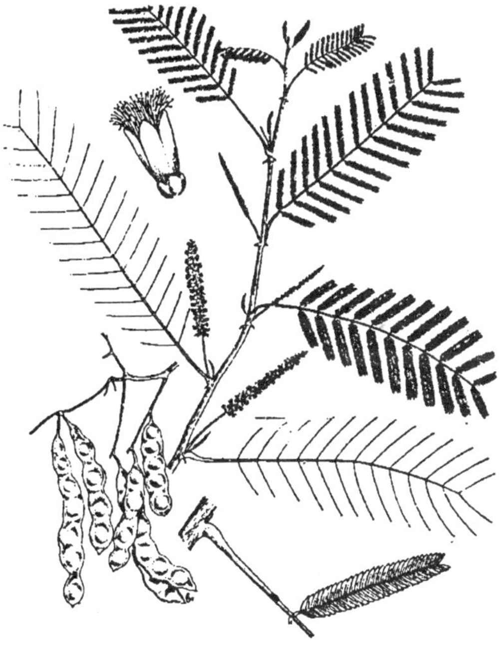

HABITAT : Konkan, Maharashtra and other parts of India.

CHEMICAL COMPOSITION : Catechin, catechutannic acid, tannin. Wood contains α, β and y catechin. l-epicatechin. Heartwood contains eight flavons. Quercetin and its derivatives. Seeds exhibited hypoglycemic activity in normal rats but not in diabetic rats.

Parts used : The whole plant, flower tops, bark and extract.

Ayurvedic properties :
Rasa - Tikta, Kaṣāya. Vīrya - Śīta.
Guṇa - Laghu, Rūkṣa. Vipāka - Kaṭu.

खदिरः शतिळो दन्त्यः कण्डूकासारुचिप्रणुत्।
तिक्तः कषायो मेदोघ्नः कृमिमेहज्वरव्रणान्।।
श्वित्रशोफामपित्ताम पाण्डुकुष्ठकफान् हरेत्।
(भावप्रकाश)

Khadiraḥśītaḷo dantyaḥ kaṇḍūkāsā rucipraṇut,
Tiktaḥ kaṣāyo medoghnaḥ kṛmimeha jvaravraṇān.
Śvitraśophāmapittāsra pāṇḍukuṣthakaphān haret.
(Bhavaprakasa)

Actions/Uses : Pittaghna, Kaphaghna, Rucivardhaka, Stambhana, Kṛmighna.

Therapeutics :

Khadir is a fairly good source of timber. The gum is of very good quality. The most important product is Catechu or Katha. Khadira bark is bitter, astringent, cooling, anthelmintic, antidysenteric and antipyretic; useful in melancholia, conjuctivitis, haemoptysis and skin diseases. Various plant parts are used in sore mouth, pain in chest, asthma, colicky pain, cancer, gravel, dysentery, phthisis, bronchitis, consumption and strangulation of intestines. Juice of bark along with asafoetida is used in haemoptysis. Mixture of flower tops, cumins, is given in gonorrhoea. Katha from heartwood is astringent, cooling and digestive and is used in relaxed conditions of throat, mouth, gums and for cough and diarrhoea. Gum resin used in masticatories.

Acacia concinna DC.

SYNONYMS : *Acacia sinuata (Lour.) Merrill.; Mimosa sinuata* Lour.; *Mimosa concinna* Willd.; *Acacia rugata (Lamk.)* Merr.

ENGLISH : Soap-pod tree.

INDIAN : Shikekaaee, Kochi.

SANSKRTA : Sātalā, Saptalā, Fenila, Carmakaśā.

HABIT : A climbing prickly scandent shrub.

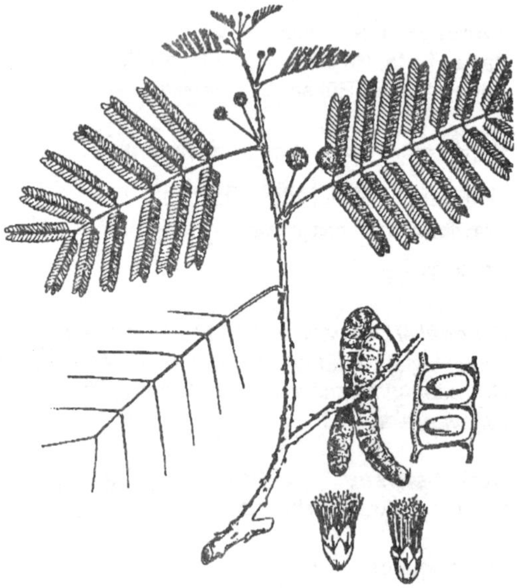

HABITAT : Tropical jungles of India.

CHEMICAL COMPOSITION : Tartaric, citric, oxalic, succinic and ascorbic acids, tartaric racimase, rutin, tannin along with alkaloids-colycotomine and nicotine from leaves. Pods contain saponins (20.8 %). Seeds contain saponin which possesses β-amrine, acacinin A & B, carbohydrate glucose, arabinose, xylose, fucose and rhamnose.

Parts used : Aerial parts leaves, bark and pods.

Ayurvedic properties :
Rasa - Tikta, Kaṣāya. Vīrya - Śīta.
Guṇa - Laghu. Vipāka - Kaṭu.

सातला कफपित्तघ्नी लघु तिक्ता कषायिका।
विसर्पकुष्ठ विस्फोटव्रणशोफनिकृन्तनी।।
(राजनिघण्टु)

Sātalā kaphapittaghnī laghu tiktā kaṣāyikā,
Visarpakuṣṭhavisphotavraṇa śophanikṛntanī.
(Rajanighantu)

Actions/Uses : Fruits : Uttejaka, Kaphagna, Vāmaka, Anulomika.

Leaves : Sour, Rocaka, Virecana stimulating liver.

Therapeutics :

Pods of Satala are cooling, deobstruent, detergent, purgative, anthelmintic, antidiarrhoeal. They improve appetite but cause vaata; cure kapha, biliousness, diseases of blood, erysepelas, leucoderma, stomatitis, ascites, piles, cardiotonic. Pods are aperient, expectorant and emetic. Decoction of pods relieves biliousness and acts as purgative. Ointment prepared from ground pods, used in skin diseases. A decoction used to remove dandruff. Pod decoction used to kill lice and skin diseases. Extract of bark used in leprosy. Leaves are cathertic and cure billiousness. A chutney made of tender leaves, salt, tamarind and chillies given in billious affections such as jaundice. Alcoholic extract of aerial parts is hypotensive, spasmolytic and diuretic.

Acacia farnesiana Willd.

ENGLISH : Cassie flower.

INDIAN : Devabaabhulla, Gandha babul, Kankar, Vilaayati kikar.

SANSKṚTA : Arimeda, Vitkhadira, Irimeda, Bālapatra, Bahuśara, Vakrakanṭa, Suśalya, Kuṣṭhāri, Yajnyānga, Gāyatrī, Dantadhāvana, Bahuśalya, Kṣatakṣama.

HABIT : A thorny bush or small tree.

HABITAT : Native to tropical America. Grown throughout India, often planted in gardens.

CHEMICAL COMPOSITION : N-acetyl-L-djenkolic acid from seeds. Isorhamnetin-3-7-glucorhamnoside, d-pinitol from flowers. Gallic, ellagic, m-digalic acids; methyl gallate, kaempferol, cyanogens; linamarin, lotaustralin, aromadendrin, naringenin, etc. isolated. Tannins and alkaloids, also rutin and apigenin-6,8-bis-C glucopyranoside from leaves. Essential oil. Pods 23 % tasnnin. Ripe pods contain tannin and several polyphenolic compounds.

Parts used : Various plant parts, bark and leaves.

Ayurvedic properties :
Rasa - Kaṣāya. Vīrya - Śīta.
Guṇa - Laghu, Rūkṣa. Vipāka - Kaṭu.

विट्खदिरः कटुरुष्णस्तिक्तो रक्तव्रणोत्थदोषहरः।
कण्डूतिविषविसर्पज्वर कुष्ठोन्मादभूतघ्नः।।
(राजनिघण्टु)

Viṭkhadiraḥ kaṭuruṣṇastikto raktavranotthadoṣaharaḥ, Kaṇḍūtiviṣavisarpajvara kuṣṭhonmādabhutaghnaḥ. (Rajanighantu)

अरिमेदः कषायोष्णो मुखदन्तगदास्त्रजित्।
हन्ति कण्डूविषश्लेष्मकृमिकुष्ठविषव्रणान्।।
(भावप्रकाश)

Arimedaḥ kaṣāyoṣṇo mukhadantagadāsrajit, Hanti kaṇḍūviṣaśleṣmakṛmikuṣṭhaviṣavraṇān. (Bhavaprakasa)

अरिमेदस्य निर्यासो मधुरस्तु बलप्रदः।
धातुवृद्धिकरश्चैव मुनिभिः संप्रभाषितः।।
(नि. र.)

Arimedasya niryāso madhurastu balapradaḥ, Dhatuvṛddhikaraścaiva munibhiḥ samprabhaṣitaḥ. (Nighanturatnakar)

Actions/Uses : Kaphapittaśāmaka, Stambhana, Snitasthāpana.

Therapeutics :

Various plant parts of Arimeda are used in madness, carbuncle, epilepsy, rabies, convulsions, delirium, sores, cholera, sterility in women, snake-bite and rinderpest. Tender leaves, bruised in a little water are prescribed for gonorrhoea. Bark is astringent and demulcent. Flowers are the source of much valued Cassie perfume.

Acacia pennata (Linn.) Willd.

SYNONYMS : *Mimosa pennata* Linn.

INDIAN : Shemba, Ari, Aila.

SANSKṚTA : Ari, Sandanika, Khadirapatrikā.

HABIT : A large, thorny climbing shrub.

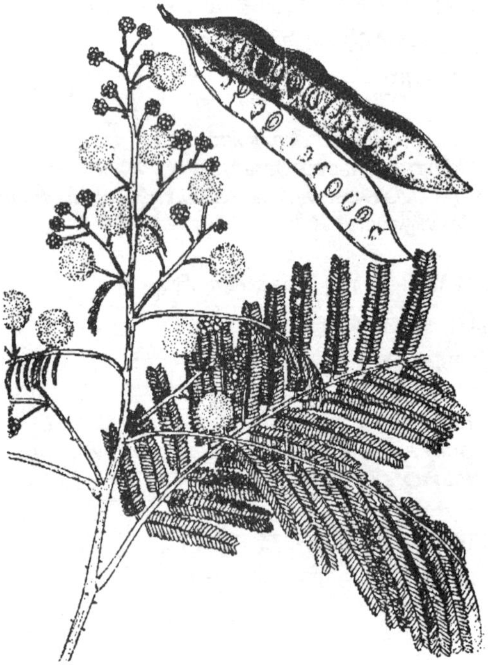

HABITAT : Bengal, Bihar, Central, Western and South India.

CHEMICAL COMPOSITION : Bark contains lupeol and a-spinasterol and stem β-sitosterol. Extract of aerial parts, hypotensive.

Parts used : Roots.

Ayurvedic properties :
Rasa - Ticta, Kaṭu, Kaṣāya. Vīrya - Ūṣṇa.
Guṇa - Laghu, Rūkṣa. Vipāka - Kaṭu.

अरिः कषायकटुका तिक्ता रक्तार्तिपित्तनुत्।
(राजनिघण्टु)

Ariḥ kaṣāya kaṭukā tiktā rakttārti pittanut.
(Rajanighantu)

Actions/Uses : Raktasangrāhaka, Tridoṣahara, Vātapittahara.

Therapeutics :

Bark of Ari is acrid, bitter, sour, and hot; cures diseases of blood, biliousness, tridosha, bronchitis, asthma. Leaf juice mixed with milk given to infants who suffer from indigestion. Leaves, chewed with sugar and cumin in bleeding gums. Decoction of young leaves given in body pain, headache and fever. Juice of bark is antidote for snake poison. Bark is used as a substitute for soap. Fruit pulp is piscicidal.

Albizzia lebbeck Benth.

SYNONYMS : *Mimosa lebbeck* Linn.; *Acacia speciosa* Willd.; *Albizzia odoratissima* Benth.

INDIAN : Shirisha, Chichola.

SANSKṚTA : Śirīśa, Śukapriya, Mṛudupuṣpaka, Śukataru, Kalinga, Madhupuṣpa, Śukrapuṣpa, Śukradruma, Śukavṛkṣa, Sitapuṣpa, Bhandika, Uddānaka, Lomaśapuṣpaka, Kapītana, Śyāmala, Śankhiṇīphala, Vṛttapuṣpa, Bhandīra, Plavangaka, Bhandī, Śyāmavarṇa, Pit-śiriśa.

HABIT : A large, erect, unarmed deciduous, spreading tree.

HABITAT : A road-side tree common throughout India.

CHEMICAL COMPOSITION : Gum, saponin and tannin. Pod saponin, labbekanin C. Seeds, a mixture of saponins lebbeddkanin A & B. Heartwood, malanoxetin, d-pinitol, okanin and leucopelargonidin. Leaves contain echinocystic acid and β-sitosterol, flavone, vicetin II. Flowers gives a colourless, sweet-smelling oil, 4.3 %.

Parts used : Bark, seeds, leaves and flowers.

Ayurvedic properties :
Rasa - Kaṣāya, Tikta, Madhura. Vīrya - Alpa-Ūṣṇa. Guṇa - Laghu, Rūkṣa, Tīkṣṇa. Vipāka - Katu.

कफपित्तहरश्चैव विषव्रणविशोधनः।
शीतवीर्यो विसर्पघ्नो शिरीषो मधुरो रसः।।
(म. नि.)

Kaphapittaharścaiva viṣavraṇaviśodhanaḥ,
Śītavīryo visarpaghno śirīṣo madhuro rasaḥ.
(Madanadinighantu)

शिरीषो मधुरोऽनुष्णस्तिक्तश्च तुवरो लघुः।
दोषशोषविसर्पघ्नो कासव्रणविषापहः।।
(भावप्रकाश)

Śirīṣo madhuro anuṣṇastiktaśca tuvaro laghuḥ,
Doṣasoṣavisarpaghnah kāsavraṇaviṣāpahaḥ.
(Bhāvaprakāṣa)

शिरीषो कटुकः शीतो विषवातहरः परः।
पामासृक्कुष्ठकण्डूतित्वग्दोषाणां विनाशकः।।
(राजनिघण्टु)

Śirīṣo kaṭukaḥ śīto viṣavātaharaḥ paraḥ,
Pāmāṣrk kuṣṭhakaṇḍūtitvagdoṣāṇām vināśakaḥ.
(Rajanighantu)

Actions/Uses : Vātaghna, Pittaghna, Kaphaghna, Śothahara, Viṣaghna, Vedanāsthāpana.

Therapeutics :
Sirisa plant is used in snake bite and scorpion sting. Root is used in hemicrania. Powder of root bark is used to strengthen gums. Leaves are good for ophthalmia and flowers for chronic cough, bronchitis and asthma. Bark is bitter, acrid, sweet, mildly thermogenic, expectrant, aphrodisiac, anodyne, anti-inflammatory, cephalic, ophthalmic, depurative, cooling, restorative, tonic, and anthelmintic, cures vaata, diseases of blood, leucoderma, itching, piles. Bark and seeds are astringent, given in piles and diarrhoea, tonic and restorative. Aqueous extract of bark used against conception in women.

Entada pursaetha DC. ssp. pursaetha.

SYNONYMS : *Entada phaseoloides auct. non (L.)* Merr., *Entada scandens auct. non (L.)* Benth.; *Mimosa entanda* Linn.

ENGLISH : Mackary bean, Ladynut.

INDIAN : Gaarambi, Gardal, Gila.

SANSKṚTA : Gilagaccha, Gārambī, Gilla.

HABIT : A gigantic climber with twisted and angled stem.

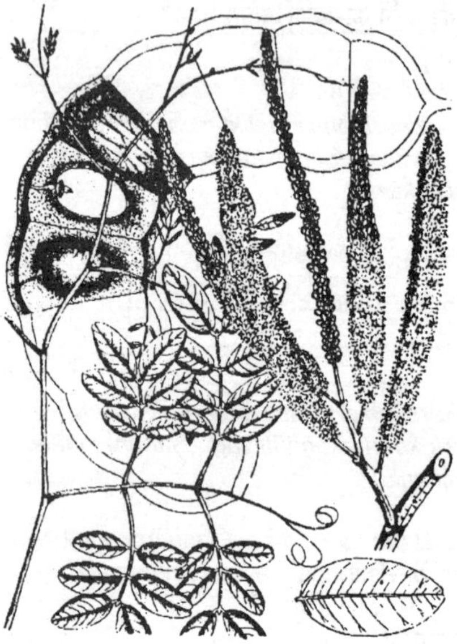

HABITAT : Ghat forests of Maharashtra, Bengal, Bihar and Orissa.

CHEMICAL COMPOSITION : Prosapognine A, lupeol, β-sitosterol, entanin an antitumor saponin from seeds. Pericarp, B-sitosterol, a-amyrin, quercetin, cyanidin chloride and gallic acid. Myristic, palmitic, stearic, arachidic, behenic, oleic, linoleic and linolenic acids from seed oil. Two amorphous saponins from seeds have a strong haemolytic action on human red blood cells and have a depressant effect on the respiratory system and inhibit movements of unstriped muscles of intestines and uterus.

Parts used : Seeds, stem and bark.

Ayurvedic properties :

Action\Uses : Pauṣṭika, Jvarahara, Śothaghna, Vraṇaropaṇa.

Therapeutics :

Seeds, stem and bark are poisonous. Seeds are fish poison, considered tonic, emetic, antiperiodic and anthelmintic. Paste prepared from seeds is applied locally for inflammatory glandular swellings. Stem, emetic. Juice of wood and bark used as external application for ulcers. Used in dropsy, anasarca, cancer, pain in loins, epilepsy, constipation and rinderpest. Seeds raw or roasted, as oral contraceptive. Powdered kernel mixed with few spices, is commonly given to women immediately after delivery, for allaying bodily pains. Seeds are used in pains of loins, in debility and in glandular swellings.

Mimosa pudica Linn.

ENGLISH : Sensitive plant, Humble plant.

INDIAN : Laajaalloo, Laajri.

SANSKṚTA : Lajjālu, Śamīpatrā, Khadirī, Anjalikārikā, Raktapādī, Samangā, Namaskarī, Sparśasankoca-parṇikā, Sparśarodanika, Samantatongati.

HABIT : A diffuse undershrub.

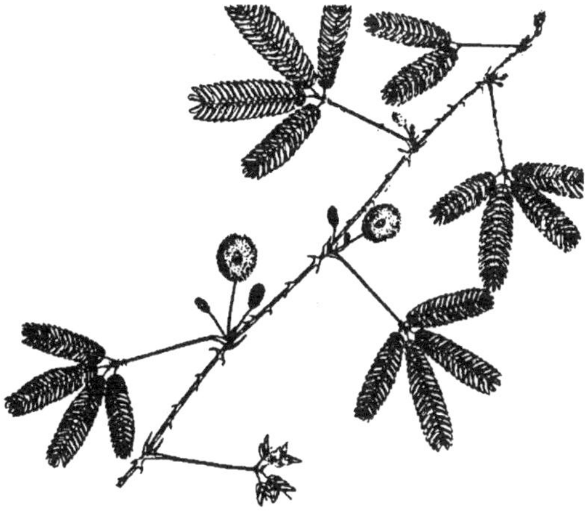

HABITAT : A native of America, naturalized throughout India.

CHEMICAL COMPOSITION : Plant contains β-amyrin, β-sitosterol and friedelin. Mucilage of seeds contain galactose and mannose. An adrenaline-like substance identified in extracts of leaves.

Parts used : The whole plant, roots and seeds.

Ayurvedic properties :
Rasa - Kaṣāya, Tikta. Vīrya - Śīta.
Guṇa - Laghu, Rukṣa. Vipāka - Kaṭu.

लज्जालुः शीतला तिक्ता कषाया कफपित्तजित्।
रक्तपित्तमतीसारं योनिरोगान् विनाशयेत्।।
(भावप्रकाश)

Lajjāluḥ śītalā tiktā kaṣāyā kaphapittajit,
Raktapittamatīsāram yonirogān vināśayet.
(Bhavaprakasa)

Actions/Uses : Kaphapittaśāmaka, Sandhāniya, Raktastambhana, Vraṇaropaṇa, Puriṣasangrahanīya, Atisāraghna, Arśaghna.

Therapeutics :

Lajjalu is astringent, cooling, antiseptic, alterative and blood purifier. It is resolvent, alterative and carminative. It is used in burning sensation of body, diarrhoea, dysentery, haemophilic conditions, leucorrhoea, morbid conditions of vagina. Root is aphrodiasiac, cooling, vulnerary; cures kapha, asthma, biliousness, vaginal and uterine complaints. Decoction of root is useful in gravellish complaints. Juice is antiseptic, alterative and blood purifier. Leaves and root are used in piles and fistula. Leaves, rubbed into paste applied to hydrocele. Leaf and stem in scorpion sting.

Family : LXII - Moraceae.

Artocarpus heterophyllus Lamk.

SYNONYMS : *Artocarpus integrifolia* Linn. *f.; Artocarpus integra auct. non* Merrill.; *Artocarpus heterophylla* Pers.

ENGLISH : Jack fruit.

INDIAN : Phannasa, Kanthal.

SANSKRTA : Panasa, Mahāsarja, Phalina, Phalavṛukṣaka, Kaṣṭaphala, Mūlaphalada, Apuṣpaphalada, Pūtaphala, Āśayaphala, Garbhakanṭaka, Kanṭakīphala, Bṛhatphala, Sthūlakanṭaphala, Campakālu, Phalasa, Sthūla, Atibṛhatphala, Campālū, Mṛdangaphala, Murajaphala, Campakoṣa, Skandaphala.

HABIT : A large evergreen tree.

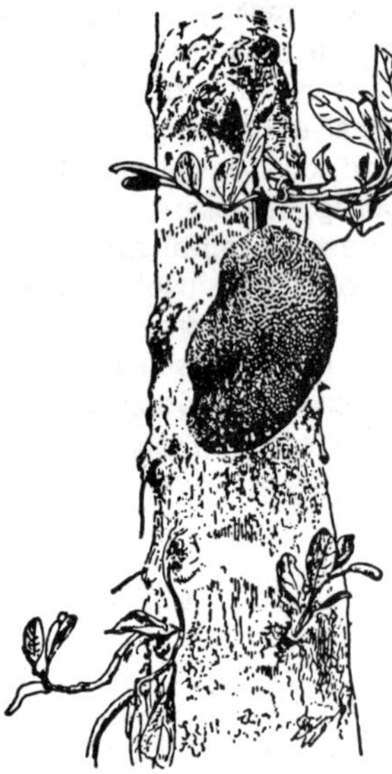

HABITAT : Indigenous to India, grows wild in Western Ghats.

CHEMICAL COMPOSITION : Fruit is a rich source of carbohydrates and essential amino acids. Latex from fruits contains cycloartenone, cycloartenol, β-sitosterol, tyrosine, valine. Cycloarteny acetate, cycloartenol and cycloartenone from bark. β-sitosterol, betulinic and ursolic acids and artoflavanone from roots. Artocarpetin, artocarpanone, artocarpin and isoartocarpin from heart-wood. Acetylcholine in seeds and leaves.

Parts used : Fruit, seeds, latex, leaves, bark and root.

Ayurvedic properties :
Rasa - Madhura, Kaṣāya. Vīrya - Śīta.
Guṇa - Gurū, Snigdha. Vipāka - Madhura.

पनसं शीतलं पक्वं स्निग्धं पित्तानिलापहम्।
तर्पणं बृंहणं स्वादु मांसळं श्लेष्मळं भृशम्।।
बल्यं शुक्लप्रदं हन्ति रक्तपित्तक्षतव्रणान्।
आमं तदेव विष्टंभि वातळं तुवरं गुरु।।
दाहकृत् मधुरं बल्यं कफमेदोविवर्धनम्।
पनसोद्भूतबीजानी वृष्याणि मधुराणी च।। (भावप्रकाश)

Panasam śītalam pakvam snigdham pittānilāpaham,
Tarpaṇam bṛnhaṇam svādu mānsaḷam śleṣmaḷam bhṛśam,
Balyam śuklapradam hanti raktapittakṣatavrāṇan,
Āmam tadeva viṣṭambhi vātaḷam tuvaram gurū.
Dāhakṛt madhuram balyam kaphamedo vivardhanam,
Panasodbhūta bījānī vṛṣyāṇi madhurāṇī ca.
(Bhāvaprakāṣa)

Actions/Uses : Vātapittaśāmaka, Śothahara, Vraṇapācana, Vṛṣya, Viṣṭambhī, Stambhana, Raktastambhana, Viṣaghna.

Therapeutics :

Latex of Panasa plant is applied to glandular swellings and abscesses to pomote suppuration. Roots are used for toothache, and are useful in skin diseases, sores and sterility in women. Decoction of roots is used internally in diarrhoea, stomach complaints and asthma. Leaves used in skin diseases and as antidote to snake bite. Ash of the leaves is useful in healing ulcers. The unripe fruit is acrid, astringent, carminative and tonic. Ripe fruit is demulcent, cooling, laxative, nuritive, fattening and useful in biliousness. The seeds are diuretic.

Ficus bengalensis Linn.

SYNONYMS : *Ficus indica* Linn.; *Urostigma benghalense (Linn.)* Gasp.

ENGLISH : Banyan tree.

INDIAN : Wadda, Wata, Bar.

SANSKṚTA : Vaṭa, Nyagrodha, Raktaphala, Bahupāda, Śṛngī, Kṣiri, Rohiṇa, Avarohī, Jatāla, Pādarohaṇa, Pakṣyāvāsa, Viṭapī, Mandalī, Nīla, Śiphāruha, Yakṣatarū, Skandharuha, Mahācchāyā, Skandhaja, Vaiśraṇāvāsa, Pādarohī, Bhāndīra, Śivāhvaya, Śunga, Vanarat, Nandi, Vṛkṣanātha, Yamapriya, Varagada, Akṣayyavaṭa.

HABIT : A very large tree with many aerial roots.

HABITAT : Sub-Himalayan tract and Western Peninsula, planted elsewhere.

CHEMICAL COMPOSITION : Bengalenoside a hypoglycemic glucoside in bark. Three methyl esters of leucoanthocyanins along with methyl ether of leucoanthocyanidin from stem bark. Friedelin, quercetin, β-sitosterol, 3-galactoside and rutin in leaves. Taraxasterol tiglate in heartwood.

Parts used : Bark, leaves, fruits, adventitious roots, latex.

Ayurvedic properties :
Rasa - Kaṣāya.
Vīrya - Śīta.
Guṇa - Guru, Rūkṣa. Vipāka - Kaṭu.

वटः कषायो मधुरः शिशिरः कफपित्तजित्।
ज्वरदाहतृषामोहव्रणशोफापहारकाः।।
(राजनिघण्टु)

Vaṭaḥ kaṣāyo madhuraḥ śiśiraḥ kaphapittajit,
Jvara dāhatṛṣāmohavraṇaśophāpahārakāh.
(Rajanighantu)

Actions/Uses : Kaphavātaśāmaka, Vedanāsthāpana, Raktastambhana, Cakṣuṡya, Śothahara, Vraṇaropaṇa, Raktapittahara, Raktaśodhana, Stambhana, Garbhasthāpana, Śukrastambhana, Mūtrasamgrahaṇīya.

Therapeutics :

Vata is widely used in the treatment of skin diseases. Plant is used in ophthalmia and other eye troubles, mouth sores, fever, madness, atrophy, emaciation or cathexy, cholera and rinderpest. Paste of root applied to scalp to grow hair long and used for menorrhagia. It cures erysepelas, burning sensation and vaginal disorders. Root fibres are used in gonnorrhoea. Leaves are applied as polutice on swellings and inflammed parts for relief. Bark is astringent, cooling and alleviates vitiated kapha and pitta. An infusion of bark cures dysentery, nervous disorders, diarrhoea, leucorrhoea, menorrhagia, and reduces blood sugar in diabetes. Milky juice is beneficial as local application in toothache, sores and ulcers, for rheumatism and lumbago and for soles of feet when cracked. Infusion of young buds is used in diarrhoea and dysentery and young tips of roots for obstinate vomiting. Juice mixed with sesamum oil is applied to burns. Latex used in genital disorders. Seeds are cooling and tonic. Powder of seed is progenitive.

Ficus carica Linn.

ENGLISH : Fig.

INDIAN : Anjira.

SANSKRTA : Anjira, Phalgū.

HABIT : A small or middle sized deciduous tree.

HABITAT : Baluchistan. Cultivated in India.

CHEMICAL COMPOSITION : Ficin in latex. Psoralen and bergapten, taraxasterol, B-sitosterol, lupeol, B-amyrin, coumarins, xanthotoxin, xanthotoxol, and marmesin in leaves. Root, guaiazulene. Umbeliferone and scopoletin in plant. Palmitic and valeric acids, p-symene, cadalene, octacosane and tricosane from essential oil.

Parts used : Fruits, roots, bark, and latex.

Ayurvedic properties :
Rasa - Madhura. Vīrya - Śīta.
Guṇa - Snigdha. Vipāka - Madhura.

अञ्जीरः शीतलः स्वादुर्गुरु रक्तरुजाहरः।
दाहं वातं च पित्तं च नाशयेतृप्तिदो मतः।।
(निघण्टु रत्नाकर)

Anjīraḥ sītalaḥ svādurgurū raktarujāharaḥ,
Dāham vātanca pittañca nāśayet tṛptido mataḥ.
(Nighanturatnakar)

Actions/Uses : Snehana, Sransana, Pācana, Kaphaghna, Pittghna, Anulomic.

Therapeutics :

Fruit is nutritive, emollient, demulcent, aperient, and laxative; used in anaemia. Latex is anthelmintic due to ficin. The fresh and dried fruits are used in constipation. Roasted figs are used as polutice for gumboils, boils and carbuncles. Milky juice from fresh green fruit is acrid and used to destroy warts.

Ficus racemosa Linn.

SYNONYMS : *Ficus glomerata* Roxb.; *Covellia glomerata* Miq.

ENGLISH : Country fig.

INDIAN : Umbara, Audumbara, Gular.

SANSKṚTA : Udumbara, Jantuphala, Yajnyānga, Hemadugdhaka, Kṣīrī, Maśakī, Sucakṣu, Kakṣāphala, Kṣīravṛkṣa, Sadāphala, Yajnayoga, Yajniya, Supratiṣṭhita, Śītavalka, Pavitraka, Saumya, Yajnasāra, Jantuvṛkṣa, Apuṣpa, Phalasambadha, Kṛmiphala, Yajnodumbara, Kāncana, Pāṇibhuk, Haritākṣi.

HABIT : A moderate-sized to large deciduous spreading tree.

HABITAT : Throughout India along the sides ravines and banks of strems.

CHEMICAL COMPOSITION : Bark, ceryl behenate, lupeol, a-amyrin and 3 unidentified compounds. Fruit, lupeol-OAc, β-sitosterol, sterol and glauanol. Leaves, gluanol-OAc, α-amyrin and β-sitosterol. A glycoside-rich fraction from leaves reported to have hypotensive and cardiac-depressive effects. Aqueous extract of bark possesses remarkable antiulcer activity against acute gastric ulcers in animals and inhibits acid secretion and stimulate gastric juice.

Parts used : Bark, fruit, latex and juice.

Ayurvedic properties :
Rasa - Kaṣāya, Madhura.Vīrya - Śīta.
Guṇa - Gurū, Rūkṣa. Vipāka - Kaṭu, Madhura.

उदुम्वरं कषायं स्यात् पक्वन्तु मधुरं हिमम्।
क्रिमिकृद् पित्तरक्तहनम् मूर्च्छादिहतृषाहम्।।
(राजनिघण्टु)

Udumbaram kaṣāyam syāt pakvantu madhram himam,
Krimikṛd pittaraktaghnam mūrchādāhatṛṣāpaham.
(Rajanighantu)

Actions/Uses : Kaphaghna, Pittaghna, Śothahara, Vedanāsthāpana, Vraṇaśodhana-ropaṇa, Varṇya, Pūriṣastambhana.

Therapeutics :

Udumbar is astringent, antiseptic, cooling and highly efficacious in threatened abortions, menorrhagia or flooding and failure of lactation, gonorrhoea, leucorrhoea, urinary diseases, haemorrhages, skin diseases and ulcers. Plant is used in smallpox, muscular pain, adenitis, scabies, spermatorrhoea, orchitis, epididymitis, hydrocele. Juice of roots is useful in measles, smallpox and chikenpox. It is also used for thirst during fever. Leaf is eaten and chewed in jaundice. A decoction of bark is used internally for menorrhagea, leucorrhoea and metrorrhagea and is useful for the growth of foetus. Decoction is used for gargle during inflammation of mouth. Root is used in dysentery, sap of root in diabetes. Leaves and bark useful as polutice for eczema. Decoction of leaves for bronchitis. Powdered leaves mixed with honey given in billious affections. Unripe fruits are astringent, sweet, carminative, digestive, stomachic and useful in diarrhoea, dyspepsia, dysentery, haemorrhages and menorrhagia. Bark is reputed for healing ulcers, skin and vaginal diseases. It is astringent. Decoction of bark is used to wash wounds. Fruit is astringet, stomachic and carminative, given in menorrhagia and haemoptysis. Milky juice or latex is useful in piles and diarrhoea and is applied externally on wounds to relieve pain and swelling. It is taken internally with sugar as an aphrodisiac.

Ficus religiosa Linn.

SYNONYMS : *Urostigma religiosum* Gasp.

ENGLISH : Peepal tree.

INDIAN : Pimpalla, Ashwattha, Bodhi.

SANSKṚTA : Pippala, Aśvatha, Caladala, Bodhidru, Bodha, Gajāśana, Satvarajanāśana, Vipra, Sevya, Caitya, Svādubījaka, Māngalya, Kṣīrapādapa, Śyāmala, Bodhivṛukṣa, Satya, Yājnika, Gajabhakṣaka, Śrīmāna, Śucidruma, Dhanyavṛukṣa, Keśavāvāsa, Lakṣmīvān, Devātmā, Dviradāśana, Nāgabandhū, Keśavālaya, Kṛṣṇāvāsa, Harivāsa, Kunjarāśana.

HABIT : A large or medium-sized deciduous tree with spreading branches.

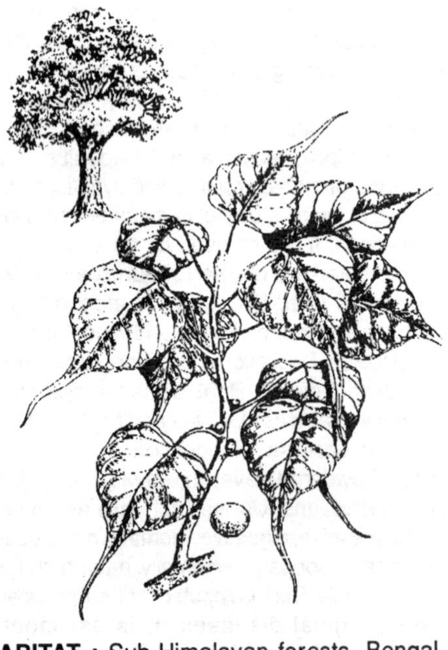

HABITAT : Sub-Himalayan forests, Bengal and Madya Pradesh. Planted elsewhere.

CHEMICAL COMPOSITION : Phytosterolin from bark, powerful CNS stimulant and hypoglycemic. β-sitosterol-D-glucoside is present in bark. Vitamin K from stem bark.

Parts used : Bark, fruit, seeds, leafbuds and latex.

Ayurvedic properties :
Rasa - Kaṣāya, Madhura. Vīrya - Śīta.
Guṇa - Gurū, Rūkṣa. Vipāka - Kaṭu.

पिप्पलः सुमधुरस्तु कषायःशीतलश्च कफपित्तविनाशी।
रक्तदाहशमनः स हि सद्यो योनिदोषहरणः किल पक्वः।।
(राजनिघण्टु)

Pippalaḥ sumadhurastu kaṣāyah,
Śītalaśca kaphapitta vināśī,
Raktadāhaśamanaḥ sa hi sadyo,
Yonidoṣaharanaḥ kila pakvaḥ.
(Rajanighantu)

Actions/Uses : Kaphapittaśāmaka, Vraṇaropaṇa, Śothahara, Varṇya, Vedanāsthapana, Raktaśuddhikara.

Therapeutics :
Various parts of Ashwattha used in otitis media suppurativa, mouth sores, atrophy, emaciation or cachexy, rheumatism, smallpox, carbuncle, rinderpeast, mucus in urine, spermatorrhoea, gravel, cholera, etc. Leaves with other ingredients is an aborticide. Leaves and young shoots are purgative. Bark is astringent and is found efficacious in gonorrhoea. Pulverised bark is applied externally on unhealthy ulcer or wounds to promote granulation. Infusion of bark given internally in scabies, ulcers and skin diseases; decoction given in gonorrhoea. It is aphrodisiac and good for lumbago. Fruit is mild laxative and digestive. Seeds are cooling, laxative and alterative. Powder of seeds taken for three days during menses sterilizes women for long time.

Morus acedosa Griff.

SYNONYMS : *Morus indica* Linn.

HABIT : A low shrub.

HABITAT : Wild in sub-Himalayan tract, cultivated elsewhere.

ENGLISH : White mulberry.

INDIAN : Toota; Shahatuta.

SANSKṚIT : Śālmali.

CHEMICAL COMPOSITION : Plant contains mulberin and mulberrochromene.

Parts used : Bark, fruit, leaves and root.

Ayurvedic properties :
Rasa - Madhura, Unripe : Amla
Vīrya - Śīta, Unripe : Ūṣṇa
Guṇa - Gurū Vipāka - Madhura

Actions/Uses : Vātapittahara.

Therapeutics :

Root of Salmali is anthelmintic and astringent. Decoction of leaves used as gargle in inflammation of vocal cords. Bark is anthelmintic and purgative. Fruit is aromatic, cooling, laxative, allays thirst in fevers.

Morus alba Linn.

INDIAN : Toota.

SANSKṚTA : Tūta, Sthula, Bramhadārū, Madhupippalī, Śahātūta, Śetūra, Tūda, Tūla.

HABIT : A low shrub.

HABITAT : Baluchistan, cultivated in various parts of India.

CHEMICAL COMPOSITION : Mulberrin, mulberrochromene, cyclomulberrin, cyclomulberrochromene from stems, roots and bark. Morusin, cyclomorusin and compound A from root bark. Mulberranol from bark and alboctalol from heartwood. Astragalin from leaves.

Parts used : Bark, fruit, leaves and root.

Ayurvedic properties :
Rasa - Madhura, Kaṣāya Vīrya - Śīta
Guṇa - Gurū, Snigdha Vipāka - Madhura

तूदं पक्वं गुरु स्वादु हिमं पित्तानिलापहम्।
तदेवामद्गुरु सरमम्लोष्णं रक्तपित्तकृत्।।
(भावप्रकाश)

Tūtam pakvam gurū svādu himam pittānilāpaham,
Tadevāmam gurū saramamloṣṇam raktapittakṛt.
(Bhavaprakasa)

तूदस्य तु फलं स्वादु बलवर्णाग्निवृद्धिकृत।
तूदं तु मधुराम्लं स्याद् वातपित्तहरं सरम्।
दाहप्रशमनं वृष्यं कषायं कफनाशनम्।।
(ध. नि.)

Tūdasya tu phalam svādu balavarṇāgni vṛddhikrt,
Tūdam tu madhurāmlam syād vātapittaharam saram,
Dāhapraśamanam vṛṣyam kaṣāyam kaphanāśanam.
(Dhanvatari)

Actions/Uses : Vātapittaghna. Leaves : Vraṇaropaṇa.

Bark : Recana, Kṛmighna. Ripe fruit : Dīpana, Anulomana.

Therapeutics :

Bark of Tuta is purgative and anthelmintic. Fruit is refrigerant in fever, used as remedy for sore throat, dyspepsia and melancholia.

Family : LXIII - Moringaceae.

Moringa oleifera Lam.

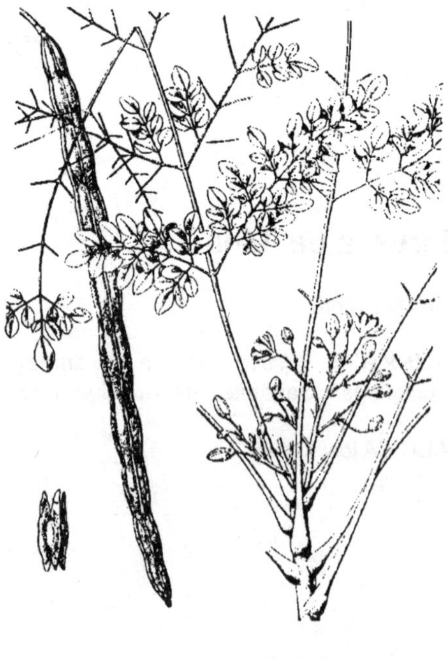

SYNONYMS : *Moringa pterygosperma* Gaertn.

ENGLISH : Drum stick, Horse radish tree.

INDIAN : Shevagaa, Mungna, Shajna, Achajhada.

SANSKṚTA : Śigru, Śobhānjana, Madhuśigru, Damṣamūla, Tīkṣṇamūla, Tīkṣṇagandha, Akṣīva, Mocaka, Haritaśāka, Śākapatra, Mūlakaparṇi, Supatraka, Upadamṣakṣama, Komalapatraka, Bahumūlā, Haritasāraka, Tīkṣṇagandhaka.

HABIT : A small or middle-sized tree.

HABITAT : Indigenous to North-West India. Cultivated throughout India.

CHEMICAL COMPOSITION : Alanine, arginine, glycine, threonine, serine, valine, glutamic and aspartic acids in flowers and fruits. Lysine, sucrose and glucose in flowers. 4-hydroxymellein, vanillin, octacosanoic acid, β-sitosterol and β-sitostenone from stems. Aldotriouronic acid from gum. Bark yields a base moringine. Pterygospermin, an antibiotic principle. Clinical studies have shown that stem bark produces significant relief in patients suffering from difficult micturition or mutrakruchcha. Abortifacient activity of leaves has been demostrated.

Parts used : Root, Root-bark, flowers and seeds.

Ayurvedic properties :
Rasa - Madhura, Tikta, Kaṭu. Vīrya - Uṣṇa.
Guṇa - Laghu, Rūkṣa, Tīkṣṇa, Sara. Vipāka - Kaṭu.

शिग्रुश्च कटुतिक्तोष्णस्तीक्ष्णो वातकफापहाः।
मुखजाड्यहरो रुच्यं दीपनो व्रणदोषनुत्॥
शिग्रुपत्रभवं शाकं रुच्यं वातकफापहम्।
कटुष्ण दीपनं पथ्यं क्रिमिघ्नं पाचनं परम्॥
(राजनिघण्टु)

Śigruśca kaṭutiktoṣṇastīkṣṇo vātakaphāpahaḥ,
Mukhajāḍyaharo rucyam dīpano vraṇadoṣanut.
Śigrupatrabhavam śākam rucyam vātakaphāpaham,
Kaṭuṣṇam dīpanam pathyam krimighnam pācanam param.
(Rajanighantu)

शिग्रुः कटुः कटुः पाके तीक्ष्णोष्णो मधुरो लघुः।
दीपनो रोचनो रुक्षः क्षारस्तिक्तो विदाहकृत्॥
संग्राहि शुक्रलो हृद्यः पित्तरक्तप्रकोपणः।
चक्षुष्यः कफपितघ्नो विद्रधिश्वयथुकृर्मान्॥
मेदोपचीविषप्लहिगुल्मगण्डव्रणान् हरेत्॥
(भा. नि.)

Śigruḥ kaṭuḥ kaṭuḥ pāke tīkṣṇoṣṇo madhuro laghuḥ,
Dīpano rocano rūkṣaḥ kṣārastikto vidāhakṛt.
Sangrāhi śukralo hṛddyaḥ pittarakta prakopaṇaḥ,
Cakṣuśyaḥ kaphapittaghno vidradhiśvayathu kṛmīn.
Medopacī viṣaplīha gulmagaṇḍavraṇān haret.
(Bhavaprakasa)

Actions/Uses : Kaphavātaghna, Dīpana, Vidāhī, Medoghna, Vraṇahara, Viṣaghna, Vidradhihara, Śirovirecana, Kṛmighna, Gulmahara.

Therapeutics :

Śigru is hot, sweet and light. It improves appetite and digestion, promotes semen and is good for heart and eye problems. Plant is antispasmodic, stimulant, expectorant and diureic; is used as stimulant in paralytic affections and intermittent fever; used in epilepsy; rubifacient in palsy and chronic rhreumatism; carminative, stomachic, abortifacient, as cardiac and circulatory tonic; in a form of a compound spirit useful in fainting, giddiness, nervous debility, spasmodic affections of bowels, hysteria and flatulence. Fresh root is acrid and vesicant, internally stimulant, diuretic and antilithic. Root bark is used as fomentation to relieve spasm. Polutice of leaves is beneficial in glandular swellings. Leaf-juice is useful in hiccough, emetic in higher doses. Mixed with honey it is applied to eyelids in eye diseases. Bark is emmenagogue and abortifacient. Fruit is used in diseases of liver and spleen, articular pains, tenesmus and paralysis. Flowers are stimulant, tonic, diuretic and useful to increase flow of bile and are also aphrodisiac. Oil from seeds is used as external application in rheumatism. Gum is used for dental caries; mixed with sesamum oil and poured into ears for relief of otalgia. Seeds are used in venereal affections. Root, bark and gum are abortifacient.

Family : LXIV - Musaceae.

Musa paradisiaca Linn.

SYNONYMS : *Musa sapientum* Linn.; *Musa paradisiaca L. var sapientum (L.)* Kuntze.

ENGLISH : Banana.

INDIAN : Kella.

SANSKṚTA : Rambhā, Kadalī, Sukumārā, Svāduphalā, Dīrghapatrā, Nissārā, Mocā, Hastiviṣāṇikā, Suphalā, Sakṛtphalā, Gucchaphalā, Kāṣṭhīrasā, Gucchadantikā, Bālakapriyā, Urustambhā, Bhānuphalā, Vanalakṣmī, Ayatacchadā, Vīrā, Granthinī, Vāraṇā, Vāraṇāmbusā, Amśamataphalā, Carmaṇavatī, Avanīsārā, Tantuvigrahā, Nagarauṣadhi, Ambusāra.

HABIT : An erect herb with underground stem.

HABITAT : Cultivated throughout India.

CHEMICAL COMPOSITION : Serotonin and norepinephrine, dopamine and an unidentified catecholamine; 14a-Methyl-9B,19-cyclo-5a-ergost-24(28)-en-2B-ol isolated. Tryptophan and indole compounds in banana ovaries.

Parts used : Root, stem, flowers, fruits and leaves.

Ayurvedic properties :
Rasa - Madhura. Vīrya - Śīta.
Guṇa - Gurū, Snigdha. Vipāka - Madhura.

रम्भापक्वफलं कषायमधुरं बल्यं च शीतं तथा।
पित्तघ्नास्त्र विमर्दनं गुरुतरं पथ्यंच मन्दानले।।
सद्यः शुक्रविवृद्धिदं क्लमहरं तृष्णापहं कान्तिदम्।
दीप्ताग्निसुखदं कफामयकरं संतर्पणं दुर्जरम्।।
(राजनिघण्टु)

Rambhāpakvaphalam kaṣāyamadhuram balyam ca śītam tathā,
Pittaghnāśravimardanam gurutaram pathyam ca mandānale.
Saddyaḥ śukravivṛddhidam klamaharam ṭṛṣnāpaham kāntidam,
Dīptāgnisukhadam kaphāmayakaram santarpaṇam durjaram.
(Rajanighantu)

Actions/Uses : Vātapittaśāmaka, Kaphavardhaka.

Ripe fruit : Balya, Raktapittaśāmaka, Jīvanīya, Brihmana, Vṛiṣya, Kṣudhāhara, Netrarogahara, Pramehahara, Tṛṣāpahā.

Green fruit : Viṣṭambhaka, Pittaraktavikārahara, Kaphakāraka, Kṣatakṣayahara, Sangrāhaka, Tṛṣāhara, Dāhahara, Vātahara.

Therapeutics :

Root of Rambha is anthelmintic. Ash of root or whole plant is anthelmintic. Flowers are astringent. Juice of stem is used in otalgia and haemoptysis. Unripe fruit is astringent. Banana fruit is mild laxative. It aids in combating diarrhoea and dysentery and promotes healing of intestinal lesions in ulcerative colitis. Useful in coeliac disease, constipation and peptic ulcer. Banana powder is effective in treatment of coeliac disease, sprue and other carbohydrate intolerance in children; is used in intestinal disorders in adults. Unripe fruit and cooked flowers are useful in diabetes. Juice of flowers used for dysentery and stems and root for disorders of blood.

Family : LXV - Myricaceae.

Myrica nagi Thunb.

SYNONYMS : *Myrica rubra* Sieb.; *Myrica farquhariana* Wall.; *Myrica esculenta* Buch.-*Ham. ex D.Don.; Myrica sapida* Wall., *Myrica nagi* Hook. f. in part non Thunb.

ENGLISH : Box myrtle, Bay-berry.

INDIAN : Kayaphalla.

SANSKRTA : Kaṭphala, Rāmapatrī, Somavalka, Kaitarya, Kumbhikā, Śriparṇikā, Kumudikā, Bhadrā, Bhadravatī, Kṛṣṇagatbha, Rohiṇī, Pracetasī, Mahākumbhī, Rāmasenaka, Ugragandha, Ranjanaka, Trutī, Laghukāśmarī, Mahāvalka, Somapādapa, Kāyaphalla, Kāyacchāla.

HABIT : A small or moderate-sized evergreen tree.

HABITAT : A native of China and Japan. Found in sub-tropical Himalayas.

CHEMICAL COMPOSITION : Plant contains myriconol, proanthocyandin, β-sitisterol, friedelin, taraxerol, myricadiol, myricetin. Also contains glycoside myricitrin.

Parts used : Bark and fruit.

Ayurvedic properties :
Rasa - Kaṣāya, Tikta, Kaṭu. Vīrya - Uṣṇa.
Guṇa - Laghu, Tīkṣṇa. Vipāka - Kaṭu.

कटफलः कटुरुष्णश्च कासश्वासज्वरापहः।
उग्रदाहहरो रुच्यो मुखरोगशमप्रदः।।
(राजनिघण्टु)

Kaṭphalaḥ kaṭuruṣṇaṣca kāsaśvāsajvarāpahaḥ,
Ugradāhaharo rucyo mukharogaśamapradaḥ.
(Rajanighantu)

कट्फलस्तुवरस्तिक्तः कटुर्वातकफज्वरान्।
हन्ति श्वासप्रमेहार्शः कण्डवामयारुचीः।।
(भावप्रकाश)

Kaṭphalastuvarstiktaḥ kaṭurvātakaphajvarān,
Hanti śvāsapramehārśaḥ kaṇḍvāmayārucīḥ.
(Bhavaprakasa)

Actions/Uses : Kaphavātaśāmaka, Dīpana, Grāhī, Hṛdya, Samdhanīya, Śirovirecana, Śūlapraśamana, Kothapraśamana, Śothahara, Arśaghna, Arucihara, Raktapittaghna, Svāsahara, Kaṇḍūhara, Pramehaghna, Galagaṇḍahara, Garbhāśayasankocaka, Mūtrasamgrahaṇīya, Vātakaphajvarahara.

Therapeutics :

Kaṭphala is aromatic and astringent, heating and stimulant. Bark is resolvent, astringent, carminative, tonic and antiseptic, useful in fever, asthma, cough; powdered and used as snuff in catarrh with headache; mixed with ginger used as a rubifacient application in cholera; fish poison. A decoction of the bark is useful in asthma, diarrhoea, fevers, lung affections, chronic bronchitis, dysentery and diuresis. Bark is chewed to relieve toothache and a lotion from it is used for washing putrid sores.
Fruits yield a wax which is used externally for healing ulcers.

Family : LXVI - Myristicaceae.

Myristica fragrans Houtt.

SYNONYMS : *Arillus myristicae; Myristica officinalis* Linn.; *Myristica moschata* Thunb.; *Myristica aromatica* Lam.

ENGLISH : Nutmeg, Mace, Nux Moschata.

INDIAN : Jaayaphalla, Jaayapatri, Jotri.

SANSKṚTA : Jātīphala, Jātikośa, Jātipatrī, Jātisasya, Putha, Mālatiphala, Majjāsāra, Jātisāra, Kusumaphala, Sumanaphala, Koṣa, Madaśaunda, Saumanaphala, Śāluka, Mālatisuta.

HABIT : A dioecious or occasionally monoecious evergreen, middle sized tree.

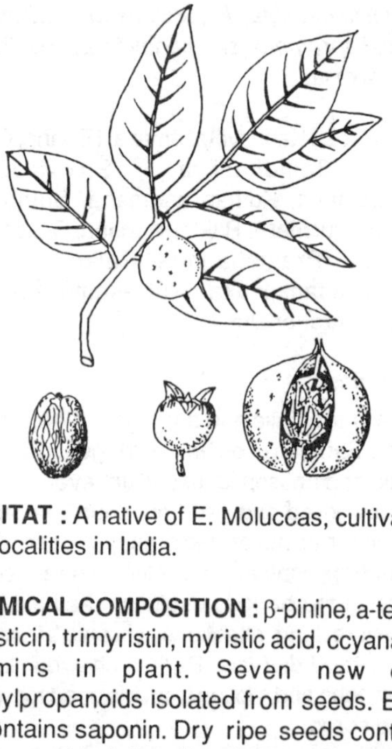

HABITAT : A native of E. Moluccas, cultivated in a few localities in India.

CHEMICAL COMPOSITION : β-pinine, a-terpinene, myristicin, trimyristin, myristic acid, ccyanadin and vitamins in plant. Seven new dimeric phenylpropanoids isolated from seeds. Essential oil contains saponin. Dry ripe seeds contain 5 to 15 per cent volatile oil and 25 to 40 per cent of fixed oil. Essential oil contains 80 per cent of a-pinene and 10 per cent myristicin, eugenol and safrol.

Parts used : Seeds and arillus.

Ayurvedic properties :
Rasa - Tikta, Kaṭu. Vīrya - Ūṣṇa.
Guṇa - Laghu, Tīkṣṇa. Vipāka - Kaṭu.

जातिफलं रसे तिक्तं तीक्ष्णोष्ण रोचनं लघु।
कटुकं दीपनं ग्राहि स्वर्यं श्लेष्मानिलापहम्।।
निहन्ति मुखवैरस्य मलदौर्गन्ध्य कृष्णताः।
कृमिकासवमिश्वास शोषपीनसहृद्रुजः।। (भावप्रकाश)

Jātiphalam rase tiktam tīkṣṇoṣṇam rocanam laghu,
Kaṭukam dipanam grahi svaryam slesmanilapaham.
Nihanti mukhavairasyam maladaurgandhya kṛṣnatāḥ,
Kṛmikāsāvamiśvāsa śosapīnasahṛdrujaḥ.
(Bhāvaprakāṣa)

Actions/Uses : Kaphaghna, Vātaghna, Dīpana, Pācana, Grāhī, Vṛṣya, Rucikara, Yakṛtottejana, Vātānulomana, Samjnyānāśaka, Svarya, Malasamgrāhī, Vedanāsthāpana, Ārtavajanana, Viṣaghna, Kṛmighna, Mukhadurgandhyahara, Vāmihara, Kāsaghna, Pinasahara, Hṛdrogahara, Śvāsaghna.

Therapeutics :

Seeds of Jatiphala are carminative, stimulating, astringent, digestive, stomachic, abortifacient and aphrodisiac; useful in flatulency, nausea and vomiting; used in tonics and electuaries. It is a constituent of preparations for dysentery, stomachache, malaria, rheumatism and sciatica. Oil from dried kernels is aperient and carminative.

HOMEOPATHIC USE : Nux moschata is a Homeopathic remedy very useful for hysterical patients. Disposition to faint, specially hysterical is a strong leading symptom. It is full of very valuable symptoms pertaining to mind, head, nervous system, female sexual organs and liver. Like females, it is a valuable remedy for children, especially in marasmus, enlarged liver, enlarged spleen, with loose, bloody stools; worse after intermittent fever. Abdomen enormously distended, flatulent distention of uterus; epilepsy, catalepsy, hysteria; Labor pains too weak or false; after delivery uterus remains uncontracted; irregularity in heart's beat, pulse small and weak are the main symptoms.

Myristica malabarica Lam.

SYNONYMS : *Myristica tomentosa* Graham.

HABIT : A mmoderate-sized tree.

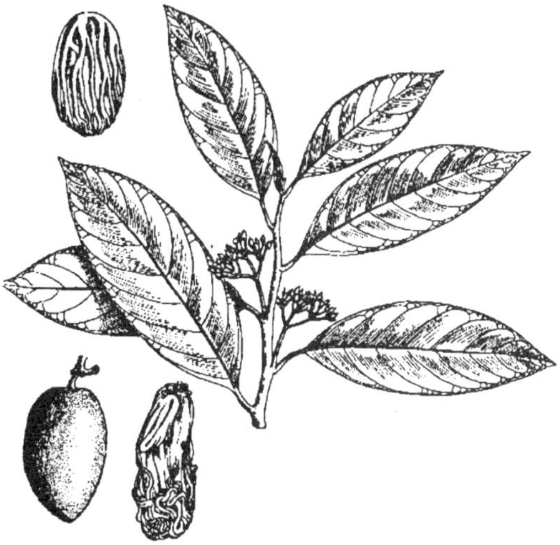

HABITAT : Everegreen forests of Konkan, Kanara and Malbar.

ENGLISH : False nutmeg, Bombay mace.

INDIAN : Raanajaayaphalla, Raampatri.

SANSKṚTA : Vanajātīphala.

CHEMICAL COMPOSITION : Plant contains myristic acid and its triglycerice. Leaves contain β-sitosterol. Fruit rind contains malabaricones, lignana, malabaricanol.

Parts used : Seeds and arillus.

Ayurvedic properties :
Rasa - Katu, Tikta. Vīrya - Ūsna.
Guṇa - Laghu, Tīksna, Snigdha. Vipāka - Katu.

Actions/Uses : Vātakaphaghna, Raktasodhaka, Kaphanisaraka, Sothahara, Varnnya, Purisavaha, Vṛṣya and Ārtavajanana.

Therapeutics :

Vanajatiphala is a local stimulant and aphrodiaisac. Seeds in form of paste are used as external application. Fat mixed with little oil applied to indolent ulcers, allays pain, cleanses the surface and establishes healthy action.

Family : LVXII - Myrsinaceae

Embelia ribes Burm.

SYNONYMS : *Embelia glandulifera* Wight.

ENGLISH : Embelia.

INDIAN : Waawadinga, Babe-rang, Kakannie, Vaivarang.

SANSKṚTA : Viḍanga, Pāvaka, Kṛumighna, Citratanḍūla, Jantunāśana, Amoghā, Kṛmihā, Vātārī, Jantughnī, Mṛugagāminī, Kairalī, Kapalī, Gahvarā, Sucitrabījā, Moghā, Kśudrataṇdula, Kṛmijita, Bhūtaghna, Vellā, Vriśanasana.

HABIT : A large scandent climbing shrub.

HABITAT : Throughout India.

CHEMICAL COMPOSITION : Fruits contain 2 to 3 per cent embelin. Berries contain vilangin.

Parts used : Root, root-bark, leaves, fruits and seeds.

Ayurvedic properties :
Rasa - Kaṭu, Kaṣāya, Tikta. Vīrya - Ūṣṇa.
Guṇa - Laghu, Rūkṣa, Tīkṣṇa. Vipāka - Kaṭu.

विडंगं कटुतीक्ष्णोष्णं रुक्षं वह्निकरं लघु।
शूलाध्मानोदरश्लेष्मकृमिवातविबंधनुत्।।
(भावप्रकाश)

Viḍangam kaṭutīkṣṇoṣṇam rūksam vahanikaram laghu,
Śūlādhmanodara śleaṣma kṛmi vāta vibandhanut.
(Bhavaprakasa)

Actions/Uses : Kaphaghna, Vātaghna, Pittakara, Dīpana, Rucya, Kṛmighna, Bhūtaghna, Jantunāśani, Viṣṭambhi, Viṣahara, Medoghna, Āmadoṣahara, Adhmānahara, Śūlahara, Mehaghna, Agnimāndyaśamanī.

Therapeutics :

Vidanga plant is used to cure pyorrhoea. Root improves digestion and cures flatulence and colic, is used as an antifertility drug. Pulp is purgative. Fresh juice is cooling, diuretic and laxative. Infusion of roots is given for coughs and diarrhoea. Paste of root-bark is applied externally to chest in pneumonia while its powder is a good remedy for toothache. Fruits are acrid, light, astringent, carminative, anthelmintic, stimulant, alterative. Its reputation is due to its action of expelling tapeworms. Fruit cures dental, oral and throat troubles except ptylism and cancer of lips. A paste of seeds is applied locally against ringworm and seed powder is used as an errhine in cold and headache. A decoction of seeds is beneficial in fevers, skin diseases and chest complaints.

Family : LXVIII - Myrtaceae

Eucalyptus globulus Labill.

ENGLISH : Eucaliptus; Blue Gum Tree.

INDIAN : Nilagiri.

SANSKṚTA : Tailaparṇī, Sugandhapatrā, Haritaparṇī, Raktaniryāsa, Malayajo, Sukṣmaparṇī.

HABIT : A tall tree.

HABITAT : Introduced and cultivated in India.

CHEMICAL COMPOSITION : Cis and trans pinocarveol derivatives. α-pinene and unsaturated a-ketone in leaves. Cystein, ornithine, asparagine, glutamic acid, threonine, a- and B- alanine and norvaline in fruits.

Parts used : Dried leaves, roots, gum, essential oil.

Ayurvedic properties :
Rasa - Kaṭu, Tikta, Kaṣāya. Vīrya - Ūṣṇa.
Guṇa - Laghu, Snigdha. Vipāka - Kaṭu.

तैलपर्णः लघुः स्निग्धः कटुतिक्तकषायकः।
वीर्योष्णः कफवातघ्नः पूतिजन्तुहरः स्मृतः।।
जीर्णकासे प्रतिश्याये स्वरभेदे च शस्यते।।
(द्रव्यगुण विज्ञान)

Tailaparṇaḥ laghuḥ snigdhaḥ kaṭutiktakaṣāyakaḥ,
Vīryoṣṇaḥ kaphavātaghnaḥ pūtijantuharaḥ smṛtaḥ,
Jīrṇakāse pratiśyāye svarabhede ca śasyate.
(Dravyagunavijnana)

Actions/Uses : Kaphavātaśāmaka, Dīpana-Pācana, Jantughna, Pūtihara, Svedajanana, Plihāsankocaka, Uttejaka.

Therapeutics :
Root of Tailaparni is mild purgative. Leaves are febrifuge, carminative, stimulant, expectorant, diaphoretic and antiseptic. They increase flow of saliva, gastric and intenstinal juices and thus increae appetite and digestion. Infusion of leaves is given in apthous ulcerations of mouth and applied locally in erysipelas. Lotion is useful for foul ulcers. Inhalation of fumes is beneficial in respiratory affections. Antimalarial properties are atributed to volatile oil. It is a powerful antiseptic and disinfectant. Gum is mild astringent, used in diarrhoea.

Melaleuca leucadendron Linn.

HABIT : An evergreen tree of small or moderate size with thick, spongy bark.

HABITAT : Native to East Indies. Indigenous to Burma, Cambodia, Thailand. Planted in Indian Gardens as ornamental.

ENGLISH : Cajuput.

INDIAN : Cajuputte, Kayaaputi.

SANSKṚTA : Śitānsu, Kāyāputi.

CHEMICAL COMPOSITION : Essential oil containing cajuputol which is identical with eucalyptol. Bark yields crystalline resinol melaleucin. Betulin, friedelin, uvaol, sitosterol, epitaraxeryl acetate and taraxastenone from leaves and stems.

Parts used : Leaves, twigs, bark and essential oil.

Ayurvedic properties :
Rasa - Kaṭu, Tikta, Kasāya. Vīrya - Ūṣṇa.
Guṇa - Laghu, Snigdha. Vipāka - Kaṭu.

शीतांशुतैलमाक्षेपशमनं वायुनाशनम्।
स्वेदनं शूलहृश्चोग्रं ज्वरघ्नं कफ्नुत परम्।।
आमवाते तथाघ्माने ज्वरेच शिरसां गदे।
दन्तरोगे च भग्ने च द्वैपेयं परियुज्यते।। (द्रव्यगुण)

*Śitāmśu tailam ākṣepaśamanam vāyunāśanam,
Svedanan śūlahṛccogram jvaraghnam kaphnut param.
Āmavāte tathādhmāne jvare ca śiraso gade.
Dantaroge ca bhagne ca dvaipeyam pariyujyate.
(Dravyaguna)*

Actions/Uses : Kaphavātaśāmaka, Dīpana-Pācana, Jantughna, Pūtihara, Svedajanana, Plihasankocaka, Uttejaka.

Therapeutics :

Cajuput oil is distilled from fresh leaves and twigs, is antiseptic, anticonvulsant, anticolic and powerful antipyretic. It is anthelmintic, stimulant, carminative, expectorant and rubifacient. It is used internally in the treatment of bronchitis, laryngitis, diarrhoea and flatulence. Externally it is applied for ear-ache, rheumatism and skin diseasess. It is stimulant, antispasmodic, diaphoretic, rubefacient. Useful for all kinds of pains, internal and external. Used for rheumatic affections, toothache, neuralgia, sprains and bruises, in choleraic diarrhoea, psoriasis, chronic pityriasis, acne and eczema.

HOMEOPATHIC USE : Oil is used to prepare Homeopathic tincture. It is useful for nightly diarrhoea, dropsy, gout and rheumatism, toothache, epilepsy. Many nervous symptoms are developed, some reflected from the female sexual organs. Persistant choking sensation of hysteria; nervous dyspmoea; nervous distension of bowels are some of the symptoms.

Myrtus communis Linn.

ENGLISH : Myrtle.

INDIAN : Vilaayati mendi, Matalai, Murada.

HABIT : An evergreen low shrub.

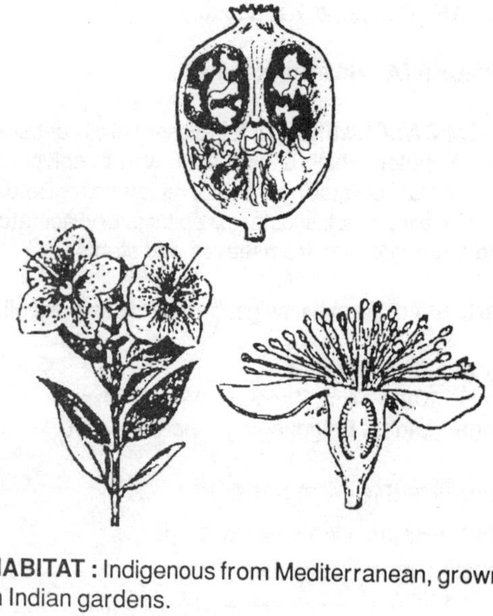

HABITAT : Indigenous from Mediterranean, grown in Indian gardens.

CHEMICAL COMPOSITION : Essential oil from leaf, myrtucommulones A and B; myrtinyl acetate, myrtenol, limonene, linalool, α-pinene, β-pinene, p-cymol, cineole, camphene, and traces of car-3-ene.

Therapeutics :

Leaves astringent, considered useful in cerebral affections, especially epilepsy, also in dyspepsia and diseases of stomach and liver. Decoction is employed as mouth wash in cases of aphthae. Fruit, carminative, given in diarrrhoea, dysentery, haemorrhage, internal ulceration and rheumatism. Leaf, applied in scorpion sting. Essential oil of leaves is antiseptic, local application in rheumatism. An injection of infusion or tincture of leaves has given excellent results for prolapsus and leucorrhoea. Powdered leaves check night sweats of phthisis and is useful in all pulmonary disorders.

HOMEOPATHIC USE : Myrt. com. is a Homeopathic remedy. It is an active antiseptic. It has a powerful action on left lung, especially the upper part. Pain in upper left lung going to right through from the front to the left scapula is the keynote. Cases of phthisis, haemoptysis, hepatisation, lung syphilis having this symptom have been cured with Myrt. com.

Psidium guajava Linn.

SYNONYMS : *Psidium pyriferum* Linn.

ENGLISH : Guava.

INDIAN : Peru, Amrud, Jamba, Tupkel, Safed safari.

SANSKRTA : Pārevata, Perkam, Dṛdhabījam, Māmsala, Pṛthaktvāca, Perāḷa, Mṛduphala, Amṛtaphala.

HABIT : An arborescent shrub or small tree.

HABITAT : Cultivated and naruralized throughout India.

CHEMICAL COMPOSITION : Richest natural source of Vitamin C and contains 4 to 10 times more of this vitamin than in citrus fruits. It also contains considerable amount of pectin. Plant yields β-sitosterol. Fruit contains arabinan, composed of galacturonic acid, galactose, arabinose with traces of xylose and rhamnose. Leaves contain β-sitosterol, maslinic acid, guijavalic acid and an unidentified acid. Pectin constituents are d-galacturonic acid, d-galactose, l-arabinose. Essential oil eugenol contains dl-limonene. Stem bark, luteic and ellagic acid, leucocyanidin, and a glycoside amritoside.

Parts used : Root-bark, fruit and leaves,

Ayurvedic properties :
Rasa - Madhura, Kaṣāya. Vīrya - Śīta.
Guṇa - Gurū, Tīksna. Vipāka - Madhura.

पारेवतन्तु मधुरं क्रिमिवातहारी।
वृष्यं तृषाज्वरविदाहहरञ्च हृद्यम्।
मूर्च्छाभ्रमश्रमविशोष विनाशकारी।
स्निग्धश्च रुच्यमुदितं वहुवीर्यदायी।।
(राजनिघण्टु)

Pārevatantu madhuram krimivātaharī,
Vṛṣyam tṛṣājvaravidāhaharñca hṛdyam.
Mūrcchābhramaśramaviśoṣa vināsakārī.
Snigdhānca rucyamuditam bahuvīryadayī.
(Rajanighantu)

Actions/Uses : Vātaghna, Pittagna, Kaphakara, Vṛṣya, Śukrala, Madanāśakara, Rucikar.

Therapeutics :
Leaves of Perukam are used as astringent for bowels and for wounds and ulcers. Young leaves are used as a tonic in the diseases of digestive functions. Their decoction and that of young shoots is prescribed as febrifuge and in antispasmodic baths. Infusion of leaves is used in cerebral affections, nephritis and cachexia. A decoction of leaves is used in cholera for arresting vomiting and diarrhoea; when gargled relieves toothache and gum boils. Pounded leaves are applied locally in rheumatism and an extract is used in epilepsy and chorea. The tincture is rubbed over the spine of children suffering from convulsions. Infusion of leaves and roots is an astringent drink. Bark is astringent and tonic and ash caustic, used in diarrhoea of children. It is generally administered in the form of decoction. Flowers are said to cool the body and are used in bronchitis. They are applied to eye sores. Perukam fruit is tonic, cooling and laxative. It is good in colic and for bleeding gums. Fruit and its conserve are astringent and used in diarrhoea and desentery.

Syzygium cumini (Linn.) Skeels.

SYNONYMS : *Eugenia jambolana* Lam.; *Myrtus cumini* Linn.;

ENGLISH : Plum jambol, Jaman, Jambolan, Blacj plum.

INDIAN : Jaambhoolla, Jaamuna, Jaman.

SANSKṚTA : Jambū, Nandī, Mahāphala, Phalendrā, Surabhipatrā, Śyāmalā, Śukapriyā, Nīlaphalā, Mahāskandhā, Rājārhā, Meghamedinī, Rājaphalā, Mahārasā, Jambula, Bahubhadra.

HABIT : A large evergreen tree.

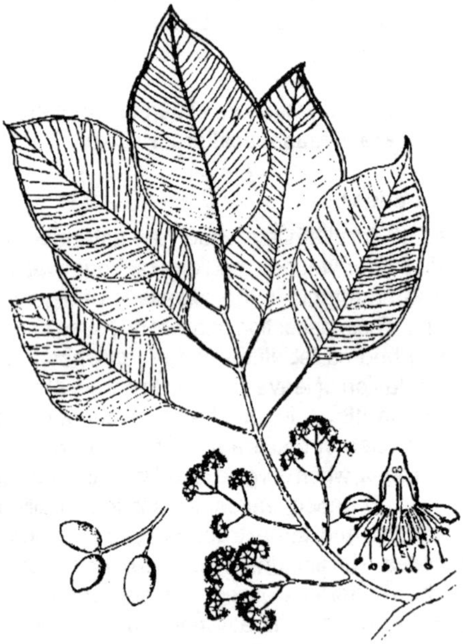

HABITAT : Throughout India.

CHEMICAL COMPOSITION : Seeds contain glycoside jambolin, ellagic acid, tannin, gallic acid, chlorophyll, fatty oil, starch, resin, sugar and traces of oil. Flowers give acetyl oleanolic acid two other triterpenoids, ellagic acid, isoquercitrin, quercetin, kaempferol and myricetin.

Parts used : Fruit, leaves, dried seeds and bark.

Ayurvedic properties :
Rasa - Kaṣāya, Madhura, Amla. Vīrya - Śīta.
Guṇa - Laghu, Rūkṣa. Vipāka - Madhura, Kaṭu.

जम्बः कषायमधुरा श्रमपित्तदाह,
कंठार्तिशोषशमनी क्रिमिदोषहन्त्री ।
श्वासातिसारकफकास विनाशिनीच
विष्टम्भिनी भवती च रोचन पाचनी च॥
(राजनिघण्टु)

Jambūḥ kaṣāyamadhurā śramapittadāha,
Kaṇṭhārtiśoṣáśamanī krimidoṣahantrī,
Śvāsātisārakaphakāsavināśinī ca,
Viṣṭambhinībhavati ca rocanipācani ca.
(Rajanighantu)

Actions/Uses : Kaphaghna, Pittaghna, Vātakara, Dīpana-Pācana, Stambhana, Grāhī, Rucya, Yakṛtottejana, Vātānulomana, Kāsaghna, Vīryaprada, Puṣṭikara, Balaprada, Śramāpaha, Raktapittahara, Dāhaśamana, Atisāraghna.

Therapeutics :
Jambu bark is astringent and is considered a specific for dysentery, it cures haemorrhages, burning sensation, dysentery, diarrhoea, diabetes, excessive thirst, dyspepsia, cough and asthma and is used in preparation of astringent decoction for gargles and washes; fresh juice given with goat's milk in diarrhoea of children. Juice of leaves is used in dysentery. Juice of ripe fruit, made into a vinegar used as a stomachic, carminative and as diuretic. Fruit is useful astringent in billious diarrhoea. Seeds used in diabetes.

Family : LXIX-Nyctaginaceae

Boerhaavia diffusa Linn.

SYNONYMS : *Boerhaavia repens* Linn.; *Boervia procumbens* Roxb.

ENGLISH : Hogweed, Patagon.

INDIAN : Punarnawaa, Raktavasu, Biskhafra.

SANSKṚTA : Punarnavā-Śveta, Punarnavā-Rakta, Viśākhā, Kathilla, Śaśivaṭīkā, Pṛthvī, Sitavarśābhū, Dīrghapatra, Śivāṭikā, Vṛścika, Mahāvarṣabhava, Kṣudrapatra, Varṣaketu, Śothagna, Śophaghni.

HABIT : A very variable, diffusely branched, pubescent or glabrous, prostrate herb. Two chief varieties are Shweta punarenava and Rakta punarnava. The two plants are sources of two different Ayurvedic drugs, Punarnava and Varhabhu, possibly with similar therapeutic effects.

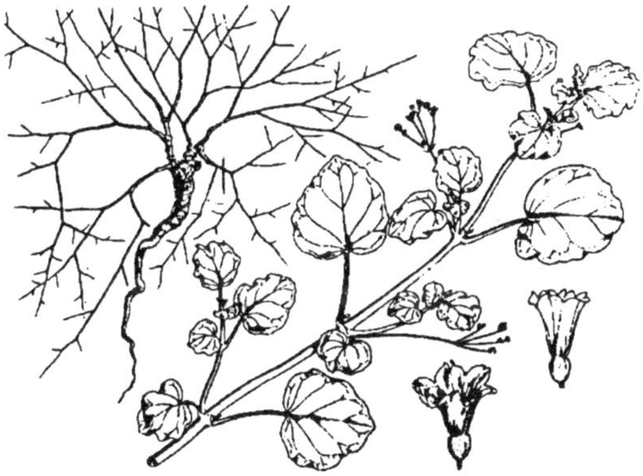

HABITAT : A weed throughout India.

CHEMICAL COMPOSITION : Hentriacontane, β-sitosterol, ursolic acid, punarnavine-1 & -2, myricyl alcohol, myristic acid, oxalic acid and alkaloids. Polysaccharide consisting of glucose, xylose, glucuronic acid, galctose, L-arabinose and L-rhamnose; and a glycoprotein. An injection of alkaloids in cats produced a distinct and persistant rise of blood pressure and marked diuresis.

Parts used : Whole herb and root.

Ayurvedic properties :
Rasa - Kaṭu, Madhura, Tikta, Kaṣāya.
Vīrya - Uṣṇa, Śīta.
Guṇa - Laghu, Rūkṣa. Vipāka - Kaṭu.

पुनर्नवावरुणा तिक्ता कटुपाका हिमा लघु।
वातला ग्राहिणी श्लेमपित्तरक्त विनाशिनी।।
(द्रव्यगुण)

Punarnavāruṇā tiktā kaṭupākā himā laghuh,
Vātalā grāhiṇī śleṣma pittarakta vināśinī.
(Dravyaguna)

कटुकषायानुरसा पाण्डुघ्नी दीपनी सरा।
शोफानिलगरश्लेष्महरी व्रण्योदरप्रणुत्।।
(भा. प्र.)

Kaṭu kaṣāyānurasā pāṇḍughnī dīpanī sarā,
Śophānilagaraśleṣmaharī vraṇyodarapraṇut.
(Bhavaprakasa)

Actions/Uses : Dīpana, Anulomana, Recana, Kāsahara, Śothahara, Hṛdya, Raktavardhaka, Rasāyana, Vayasthāpana, Mūtral, Svedopaga. White : Pittaghna, Vātaghna, Kaphaghna.

Red : Pittahara, Vātakara.

Therapeutics :

Punarnava is pungent, bitter, astringent, hot and laxative. It is cooling, stomachic, diuretic, diaphoretic, expectorant, antipyretic and cardiotonic. It stimulates function of heart and kidney and is a specific for jaundice, diabetes, general debility and oedema. It is a rejuvenative drug. It is used in epilepsy, pain in abdomen due to congestion of blood, prolapsus ani and fistula ani, dysentery, otitis media. The whole plant, fresh or dried is the source of drug Punarnava which is official in I.P. as a diuretic. Root is diuretic, laxative, anthelmintic and febrifuge. It is used as expectorant in asthma, stomachic, in oedema, anaemia, jaundice, ascites, ansarca, scanty urine and internal inflammation. Root is useful for restoration of virility in man. Its polutice mixed with palm oil is applied to boils. It is an antidote to snake venom. Roots and leaves are more diuretic and antiinflammatory than stems and whole plant. Leaf ash and roots are taken to cure night blindnesss. Alcoholic extract of roots and plant is spasmolytic. Leaves are used in ophthalmia and in eye wounds, in muscular pain; in dropsy and gonorrhoea; to purify blood and hasten delivery. Paste of leaves taken orally to check bleeding after delivery. Dry and powdered leaves mixed with Brassica oil used for external application on itch and eczema. Boiled with rice, garlic and water rubbed on body to cure rheumatism. Flowers and seeds as contraceptives in Ayurved. Diuretic and anti-inflammatory activities are maximum in samples collected in rainy season. Combination of these two activities make Punarnava a very useful drug for the treatment of inflammatory renal diseases and common clinical problems like nephrotic syndrome. It is particularly useful as a maintaining drug. It is effective in cases of oedema and ascites resulting from early cirrhosis of the liver and chronic peritonitis. It is useful in abdominal tumours and cancer.

Mirabilis jalapa Linn.

ENGLISH : Marvel of Peru; Four Oíclock plant.

INDIAN : Gulabasa.

SANSKṚIT : Trisandhi, Kṛṣnakeli, Sandhyā-rāga.

CHEMICAL COMPOSITION : Alkaloid trigonelline. Tricosan-12-one, n-hexacosanol, β-sitosterol and tetracosanoic acid isolated, while tartaric, citric acids, valine tryptophan, leucine, alanine and glycine detected in leaves. Indicaxanthin and vulgaxanthin I, Meraxanthins I, II, III and IV from flowers.

HABIT : A tall perennial herb or undershrub.

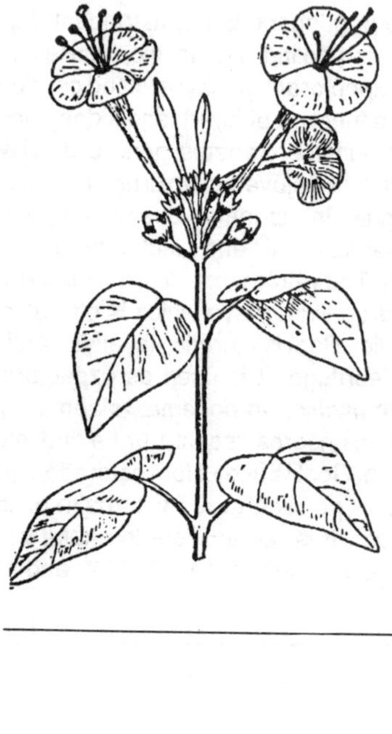

HABITAT : A native of tropical America. Cultivated or spontaneous over greater parts of India.

Parts used : Tuber, root and leaves.

Ayurvedic properties :

त्रिसन्धिस्त्रिविधा ज्ञेया रक्ता चान्या सितासिता।
कफकासहरा रुच्या त्वग्दोषशमनी परा।।   (राजनिघण्टु)

Trisandhistrividhā jneyā raktā cānyā sitāsitā,
Kaphakāsahara rucya tvagdosa samani para.
(Rajanighantu)

त्रिसंधी शीतला रुच्या तिक्ता विषविनाशिनी।
त्वग्दोषकासदद्रुघ्ना वातपित्तकफापहा।।   (निघण्टु रत्नाकर)

Trisandhī śītalā rucyā tiktā viṣa vināśinī,
Tvagdoṣa kāsa dadrughnā vātapittakaphāpahā.
(Nighanturatnakara)

Actions/Uses : Kaphahara, Kāsahara, Ṡothahara

Therapeutics :

Tuber possesses purgative properties similar to Jalap. It is used as a polutice on carbuncles. Root is a mild purgative and aphrodisiac. Dried root is nutrient. Powdered and fried in ghee with spices it is given in milk as a nourishing and strengthening medicine. Rubbed with water it is applied externally in contusions. Leaves, maturant, lessen inflammation, applied to boils, phlegmons and whitlow. Fresh leaf-juice is demulcent and found useful when applied to the body in urticaria.

Family : LXX-Nymphaeaceae

Nelumbo nucifera Gaertn.

SYNONYMS : *Nelumbium speciosum* Willd.; *Nelumbo nelumbo* Druce.

ENGLISH : Lotus, Chinese water lily.

INDIAN : Kamalla, Kanwal.

SANSKṚTA : Kamala, Padmini Padma, Nalina, Aravinda, Mahotpala, Pankeruha, Tāmarasa, Sahasrapatra, Jalaja, Sārasa, Bisaja, Rājīva, Puṣkara, Ambhoruha.

HABIT : A handsome aquatic herb.

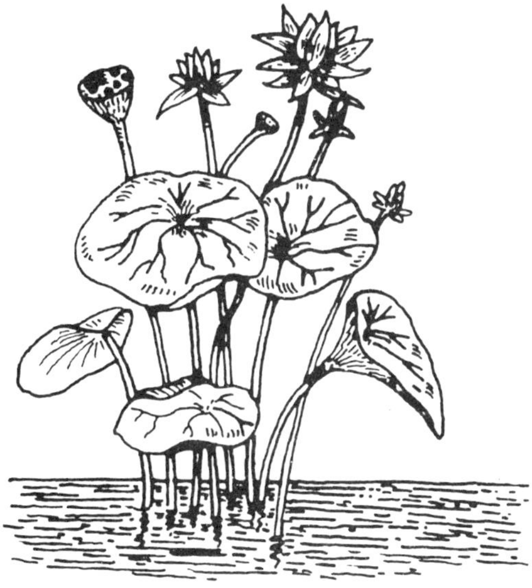

HABITAT : Native of China, Japan and possibly India. Throughout the warmer parts of India.

CHEMICAL COMPOSITION : Plant contains a number of alkaloids. Raffinose and stachyose from rhizomes. Nelumboside from leaves. Nornuciferine, Pronuciferine, Roemerine, nuciferine, anonaine etc. from leaves and seeds. Nuphrarine causes lasting stimulation of respiration; impaired respiration is restored and stimulated in animal experiments.

Parts used : Rhizomes, flowers, stalk, leaves, seeds and filaments.

Ayurvedic properties :
Rasa - Madhura, Kaṣāya, Tikta. Vīrya - Śīta.
Guṇa - Gurū, Snigdha, Picchila. Vipāka - Madhura.

कमलं शीतलं वर्ण्यं मधुरं कफपित्तजित्।
तृष्णादाहास्त्रविस्फोट विषवीसर्पनाशनम्।।
पद्मबीजं हिमं स्वादु कषायं तिक्तकं गुरुः।।
(भावप्रकाश)

Kamalam śītalam varṇyam madhuram kaphapittajit,
Tṛṣṇādāhāsravisphoṭa viṣavīsarpanāśanam.
Padmabījam himam svādu kaṣāyam tiktakam guruḥ.
(Bhavaprakasa)

पद्मिनी द्वितीयम् शीतम मधुरम् लवणम् गुरु।
बलासरक्तपित्तघ्नम् रुक्षा विष्टंभी वातलभ्।।
(कैयदेव)

Padminī dvitīyam śītam madhuram lavanam gurū,
Balāsaraktapittaghnam rūkṣo viṣṭabhī vatalam.
(Kaiyadeva nighantu)

Actions/Uses : Kaphapittaghna, Dāhapraśamana, Mūtravirajanīya, Samgrāhī, Raktapittasāmaka, Tṛṣṇāhara, Viṣaghna, Visphotahara, Visarpahara, Varṇya.

Therapeutics :
Kamala root is bitter, it cures cough and biliousness, allays thirst and is cooling body, in powder form it is prescribed in piles as demulcent, also for dysentery and dyspepsia; used as a paste in skin affections and ringworm. The whole plant is sweet, cool, slightly bitter; gives tone to the breast, removes worms; allays thirst, fever, biliousness, vomiting and strangury. Flowers are used in diarrhoea, cholera, fever and diseases of liver and heart. Flowers and filaments are cooling, refrigerant, astringent, sedative, cholagogue, diuretic, bitter, and expectorant. Filaments are used in burning sensation of body, bleeding piles and menorrhagia. Fruit removes kapha and pitta and foul breath. Seeds are demulcent and nutritive.

Family : LXXI - Oleaceae

Jasminum grandiflorum Linn.

SYNONYMS : *Jasminum officinale* Linn.; *Jasminum officinale* Linn. *var. grandiflorum* Bailey.

ENGLISH : Common jasmine, Spanish Jasmine.

INDIAN : Chameli, Jati.

SANSKṚTA : Mālatī, Jātī, Sumanā, Bahupuṣpī, Gaṇikā, Ambaṣṭhā, Hemapuṣpikā, Manojnyā, Rājaputrī, Priyavadā, Hṛdyagandhā, Cetikā, Telabhāvinī, Cambeli, Cetaki, Somanasyāyanī, Cameli, Malligāī, Yāsamina.

HABIT : A large scrambling or twining shrub.

HABITAT : Native of N.W. Himalayas, wildly grown throughout India.

CHEMICAL COMPOSITION : Salicylic acid in leaves. Essential oil in flowers contains lactone IV. Pyridine and nicotinate derivatives.

Parts used : Whole plant, root, leaves and flowers.

Ayurvedic properrties :
Rasa - Tikta, Kaṣāya. Vīrya - Ūṣṇa.
Guṇa - Laghu, Mṛudu, Snigdha, Śīta. Vipāka - Kaṭu.

मालती शीता तिक्ता स्यात् कफघ्नी मुखपाकनुत्।
कुड्मलं नेत्ररोगघ्नं व्रणविस्फोटकुष्ठनुत्।।
(राजनिघण्टु)

Mālatī śīta tiktā syāt kaphaghnī mukhapākanut,
Kuḍmalam netrarogaghnam
vraṇavisphotakuṣṭhnut.
(Rajanighantu)

जातियुग्मं तिक्तमुष्णं तुवरं लघु दोषजित्।
शिरोऽक्षिमुखदन्तार्तिविषकुष्ठव्रणास्त्रजित्।।
(भावप्रकाश)

Jātiyugmam tiktamuṣṇam tuvaram laghu doṣajit,
Śiro akṣimukhadantārtiviṣakuṣṭhavraṇāsrajit.
(Bhavaprakasa)

Actions/Uses : Tridoṣahara, Snigdha, Anulomana, Mūtrajanana, Raktaprasādana, Vājikaraṇa, Ārtavajanana, Śirorogahara, Dantarogahara, Kuṣṭhahara, Akṣimukharogahara.

Therapeutics :

Malati is good for healing chronic ulcers, skin diseases and poisonous affections. It is bitter, astringent, anthelmintic, diuretic and emmenagogue. Root is used in the treatment of ringworm. Leaves chewed as a treatment for ulceration or eruptions in mouth. Fresh juice applied to corns. Flowers and leaves are aphrodisiac and overcome menstrual irregularities. Flowers as well as a poluticee of leaves are applied as a plaster to loins, genitals and pubes as an aphrodisiac. Flowers are used as an application in skin diseases, headache and weak eyes. Flowers and leaf juice are used in various kinds of tumors. Oil prepared with juice of leaves used in otorrhoea. Alcoholic extract of aerial parts is hypotensive and anticancer. Shrub used in burns.

Jasminum sambac Ait.

ENGLISH : Double Jasmin, Arabian Lily, Tuscan jasmine.

INDIAN : Mogaraa, Bat-mogara, Motiaa, Banamallikaa, Chambaa.

SANSKṚTA : Mallikā, Mogarā, Motiā.

HABIT : A scandent or sub-errect woody perennial shrub.

HABITAT : Cutivated as an ornamental shrub throughout India.

CHEMICAL COMPOSITION : Pyridine and nicotinate derivatives. The flowers contain a yellcw pigment, used as a substitute for saffron.

Parts used : Roots, leaves and flowers.

Ayurvedic properties :
Rasa - Tikta, Kaṣāya. Vīrya - Ūṣṇa.
Guṇa - Laghu, Mṛdu, Snigdha. Vipāka - Kaṭu.

नेत्ररोगापहन्त्री स्यात् कटुष्णा वृत्तमल्लिका।
व्रणघ्नी गन्धवहुला दारयत्यास्यजान् गदान्।।
(राजनिघण्टु)

Netrarogāpahantrī syāt kaṭuṣṇa vṛttamallikā,
Vraṇaghnī gandhabahulā dārayatyāsya Jāngadān.
(Rajanighantu)

मल्लिका कटुता तिक्ता लघ्वीकोष्णा च शुक्रला।
चक्षुष्या कुष्ठविस्फोट मुख रुक्कण्डुतापहा।।
विषदोषम् व्रणम् वहतम् पित्तम् रक्करुजमूतथा।
हृद्रोगारुचि दुर्नामनाशिनी परिकीर्तिता।
(नि. र.)

Mallikā kaṭutātiktā laghvī coṣṇā ca śukralā,
Cakṣuṣyā kuṣṭha visphota mukharukkaṇḍu tāpahā.
Viṣadoṣam vraṇam vātam pittam raktarujam tathā,
Hṛdroga ruci durnāma nāśinī parikīrtitā.
(Nighanturatnakara)

Actions/Uses : Tridoṣaghna, Śothaghna, Vraṇaropaṇa, Stanyanāśana, Garbhāśayauttejaka.

Therapeutics :
Mallika is cooling, used in cases of insanity, weakness of sight and affections of mouth. It is used as a deodorant in foul smelling ear and nose diseases. Root is emmenagogue. Flowers are lactifuge, applied unmoistened to breasts to arrest secretion of milk in puerperal state in cases of threatened abscess. Leaves if boiled in oil exude a balsam which is used for anointing the head in eye complaints and to strengthen vision. It is also used as a remedy in cases of insanity. Dried leaves soaked in water and made into a polutice used in indolent ulcers. Alcoholic extract of aerial part is CNS depressant and hypotensive.

Nyctanthes arbor-tristis Linn.

ENGLISH : Indian mourner, Night jasmine, Coral jasmine.

INDIAN : Paarijaata, Praajakta, Harasinghar, Seoli, Khurasli.

SANSKṚTA : Śephāli, Pārijāta, Mandāra, Prājakta, Rāgapuṣpī, Hariśṛngārapuṣpaka, Nālakunkūmaka, Kharapatraka.

HABIT : A hardy large shrub or small tree.

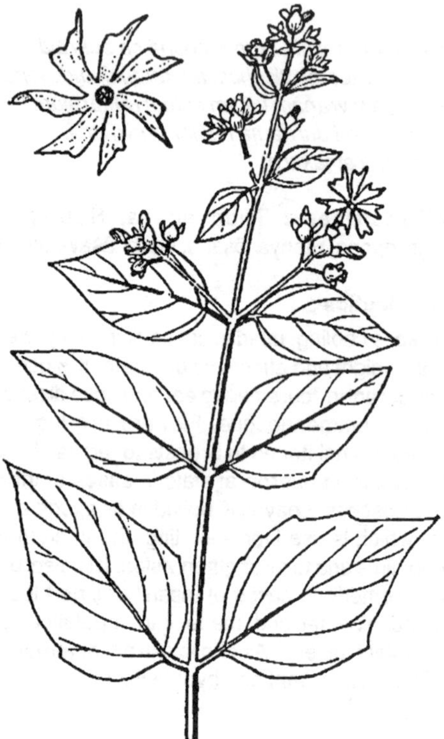

HABITAT : Native to India and cultivated in many parts.

CHEMICAL COMPOSITION : Leaves give manittol, free glucose and fructose benzoic acid, β-amyrin, β-sitosterol, hentriacontane, astragalin and nicotiflorin from leaves. Nyctanthoside, crocin-1, crocin-3 and D-mannitol from flowers. β-sitosterol and glycoside naringenin from stem. Essential oil.

Parts used : Whole plant, seeds, leaves and bark.

Ayurvedic properties :
Rasa - Tikta. Vīrya - Uṣṇa.
Guṇa - Laghu, Rūkṣa. Vipāka - Kaṭu.

शेफालिः कटुतिक्तोष्णा रुक्षा वातक्षयापहा।
स्याद्ङ्ग सन्धिवातघ्नी गुद वातादिदोषनुत्।।
(राजनिघण्टु)

Śephāliḥ kaṭutiktoṣṇā rūkṣā vātakṣayāpahā,
Syādanga sandhivātaghno gudavātādi doṣanut.
(Rajanighantu)

Actions/Uses : Kaphavātaśāmaka, Pittaśodhana, Anulomana, Dīpana, Kṛmighna, Mūtrala, Viṣaghna, Vedanāsthāpana, Gridhrasihara, Balya, Sandhirogahara, Tvagrogahara.

Therapeutics :

Sephali or Parijataka plant is cholagogue, anthelmintic and laxative. It is used in dysentery, menorrhagia, sores and ulcers. Leaves are useful in fever and rheumatism; fresh juice with honey is given in chronic fever. Expressed juice of leaves is a mild bitter tonic, colagogue, laxative, anthelmintic, given with little sugar to children as a remedy for intestinal worms. Decoction of leaves prepared over a gentle fire, is recommended as a specific for obstinate sciatica.

Family : LXXII - Oxalidaceae

Oxalis corniculata Linn.

ENGLISH : Indian Sorrel.

INDIAN : Ambutee, Amrul sak, Chuka tripati, Anjati, Bhinsarpati.

SANSKṚTA : Cāngerī, Āmlapatrikā, Āmlaparṇikā, Triparṇī, Ambaṣṭha, Amlika, Śuklika, Cukrita, Āmalonikā.

HABIT : A small annual or perennial procumbent or erect herb.

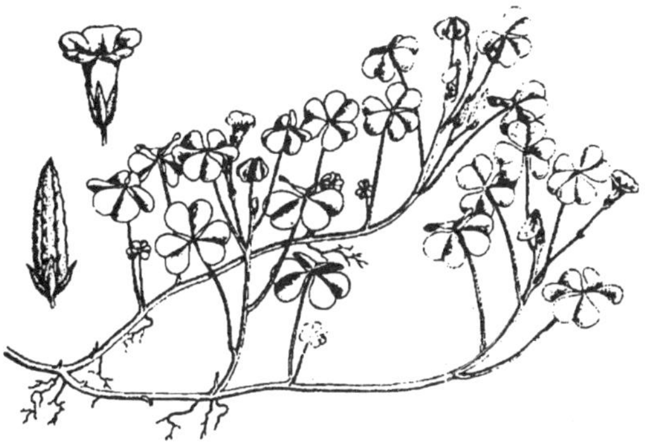

HABITAT : Common in moist and cultivated places and warmer parts of India.

CHEMICAL COMPOSITION : Malic, tartaric and citric acids from leaves and stem. Acid potassium oxalate. Leaves are a good source of vitamin C. Alcoholic extract of leaves is antibacterial.

Parts used : Whole plant.

Ayurvedic properties :
Rasa - Amla, Kaṣāya. Vīrya - Uṣṇa.
Guṇa - Laghu, Rūkṣa. Vipāka - Amla.

चाङ्गेरी दीपनी रुच्या रुक्षोष्णा कफवातनुत्।
पित्तलाम्ला ग्रहण्यर्शः कुष्ठातिसारनाशनी।।
(भावप्रकाश)

Cangeri dipani rucya ruksosna kaphavatanut,
Pittalamla grahanyarsah kusthatisaranasini.
(Bhavaprakasa)

Actions/Uses : Kaphavātaśāmaka, Pittavardhaka, Rucikara, Grāhī, Dīpana-Pācana, Vedanāsthāpana, Yakṛtottejaka.

Therapeutics :

Cangeri plant possesses astringent, vermifuge, emmenagogue and antiseptic properties. Boiled with buttermilk, it is a remedy for indigestion and diarrhoea in children. Fresh juice of plant cure dyspepsia, piles, anaemia and tympanitis. Leaves, coolling, refrigerating, stomachic, antiscorbutic and appetizing. They are used in fevers, dysentery, scurvy and billiousness and for removing corns, warts and other excrescences of the skin. An infusion of leaves is used to remove opacities of cornea.

Family : LXXIII - Pandanaceae

Pandanus tectorius Soland. ex Parkinson.

SYNONYMS : *Pandanus odoratissimus* Roxb.; *Pandanus fascicularis* Lamk.; *Pandanus laevis* Kunth.; *Pandanus variegatus Miq.; Pandanus latifolius* Hassk.; *Pandanus amaryllifolius* Roxb.

ENGLISH : Screw pine.

INDIAN : Kewadaa, Ketaki, Keura, Gagandhul.

SANSKṚTA : Ketakī.

HABIT : A densely branched shrub rarely erect.

HABITAT : Seacoast of Indian Peninsula.

CHEMICAL COMPOSITION : Essential from outer part of flower containg 70 % methyl ether of β-phenylethyl alcohol. Blossoms yield essential oil containing benzyl benzoate, benzyl salicylate, benzyl alcohol, geraniol, linalool, linalyl acetate, bromostyene, guaiacol, phenylethyl alcohol and aldehydes.

Parts used : Flower, root and oil.

Ayurvedic properties :
Rasa - Tikta, Madhura, Kaṭu. Vīrya - Ūṣṇa.
Guṇa - Laghu, Snigdha. Vipāka - Kaṭu.

केतकः कटुकः स्वादुर्लघुस्तिक्तः कफावहः।
(भावप्रकाश)

Ketakaḥ kaṭukaḥ svādurlaghustiktaḥ kaphāvahaḥ.
(Bhavaprakasa)

केतकी कुसुमं वर्ण्यं केशदौर्गन्ध्यनाशनम्।
हेमाभं मदनोन्मादवर्धनं सौख्यकारि च।।
तस्य स्तनोऽतिशिशिरः कटुः पित्तकफापहः।
रसायनकरो बल्यो देहदार्ढ्यकरः परः।।
(राजनिघण्टु)

Ketakī kusumam varṇyam
keśadaurgandhyanāśanam,
Hemāmbha madanonmādavardhanam
saukhyakāri ca.
Tasya stano atiśiśiraḥ kaṭuḥ pittakaphāpahaḥ,
Rasāyanakaro balyo dehadāṛdhyakaraḥ paraḥ.
(Rajanighantu)

Actions/Uses : Tridoṣaśāmaka, Dīpana-Pācana, Anulomana, Hṛdya, Raktaprasādana, Prajasthāpana, Vṛṣya, Svedajanana, Kuṣṭhaghna, Kaṭupauṣṭika, Varṇya, Vedanāsthāpana, Keṣya, Saumanasyajanana, Ākṣepahara, Daurgandyahara, Vraṇaropaṇa.

Therapeutics :

Ketaki is bitter, cooling, aromatic, antiseptic, aphrodisiac, carminative, stomachic and abortifacient. It cures earache, headache, eye diseases, ulcers and scabies. It strengthens heart and liver. It is indicated in constipation, hysteria, debility, sterility and burning sensation of body. Leaves are bitter, pungent, aromatic, used in leprosy, smallpox, syphilis, scabies and leucoderma. Oil of bracts, stimulant, antispasmodic, administerd for headache and rheumatism.

Family : LXXIV-Papaveraceae

Argemone mexicana Linn.

ENGLISH : Mexican prickly poppy, Yellow thistle.

INDIAN : Dhotraa-piwallaa, Kaate Dhotraa.

SANSKRTA : Svarṇakṣirī, Kancanakṣirī, Kaṭuparṇī, Pītadhattura, Pītadugdhā, Dārudī, Satyanāsī, Svarṇadugdhā, Svarṇāhvā, Rukmiṇī, Hemakṣīrī, Kancanī, Suvarṇakṣirī, Hemāngā, Kanakakṣīrī, Pisollā, Kṣiragarbhā, Cokaka, Tiktadugdhikā, Hemavatī, Bhrahmadanti, Paṭuparṇi.

HABIT : An erect, prickly annual herb, 30 cm. to 1.2 m tall.

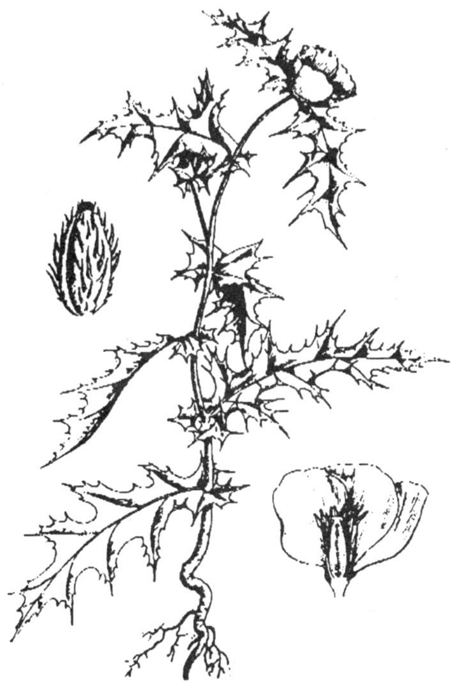

HABITAT : A native of Mexico. Naturalized and cultivated throughout India. It is cultivated as a crop in Egypt, USSR, Turkey, Yugoslavia, France and Australia.

CHEMICAL COMPOSITION : Protopine nitrate, berberine nitrate, ceryl alcohol, β-sitisterol, succinic, citric, tartaric and maleic acids, glucose and fructose. Argemone oil. Myristic, palmitic, oleic and linoleic acids and hydroxy and keto fatty acids in seed oil. Prolonged use of oil produces toxic effects resembling those of epidemic dropsy.

Parts used : Root, Seeds, milky juice and oil.

Ayurvedic properties :

Rasa - Tikta.	Vīrya - Uṣṇa, Śīta.
Guṇa - Laghu, Rūkṣa.	Vipāka - Kaṭu.

स्वर्णक्षीरी हिमा तिक्ता क्रिमिपित्तकफापहा।
मूत्रकृच्छ्राश्मरीशोफदाहज्वरहरा परा।।
(राजनिघण्टु)

Svarṇakṣīrī hima tiktā krimipittakaphāpahā,
Mūtrakṛcchāśmarīśophadāha jvaraharāparā.
(Rajanighantu)

Actions/Uses : Kaphapittaghna, Bhedanīya, Virecana, Śothahara, Mūtrala, Raktaśodhaka, Vraṇaśodhana, Ropaṇa, Viṣaghna, Vibandha, Kuṣṭhaghna, Viṣamajvaraghna, Vedanāsthāpana, Anaha, Kṛimighna.

Therapeutics :

Svarnaksiri plant is used as anthelmintic, antileprotic, tonic, diuretic and as antidote. Root is alterative, used in chronic skin diseases. Roots used for syphilitic infections and tapeworms. Roots is burnt to provide heat in treatment of piles; given in pain for severe stomachache. Seeds are laxative, emetic, narcotic, expectorant and demulcent, used in abdominal colic, diarrhoea and dysentery, are poisonous in large quantity. Seeds are pounded in Mahua oil and applied to eczema and itching. Ground seeds mixed with Brassica oil is used for itch. Yellow juice of plant used for dropsy, jaundice and cutaneous affections. Oil is purgative, used for cutaneous affections. Latex of plant is used for treatment of warts, tumors, cancer syphilis and skin diseases. Crude opium is still used to a limited extent in modern medicine as well as traditional medicine. Its main uses lie as narcotic, sedative, anodyne, antispasmodic, hypnotic and sudorific. It is used as a paregogic in controlling dysentery and diarrhoea.

Family : LXXV - Pedaliaceae

Papaver somniferum Linn.

ENGLISH : Poppy, Mawseed, White poppy, Mexican poppy.

INDIAN : Aphoo, Aphima, Kashkash, Posta.

SANSKṚTA : Aphūka, Ahiphena, Aphana, Niphena, Nāgaphena, Khasa, Nayaphena, Khaskhasarasa, Cosa.

HABIT : An erect, rarely branched, usually glaucous annual herb.

HABITAT : A native of West Asia, now grown in Uttar Pradesh, Punjab, Rajasthan and Madhya Pradesha. Wildly cultivated in gardens as ornamental plant.

CHEMICAL COMPOSITION : Sap contains oxalic acid and opium contains some 25 alkaloids, the chief being morphine, codeine, thebaine, narcotine, narceine, noscapine and papaverine. Poppy seed oil is a drying oil. It dries much slower and more uniformly than linseed oil. Meconic acid 3-5 %. Opium contains more than two dozen alkaloids. The main are morphine, codein, nascopine, papaverine and thebaine. Minor alkaloids include aporeine, codamine, cryptopine, gnoscopine, hydrocotarnine, laudanine, narcotoline, neopine, oxynarcotine and papaveramine. Alkaloids berberine, protopine and allocryptonine. Prolonged use of oil produces in man toxic effects resembling occuring in epidemic dropsy.

Parts used : Seeds, milky juice, fruit bark.

Ayurvedic properties :
Rasa - Tikta, Kaṣāya. Vīrya - Uṣṇa.
Guṇa - Laghu, Rūkṣa, Vikāsi. Vipāka - Kaṭu.

आफूकं शोषणं ग्राहि श्लेष्मघ्नं वातपित्तलम्।
आक्षेपशमनं निद्राजननं मदकारि च।।
स्वेदनं वेदनाहृच्च मूत्रातिसारणूत् परम्।
कासश्वासाति सारघ्नं शोणितस्त्रूतिवारणम्।।
(भावप्रकाश)

Āphukam śoṣaṇam grāhiśleṣmaghnam vātapittalam.
Ākṣepaśamanam nidrājananam madakāri ca.
Svedanam vedanāhṛcca mūtrātisāraṇūt param,
Kāsaśvāsātisāraghnam śoṇitasrūtivāraṇam.
(Bhavaprakasa)

Actions/Uses : Kaphaghna, Vātaśāmaka, Pittaprakopī, Stambhana, Pūriṣa, Śūlapraśamana, Pralāpanāśana, Nidrājanana.

Therapeutics :
Ahiphena or opium is used as narcotic, sedative, anodyne, hypnotic, antispasmodic and sudorific. It has more marked diaphoretic action than morphine but its hypnotic and analgesic effects are less certain. Opium is the inspissated milky juice from immature capsules and is a narcotic. It is useful for cough and bronchitis and in cases of diarrhoea and dysentery. Morphine is used to relieve pain, anxiety and sleeplessness due to pain.

Pedalium murex Linn.

SIMILAR : *Tribulus terrestris* Linn.

ENGLISH : Burra Gokhru.

INDIAN : Bhar-Gokharu, Selusaran, Gokshura, Hatticharatte.

SANSKṚTA : Gokśūra, Suryā, Gokanṭaka, Svādanstra, Svādukanṭaka, Trikanṭaka, Vanaśṛugāta, Caṇadruma, Ikṣūgandhikā, Palankadhā, Kanṭaphala, Bhakṣyataka, Vyāladanśtraka, Kṣūrānga, Bhadrakanṭa, Kṣūraka, Duh cakrama, Mahāngaka.

HABIT : A diffuse, more or less succulent herb.

HABITAT : Many parts of India.

CHEMICAL COMPOSITION : Pedalitin. diosmetin, dinatin and a gum isolated from leaves. Young branches contain mucilage and root, alkaloids.

Parts used : Fruits, stem, roots and leaves.

Ayurvedic properties :
Rasa - Madhura. Vīrya - Śīta.
Guṇa - Gurū, Snigdha. Vipāka - Madhura.

गोक्षुरः शीतलः स्वादुर्बलकृद्बस्तिशोधनः।
मधुरों दीपनों वृष्यः पुष्टिदश्चाश्मराहरः।
प्रमेहश्वासकासार्शः कृच्छ्रहृद्रोंगवातनुत्।।
(भावप्रकाश)

Gokṣuraḥ śītalaḥ svādurbalakṛd bastiśodhanaḥ,
Madhuro dīpano vṛṣyaḥ puṣṭidaścāśmarīharaḥ,
Pramehaśvāsakāsārśaḥ kṛcchahṛdrogavātanut.
(Bhavaprakasa)

गौक्षुरः मूत्रविरेचनीयानाम्।
(चरक)

Gokṣuraḥ mūtrāvirecanīyānām.
(Caraka)

Actions/Uses : Vātaghna, Pittaśāmaka, Kaphakara, Anulomana, Balya, Hṛdrogahara, Vedanāsthapāna, Mūtravaha, Baśtisodhana, Kaphanissāraka.

Therapeutics :

Decoction of root of Goksura is antibilious. Infusion of leaves and stems is used in gonorrhoea and dysuria. Fruit and seeds are antispasmodic, aphrodesiac, demulcent and diuretic; in decoction given for incontinence of urine, gleet, nocturnal emmissions, spermatorrhoea, impotence and irritation of urinary organs. Juice of fruit is emmenagogue, used in puerperal diseases and to promote lochial discharge. Properties and used are the same as Tribulus terrestris.

Sesamum orientale Linn.

SYNONYMS : *Sesamum indicum* Linn.

ENGLISH : Sesame, Gingelly.

INDIAN : Teella.

SANSKṚTA : Tila, Homadhānya, Pavitra, Pitṛtarpaṇa, Pāpaghna, Pūtadhānya, Jartila, Vanodbhava.

HABIT : An erect branched or unbranched annual.

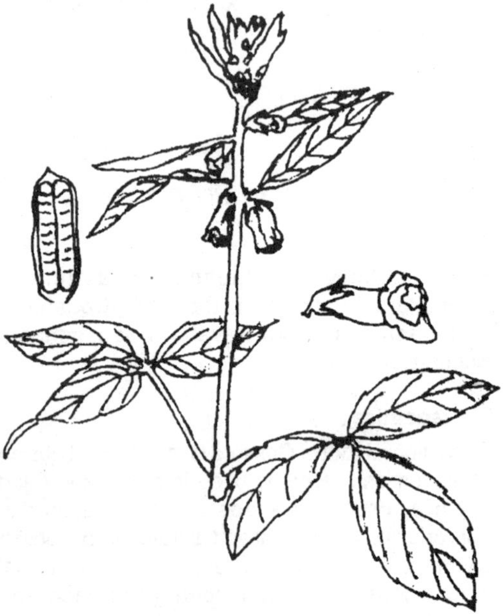

HABITAT : Cultivated all over India.

CHEMICAL COMPOSITION : Seeds are fairly rich in thiamine and niacin. The principal protein is a globulin. Seeds contain a fixed oil and leaves contain a gummy matter. A flavonoid glucoside pedalin isolated from leaves.

Parts used : Seeds, leaves, roots and oil.

Ayurvedic properties :
Rasa - Madhura, Anurasa-Kaṣāya. Vīrya - Ūṣṇa.
Guṇa - Gurū, Snigdha. Vipāka - Madhura.

स्निग्धोष्णो मधुररस्तिक्तः कषायः कटुकरस्तिलः।
त्वच्यः केश्यश्च बल्यश्च वातघ्नः कफपित्तकृत्॥
(च. सू.)

Snigdhoṣṇo madhurastiktaḥ kaṣāyaḥ kaṭukastilaḥ,
Tvacyaḥ keśyaśca balyaśca vātaghnaḥ kaphapittakṛt.
(Caraka Sutra 27)

Actions/Uses : Vātaśāmaka, Kaphapittaprakopī, Sandhānīya, Keṣya, Snehana, Vṛṣya, Vraṇaśodhana-ropaṇa, Dantya, Dīpana, Balya, Grāhī, Śūlapraśamana, Mūtrasamgrahaṇīya, Vājikarṇa, Pūyaghna, Ārtavajanana.

Therapeutics :

Tila seeds are laxative, emolient, tonic, diuretic, lactagogue. They are useful in piles, are used as a nourishing food and also as a flavouring agent. A paste of seeds mixed with butter is applied externally in bleeding piles. Seeds in form of a decoction are used as emmenagogue; as a polutice applied to ulcers. Seeds and oil are used as demulcent in dysentery and urinary complaints in combination with other medicines. Leaves in form of an infusion used in affections requiring demulcents. An infusion of fresh leaves in cold water is much used in North America as a demulcent drink in cholera infantum and other disorders of bowels in children. It is also employed in catarrh and affections of the urinary organs.

Family : LXXVI - Piperaceae

Piper betle Linn.

SYNONYMS : *Chavica betle* Miq.

ENGLISH : Betel.

INDIAN : Paanawela, Naagawela, Tambuli.

SANSKṚTA : Tāmbula, Nāgavallī, Tāmbulavalai, Nāgavallarī, Pātālavāsini.

HABIT : A perennial dioecious creeper.

1

HABITAT : Native of Malaysia. Cultivated in hotter and damper parts of India.

CHEMICAL COMPOSITION : Plant contains a-terpinene, p-cymene, carvacrol, chavicol and its derivatives, allyl catechol, eugenol, estragol, oxalic acid, malic acid and amino acids. Leaves contain good amount of vitamins particulary nicotinic acid, ascorbic acid and carotin. They also contain significant amounts of all the essential amino acids except lysine, histidine and arginine. Large concentrations of asparagine are present while glycine and proline occur in good amount. Essential oil of leaf gives it the aromatic flavour. β-sitosterol from roots.

Parts used : Leaf, root and fruit.

Ayurvedic properties :
Rasa - Kaṭu, Tikta, Kaṣāya. Vīrya - Uṣṇa.
Guṇa - Laghu, Rūkṣa, Tīkṣṇa, Viṣada.
Vipāka - Kaṭu.

ताम्बुलं विशदं रुच्यम् तिक्तोष्णं तुवरं सरम्।
वश्यं तिक्तं कटु क्षारं रक्तपित्तश्रमापहम्।।
वल्यं श्लेष्मास्यदौर्गन्ध्यमलवातश्रमापहम्।।
नक्तान्धशमनं कामदीपनं क्षतरोपणम्।।
(द्रव्यगुण)

Tāmbulam viśadam rucyam tikoṣṇam tuvaram saram,
Vaśyam tiktam kaṭu kṣāram raktapittaśramāpaham,
Balyam śleṣmāsyadaurgandhyamalavātaśramāpaham.
Naktāndhaśamanam kāmadīpanam kṣataropaṇam.
(Dravyaguna)

Actions/Uses : Vātaghna, Raktapittakara, Śleśmahara, Rucya, Balya, Dīpana, Svarya, Sramsana, Kāsahara, Asyadaurgandhyahara, Sramahara, Pittaprakopana, Kaṇḍūhara, Kṛmihara, Pīnasajit, Kāmāgni, Kāmajananī, Sandīpana.

Therapeutics :
Tambula is hot, acrid, bitter astringent, aromatic, carminative, stimulant, aphrodisiac and antiseptic. Root is used to prevent child bearing. Leaf is aromatic, carminative, stimulant, astringent used as a preventive of worms and in snakebite. Juice of leaves is dropped into eyes in painful affections and in night blindness; also used to allay thirst and to relieve cerebral congestion and satyriasis. Essential oil from leaves is used in respiratory catarrhs and as antiseptic. Fruit is employed with honey as remedy for cough.

Piper cubeba Linn. f.

SYNONYMS : *Cubeba offininalis* Miq.; *Litsea citrata* Blume.; *Litsea cubeba* Pers.

ENGLISH : Cubeba, Tailed pepper.

INDIAN : Kankola, Kabaabchini, Himsi mire, Ckini.

SANSKṚTA : Kankola, Sugandha-murica, Surapriya, Vṛttaphala.

HABIT : A deciduous aromatic climbing shrub or a small tree.

HABITAT : Indeginous to Java and Sumatra, cultivated to some extent in India.

CHEMICAL COMPOSITION : Thirteen compounds identified in oil from fresh bark; citral was major constituent. Essential oil from fruits contained cubebin, limonene, sabinene, methyltheptenone, citronellal and citrals.

Parts used : Fruit and oil.

Ayurvedic properties :
Rasa - Kaṭu, Tikta. Vīrya - Uṣṇa.
Guṇa - Laghu, Tīkṣṇa, Rūkṣa. Vipāka - Kaṭu.

सुरप्रियं वृत्तफलं तद्वायुशमनं मतम्।
श्लेष्मोत्सारणमाग्नेयं मूत्रवृद्धिकरं तथा।।
औपसर्गिकमेहश्च शुक्रमेहं सुदारुणम्।
श्वेतप्रदरमर्शांसि कृच्छ्रश्चापि विनाशयेत्। (द्रव्यगुण)

Surapriyam vṛttaphalam tadvāyuśamanam matam,
Śleṣmotsāraṇamāgneyam mūtravṛdhikaram tathā.
Aupasargukamehañca śukrameham sudāruṇam,
Śvetapradaramarsamsi kṛcchañcāpi vināśayeta.
(Dravyaguna)

कङ्कोलं कटुतिक्तोष्णं वक्त्रजाड्यहरं परम्।
दीपनं पाचनं रुच्यं कफवातनिकृन्तनम्।। (राजनिघण्टु)

Kaṅkolam kaṭutiktoṣṇam vaktrajāḍyaharam param,
Dīpanam pācanam rucyam kaphavātanikṛntanam.
(Rajanighantu)

Actions/Uses : Kaphavātaśāmaka, Raktotkleśakara, Uttejaka, Hṛdya, Śothahara, Durgandhināśaka, Kṛimighna, Vraṇaropaṇa, Dīpana, Pācana, Rucikara, Anulomana, Vājikara, Mūtrala, Ārtavajanana, Kaphanissāraka.

Therapeutics :

Kankola or Cubebs are stimulant, carminative, local irritant, diuretic and expectorant. It cures asthma. Unripe fruit is aromatic, diuretic and expectorant. Fruit is edible, aromatic and carminative, used for dizziness, headache, hysteria, paralysis and loss of memory. A valuable remedy in cases of gleet, internal inflammations and catarrh; used in coughs, bronchitis and lung troubles in general. Oil is used in genito-urinary diseases like cystitis, gonorrhoea and gleet.

HOMEOPATHIC USE : Cubeba is a Homeopathic remedy. It affects mucous membrane of nose, air passages and intestines, as well as that of urinogenital tract. Some of the characteristic symptoms are burning in throat with drynes, constant need to swallow saliva, burning in stomach; in abdomen; in rectum; in urethra; in fossa navicularis. It is of use in gonorrhoea after the inflammatory stage is passed under the usual remedies, if there remains burning in urethra after micturition, and the discharge is bland.

Piper longum Linn.

SYNONYMS : *Chavica roxburghii* Miq.

ENGLISH : Indian long peppere.

INDIAN : Pimpallee, Pipal, Pipli, Piplamul.

SANSKṚTA : Pippalī, Māgadhī, Kṛṣṇā, Vaidehī, Kolā, Capalā, Kaṇā, Ūṣaṇā, Upakullyā, Śaundī, Tīkṣṇatanḍulā, Kaṭubijā, Korāngī, Tiktatanḍulā, Syāmā, Dantaphala, Maghādodbhava.

HABIT : A slender aromatic climber with perennial woody roots.

HABITAT : Hotter parts of India.

CHEMICAL COMPOSITION : Plant contains essential oil consisting of long-chain hydrocarbons, mono- and sesquitertenes. Piperlongumine, piperlonguminine, sesamin, piperine, and methyl 3,4,5-trimethoxycinnamate from roots.

Parts used : Root and fruit.

Ayurvedic properties :
Rasa - Kaṭu, Tikta. Vīrya - Anūṣṇa, Śīta when fresh. Guṇa - Laghu, Snigdha, Tīkṣṇa. Vipāka - Madhura, Kaṭu.

पिप्पली रेचनी हन्ति श्वासकासोदरज्वरान्।
कुष्ठप्रमेहगुल्मार्शः प्लीहा शूलाममारुतान्।।
पिप्पली दीपनी वृष्या स्वादुपाका रसायनी।
अनुष्णा कटुका स्निग्धा वातश्लेष्महरी लघुः।।
दीपनं पिप्पलीमूलं कटुष्णं पाचनं लघु।
रुक्षं पित्तकरं भेदि कफवातोदरापहम्।।
आनाहप्लीहगुल्मघ्नं कृमिश्वासक्षयापहम्।। (भावप्रकाश)

Pippalī recanī hanti śvāsakasodarajvarān,
Kuṣṭhapramehagulmārśaḥ plīhā śūlāmamārutān.
Pippalī dīpanī vṛṣyā svādupākāh rasāyanī,
Anūṣṇā kaṭukā snigdhāvātaśleṣmaharī laghuḥ.
Dīpanam pippalīmūlam kaṭūṣṇam pācanam laghu,
Rūkṣam pittakaram bhedi kaphavātodarāpaham,
Ānāhaplīhagulmaghnam kṛmiśvāsakṣayāpaham.
(Bhavaprakasa)

Actions/Uses : Kaphaghna, Vātaghna, Dīpana, Pāchan, Vṛṣya, Rasāyana, Recana, Medhya, Śleṣmāhara, Jvaraghna, Kuṣṭhaghna, Pramehaghna, Gulmahara, Plīhāghna, Arśahara, Śūlahara, Medohara, Śvāsaghna, Kāsaghna, Kṛmighna, Pānḍuhara, Anahahara, Kṣāyahara, Jirṇajvarahara.

Therapeutics :
Pippali is capable of improving intellect and memory power and also to regain health by dispelling disease. It is acrid, hot, light, digestive, appetiser, aphrodisiac and tonic. It cures cough, dyspnoea, ascites, leprosy, diabetes, piles, colic, anemia, indigestion, and dispels cardiac and spleenic disorders, chronic fever. Fruits as well as roots are attributed with numerous medicinal properties and are used for diseases of respiratory tract, viz. cough, bronchitis, asthma etc; as counter-irritant and analgesic when applied locally on muscular pains and inflammation; as snuff in coma and drowsiness and internally as carminative; as sedative in insomnia and epilepsy; as general tonic and haematinic; as cholagogue in obstruction of bile duct and gall bladder; as an emmenagogue and abortifacient; and for miscellaneous purposes as anthelmintic and in dysentery and leprosy. Properties and uses of P. longum are similar to those of black pepper.

Piper nigrum Linn.

SYNONYMS : *Piper trioicum* Roxb.

ENGLISH : Pepper, Black pepper.

INDIAN : Miree, Kalimirch, Golmorich.

SANSKṚTA : Marica, Palita, Śyāma, Pallijabhūṣaṇa, Yavaneṣṭa, Dharmapatana, Vṛttaphala, Śākānga, Kola, Kaṭūka, Sarvahita, Kṛuṣṇa, Carmabandhana, Ūṣṇa, Hapuśa, Cātakāśira, Carapriya, Kṛmihara, Suvṛtta.

HABIT : A branching, climbing perennial shrub.

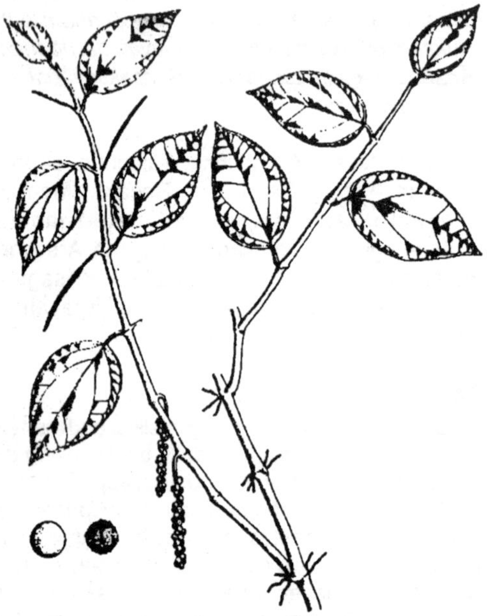

HABITAT : Cultivated in hot and damp parts of India.

CHEMICAL COMPOSITION : Alkaloids chavicine, piperine, hentriacontane-16-one, hentriacontane and β-sitosterol from stems. a-cis-bergamotene and β-cis-bergamotene from essential oil. Pipercide from fruits.

Parts used : Fruits.

Ayurvedic properties :
Rasa - Kaṭu. Raw fruit : Madhura.Vīrya - Ūṣṇa.
Guṇa - Laghu, Tīkṣṇa, Rūkṣa. Vipāka - Kaṭu.

मरिचं कटुकं तीक्ष्णं दीपनं कफवातजित्।
उष्णं पित्तकरं रुक्षं श्वासशूल कृमीन्हरेत्।।
(भावप्रकाश)

Maricam kaṭukam tīkṣṇam dīpanam kaphavātajit,
Ūṣṇam pittakaram rūkṣam śvasāśūlakṛmīnharet.
(Bhavaprakasa)

Actions/Uses : Vātaghna, Kaphaghna, Pittakar, Dīpan, Śvāsahara, Śūlahara, Kṛmighna, Kasahara, Kuṣṭahara, Raktotkleṣakara, Avṛṣya, Lekhana.

Therapeutics :

Marica fruit is acrid, bitter, hot, light, alterative and is used as aromatic, stimulant, carminative and is a valuable gastro-intestinal stimulant, of great service as a stomachic in dyspepsia and flatulence, congestive chills and indigestion. Its action as a stimulant is more especially evident on mucous membranes of rectom and urinary organs. It is useful in cholera, in weakness after fevers, vertigo, coma and as an antiperiodic in malarial fevers; as an alterative in paraplegia and arthritic diseases. Externally applied it is rubifacient, and as a local application for relaxed sore throat, piles and skin diseases. It is regarded as a useful remedy in haemorrhoidal affections and in relaxed conditions of rectum attended with prolsapsus. It is used as a substitute for cubebs in gonorrhoea.

HOMEOPATHIC USE : Piper nigrum is a Homeopathic remedy. Full, heavy headache, sensation of burning occur almost everywhere. Pressure in nasal bones, in temples and in facial bones; contraction of uterus and a sensation as if something about to enter it, sensation of a foreign round body rising to stomach are the chief symptoms during proving.

Family : LXXVII-Plantaginaceae

Plantago ovata Forsk.

ENGLISH : Blond psyllium, Spogel seeds, Ispaghula.

INDIAN : Isabagola.

SANSKṚTA : Aśvagola, Aśvakarṇabīja, Iśadgola, Śītabīja.

HABIT : A stemless or sub-caulescent softly hairy or wooly annual herb.

HABITAT : Indigenous to Mediterranean and West Asia. Grown in Punjab, Sind and Baluchistan.

CHEMICAL COMPOSITION : Mucilage. Seeds contain holoside planteose. Seed oil has 50 % linoleic acid, prevents atherosclerosis. Oil is more active than safflower oil, reduces serum cholesterol level in ribbits. The husk of the seed contains a colloidal mucilage (polysaccharide), mainly consisting of xylose, arabinose, galacturonic acid with rhamnose and galactose.

Parts used : Seeds.

Ayurvedic properties :
Rasa - Madhura. Vīrya - Śīta.
Guṇa - Snigdha, Gurū, Picchila. Vipāka - Madhura.

अश्वगोलं गुरु स्वादु स्निग्धं शीतंच पिच्छिलम्।
स्नेहनं मूत्रजननं श्लेष्मनिः सारणं परम्।।
दाहतृष्णाहरं बल्यं ज्वरघ्नं चाथ शस्यते।
प्रवाहि कातिसारामदाह तृष्णा ज्वरादिषु।
वातपित्तामये कासे दौर्बल्ये मूत्रकृच्छ्रजे।।

Aśvagolam guru svādu snigdham śītam ca picchilam,
Snehanam mūtrajananam śleṣmaniḥ sāraṇam param.
Dāhatṛṣṇāhāram balyam jvaraghnam cātha śasyate,
Pravāhikātisārāmadāha tṛṣṇā Jvarādiṣu,
Vātapittāmaye kāse daurbalye mūtrakrcchraje.

Actions/Uses : Vātapittaghna, Mūṭrajanana, Śleṣmānissāraka, Dāhahara, Balya, Jvaraghna, Mūtrakṛcchraghna, Daurbalyahara, Kāsahara, Pravāhikaghna, Atisāraghna.

Therapeutics :

Seeds of Asvakarnabija is demulcent, cooling, diuretic and laxative. Used in inflammatoy conditions of mucous membrane of gastro-intestinal and genito-urinary tracts and in chronic dysentery, diarrhoea and constipation. Infusion or decoction of the whole seed is used to control chronic constipation. In addition to its medicinal use, it is employed as a stabilizer in ice creams and as an ingredient in chocolates and other food materials. It is also used for sizing textiles and as a base for cosmetics.

Family : LXXVIII-Plumbaginaceae

Plumbago indica Linn.

SYNONYMS : *Plumbago rosea* Linn.

ENGLISH : Fire plant.

INDIAN : Laala chitraka, Chitra, Rakta-chitra.

SANSKṚTA : Raktacitraka, Citraka, Agni, Vahni, Dāhanah, Analah.

HABIT : A woody shrubby red-flowered perennial.

HABITAT : An ornamental plant cultivated in gardens throughout India.

CHEMICAL COMPOSITION : Plubbagin, sitosterol glucoside.

Parts used : Root, root-bark, milky juice.

Ayurvedic properties :
Rasa - Kaṭu. Vīrya - Ūṣṇa.
Guṇa - Laghu, Rūkṣa, Tīkṣṇa. Vipāka - Kaṭu.

स्थूलकायकरो रुच्यो कुष्ठघ्नो रक्तचित्रकः।
रसेनियामको लेहे बेधकश्च रसायनः।।
(राजनिघण्टु)

Sthūlakāyakaro rucyo kuṣṭhaghno raktacitrakaḥ,
Rase niyāmako lohe vedhakaśca rasāyanaḥ.
(Rajanighantu)

Actions/Uses : Rucya, Rasāyana, Śūlahara, Śothahara, Vanhikṛita, Visphotajanana, Kuṣṭhahara, Arśoghna, Arśahara, Pānduhara.

Therapeutics :

Rakta Citraka is an esteemed remedy for leucoderma and other skin diseases. Drug is used only after adequate curing and purification. Root is caustic, causing blisters on the skin. It is alterative, gastric stimulant and appetiser; in large doses it is acronarcotic poison. It has a specific action on uterus. It is acrid, abortifacient, vesicant and stimulant. It is used in the treatment of certain types of leucoderma. Tempered with little bland oil is used as external application in rheumatism and paralytic affections; also prescribed internally for these complaints. It is a powerful sialogogue, remedy for secondary syphilis and leprosy. Milky juice is useful in ophthalmia and application to scabies.

Plumbago zeylanica Linn.

ENGLISH : Ceylone Lead Wort.

INDIAN : Chitraka, Chita, Chitramula.

SANSKṚTA : Citraka, Agni, Anala, Dahana, Pāthī, Vanhisamjnaka, Vyāla, Ūṣṇa, Jyotī, Pāvaka, Śabala, Aruṇa, Hutāśana, Śūra, Śaṭha, Jvāla, Śārdūla, Citrapālī, Śikhī, Kṛśānudahana, Vāruṇa, Citrānga, Dvīpī, Jyotiśka, Hutabhunka.

HABIT : A perennial, sub-scanent shrub with white flowers.

HABITAT : indigenous to South-East Asia. Found throughout India.

CHEMICAL COMPOSITION : Contains plumbagin which externally is a strong irritant and a powerful germicide, stimulates muscular tissue in smaller doses and paralyses in larger ones; stimulates contraction of the muscular tissue of heart, intestines and worms; stimulates secretion of sweat, urine and bile and has stimulant action on nervous system. Roots contain plumbagin, 3-chloroplumbagin, 3,3-biplumbagin, chitranone, zeylinone, isozoeylinone, elliptinone and droserone. Plant extract prevented 100% ovulation and implantation in female rats.

Parts used : Root and fresh root-bark.

Ayurvedic properties :
Rasa - Kaṭu. Vīrya - Ūṣṇa.
Guṇa - Laghu, Rūkṣa, Tīkṣṇa. Vipāka - Kaṭu.

चित्रकोग्निशमः पाके कटुः शोफकफापहः।
वातोदरार्शः ग्रहणीक्रिमिकण्डूतिनाशनः।।
(राजनिघण्टु)

Citrakognisamaḥ pāke kaṭuḥ śophakaphāpāhaḥ,
Vātodarārśaḥ grahaṇī krimi kaṇḍūti nāśanaḥ.
(Rajanighantu)

चित्रकः कटुका पाके वह्निकृत् पाचनो लघुः।
रुक्षोष्णो ग्रहणीकुष्ठशोथार्शः कृमिकासनुत्।।
वातश्लेमहरो ग्राहि वातर्शः श्लेष्मपित्तहृत्।।
(भावप्रकाश)

Citrakaḥ kaṭukaḥ pāke vahnikṛt pācano laghuḥ,
Rukṣoṣno grahaṇīkuṣṭhaśotharsaḥ kṛmikāsanut,
Vātaśleṣmaharo grāhī vātarśaḥ śleṣmapittahṛt.
(Bhavaprakasa)

Actions/Uses : Kaphaghna, Vātaghna, Tṛptighna, Lekhana, Bhedana, Dīpana, Pācana, Grāhī, Kṛmighna, Grahaṇīhara, Rasāyana, Śūlahara, Śothahara, Vanhikṛta, Visphotajanana, Kuṣṭhahara, Arśoghna, Arśahara, Pānduhara, Vātodaraghna, Gudaśothahara.

Therapeutics :
Citraka root is powerfully poisonous. It is abortifacient and vesicant. It is diuretic, caustic and expellent of phlegmatic humours and is useful in rheumatism. It is an appetizer, used in skin diseases, diarrhoea, dyspepsia, piles, anasarca. Made into a paste with vinegar, milk or salt and water it is applied externally in leprosy and other skin diseases. Tincture of root bark is a powerful sudorific and antiperiodic. Milky juice is used as application in scabies and unhealthy ulcers.

Family : LXXIX - Poaceae

Bambusa arundinacea Willd.

SYNONYMS : *Bambusa bambos* Druce.; *Bambusa arundinacea* Retz.; *Arundo bambos* Linn.

ENGLISH : Spiny bamboo.

INDIAN : Kate-Baamboo, Kallaka.

SANSKṚTA : Vamśa, Tvakṣāra, Tṛṇadhvaja, Karmāra, Veṇu, Kanṭālū, Śataparva, Yavaphala, Tṛṇaketu, Maskara, Kanṭakī, Mahābala, Dṛudhagranthi, Danḍapatra, Bahuparvaka, Dhanurdruma, Dhanuśya, Śatpadālaya, Śabdamāla, Tejana.

HABIT : A graceful spinous perennial erect herb.

HABITAT : Wild throughout the country especially in hill forests.

CHEMICAL COMPOSITION : Cholin, betain, nuclease, urease, proteolytic enzyme, diastatic and emulsifying enzyme; bangsolochan, tabashir. Young roots contain a cyanogenetic glucoside and are poisonous. Young shoots contain HCN, benzoic acid, reducing sugars and waxes.

Parts used : Roots, leaves, Patraankura, fruits and Vamshlochana.

Ayurvedic properties :
Rasa - Madhura, Kṣāya. Vīrya - Śīta.
Guṇa - Rūkṣa, Laghu, Tīkṣṇa, Sara.
Vipāka - Madhura.

वंशः सरो हिमः स्वादुः कषायो बस्तिशोधनः।
छेदनः कफपित्तघ्नः कुष्ठास्त्रव्रणशोफजित्॥
तत्करीरः कटुः पाके रसे रुक्षो गुरुः सरः।
कषायः कफकृत् स्वादुर्विदाही वातपित्तलः॥
(भावप्रकाश)

Vanśaḥ saro himaḥ svāduḥ kaṣāyo bastiśodhanaḥ,
Chedanaḥ kaphapittaghnaḥ
kuṣṭhāsravraṇaśophajit.
Tatkarīraḥ kaṭuḥ pāke rase rūkṣo guruḥ saraḥ,
Kṣāyaḥ kaphakṛt svādurvidāhi vātapittalaḥ.
(Bhavaprakasa)

Actions/Uses : Balya, Brihmana, Bastiśodhaka, Cedinī, Kamalaghna, Kuṣṭhaghna, Vraṇaghna, Śothaghna, Badhamūtraghna, Vṛṣya, Dīpana, Pācana, Medoghna, Tṛsnāhara, Kāsaghna, Jvaraghna, Śvāsaghna, Viṣahara, Kṣāyaghna, Raktapittahara, Pāndughna, Kaṣṭārtavahara, Garbhāśayaśodhana, Kṛmighna.

Therapeutics :

Roots of Vamsa are sweet, astringent, cooling, tonic, laxative and diuretic. They are used in skin diseases, leprosy, ringworm burning sensation, strangury and general debility. Leaves are sweet, cooling, astringent, emmenagogue, ophthalmic, constipating and febrifuge, used in amenorrhoea, dysmenorrhoea, in eye troubles, lumbago, haemorrhoids, diarrhoea, gonorrhoea, haemetemesis, skin diseases, fever, and vetrinary practice. Sprouts are used as laxative, anti-inflammatory, digestive, carminative and anthelmintic. Mixed with black pepper and common salt, are used to check diarrhoea in cattle. Bamboo manna which is the silicious secretion found in internodes of stems is tonic, useful in fever, cough and in snake bite.

Cymbopogon citratus (DC.) Stapf.

SYNONYMS : *Andropogon citratus DC.*

HABIT : A tufted perennial erect herb with a short rhizome.

HABITAT : Grown in gardens.

SIMILAR : Cymbopogon schoenanthus (Linn.) Spreng.; Andropogon schoenanthus Linn.; Cymbopogon martini.

ENGLISH : Lemon grass.

INDIAN : Gawati chahaa, Hirva chaa, Ole Cha. Gandhatrina.

SANSKṚTA : Rohiśa, Sugandhabhūtṛṇa, Kat-ṭṛṇa, Kuṭṛṇa, Pūti, Bhūtī, Bhūtika, Śyāmaka, Dhyāmaka, Devadagdhaka, Sakala, Paura, Pātala, Saugandhika, Devadadaka, Binducita, Bhustraṇa, Devajagdhaka.

CHEMICAL COMPOSITION : Cymbopogone from leaf wax. Cymbopogonol. Lemmongrass oil contains citral.

Parts used : Stem, leaves, flowers and oil.

Ayurvedic properties :
Rasa - Kaṭu, Tikta. Vīrya - Ūṣṇa.
Guṇa -Laghu, Rūkṣa, Tīkṣṇa. Vipāka - Kaṭu.

रोहिषं तुवरं तिक्तं कटुपाकं व्यपोहति।
हृटकण्डूव्याधिवातास्त्रशूलकासकफज्वरान्।।
(भावप्रकाश)

Rohiṣam tuvaram tiktam kaṭupākam vyapohati,
Hṛtkaṇḍūvyādhivātāsraśūlakāsakaphajvarān.
(Bhavaprakasa)

Actions/Uses : Kapha-vātaśāmaka.

Therapeutics :
Herb Rohisa is stimulant, diaphoretic, anti-spasmodic. Oil is carminative in cholera and tonic. Leaves are chewed in Samoa for sore gums, also used for elephantiasis. Infusion of leaves is sudorific, stimulant and antiperiodic, useful in catarrh. Juice applied in headache.

Cymbopogon jwarancusa Schult.

SYNONYMS : *Andropogon jwarancusa* Jones.

ENGLISH : Schaenanthe officinale, Squinauch.

INDIAN : Waalla-piwalla.

SANSKṚTA : Lāmajjaka, Gardhabhapriya, Uśtrapriya.

HABIT : A highly aromatic grass.

HABITAT : Various parts of India.

CHEMICAL COMPOSITION : Essential oil, antimicrobial. Piperitone, β-caryophyllene, p-cymene, piperitol, perillyl alcohol, farnesol and pamitic acidin essential oil.

Parts used : Stem, leaves, flowers and oil.

Ayurvedic properties :
Rasa - Kaṭu, Tikta. Vīrya - Ūṣṇa.
Guṇa - Laghu, Rūkṣa, Tīkṣṇa. Vipāka - Kaṭu.

लामज्जकं तु मधुरं तिक्तं शीतंच पाचकम्।
स्तम्भनं लघु पित्तघ्नं वाततृट दाहनाशनम्।।
त्रिदोषश्रममूर्च्छास्त्रशूलवान्ति विनाशनम्।
ज्वरं च रक्तदोषं च स्वेदं कृच्छं मदं कफम्।।
व्रणं विषं विसर्पंच नाशयेदिन्ि कीर्तितम्।

Lāmajjakam tu madhuram tiktam śītam ca pācakam,
Stambhanam laghu pittaghnam vātatṛtadāhanāśanam.
Tridoṣaśramamurcchāsrtra śūla vānti vināśanam.
Jvaram ca raktadoṣam ca svedam kṛccham madam kapham,
Vraṇam viṣam visarpañca ñāśayediti kīrtitam.
(Nighanturatnakara)

Actions/Uses : Kaphavātaśāmaka, Raktotkleṣakara, Mūtrajanana, Vedanāsthāpana, Rucikara, Dīpana, Pācana, Anulomana, Kṛmighna, Raktaśodhana, Hṛdayottejaka, Kaphanissāraṇa, Stanyajanana.

Therapeutics :

Lamajjaka grass is carminative, stimulant and emmenagogue. It is used as a stimulant diaphoretic in gout, chronic rheumatism and intermittent fever; also to purify blood and in coughs, and cholera aromatic, tonic in dyspepsia.

Cynodon dactylon (Linn.) Pers.

SYNONYMS : *Panicum dactylon* Linn.

ENGLISH : Dog's Teeth grass.

INDIAN : Durwaa, Hariyaali.

SANSKṚTA : Dūrvā, Śataparṇa, Śatavīrya, Golomī, Vṛkṣādāni, Latā, Payasya, Harita, Śāmbhavī, Śāntā, Pūtā, Anuṣṇavallikā, Śiveṣṭā, Mangalā, Anantā, Durbharā, Pracandā, Sukhavallabhā, Śītalatā, Vijayā, Gaurī, Śītalā, Bhadrā, Bhārgavī, Vighnā, Śvetā.

HABIT : A hardy perennial with creeping culms, rooting at nodes.

HABITAT : Throughout India.

CHEMICAL COMPOSITION : An alkaloid from the plant caused a slowing of blood flow in mesenteric capillaries of rats and mice. A glycoside from the plant caused hypotension in cat.

Parts used : Whole plant.

Ayurvedic properties :
Rasa - Madhura, Kṣāya, Tikta.Vīrya - Śīta.
Guṇa - Laghu, Snigdha. Vipāka - Madhura.

दूर्वाः कषायाः मधुराश्च शीताः पित्ततृषारोचकवान्ति हन्त्यः।
सदाह मूर्च्छा ग्रह भूत शान्ति श्लेष्मश्रमध्वंसन तृप्तिदाश्च।।
(राजनिघण्टु)

Dūrvāḥ kaṣāyāḥ madhurāśca śītaḥ
pittaṭṛṣā rocaka vānti hantryaḥ,
Sadāha mūrchā graha bhūta śānti,
Śleṣmaśramadhvamsana trptidāśca.
(Rajanighantu)

Actions/Uses : Kaphapittaśāmaka, Vraṇaropaṇa, Raktapittahara, Stambhana, Dāhapraśamana, Tṛṣṇāghna, Arocakaghna, Vāntihara, Mūrcchaghna, Śramahara, Kaṇḍūhara, Tvagrogaghna, Garbhasrāvahara, Varṇya, Viṣarpaghna.

Therapeutics :

Durva is reputed as a remedy in epitaxis, haematuria and scabies. Plant is used in inflammed tumors, whitlows and fleshy excrescences. Juice of plant is astringent, used as application to fresh cuts and wounds; diuretic used in dropsy and ansarca, in hysteria, epilepsy, insanity;.astringent in chronic diarrhoea and dysentery; useful in catarrhal ophthalmia. Infusion of root is used for stopping bleeding from piles. Decoction of roots is diuretic and used in dropsy and secondary syphilis. Crushed roots mixed with curds used in chronic gleet. Rhizome is cooling, astringent, diuretic, demulcent, aperient, ophthalmic, haemostatic and suppurative. It checks bleeding from cuts and wounds and is useful in fever, burning sensation, chronic diarrhoea, dysentery, anasarca, dropsy, catarrhal ophthalmia, dysuria, bleeding piles, eye affections, epilepsy, hysteria and insanity. Used in urinary and bladder complaints, cystitis, nephritis. Also recommended in gout and rheumatism.

Desmostachya bipinnata Stapf.

SYNONYMS : *Poa cynosuroides* Retz.; *Eragrostis cynosuroides* Beauv.

INDIAN : Darbha, Kusha.

SANSKṚTA : Darbha, Śucicira, Yajniyapatraka, Bramhapavitra, Puṇyaṭruṇa, Vanhiputaṭruṇa, Kuśa, Sucyagra, Yajnyabhūṣaṇa, Pūta, Vajra.

HABIT : A tall tufted perennial grass.

HABITAT : Throughout India in hot and dry places.

CHEMICAL COMPOSITION : Crude protein, 6.75; crude fibre, 40.30% on dry basis.

Parts used : Roots.

Ayurvedic properties :
Rasa - Madhura, Kaṣāya. Vīrya - Śīta.
Guṇa - Laghu, Snigdha. Vipāka - Madhura.

दर्भद्वयं त्रिदोषघ्नं मधुरं तुवरं हिमम्।
मूत्रकृच्छ्राश्मरी तृष्णाबस्तिरुक् प्रदरोस्त्रजित्।।
(भावप्रकाश)

Darbhadvayam tridoṣaghnam madhuram tuvaram himam,
Mūtrakṛcchāśmarī tṛṣṇā bastiruk pradarāsrajit.
(Bhavaprakasa)

मृदुदर्भः कुशो बर्हिः शुचिचीरः सुवृत्तकः।
खरोऽन्य पृथुलः शीरी गुन्द्रा च नीरजः स्मृतः।।
द्रर्भयुग्मं पवित्रं स्यान्मूत्रकृच्छ्रघ्नशीतलम्।
रक्तपित्त प्रशमनं केवलं पित्तनाशनम्।।

Mṛdudarbhaḥ kuśo barhiḥ śucircāraḥ suvṛttakaḥ,
Kharo anyaḥ pṛthulaḥ sīrrā gundraḥ ca nīrajaḥ smṛtaḥ.
(Dhanvantan)

Actions/Uses : Tridoṣaśāmaka, Stambhana, Śītala, Mūtrala, Tṛsnāghna, Raktaskandana, Vraṇaropaṇa.

Therapeutics :

Roots of Darbha is cooling, sweet, astringent, diuretic and galactagogue. It rehabilitates morbid vaata, pitta and kapha, useful in dysentery, diarrhoea, thirst, urinary calculi, dysuria and other diseases of bladder, raktapitta and skin diseases. It is litholytic and is usually an ingredient in preparations for dysentery and menorrrhagia. Culms are diuretic, stimulant, used in dysentery and menorrhagia.

Hordeum vulgare Linn.

SYNONYMS : *Hordeum sativum* Pers.

ENGLISH : Barley.

INDIAN : Jawasa, Jau, Jav.

SANSKṚTA : Yava, Java, Divya.

HABIT : An annual erect, stout tufted herb.

CHEMICAL COMPOSITION : Fat, starch, protein. Alkaloid hordenine. Luteolin glycoside, orientoside and orientin. Hordenine and its methyl ether have bronchodialtory properties.

Parts used : Fruit and seed.

Ayurvedic properties :
Rasa - Madhura. Vīrya - Śīta.
Guṇa - Gurū, Rūkṣa. Vipāka - Kaṭu.

रुक्षः शीतो गुरुः स्वादुः कषायो मधुरो यवः।
वृष्यो ग्राहि कफघ्नश्च स्यात्पित्तश्वासकासनुत्।।

Rūkṣaḥ śīto guruḥ svāduḥ kaṣāyo madhuro yavaḥ,
Vṛṣyo grāhi kaphaghnsca syātpītta śvāsakāsanut.
(Nighanturatnakara)

Actions/Uses : Lekhana, Medhya.

Therapeutics :

Decorticated seeds are nutritive and demulcent, easy to digest, used in diatary of sick. Barley-water is a demulcent food beverage for children suffering from diarrhoea, catarrhal inflammation of bowels. Parched and powdered much employed in the form of gruelling cases of painful and atonic dyspepsia. Barley malt obtained from germinated seeds is mostly used for the manufacture of beer and whiskey.

Saccharum officinarum Linn.

HABIT : An erect perennial grass.

HABITAT : Cultivated in hotter parts of India.

Similar : *Saccharum munja* Roxb.; *Saccharum cilliare* Anders.

ENGLISH : Sugarcane.

INDIAN : Oosa, Pundia.

SANSKṚTA : Mūnja, Sara, Ikṣhu, Kāṇḍekṣu, Sitapuṣpaka.

CHEMICAL COMPOSITION : Sugar and calcium oxalate. 5-O-methyapigenin etc. from flowers.

Parts used : Stem.

Ayurvedic properties :
Rasa - Madhura, Tikta.   Vīrya - Śīta.
Guṇa - Laghu, Snigdha.   Vipāka - Madhura.

मुञ्जद्वयं तु मधुरं तुवरं शिशिरं तथा।
दाहतृष्णा विसर्पास्त्र मूत्रकृच्छ्र क्षिरोगजित्।
दोषत्रयहरं वृष्यं मेखलासूपयुज्यते।।
(भावप्रकाश)

Muñjadvayam tu tuvaram śiśiram tathā,
Dāhaṭṛṣnāvisarpāsramutrakṛcchkṣirogajit,
Doṣatrayaharam vṛṣyam mekhalāsūpayujyate.
(Bhavaprakasa)

Actions/Uses : Tridoṣaśāmaka, Tṛṣnaghna, Raktaśodhana, Mūtrala, Tiktapittahara, Stanyajanana, Vṛṣya.

Therapeutics :
Stems of Sara are sweet, laxative, diuretic, cooling and aphrodisiac. Root, demulcent, cooling and diuretic.

Sorhgum bicolor (Linn.) Moench.

SYNONYMS : *Andropogon sorghum* Brot.; *Sorghum vulgare (Linn.)* Pers.

HABIT : A stout annual grass with culms.

HABITAT : Widely cultivated in India.

ENGLISH : Sorgho.

INDIAN : Jondhallaa, Jowaari.

SANSKṚTA : Yāvanāla.

CHEMICAL COMPOSITION : Hexacosanol, β-sitosterol and glucose from roots. Dhurrin and leuteforol isolated. Cyanogenetic glucoside in aerial shoots. Leaves contain HCN.

Parts used : Seeds.

Ayurvedic properties :
Rasa - Madhura. Vīrya - Śīta.
Guṇa - Gurū, Rūkṣa. Vipāka - Madhura.

यारनालो गुरु शीतो रुक्षो ग्राही रुचिप्रदः।
वृष्यो मलस्तम्भकरः स्वादुः पित्तकफापहः।।
रक्तरोगप्रशमनो मुनिभिः पूर्वमीरितः।

Yāvanālo guru śīto rūkṣo grāhī rucipradaḥ,
Vṛṣyo malastambhakaraḥ svāduḥ pittakaphāpaḥ.
(Nighanturatnakara)

Actions/Uses : Kaphahara, Pittahara, Grāhī, Malastambhakara, Vīryavardhaka.

Therapeutics :
Seeds are diuretic, demulcent and aphrodisiac.

Vetiveria zizanioides (Linn.) Nash.

SYNONYMS : *Andropogon muricatus* Retz.; *Andropogon squarrosus* Hook. *f.*; *Phalaris zizanoides* Linn.

HABIT : A densely tufted grass.

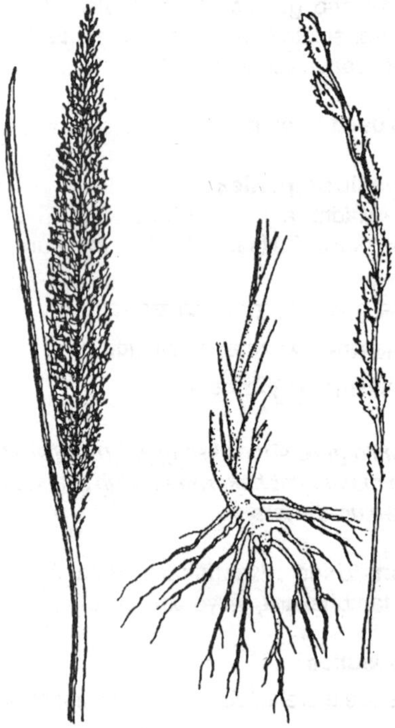

HABITAT : Practically over the whole of India.

ENGLISH : Cuscus grass, Vetiver, Vitivayr, Khas-khas.

INDIAN : Waalla, Khasa.

SANSKṚTA : Uśīra, Nalada, Sevya, Amruṇāla, Samagandhaka, Ambu, Jalavāsa, Varitara, Sugandhika, Gandhadhya, Vīratara, Bahūmulaka, Vāla, Hrīvera, Brahiṣṭa, Udecya, Keśa, Lalanapriya, Kuntalosira, Reśira, Sugandhimūla.

CHEMICAL COMPOSITION : Essential oil containing β-vetivone. Isobisabolene, khusol and azulene from oil.

Parts used : Roots

Ayurvedic properties :
Rasa - Tikta-Madhura. Vīrya - Śīta.
Guṇa - Laghu, Rūkṣa. Vipāka - Kaṭu.

उशीरं पाचनं शीतं स्तंभनं लघु तिक्तकम्।
मधुरं ज्वरहृद्वान्तिमदजित् कफपित्तनुत्।
तृष्णा सविषवीसर्पदाह कृच्छ्रव्रणापहम्।।
(भावप्रकाश)

Uśīram pācanam śītam stambhanam laghu tiktakam,
Madhuram jvrahṛd vāntimadajit kaphapittanut,
Tṛṣṇā saviṣavisarpadāhakṛcchvraṇāpaham.
(Bhavaprakasa)

Actions/Uses : Pittaghna, Kaphaghna, Vātakar, Śītala, Varṇya, Vedanāsthāpana, Dāhaśāmaka, Dīpana, Pācana, Tṛsṇāghna.

Therapeutics :
Usira roots are tonic, diuretic, refrigerant, stomachic, stimulant, antispasmodic, diaphoretic and emmenagogue. Infusion of roots is refrigerent, febrifuge, diaphoretic, stimulant and emmenagogue; pulvereized and made into paste in water is used as a cooling external application in fevers; their essence used as tonic.

Family : LXXX-Polygonaceae

Rheum officinale Baillon.

SYNONYMS : *Rheum rhaponticum* Willd.; *Rheum palmatum* Linn. *Rheum emodi Wall. ex Meissn.*

ENGLISH : Rhubarb.

INDIAN : Revand-chini.

SANSKṚTA : Gandhini, Pita, Banglacini.

HABIT : A stout herb with rhizomes.

HABIT : Central and Western China. Found in Himalayas from Kashmir to Sikkim. Cultivated in Assam.

CHEMICAL COMPOSITION : Anthracene derivatives 2-3 %. Aloe emodin (Physion), Rutin, rhein, emodin anthrone and rhein antrnone and five glucosides isolated from rhizome. Tannins, Glucogallin and tetrarin. Resins and fatty oil.

Parts used : Rhizime and roots.

Ayurvedic properties :
Rasa - Kaṭu, Tikta. Vīrya - Uṣṇa.
Guṇa - Tīkṣṇa. Vipāka - Kaṭu.

गन्धिनी पीतमूली च वल्या सा मृदुरेचनी।
हन्त्यजीर्णमतिसारं वन्हिमान्धमरोचकम्।।
विट्संग शीतपित्तश्च दुष्टव्रणविरोहिणी।। (द्रव्यगुण)

Gandhinī pītamulī ca balyā sā mṛdurecanī,
Hantyajīrṇamatisāram vanhimāndyamrocakam.
Viṭsangam śītapittañca duṣṭavraṇavirohiṇī.
(Dravyaguna)

Actions/Uses : Balya, Hṛdya, Bhedanīya, Dīpanīya, Anulomika, Yakṛtottejaka, Vātaśleṣmāhara, Śvāsahara.

Therapeutics :
Gandhini is a tonic and mild laxative; useful in the treatment of dyspepsia, diarrhoea, anorexia, constipation, urticaria and septic wounds. Roots of Gandhini are bitter, astringent, purgative, stomachic, aperient, especially useful in infantile stomach troubles. In small doses it acts as a stomachic and slightly astringent and in large doses as a purgative. Their stimulating effect combined with aperient properties find useful application of roots in atonic dyspepsia. Under the use of it the secretions, particularly urine becomes coloured and cutaneous secretion, especially of the arm-pits and the milk also becomes coloured. It is a purgative of much value in diarrhoea depending upon the presence of irritaing matter in the alimentary canal, by first causing its evacuation and afterwards acting as an astingent. Externally powdered root is used with good effect to indolent ulcers and sloughing ulcers.

HOMEOPATHIC USE : Rheum is a homeopathic remedy, frequently indicated in children. Bad smell is a great characteristic. It may be prescribed without hesitation in sour stool of children. Screaming, weeping of children; restlessness with weeping. Best remedy in teething infants, especially for difficult teething with diarrhoea. It affects the eyes; digestive system very profoundly; and the females. Milk of nursing women, yellow and bitter; infant refuses breast. After sleeping eyes are sealed up with gum. Foetid breath, foul taste. Child smells sourish, even if washed or bathed every day.

Rumex crispus Linn.

SIMILAR : *Rumex dentatus* Linn.

ENGLISH : Curled Dock, Yellow Dock.

INDIAN : Amla-betasa.

SANSKṚTA : Cukram, Cukrikā, Patrāmla, Rocanī, Śatavedhanī.

HABIT : A perennial, erect, glabrous herb.

HABITAT : Europe and North Asia.

CHEMICAL COMPOSITION : Chrysophanol, chrysophanic acid, emodin and β-sitosterol and essential oil.

Ayurvedic properties :

Rasa - Amla	Vīrya - Ūṣṇa
Guṇa - Laghu, Rūkṣa	Vipāka - Amla

Parts used : Whole plant, leaves.

चुक्रम् स्यादम्लपत्रन्तु लघूष्णं वातगुल्मनुत्।
रुचिकृद् दीपनं पथ्यमीषत् पित्तकरं परम्।। (राजनिघण्टु)

Cukram syādamlapatrantu laghūṣṇam vātagulmanut,
Rucikṛd dīpanam pathyamīṣat pittakaram param.
(Rajanighantu)

Actions/Uses : Vātaśāmaka, Kaphapittaprakopī, Rucikara, Dīpana, Bhedana, Yakṛtottejaka, Snehana, Grāhī, Hṛdya.

Therapeutics :

Cukram root is mildly laxative and astringent. It causes irrritation of the skin and mucous membrane, also of the digestive system.

HOMEOPATHIC USE : Rumex is a Homeopathic remedy, pre-eminently a cough remdy. It is one of the most quick-acting remedies. Main symptoms are, violent, incessant, and fatiguing cough with little expectoration; attacks of barking hoarse cough at 11 P.M. and 2 and 5 A.M. Bowels get so much irritated in early morning hours that they produce sudden urging to stool.

Rumex vesicarius Linn.

ENGLISH : Sorrel, Bladder dock.

INDIAN : Chaakawata, Aambatchukaa, Chuka, Ambari, Palak.

SANSKṚTA : Cukā, Cukra, Patrāmla, Hilamocikā, Śākarāja, Rājaśāka, Cakravartī, Prasādaka, Vīraśāka, Śākavīra, Kṣārapatra, Vāstuka, Śākarāt.

HABIT : An annual monoeocious, glabrous herb.

HABITAT : Widely cultivated as a vegetable in Tripura, West Bengal and Bihar.

CHEMICAL COMPOSITION : Leaves contain vitamins and amino acids. An anthraquinone glucoside, acid potassium oxalate and tartaric acid, cystine, glutamic acid, proline, phenylalanine and histidine are also detected.

Parts used : Whole plant, leaves.

Ayurvedic properties :
Rasa - Amla. Vīrya - Uṣṇa.
Guṇa - Laghu, Rūkṣa. Vipāka - Amla.

चुकोऽग्निद्दीपनश्चोष्णो रुचिकारी लघुः स्मृतः।
पित्तलः सारकः पथ्यो अत्यम्लः शूलनाशकः।।
गुल्माग्निमान्घहृत्पीडा बद्धविट् कत्वनाशनः।
आमवात तृषावान्तिकफ गुल्मापहो मतः।।
वातं च मुखवैरस्यं नाशयेदिति कीर्तितः।
(निघण्टु रत्नाकर)

Cuko agniddīpanaścoṣṇo rucikarī lghuḥ smrtaḥ,
Pittalaḥ sārākaḥ pathyo atyamlaḥ śūlanāsakaḥ.
Gulmāgnimāndya tṛpīdā baddhaviṭkaṭva nāśanaḥ,
Āmavāta ṭṛsā vānti kapha gulmāpaho mataḥ.
Vātam ca mukhavairasyam nāśayediti kīrtitaḥ.
(Nighanturatnakara)

Actions/Uses : Vātaśāmaka, Kaphapittakara, Dīpana, Śōthahara, Yakṛtottejaka, Rocana, Pācana, Bhedana, Dāhahara, Viṣaghna, Vedanāsthāpana.

Therapeutics :

Cuka is astringent and sedative like Gandhini and Sarsaparilla; given in disorders of lymphatic and glandular system. Useful in scurvy and chronic skin eruptions. Herb is very sour, laxative, stomachic; useful in asthma, bronchitis, hiccough, pain, piles, flatulence, dyspepsia, vomiting, constipation, alcoholism, tumours, diseases of spleen and in heart troubles. Fresh plant juice is refrigerent, diaphoretic, diuretic and antiscorbutic.
Leaves are used as aperient, diuretic and also in snake-bite. Leaf juice allays toothache. Roasted seeds are useful in dysentery and scorpion-bite. In Europe herb is used for urinary and renal troubles.

Family : LXXXI - Primulaceae

Anagallis arvensis Linn.

ENGLISH : Scarlet Pimpernel, Weather-glass, Red Chickweed.

INDIAN : Jonkmaaree.

SANSKṚTA : Jonkamarī.

HABIT : A small, much-branched, gland-dotted, annual herb.

HABITAT : Europe, Asiatic Russia and Great Britain. More or less throughout India as a weed.

CHEMICAL COMPOSITION : A pungent, acrid, volatile oil with a peculiar smell is present in the herb. The herb contains anagalligenone B, glucofrucoside and tannin. Roots contain saponins of triterpene series. Aerial parts yield alkaloids; n-hecacosane, rutin, β-sitosterol, stigmasterol, β-amyrin, lacceric acidanagalllgenin, glucose, arabinose and xylose. Flowers, stigmasterol, β-sitosterol, α-spinasterol 3-glucosides, kaempferol, quercetin and rutin.

Parts used : Entire herb.

Ayurvedic properties :
Rasa - Kaṭu, Tikta. Vīrya - Śīta.
Guṇa - Laghu, Rūkṣa. Vipāka - Kaṭu.

Actions/Uses : Anulomic, Śothaghna, Vedanāsthāpan, Viṣahara, Vraṇaropaṇa, Avasādaka.

Therapeutics :
Jonkamri is used in Europe as diuretic, diaphoretic, expectorant, in dropsy, rheumatism and in hepatic and renal complaints; while in India used in gout, cerebral affections, hydrophobia, leprosy, dropsy, epilepsy, mania, as a fish poison and in snake-bite.

HOMEOPATHIC USE : Anagallis is a Homeopathic remedy. It has marked action on skin, charecterized by great itching and tingling everywhere. An old medicine for hydrophobia and dropsy. Possesses power of softening flesh and destroying warts. Itching; dry, bran-like eruption especially on hands and fingers; palms especially affected. Vesicles in groups. Ulcers and swellings on joints are some of the main symptoms.

Family : LXXXII - Punicaceae

Punica granatum Linn.

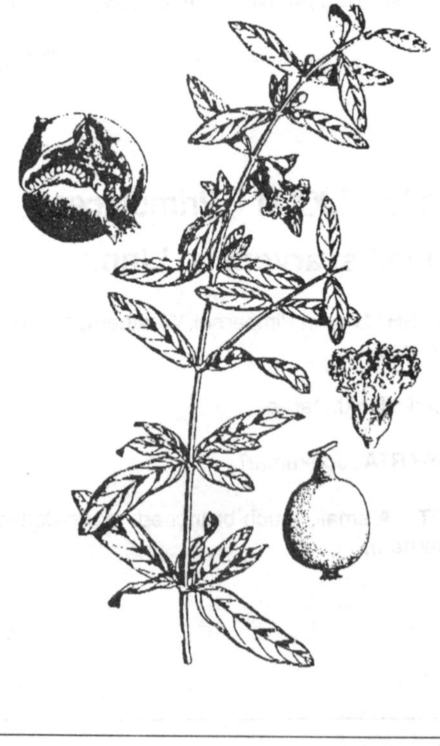

ENGLISH : Pomegranate.

INDIAN : Daallinba, Anaara.

SANSKṚTA : Dādima, Dantabīja, Raktapuṣpa, Lohitapuṣpaka, Karaka, Dādimīsāra, Kuttima, Phalaśādava, Raktabīja, Suphala, Madhubīja, Kucaphala, Rocana, Śukavallabha, Maṇibīja, Valkaphala, Sunīla, Vṛttaphala, Raktakusuma, Madurāmla, Śukapriya, Daśanabīja, Nīlapatra, Parvarut, Piṇḍapuṣpa, Piṇḍīra.

HABIT : A large deciduous shrub or small tree.

HABITAT : Cultivated in many parts of India, wild in the north-west regions.

CHEMICAL COMPOSITION : Malvidin pentose glucosides, tannin, ursolic acid in various parts, Bark yields alkaloids pelletierine etc. Root bark contains four aikaloids, pseudo-pelletierine, pelletierine, isopelletierine and methylpelletierine. Sitosterol; two tannins-punicalagin and punicalin from peals. Pectin from fruits contained mannose, galactose, rhamnose, arabinose and glucose. Principal sugar acid was galacturonic acid.

Parts used : Flowers, fruits, dried fruit-bark, leaves, root-bark.

Ayurvedic properties :
Rasa - Madhura, Kaṣāya, Amla.
Vīrya - Ūṣṇa, Anūṣṇa.
Guṇa - Laghu, Snigdha. Vipāka - Madhura.

तत्तुस्वादु त्रिदोषघ्नं तृड्दाहज्वरनाशनम्।
हृत्कण्ठमुखदोषघ्नं तर्पणं शुक्रलं लघु।।
कषायानुरसं ग्राहि स्निग्धं मेधाबलप्रदम्।
स्वाद्वम्लं दीपनं रुच्यं किञ्चित्पित्तकरं लघु।।
अम्लं तु पित्तजनकं आमवात कफापहम्।।
(भावप्रकाश)

Tattu svādu tridoṣaghnam tṛḍ dāhajvranāśanam,
Hṛtkaṇṭhamukhadoṣaghnam tarpaṇam śukralam laghu.
Kaṣāyānurasam grāhi snigdham medhābalapradam,
Svādvamlam dipānam rucyam kīncitpittakaram laghu.
Amlam tu pittajanakam āmavātakaphāpaham.
(Bhavaprakasa)

दाड़िमं मधुरमम्लकषायं कासवात कफपित्तविनाशी।
ग्राहि दीपनकरञ्च लघुष्णं शीतलं श्रमहरं रुचिदायी।।
(राजनिघण्टु)

Dāḍimam madhurmanlakaṣāyam,
Kāsa vātakaphapittavināsī,
Grāhi dīpanakarānca laghuṣṇam,
Śītalam śramaharam rucudayī.
(Rajanighantu)

Actions/Uses : Pittaghna, Kaphavātaghna, Vātaśāmaka, Tṛsṇāhara, Tarpaṇa, Śūkrala, Balakara, Medhya, Grāhī, Jantughna, Śothahara, Hṛdya, Vraṇaropaṇa-Śodhana, Dāhaghna, Jvaraghna, Mukhadoṣahara.

Therapeutics :

Root and stem bark of Dāḍima are astringent, anthelmintic, specific in tapeworm. Grains and its alkaloids are astringent, anthelmintic and taenifuge. Flowers, bark of the tree, and rind of fruit are astringent and stomachic. Fresh juice is cooling and refrigerant. Rind of fruit combined with aromatics like cloves etc. useful in diarrhoea and dysentery. Seeds are stomachic; pulp is cardiac and stomachic.

Family : LXXXIII-Ranunculaceae

Aconitum heterophyllum Wall (ex Royle).

INDIAN : Ativiasha, Atis.

SANSKṚTA : Viṣā-1, Ativiṣā, Śuklakandā, Bhrangurā, Amrtā, Ghuṇavallabhā, Kāśmīrā, Śiśubhaiṣajyā, Ghuṇapriyā.

HABIT : A tall perrennial erect showy herb with tubers.

HABITAT : Common in sub-alpine and alpine zones of western Himalayas.

CHEMICAL COMPOSITION : Diterpenoid alsoloid, atisine is the main constituent(0,4 %). Others include atidine, histsine, isoatisine, atisinine, isoasitisine, dihydroatisine, heteratisine, hetidine, hetisine, hetidine, hetisinone heterophyllisine, heterphylline, heterophyllidine.

Parts used : Dried tuberous roots

Ayurvedic properties :
Rasa - Tikta, Kaṭu. Vīrya - Ūṣṇa.
Guṇa - Laghu, Rūkṣa. Vipāka - Kaṭu.

विषा सोष्णा कटुस्तिक्ता दीपनी पाचनी हरेत्।
कफपित्तातिसाराम विषकासचमिकृमीन्।।
(भावप्रकाश)

Viṣā sokṣṇā kaṭsṭiktā dīpanī pācanī haret,
Kaphapittātisārām viṣakāsa vami kṛmīn.
(Bhavaprakasa)

त्रिप्रकारं चातिविषं किञ्चिदुष्णं च तिक्तकम्।
अग्निदीप्तिकरं ग्राहि त्रिदोषां च पाचकम्।।
रक्तपित्तज्वरामाति सारकासविषापहम्।
यकृद्वान्तितृषां चैव कृमिनर्शांश्च पीनसम्।।
पित्तोदरं चातिसारं हर्षव्याधिहरं मतम्।
(नि. र.)

Triprakāram cātiviṣam kiñciduṣṇam ca tiktak1am,
Agnidīptikaram grāhi tridoṣānam ca pācakam.
Raktapittajvarāmātisārakāsaviṣāpaham,
Yakṛdvāntitṛṣām caiva kṛminārsāmśca pīnasam.
Pittodaram cātisāram harṣavyādgiharam matam.
(Nighanturatnakara)

Actions/Uses : Pittahara, Kaphaghna, Vātaghna, Cchardinigrahaṇa, Dīpana-Pācana, Grāhī, Arśoghna, Kṛmighna, Śophaghna, Kāsaghna, Viṣahara.

Therapeutics :

Viṣā is beneficial in deranged kapha and pitta. It is expectorant, antipyretic, antidysenteric, antidiarrhoeal and antiemetic. Tubers and root are bitter, tonic, stomachic, astringent, antiperiodic, aphrodisiac, useful in diarrhoea and dysentery, acute inflammation, dyspepsia, cough, hysteria, loss of memory, piles and throat diseases. Root powder is useful in splenic fever and gastric troubles of children suffering from cough, coryza and vomiting. Root is considered to be aphrodisia, digestive, valuable febrifuge and an antiferility agent.

Aconitum napellus Linn.

ENGLISH : Aconite, Wolfsbane, Monkshood.

INDIAN : Dudhiyaa Bachanaaga.

SANSKṚTA : Viṣā-2, Vatsanābha.

HABIT : A herb.

HABITAT : Indigenous to Alps and Pyrenees; widely spread all over the world.

CHEMICAL COMPOSITION : Diterpenoid alkaloids. Aconitine is the principal alkaloid (0.4 %), others are Aconine, Picroacontine. Neopelline, Neolin, Napellin, Mesaconitine, Hypaconitine in traces.

Parts used : Green leaves and root.

Ayurvedic properties :

Rasa - Tikta, Kaṭu.	Vīrya - Ūṣṇa.
Guṇa - Laghu, Rūkṣa.	Vipāka - Kaṭu.

विषं प्राणहरं प्रोक्तं व्यवयि च विकाषि च ।
आग्नेयं वातकफहृत् योगवाहि मदावहम्।।
तदेव युक्तियुक्तं तु प्राणदायि रसायनम्।
योगवाहि त्रिदोषघ्नं बृंहणं वीर्यवर्द्धनम्।।
(भावप्रकाश)

Viṣam prāṇaharam proktam vyavayi ca vikaṣi ca,
Āgneyam vātakaphahṛt yogavāhi madāvaham.
Tadeva yuktiyuktam tu prāṇadāyi rasāyanam,
Yogavāhi tridoṣaghnam bṛumhaṇam vīryavarddhanam.
(Bhavaprakasa)

Actions/Uses : Pittahara, Kaphaghna, Vātaghna, Cchardinigrahaṇa, Dīpana-Pācana, Arśoghna, Kṛmighna, Śophaghna, Kāsaghna, Grāhī, Viṣahara.

Therapeutics :

Visa is extremely toxic in large doses. In small doses it acts as a tonic, remedy for fever, nervous debility, rheumatism and cardiac diseases. It is sedative, anodyne and febrifuge. Tincture aconite is used in neuralgia and in rheumatism. Given internally in a very small dose it lowers temperature in fever, relieves pain, cough irritation and useful in cystitis and diarrhoea. It is useful in all febrile and inflammatory conditions such as acute catarrh, tonsilitis and croup, scarlatina, gastritis and facial neuralgia. In palpitation of the heart spasm, it has been used with success. Alkaloid is a cardiac irritant.

HOMEOPATHIC USE : Aconite is a Homeopathic remedy. It is a short acting remedy and there is hardly any acute disease in which it is not more or less called for. The characteristic notes of the remedy are suddenness, intesnity, anxiety, restlessness, fear causation, fright, intense dry cold heat, dry, hot skin. Symptoms come like a storm. It is a great febrifuge. Frequently indicated in stiff-neck, acute lumbago and in primary stage of rheumatic ophthalmia.

Clematis triloba Heyne ex Roth.

INDIAN : Morawela, Ranjani, Murhari.

SANSKRTA : Moraṭā, Mūrvā, Madhurasā, Devī, Tejanī, Tiktā, Sruvā, Murangīkā, Madhulikā, Madhuśreṇī, Pīluparṇī, Divyalatā, Triparṇī, Tejasvinī, Bhinnadalī, Madhumatī, Pṛthakparṇī, Gokarṇī, Jvalini, Lalukarṇikā, Dahanī, Devaśreṇī, Gopavalli, Svādurasā, Vṛsatvacā, Abhirasā, Vṛṣaśreṇī, Pṛthu, Laghuparṇī.

HABIT : An extensive climber.

HABITAT : Konkan and Western Ghats.

CHEMICAL COMPOSITION : Anemonin. Alcoholic extract of plant is hypotensive and spamolytic.

Parts used : Roots.

Ayurvedic properties :
Rasa - Madhura. Vīrya - Ūṣṇa.
Guṇa - Sara, Gurū, Kāṣaya. Vipāka -Kaṭu.

मोरटा तुवरा तिक्ता स्वाद्विचोष्णा गुरुः स्मृता।
पाककाले तु कटुका सारकाच त्रिदोषहा।।
रक्तदोष मेदोरोगं कुष्ठं मेहं ज्वरं तथा।
वान्तिंच मुखशोषं च भ्रमं कण्डूं तृषां तथा।।
हृद्रोगं च कफं पित्तं वातंच विषमज्वरम्।
नाशयेदिती तैरुक्तं कन्दोऽस्याः कृमिनाशकः।।
कृमिकलिकरोगं च विषदोषं च नाशयेत्।
(निघण्टु रत्नाकर)

Moraṭā tuvarā tiktā svādvi coṣṇā guruḥ smṛtā,
Pākakāle tu kaṭukā sārakā ca tridoṣahā.
Raktadoṣa medorogam kuṣṭham meham jvaram tathā,
Vantiñ ca mukhaśoṣañ ca bhramam kaṇḍūm tṛṣām tathā.
Hṛdrogañ ca kapham pittam vātañ ca viṣamajvaram,
Nāśayeditīm tairuktam kando asyāḥ kṛmināśakaḥ.
(Nighanturatnakara)

Actions/Uses : Tridoṣaśāmaka, Kaphavātanāśaka, Raktapittahara, Mehaghna, Tṛuṣṇāhara, Hṛdrogahara, Kaṇḍūhara.

Therapeutics :

Morata plant is applied to boils and itch, used in leprosy, blood diseases and fevers. Used in snake-bite. Leaves are alterative, acrid and sedative. Infusion of leaves is employed in blood-diseases such as syphilis, scrofula, leprosy and in chronic fevers. Leaf juice mixed with that of Holarrhena antidysenterica is a remedy for conjunctivitis

Helleborus niger Linn.

ENGLISH : Hellebore-black, Christmas Rose.

INDIAN : Katukee, Kaddu.

SANSKṚTA : Katukī, Rohiṇī, Matsyapittā, Matsyabhedī, Bahulā, Sakulā.

HABIT : A herb with rhizomes.

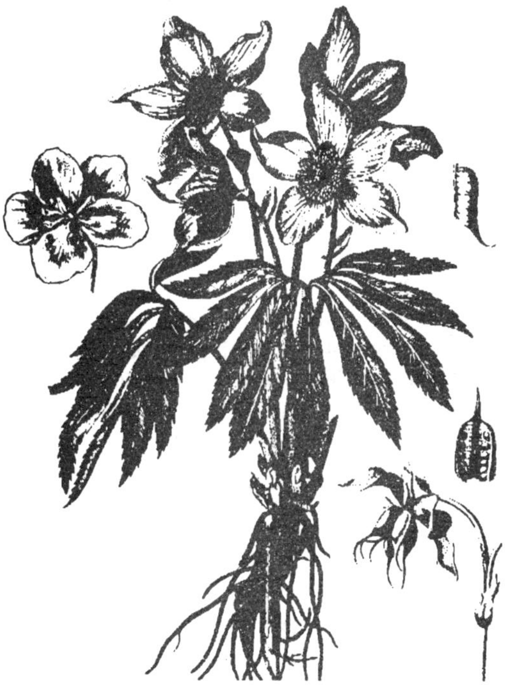

HABITAT : A native of Cental and Southern Europe occuring in Himalayas from Kashmir to Sikkim.

CHEMICAL COMPOSITION : Helleborin, hellebrin and helleborein. Dried rhizome also contains a saponin and a volatile oil. Entire plant and especially rhizome contains highly active glycosides that regulate blood flow in arteries and have a diuretic action.

Parts used : Rhizome, roots and whole plant.

Ayurvedic properties :
Rasa - Tikta. Vīrya - Śīta.
Guṇa - Laghu, Rūkṣa. Vipāka - Kaṭu.

कटुका कटुका पाके तिक्ता रुक्षा हिमा लघुः।।
भेदनी दीपनी हृद्या कफपित्तज्वरापहा।
प्रमेह श्वासकासास्त्रदाहकुष्ठकृमिप्रणुत्।।
(भावप्रकाश)

Kaṭukā kaṭukā pāke tiktā rūkṣā himā lahguḥ.
Bhedanī dīpanī hṛddyā kaphapittajvarāpahā,
Prameha śvāsakāsāsra dāha kuṣṭha kṛmipraṇut.
(Bhavaprakasa)

कटुका पित्तजित्तिक्ता कटुः शीतास्त्र दाहजित्।
बलासारोचकान् हन्ति विषमज्वर नाशिनी।।
(ध. नि.)

Kaṭukā pittajittiktā kaṭuḥ śītāsra dāhajit,
Balāsārocakān hanti viṣamajvara nāśinī.
(Dhanvantari nighantu)

Actions/Uses : Kaphapittaśodhana. Bhedanīya, Lekhanīya.

Therapeutics :
Plant is hydrogogue, cathartic, emmenagogue and anthelmintic; in large doses acro-narcotic poison. In small doses it acts like digitalis. It strengthens heart and helps to improve circulation. It acts on kidneys and increases urination. Due to its mild laxative action its use in fever proves beneficial. Rhizome is diuretic, cathartic, anthelmintic and emmenagogue; used as local anaesthetic, cardiac tonic like digitalis, in apoplexy and skin diseases. Has been used in dropsy, chlorosis and amenorrhoea. It is of value in nervous disorders, hysteria and melancholia.

HOMEOPATHIC USE : Helleborus nigre is a Homeopathic remedy. It is a remedy which is often called for in serious diseases like meningitis, typhoid fever and hydrocephalus in their late stages when the patient is in a low state of vitality with blunt sensibility. Stupefaction and automatic action are the two key words to depict the action of this great remedy with very limited range. It has proved useful in injuries of the brain.

Family : LXXXIV-Rhamnaceae

Ventilago madraspatana Gaertn.

SYNONYMS : *Ventilago maderaspatana* Wight.; *Ventilago denticulata* Willd; *Ventilago calyculata* Tulsane.; *Ventilago madraspatana var. calyculata (Tul.)* King.

HABIT : A large woody climber.

HABITAT : Maharashtra, Western Ghats and hotter parts of India.

ENGLISH : Red creeper.

INDIAN : Lokhanddee; Khandavel, Pitti, Kanvel.

SANSKṚIT : Raktavallī.

CHEMICAL COMPOSITION : Trihydroxy-methyl-anthranolmonomethyl ether; emodinmonomethyl ether in root bark. Lupeol, B-sitosterol and its glucosidee in fruits, leaves and stem. Friedelin in stem bark.

Parts used : Bark.

Ayurvedic properties :

Actions/Uses : Grāhī.

Therapeutics :

Root bark is carminative, stomachic, stimulant and tonic. In the form of powder it is prescribed in atonic dyspepsia, mild fever, and debility. Stem bark is powdered, mixed with gingelly oil and used as application for skin diseases and itch.

Ziziphus mauritiana Lamk.

SYNONYMS : *Zizyphus jujuba* Lam.; *Rhamnus jujuba* Linn.

ENGLISH : Jujube, Indian Cherry Plum.

INDIAN : Bora, Ber.

SANSKṚIT : Badara, Unnāva, Badarī, Karkadhū, Phenila, Suphala, Kola, Sauviraka, Gudaphala, Baliṣṭa, Phalaśaiśira, Vakrakantaka, Dṛḍhabīja, Vṛttaphala, Kaṇṭakī, Svaccha, Rāṣtravṛudhīkarī, Subīja, Gṛndhranakhī, Kośaphala, Kuvala, Svāduka.

HABIT : A small tree or large shrub usually armed.

HABITAT : Common in hotter parts of India, cultivated in gardens or found wild in waste places.

CHEMICAL COMPOSITION : Carbohydrates, fat, protein, amino acids, anthocyanins from fruit, seeds and leaves. Leucocyanidin from bark. Leucopelargonidin, betulinic and ceabothic acids from wood. Rutin from leaves. Mauritines A, B, C, D, E and F, frangufoline and amphbines B, D and F.

Parts used : Fruits and leaves.

Ayurvedic properties :
Rasa - Madhura, Amla, Kaṣāya. Vīrya - Śīta.
Guṇa - Gurū, Snigdha, Picchila. Vipāka - Madhura.

वदरं मधुरं कषायमम्लम् परिपक्वं मधुराम्लमुष्णमेतत्।
कफकृत् पचनातिसाररक्तश्रमशोषार्ति विनाशनञ्च रुच्यम्।।
(राजनिघण्टु)

Badaram madhuram kaṣāyamamlam,
Paripakvam madhurāmlamuṣṇametat,
Kaphakṛt pacanātisāra rakta-
Śramaśoṣārti vināśañca rucyam.
(Rajanighantu)

Actions/Uses : Vātapittaghna, Dīpana, Anulomana, Dāhaśāmaka, Pācana, Tṛṣṇānigrahaṇa, Plīhāvṛdhihara, Vraṇaropaṇa.

Therapeutics :

Badara root is bitter and cooling; cures kapha; biliousness and headache. Decoction of roots is used in fever and as powder applied to old wounds and ulcers. Bark, a remedy in diarrhoea; cures boils. Leaves are bitter and cooling; cure kapha; biliousness and diarrhoea. They are antipyretic; reduce obesity. They form a plaster in strangury. Fruit is mucilaginous, pectoral, styptic, considered to purify blood and aid digestion. They are cooling, aphrodisiac, anodyne, tonic, laxative, and invigorating. The preparation JOSHANDA made from them is used in chest complaints. Kernels are sedative and are recommended as a soporific and are prescribed to stop nausea and vomiting and for relief from abdominal pain in pregnancy.

Family : LXXXV - Rosaceae

Cydonia oblonga Mill.

SYNONYMS : *Cydonia vulgaris* Pers.; *Pyrus cydonia* Linn.

ENGLISH : Quince.

INDIAN : Amrutaphalla; Beehi.

SANSKṚTA : Pātalā, Amṛutaphala, Simbitikā.

HABIT : A shrub or a small tree.

HABITAT : Native of Persia, grown in England and cultivated in Punjab, Kashmir.

CHEMICAL COMPOSITION : Fruit skin flavonols, hyperin, quercetin and isoquercetin; terpemoid lactones.

Parts used : Fruits, seeds and mucilage.

Ayurvedic properties :
Rasa - Madhura. Vīrya - Śīta.
Guṇa - Gurū, Snigdha. Vipāka - Madhura.

पाटला पिच्छिला प्रोक्ता सा स्निग्धा कासवारिणी।
शिश्नयोन्योहरेद्दाहं व्रणदाहनिवारणी।। (द्रव्यगुण)

Pāṭalā picchilā proktā sā snigdgā kāsavāriṇī,
Śiśnayonyo hareddāham vraṇadāhanivāraṇī.
(Dravyaguna)

Actions/Uses : Kaphaghna, Grāhī, Snehana, Mūtrajanana, Pauṣṭika. Kāsahara, Vraṇadāhahara.

Therapeutics :

Leaves, buds and bark of Patala are astringent. Fruits are sub-acid or sour in taste. They are astringent, expectorant and cardiac tonic. Seeds are mucilaginous, demulcent and used against fever, in dysentery, diarrhoea, sore throat; externally also, in eye dieases, as a soothing lotion and to inflammed skin as soothing and protective, useful in soreness of mucous membrane.

Prunus amygdalus Batsch.

SYNONYMS : *Prunus communis* Arcang.; *Prunus amygdalus* Baill.; *Amygdalus communis* Linn.; *Prunus dulcis (Mill.) D.A.* Webb.

HABIT : A middle sized tree.

HABITAT : Native of Asia Minor and Persia; cultivated in cooler parts of Punjab and Kashmir.

ENGLISH : Amygdala amara, Bitter Almond.

INDIAN : Badaama.

SANSKṚTA : Vātāda, Badāma, Vātavairī, Netropamaphala, Vātāma, Anilāntaka, Yugmaśuktiputopana, Suphala.

CHEMICAL COMPOSITION : Chief protein is a globulin, amandin . A fatty oil known as Sweet Almond Oil. Prunasin, decosterin and sitosterol from bitter seeds. HCN glucosides.

Parts used : Seed and oil.

Ayurvedic properties :
Rasa - Madhura. Vīrya - Ūṣṇa.
Guṇa - Gurū, Snigdha. Vipāka - Madhura.

वाताद्मज्ञा मधुरो वृष्यः पित्तानिलापहः।
स्निग्धोष्णः कफकृन्नेष्टो रक्तपित्तविकारिणाम्।।
(भावप्रकाश)

Vātādamajjā madhuro vṛṣyaḥ pittānilāpahaḥ,
Snigdhoṣṇaḥ kaphakṛnneṣṭo raktapittavikāriṇām.
(Bhavaprakasa)

Actions/Uses : Vātapittaśāmaka, Kaphapittavardhaka, Kṛmighna, Tvagrogahara, Kaṇḍūghna.

Therapeutics :

Vatada, Badama or Sweet almonds are emollient

and nutritive. Principal use of sweet almonds is for making demulcent and emollient mixtures for pulmonary affections. Oil is emollient when applied locally and nutritive when taken in small quantity and laxative in large doses. Seeds are highly nutritious, demulcent, stimulant and nervine tonic. They are lithontriptic and diuretic and their polutice is useful for irritable sores and skin eruptions. Seeds are valuable in diets for peptic ulcers. Unripe fruit is given as an astringent application to gums and mouth. Bitter almonds produce analogous effects to those of hydrocyanic acid.

HOMEOPATHIC USE : Amygdalae amara aqua is a Homeopathic remedy useful in asthma, coma, epilepsy, urticaria and headache. Symptoms are not very distinguishing from those of Hydrocyanic acid.

Prunus armeniaca Linn.

ENGLISH : Common apricot.

INDIAN : Jardaalloo, Khubani, Chuari, Kushmiaru.

SANSKṚTA : Urumāṇa, Pītālu, Āluka.

HABIT : A moderate-sized tree.

HABITAT : Native of China, Asia Minor and Persia. Commonly cultivated in N. India.

CHEMICAL COMPOSITION : A good source of sugars and vitamin A, and contains thiamine and iron. Seeds contain fatty oil, ethereal oil, vitamins, enzyme and amygdalin. Leaves contain quercitin, cyanidin, kaempferol, caffeic acid, p-coumaric acid.

Parts used : Seed and oil.

Ayurvedic properties :
Rasa - Madhura. Vīrya - Ūsṇa.
Guṇa - Gurū, Snigdha. Vipāka - Madhura.

पित्तश्लेष्महराण्याहुः स्निग्धोष्णानि गुरुणिच।
बृहणान्यनिघ्नाति बल्यानि मधुराणीच।।
(निघण्टु आदर्श बापालाल वैद्य (पूर्वार्ध))

Pittaślesmaharāṇyāhuḥ snigdhoṣṇāni gurūṇi ca,
Bṛhaṇanyānighnāni balyañi madhurāṇi ca.
(Susrta)

गुरुष्णस्निग्ध मधुर सोरुमान बलप्रदः।

Gurūṣṇasnigdha madhura sorumāna balapradaḥ.

Actions/Uses : Tridoṣahara, Balya, Brunhana.

Therapeutics :

Fruit of Urumana is sweet, antidiarrhoeal, antipyretic and emetic; allays thirst; but is bad for old people. Dried fruit is nutritive, used as laxative and refrigerent in fevers. Seeds are nutritive, tonic and anthelmintic; used in diseases of liver, piles, earache and deafness.

Prunus persica Batsch.

SYNONYMS : *Amygdala persica* Linn.

ENGLISH : Peach, Plum, Nectarine.

INDIAN : Aaluka, Aru, Shaftalu.

SANSKRTA : Āruka.

HABIT : A large deciduous shrub or small tree with glabrous twigs.

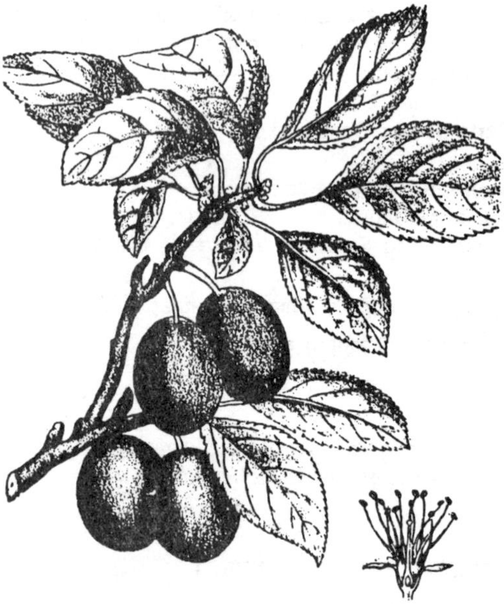

HABITAT : Native of China. Cultivated in Himalayas; grows in tropical and semitropical regions and in temperate climates usually under grass.

CHEMICAL COMPOSITION : HCN. Hentriacontane, hentrianontanol, β-sitosterol and ursolic acid from leaves. Multinoside A, kaempferol-3-rhamnoside, quercetin etc from flowers.

Parts used : Flowers, frruits, leaves and bark.

Ayurvedic properties :

Rasa - Madhura.	Vīrya - Ūṣṇa (alpa).
Guna - Gurū	Vipāka - Madhura.

नात्युष्णं गुरु संपक्वं स्वादुप्रायं मुखप्रियम्।
बृंहणं जीर्यति क्षिप्रं नातिदोषलमारुकम्।।
द्विविधं शीतमुष्णं च मधुरं चाम्लमेव च।।
गुरु पाके च तज्ज्ञेयं अरुच्यत्यग्निनाशनम्।।
रुच्यंमत्यग्निशमनमम्लं मधुरमारुकम्।
पक्वमाशु जरां याति नात्युष्णं गुरु दोषलम्।।
(निघण्टु रत्नाकर)

Nātyuṣṇam gurū sampakvam svāduprāyam mukhapriyam,
Bṛmhaṇam jīryati kṣipram nātidoṣalamārukam.
Dvividham śītamuṣṇam ca madhuram cāmlameva ca,
Gurū pāke ca tat jneyam arucyatyagnināśanam.
Rucyamtyāgniśamanamamlam madhuramārukakam.
Pakvamāśu jarām yāti nātyuṣṇam gurudoṣalam.
(Nighantuadarsa)

आरुकं तुवरं भेदि गुरुष्णं कफपित्तलम्।
पक्वंतु मधुरं नाति गुरुष्णं कफपित्तलम्।।
बृंहणं रोचनं शीघ्रं जरा. मारुतनाशनम्।
(कैयदेव निघण्टु)

Ārukam tuvaram bhedi gurūṣṇam kaphapittalam.
Pakvamtumadhuram nātigurūṣṇam kaphapittalam,
Bṛnhaṇam rocanam śīghram jarāmāruta nāśanam.
(Kaiyadevanighantu)

Actions/Uses : Brimhana, Mukhapriya, Hṛdya, Mehaghna, Arśoghna, Gulmahara, Raktadoṣahara.

Therapeutics :

Aruka is sedative, diuretic and expectorant. Most useful for irritation and congestion of the gastric surfaces. Also used in cough and chronic bronchitis. Fruit is stomachic, demulcent, antiscorbutic; useful as ascaricide.

HOMEOPATHIC USE : Amygdalus persica is a most valuable Homeopathic remedy for vomiting of various kinds, morning sicknes. Gastric irritaion of children; no form of food is tolerated ; constant nausea and vomiting are chief indications.

Rosa damascena Mill.

ENGLISH : Persian Rose, Damask rose.

INDIAN : Gulaaba, Bussorah, Falsi gulab.

SANSKRṚTA : Śatapatrī, Taruṇī, Karṇikā, Cārukeśarā, Rāmataraṇī, Bhṛungeṣṭā, Sahā, Kumārī, Gandhādhyā, Bhṛungavallabhā, Bahupatrā, Kṛpakā, Taraṇīvallī, Alikulapriyā, Kaṇṭakapravṛtā, Yātalā, Dhīrā, Kaṇṭakādhyā, Lāksā.

HABIT : An erect woody perennial shrub.

HABITAT : Cultivated all over India.

CHEMICAL COMPOSITION : Essential oil. Quercetin, kaempferol and cyanidin from whole plant; cyanidin3,5-diglucoside from petals. Detection of citronell, nerol, geraniol and phenylethanol in essential oil. Flowers contain a bitter principle, tanning matter, fatty oil and organic acids. Hips contain the pigments lycopene, β- and y_carotenes, rubixanthin, zeaxanthin, xanthophyll and taraxanthin.

Parts used : Flowers.

Ayurvedic properties :
Rasa - Tikta, Kaṣāya, Madhura. Vīrya - Śīta.
Guṇa - Laghu, Snigdha. Vipāka - Madhura.

शतपत्री हिमा तिक्ता कषाया कुष्ठनाशनी।
मुखस्फोटहरा रुच्या सुरभिः पित्तदाहनुत्।।
(राजनिघण्टु)

Śatapatrī himā tiktā kaṣāyā kuṣṭhanāsanī,
Mukhasphoṭaharā rucyā surabhiḥ pittadāhanut.
(Rajanighantu)

Actions/Uses : Vātapittaśāmaka, Hṛdya, Dīpana, Pācana, Varṇya, Vātānulomana, Durgandhināśana, Śothahara, Grāhī, Mṛduvirecana, Śonitasthāpana, Medhya, Vājikaraṇa, Dhātuvardhaka.

Therapeutics :

Petals of Satapatri applied externally as astringent. Made into a conserve with equal parts of sugar, known as Gulkand, used as tonic and fattening. Flower buds are astringent, considered aperient, cardiacal, tonic cephalic, removing bile and cold humours.

Family : LXXXVI - Rubiaceae

Gardenia gummifera Linn.

SYNONYMS : *Gardenia lucida* Roxb.

ENGLISH : Gambi resin.

INDIAN : Dikemaali.

SANSKṚTA : Nāḍihingu, Hinguśivātikā, Hingupatrī, Jantukā, Piṅgā, Rāmaṭhī, Vanśapatī, Piṇḍāhvā, Śūvīryā, Hingunādikā, Veṇupatrī, Vanśadalā, Nāḍīhinguphalā.

HABIT : A large handsome shrub or a small tree.

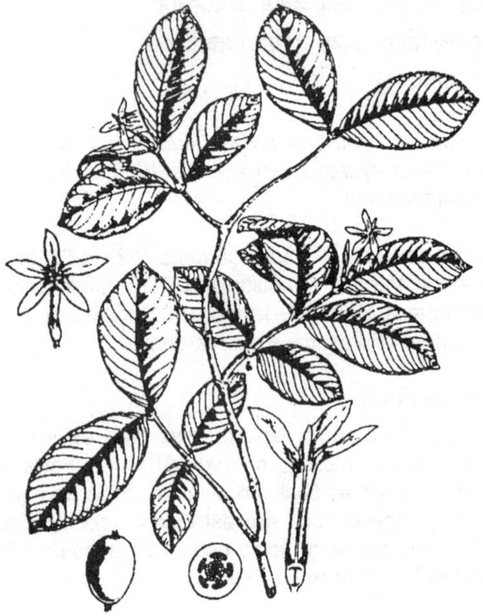

HABITAT : Various parts of India.

CHEMICAL COMPOSITION : Oleanonic aldehyde, sitosterol, D-mannitol, erythrodiol, and 5 flavons including gardenin, de-Me-tangeretin and nevadensin. Fruits, nonacosane, β-sitosterol and d-mannitol.

Parts used : Gum

Ayurvedic properties :
Rasa - Kaṭu, Tikta. Vīrya - Ūṣṇa.
Guṇa - Laghu, Rūkṣa, Tīksna. Vipāka - Katu.

नाडीहिङ्गु कटूष्णच कफवातार्तिशान्तिकृत्।
विष्ठाविबंन्धदोषघ्नमानाहामयहारि च॥
(राजनिघण्टु)

Nāḍihiṅgu katūṣṇam ca kaphavātārtiśāntikṛt,
Viṣṭhāvibandhadoṣaghnamānāhāmayahāri ca.
(Rajanighantu)

Actions/Uses : Kaphavātaśāmaka, Kaphanissāraka, Rucikara, Dīpana, Pācana, Anulomana, Kṛmighna, Śodhana, Jantughna, Ślesmāpūtihara, Vraṇaropaṇa.

Therapeutics :

Nadihingu Gum is antiperiodic, cathartic, alterative, anthelmintic, antispasmodic, carminative and stimulant in dyspepsia. Externally it is antiseptic and stimulant.

Randia dumetorum Lam.

SYNONYMS : *Xeromphis spinosa (Thunb.)* Keay.; *Randia brandisii* Gamble.; *Randia longispina W.A.; Randia tomentosa W.A.; Xeromphis spinosa Keay.; Randia dumetorum* Poir. *Randia spinosa.*

ENGLISH : Emetic nut.

INDIAN : Gel-phalla, Mainphal, Wagatta, Mindhal.

SANSKRTA : Madana, Madanaphala, Chardana, Piṇḍī, Śalyaka, Nata, Viṣpuṣpaka, Piṇḍītaka, Dhārāphala, Marūbaka, Rāṭhā, Ghanṭākhya, Viṣamuṣṭi, Kanṭāla, Ghanṭāla, Harṇya, Bastiśodhana, Viṣamuśkaka, Gālava, Śvasana, Tagara, Mātula.

HABIT : A deciduous thorny shrub or small tree.

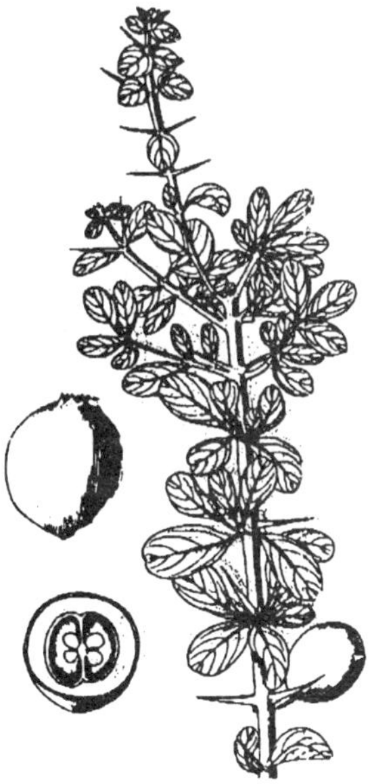

HABITAT : Sub-Himalayan tract, Konkan and Peninsular India.

CHEMICAL COMPOSITION : Six saponins-dumetoronins A,B,C,D,E and F- isolated, all of them contained oleanolic acid as glycone.

Parts used : Fruit.

Ayurvedic properrties :
Rasa - Madhura, Tikta, Kaṣāya, Kaṭu.
Vīrya - Ūṣṇa.
Guṇa - Laghu, Rūkṣa. Vipāka - Kaṭu.

मदनो मधुरस्तिक्तो वीर्येष्णो लेखनो लघुः।
वान्तिकृद्विद्रधिहरः प्रतिश्यायव्रणान्तकः।
राक्षः कुष्ठकफानाहशोथगुल्मव्रणापहः।।
(भावप्रकाश)

Madano madhurastikto viryeṣno lekhano laghuḥ,
Vāntikṛdvidradhiharaḥ pratiśyāyavraṇāntakaḥ,
Rukṣaḥ kuṣṭhakaphānāhaśothagulmavraṇāpahaḥ.
(Bhavaprakasa)

वमन द्रव्याणां मदनफलानि श्रेष्ठानि अनपायित्वात्।
(च. क.)

Vamana dravyāṇām madanaphalāni śreṣṭhāni anapāyitvāt.
(C.kal.1)

Actions/Uses : Kaphaghna, Vātaghna, Vedanāsthāpana, Vāmaka, Śothahara, Svedajanana, Lekhana, Kuṣṭhaghna, Ārtavajanana.

Therapeutics :

Aqueous extract of root bark of Madanaphala is actively insecticidal. Fruit is irritating emetic, purgative. carminative, aphrodisiac, useful in bronchitis, pain in muscles, paralysis; also used as fish poison. Pulp of fruit, dried and powdered is a valuable emetic. In small dose it is nauseant, expectorant and diaphoretic. The fruit is a nervine calmative and antispasmodic. It is a domestic remedy for ailments of children during teething. It is used in dysentery, and as anthelmintic, abortifacient; ground to coarse powder applied to the tongue and palate for fevers and incidental ailments of children during teething. Bark is astringent, given internally and also applied externally when bones ache during fever; externally applied as anodyne in rheumatism.

Rubia cordifolia Linn.

SYNONYMS : *Rubia munjista* Roxb.

ENGLISH : Indian Madder.

INDIAN : Manjishttha, Manjit, Manjestha, Manjith.

SANSKṚTA : Manjiṣṭhā, Vikasā, Yojanavallī, Raktā, Samangā, Rohiṇī, Tāmravallī, Vastraranjinī, Raktapuṣpī, Kala-meśhika.

HABIT : A very variable, prickly creeper or climber.

HABITAT : Throughout India in hilly districts, Konkan.

CHEMICAL COMPOSITION : Colouring matter in the roots is a mixture of purpurin and munjistin. Roots also contain small amounts of xanthopurpurin and pseudopurpurin. Munjistin, alizarin and its glucoside in plant.

Parts used : Roots.

Ayurvedic properties :
Rasa - Tikta, Kaṣāya, Madhura. Vīrya - Uṣṇa.
Guṇa - Gurū, Rūkṣa. Vipāka - Kaṭu.

मञ्जिष्ठा मधुरा तिक्ता कषाया स्वरवर्णकृत्।
गुरुरुष्णा विषश्लेष्मशोथयोन्यक्षिकर्णरुक्।।
रक्तातिसार कुष्ठास्त्रवीसर्पव्रणमेहनुत्।।
(भावप्रकाश)

Mañjiṣṭhā madhurā tiktā kaṣāya svaravarṇakṛt,
Gururūṣṇā viṣaśleṣmaśothayonyakṣikarnaruk,
Raktātisārakuṣṭhāsravīsarpavraṇamehanut.
(Bhavaprakasa)

Actions/Uses : Kaphaghna, Pittaghna, Vātaghna, Dīpana, Śothahara, Vraṇaśodhana-ropaṇa, Kuṣṭhaghna, Kṛmighna, Garbhāṣayottejaka, Pācana, Raktaprasādana.

Therapeutics :

Manjiṣṭhā is an efficient blood purifier and extensively used against blood, skin and urinary diseases. It is alterative, antiseptic, astringent, bitter and tonic.It is useful in diseases of blood, dysentery, ear and eye diseases, ulcers, inflammations and swellings, leprosy and urinogenital disorders. Dried root is emmenagogue, astringent, diuretic; much used in dropsy, paralysis, jaundice, amenorrhoea and visceral obstruction. Roots are tonic, alterative, astringent antidysenteric, deobstruent and antiseptic. They are used in rheumatism and form an ingredient of several Ayurvedic preparations. They are active against Staphylococcus aureus and are made into paste and applied into ulcers, inflammations and skin troubles. Stem is used in cobra-bite and scorpion sting.

Family : LXXXVII - Rutaceae

Aegle marmelos Corr.

ENGLISH : Bael fruit tree.

INDIAN : Bel.

SANSKṚTA : Bilva, Śalātu, Hṛdyagandha, Karkaṭa, Samirasāraka, Śivadruma, Triśikha, Śaivapatra, Subhuntika, Śailuśa, Śiveṣhṭa, Gandhapatra, Dūrāruha, Triśakhapatra, Lakṣmīphala, Gandhaphala, Sadāphala, Pūtimāruta, Sadāruha, Gandhagarbha, Śalya, Satyakarmā, Arimeda, Kanṭakādhyā, Sitānana, Mahākapithya, Nīlamalikā, Mālūra, Śrī, Pūtivāha, Goharītakī, Mangalya, Śāndilya, Śailūsa.

HABIT : A middle sized slender aromatic armed tree.

HABITAT : Wild in Çentral and Southern India, cultivated all over India.

CHEMICAL COMPOSITION : Umbelliferon, skimmianine, marmin, B-sitosterol, lupeol and y-sitosterol from immature bark and roots. Fruit contains psoralein and tannic acid; aegelinol, furocoumarins, furanocoumarin, marmelosin, marmelide. Ripe fruit, xanthotoxol, marmesin etc. The pulp contains mucilage, pectin, reducing sugars, tannin, a volatile voil, bitter principle. Fresh leaves yield a yellowish green oil. Extract of fruits lowered blood sugar level in normal rabbits but in diabetic rabbits reduction was insignificant. Essential oil is antifungal.

Parts used : Fruits both ripe and unripe, root-bark, leaves, rind of ripe fruit and leaves.

Ayurvedic properties :
Rasa - Kaṣāya, Tikta. Vīrya - Ūṣṇa.
Guṇa - Laghu, Rūkṣa. Vipāka - Kaṭu.

विल्वस्तु मधुरो हृद्यः कषायः पित्तजित् गुरुः।
कफज्वरातिसारघ्नो रुचिकृद्दीपनः परः।।
विल्वमूलं त्रिदोषघ्नं मधुरं लघु वातनुत्।
फलंन्तु कोमलं स्निग्धं गुरु संग्राहि दीपनम्।।
तदेव पक्कं विजेयं मधुरं सरसं गुरु।
कटु तित्तकषायोष्णं संग्राहि च त्रिदोषजित्।।
(राजनिघण्टु)

Bilvastu madhuro hṛdyaḥ kaṣāyaḥ pittajit guruḥ,
Kaphajvarātisāraghno rucikṛd dīpanaḥ paraḥ,
Bilwamūlam tridoṣaghnam madhuram laghu vātanut,
Phalam tu komalam snigdham guru sangrāhi dīpanam.
Tadeva pakvam vijneyam madhuram sarasam gurū,
Kaṭutiktakaṣāyoṣṇam sangrāhi ca tridoṣajit.
(Rajanighantu)

Actions/Uses : Kaphaghna, Vātaghna, Pittakar.

Leaves : Śothahara, Vedanāsthāpana, Svedakara.

Raw fruit : Dīpana, Pācana, Grahī, Kṛmighna, Atisāraghna.

Ripe fruit : Mṛdurecana, Arśoghna, Grahaṇīrogahara, Āmapacana.

Therapeutics :

Bilva is astringent, cooling, carminative, restorative, laxative, febrifuge and stomachic. It is used in colitis, colic, dysentery, diarrhoea, flatulence, difficult micturition, fever and vomiting. Root bark is used in intermittent fever and is useful in hypochondriasis, melancholia and palpitation of heart. Alcoholic extract of fruit or root is hypoglycemic and spasmogenic. Leaves are febrifuge. Leaf juice is applied externally in abscess. Unripe or half-ripe fruit is astringent, digestive, stomachic, used in diarrhoea. Pulp of ripe fruit is aromatic, cooling and laxative. Seed oil is antibacterial. Ash is used to kill worms and for injuries caused by animals. Various parts of plant are used in thirst, stomach pain, constipation, diarrhoea, dysentery, cholera, nausea, night fever, convulsions, cramps, nausea, puerperal fever, postnatal complications, breast pain, suppuration and snake-bite. Bark, leaves and fruits used for treatment of intestinal diseases.

HOMEOPATHIC USE : A short Homeopathic proving of a tincture from leaves was done by Biswas of Patna. It is found to destroy phlegm and useful in fever associated with catrrhal symptoms. In chronic gastro-intestinal catarrh where patients suffer from flatulent colicky pains it has a decided effect to check them. In seminal weakness and impotency it is a grand remedy.

Atalantia monophylla Corr.

SYNONYMS : *Atalantia floribunda* Wight.; *Limonia monophylla* Roxb. *Atalantia malabarica (Rafin.)* Tanka.

HABIT : A large, thorny shrub or small tree.

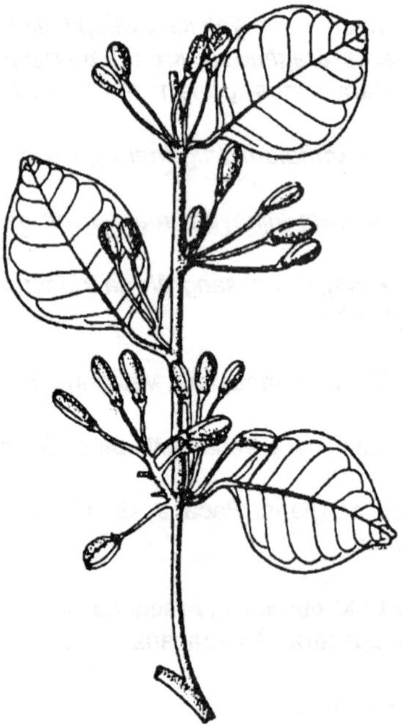

HABITAT : Konkan and Western Ghats.

ENGLISH : Wild lime.

INDIAN : Ran-limba; Maakadee, Maakadanimboo.

SANSKṚTA : Jambīra, Atavi-jambīra.

CHEMICAL COMPOSITION : Essential oil from leaves is antitubercular and antifungal. Root bark contains a tetranortriterpenoid, atalantin; stigmasterol, xanthyletin, marmesin and y-sitosterol. It also contains acridone alkaloids. Leaves yield friedelin and epifriedelanol and a mixture of stigmasterol and y-sitosterol. Fruit alkaloid serverine.

Parts used : Berries, leavea and root.

Ayurvedic properties :
Rasa - Madhura, Amla. Vīrya - Śīta, Ūṣṇa.
Guṇa - Sara, Gurū, Snigdha, Viṣada.
Vipāka - Madhura.

जम्बीरस्यफलं रसेम्लमधुरं वातापहं पित्तकृत्।
पथ्यं पाचनरोचनं वलकरं वत्ह्नेर्विवृद्धिप्रदम्।।
पक्वञ्चेद् मधुरं कफार्ति शमनं पितास्त्रदोषा पणुत्।
वर्ण्यं वीर्यविवर्द्धनंच रुचिकृत् पुष्टिप्रदं तर्पणम्।।
(राजनिघण्टु)

Jambīrasya phalam rasemlmadhuram vātāpaham pittakṛt,
Pathyam pācanarocanam balakaram vanhervivṛddhipradam.
Pakvañced madhuram kaphārttiśamanam pittāsradoṣapaṇut,
Varṇyamvīryavivarddhanam ca rucikrt

puṣṭi. pradam tarpaṇam.
(Rajanighantu)

Actions/Uses : Kaphaghna, Vātaghna, Pittakar. Pācana, Rocana, Balya, Vīryavardhaka.

Therapeutics :

Root of Jambira is antiseptic, antispasmodic and stimulant; useful in rheumatism and swellings. Decoction of leaves is beneficial in itches and other cutaneous complaints. Leaf juice is an ingredient of a compound liniment used in hemiplegia. Leaves are used in snake bite. Berries are made into pickle which forms a useful curry-diet in fevers and ailments attended with loss of taste and appetite. Oil from berries is used externally in chronic rheumatism and paralysis.

Citrus reticulata Blanco.

SYNONYMS : *Citrus nobilis* Lour., *Citrus chrysocarpa* Lushington., *Citrus aurantium* Linn. *var. aurantium. Citrus aurantium L. var. bigaradia Hook.f.; Citrus vulgaris* Risso.

HABIT : A medium-sized thorny tree.

HABITAT : Indigenous to India and cultivated throughout India.

ENGLISH : Sour-bitter orange; Seville or bigarade orange.

INDIAN : Santre, Naaringa, Khatta.

SANSKṚTA : Nāranga, Nāgaranga, Nārangaka, Tvak-sugandha, Yogasāra, Mukhapriya, Yaugika, Pārāvata, Sudha.

CHEMICAL COMPOSITION : Leaves and tender twigs yield an essential oil known as petitgrain oil, Essential oil contains 7-hydroxycoumarin geranyl ether. Dried peels give pectin and glucosides. Nobilitin and citrantin identical with hesperidin. Flowers gycosides neopesperidin and aringin. Seeds bitter compounds, limonin and de-Acnomilin. Roots, xanthyletin.

Parts used : Rind, flowers and essential oil.

Ayurvedic properties :
Rasa - Madhura, Amla. Vīrya - Śīta, Uṣṇa.
Guṇa - Sara, Gurū, Snigdha, Viṣada.
Vipāka - Madhura.

नारङ्गं मधुरं चाम्लं गुरुष्णः चैव रोचनम्।
वातामक्रिमिशूलघ्नं श्रमहृद् वलरुच्यदम्।।
(राजनिघण्टु)

Narangam madhuram camlam gurusnam caiva rocanam,
Vatamakrimisulaghnam sramahrd balarucyadam.
(Rajanighantu)

Actions/Uses : Rucya, Dīpana, Pācana, Tṛṣṇāhara, Hṛdya, Balya, Raktapittahara, Āmadoṣahara, Kṛmighna, Udaraśūlahara.

Therapeutics :

Naranga plant used against enlarged spleen, stomach-ache and menorrhagia. Leaves used as tea in headache and stomach pain. Fruit is bitter, tonic, carminative laxative and stomachic.

Citrus limon (Linn.) Burm. f.

SYNONYMS : *Citrus medica* Linn. *var. limonum* Linn.; *Citrus medica* Linn. *var. limonum Hook.f.; Citrus limonis (Linn.) Burm f.; Citrus pseudolimon* Tanaka.; *Citrus medica ssp. limon* Linn.

SIMILAR : *Citrus jambhuri* Lush.

ENGLISH : Lemon.

INDIAN : Limbu, Kagadi limbu, Idalimbu.

SANSKṚTA : Nimbuka, Amlajambīra, Vanhi, Dīpya, Śodhana, Jambīra, Rocana, Vanhibīja, Amlasāra, Dantāghrāta, Jantumārī, Rājanimbuka.

HABIT : A low shrub.

HABITAT : Cultivated all over India.

CHEMICAL COMPOSITION : Oil from peel contain d-limonene, d-a-pinene, camphene, linalool. Juice is bactericidal and contains an antipneumonia factor. Peel contains bitter principle, essential oil, hesperidin, citronellal, bergamotine, eriocitrin. apigenin, luteolin, chrysoseriol, quercetin, isorhamnetin, limocitin and limocitriol. Seeds, bitter principles, obacunone and limonin. Roots contain xanthyletin, stigmasterol and B-sitosterol.

Parts used : Fruit.

Ayurvedic properties :
Rasa - Amla. Vīrya - Anūṣṇa.
Guṇa - Laghu, Rūkṣa. Vipāka - Madhura.

निम्वुफलं प्रथितमम्लरसं कटूष्णं।
गुल्माम वातहरमग्निवृद्धिकारि।।
चक्षुष्यमेतदय कास कफार्त्ति कण्ठ।
विच्छर्दिहारि परिपक्वमतीव रुच्यम्।।
(राजनिघण्टु)

Nimbuphalam prathitamamlarasam kaṭūṣṇam,
Gulmāma vātaharamagnivṛddhikāri,
Cakṣuṣyametadatha kāsa kaphārtti kaṇṭha,
Viccharddihari paripakvamatīva rucyam.
(Rajanighantu)

Actions/Uses : Pittaghna, Vātānulomana, Āmapācana, Anulomana, Agnidipana, Viṣamajvaraghna.

Therapeutics :
Rind of ripe fruit of Nimbuka is stomachic and carminative. Juice is antiscorbutic, refrigerative in scurvy. Used in cough, cold, dysentery, diarrhoea, febrile diseases and in rheumatism, Leaves and stems are antibacterial.

Citrus Maxima (Burm.) Merr.

SYNONYMS : *Citrus decumana* Linn.; *Citrus grandis (Linn.)* Osbeck. *Citrus paradisi* Macf.

HABIT : A low shrub.

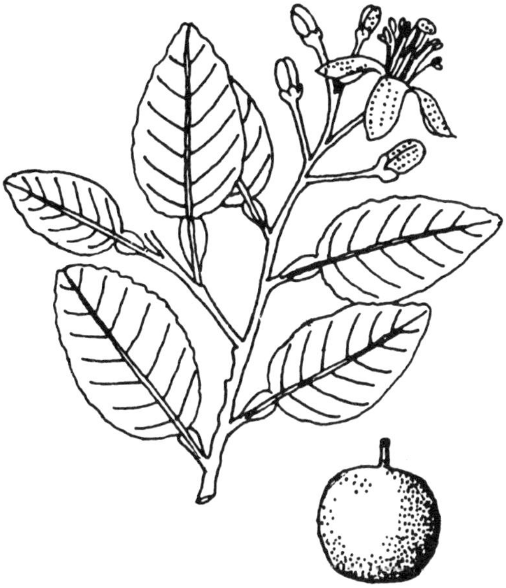

HABITAT : Grown in various parts of India.

ENGLISH : Grape fruit, Papanasa.

INDIAN : Papanasa, Chakotra.

SANSKṚTA : Madhukarkaṭi.

CHEMICAL COMPOSITION : Naringin and poncirin in peels. Oil frrom peels, d-limonene, a-pinene, linalool, geraniol etc.

Parts used : Leaves and fruit.

Ayurvedic properties :
Rasa - Madhura. Vīrya - Śīta.
Guṇa - Gurū, Śītala. Vipāka - Madhura.

मधुकर्कटिका स्वाद्वी रोचनी शीतला गुरुः।
रक्तपित्तक्षयश्वास कासहिक्काभ्रमापहा।।
(राजनिघण्टु)

Madhukarkaṭikā svādvī rocanī śītalā guruḥ,
Raktapittakṣayaśvāsa kāsahikkābhramāpahā.
(Rajanighantu)

Actions/Uses : Svādu, Agnidīpana, Raktapittahara, Kāsahara.

Therapeutics :

Madhukarkati fruit is nutritive, cardiotonic and refrigerent. Leaves are useful in epilepsy, chorea and convulsive cough. It is used in the treatment of haemorrhagic diseases, cures bronchial troubles, cough, hiccup and vertigo.

Citrus medica Linn.

SYNONYMS : *Citrus medica var. acida* Brandis. *Citrus medica var. medica* Watt.; *Citrus medica Linn var. limetta Weight & Arn.; Citrus acris* Mill.

ENGLISH : Citrous, Adam's apple, Lime fruit.

INDIAN : Mahaallunga.

SANSKṚTA : Bījapūra, Mahālunga, Bījapuraka, Mātulaka, Ruçaka, Phalapuraka, Lunga, Bījāhva, Amlakeśara, Sumanīphala.

HABIT : A low shrub.

HABITAT : Cultivated in warm moist regions in India.

CHEMICAL COMPOSITION : Bitter principle, rutaevin. Xanthyletin from roots. Essential oil microbial, contains a-Pinene, β-pinene, myrcene, camphene, limonene, a-phellandrene, 3-carene, p-cymene, y-terpinene, terpinolene, nonyl aldehyde and citronellal oil.

Parts used : Fruit, Leaves, seeds.

Ayurvedic properties :
Rasa - Amla, Madhura. Vīrya - Śīta.
Guṇa - Tikṣṇa, Snigdha. Vipāka - Madhura/Amla.

वीजपुरफलं स्वादु सरसेम्लं दीपनं लघु।
रक्तपित्तहरं कण्ठजिह्वाहृदय शोधनम्।।
श्वासकासारुचिहरं हृद्यम् तृष्णाहरं स्मृतम्।।
(भावप्रकाश)

Bījapuraphalam svādu rasemlam dīpanam laghu,
Raktapittaharam kaṇthajihvāhṛdaya śodhanam,
Śvāsakāsāruciharam ḥrdyam tṛṣṇāharam smṛtam.
(Bhavaprakasa)

Actions/Uses : Kaphavātaśamaka, Vātapittaśāmaka.

Therapeutics :

Bijapura is light, cooling, digestive, carminative, anthelmintic, stomachic, tonic and is useful in asthma, bilious vomiting, blood purification, cold, cough, fever, dyspepsia, thirst, hiccough, lumbago and sciatica. It cures flatulence, piles, dysentery, diarrhoea, cardiac and menstrual disorders. Root anthelmintic, useful in constipation, vomiting and urinary calculus. Flowers and buds are stimulant and astingent. Ripe fruit is stimulant and tonic. Fruit is expellant of poisons. Yellow pulp is aromatic and stomachic. Rind is aromatic, stimulant, hot, dry and tonic and also antiscorbutic. Fruit juice is astringent, digestive and refrigereting. Distilled water of the fruit is sedative.

Citrus limetta Risso.

SYNONYMS : *Citrus sinensis (*Linn.*) Osbeck. Citrus medica* Linn *var. limetta (Risso) Hook. f.; Citrus limettioides* Tanaka.

ENGLISH : Sweet orange, Sweet lime.

INDIAN : Mosambi. Mithanimbu.

SANSKṚTA : Miṣṭanimbuphala.

HABIT : A low shrub.

HABITAT : Cultivated in many parts of India.

CHEMICAL COMPOSITION : Essential oil. Nobiletin from peels. Fruits, limonoic acid; coumarins. Frrit flavone, rutinosides, naringin. Roots xanthyletin, stigmasterol and β-sitosterol.

Ayurvedic properties :
Rasa - Amla, Madhura. Vīrya - Śīta.
Guṇa - Tīkṣṇa, Snigdha. Vipāka - Madhura/Amla.

मिष्टनिम्बुफलं स्वादु गुरु मारुतपित्तनुत्।
गलरोगविषध्वंसी कफोत्क्लेशी च रक्तहृत्।
शोषारुचितृषाच्छर्द्दिहरं बल्यच बृंहणम्।। (द्रव्यगुण)

Miṣtanimbuphalam svādu gurū mārutapittanut,
Galarogaviṣadhvansī kaphotkleśī ca raktahṛt,
śoṣārucitṛṣācchardiharam balyancya bṛrmhaṇam.
(Drayaguna)

Actions/Uses : Kaphavataśāmaka, Vatapittaśāmaka.

Therapeutics :
Mistanimbuphala purifies blood, allays thirst in fevers, cures catarrh, improves appetite. Fruit juice useful in bilious affections and bilious diarrhoea. Rind is carminative and tonic. Fresh rind is rubbed over face as remedy for acne.

Feronia elephantum Correa.

SYNONYMS : *Feronia limonia (Linn.)* Swingle.; *Limonia elephantum (Correa.)* Panigrahi.; *Limonia acidissima* Linn.; *Schinus limonia* Linn.

ENGLISH : Wood apple, Elephant apple.

INDIAN : Kawattha, Kapittha, Bilin, Kait, Kovit.

SANSKṚTA : Kapittha, Dadhistha, Surabhicchada, Kapipriya, Dadhi, Puṣpapahala, Dantasátha, Phalasugandhika, Cirapākī, Karabhithū, Kanṭī, Gandhapatra, Grāhiphala, Kaṣāyāmlaphala.

HABIT : A tall deciduous tree.

HABITAT : Native of India and Ceylon. Cultivated in many parts of India.

CHEMICAL COMPOSITION : Leaves, stigmasterol and bergapten. Unripe fruit, stigmasterol. Root-bark, feronialactones. Bark, marmesin. Ursplic acid and a flavone glucoside from heartwood. Seed and fruit contain oil and protein. Oil is composed of palmitic, oleic, linoleic and linolenic acids.

Parts used : Fruit, gum, leaves and bark.

Ayurvedic properties :
Rasa -Madhura, Amla, Kaṣāya. Vīrya - Śīta.
Guṇa - Laghu, Rūkṣa.
Vipāka - Madhura, Amla, Kaṭu.

कपित्थमामं संग्राहि कषायं लघु लेखनम्।
पक्वं गुरु तृषाहिक्काशमनं वातपित्तजित्।
स्यादल्पतुवरं कण्ठशोधनं ग्राहि दुर्जरम्।।
(द्रव्यगुण)

Kapitthamāmam sangrāhi kaṣāyam laghu lekhanam,
Pakvam guru tṛṣāhikkāśamanam vātapittajit,
Syādalpatuvaram kaṇthaśodhanam grāhi durjaram.
(Dravyaguna)

Actions/Uses : Hikkāghna, Cardihara, Atisāraghna, Grāhī, Lekhana, Śvāsaghna, Kāsaghna, Arucihara, Vṛṣya, Sangrāhī, Mūtradoṣahara, Viṣghna.

Roots : Uttejaka, Sansrana.

Leaves : Grāhī, Pācana, Hikkāghna, Atisāraghna.

Flowers : Akhaviṣaghna.

Unripe fruit : Kaṣāya, Grāhī, Lekhana.

Ripe Fruit : Śvāsaghna, Kāsaghna, Vṛṣya, Viṣaghna.

Seeds : Kapalaviṣarpaghna.

Therapeutics :
Leaves of Kapittha are aromatic and carminative. Fruit is tonic, refreshing, cardiacal, antiscorbutic, alexiformic, astringent when unripe, stomachic and stimulant. It is used as a substitute for bael in the treatment of diarrhoea and dysentery. The pulp is used for affections of gums and throat. It is applied externally as a remedy for bites of vnenomous insects and reptiles. Bark is prescribed for biliousness.

Glycosmis pentaphylla Correa.

SYNONYMS : *Glycosmis cochinchinensis Pierre ex* Engler., *Glycosmis arborea (*Roxb.*) DC., Limonia pentaphylla* Retz., *Atista indica.*

ENGLISH : Ash sheora.

INDIAN : Bana-nimbu, Menki, Kimiro.

SANSKṚTA : Vadadru, Aśvaśakoṭa.

HABIT : A low shrub.

HABITAT : Comman in waste places, road-sides throughout India.

CHEMICAL COMPOSITION : Contains glycosmin identical with veratroylsalicin. Stem and leaves contain alkaloids arborinine. Leaves, arborinone. Root bark, skimmianine, y-fagarine and dictamine.

Parts used : Whole plant, leaves, twigs and roots.

Ayurvedic properties : Kaphahara, Pittaghna, Vatakara, Kṛimighna.

वदद्रुश्चास्यशाखोटः सपित्तकफनाशनः।
वातलश्च क्रिमीन् हन्ति पाण्डुताज्वरकामलाः।।
(द्रव्यगुण)

Vadadruscāśyáśākhoṭaḥ sapittakaphanāśánaḥ,
Vātalaśca krimīn hanti pāṇḍutājvarakamālāḥ.
(Dravyaguna)

Therapeutics :

Vadadru plant is used in indigenous medicine for cough, rheumatism, anaemia and jaundice. It is bitter, astringent, vermifuge, anti-inflammatory and expectorant. Roots, pounded and mixed with sugar given in low fever. Decoction of roots is helpful in facial inflammation. Juice of leaves is used in fever and liver complaints. Paste of leaves is applied to eczema and other skin diseases. Twigs are used as tooth-brush. Wood used in snake bite. Alcoholic extract of aerial parts is spasmolytic and diuretic.

HOMEOPATHIC USE : Glycosmis pentaphylla or atista indica is a Homeopathic remedy. Leaves are used to prepare tincture. It is used as an antimalarial tonic and in biliary colic, worm-colic, diarrhoea and dysentery.

Murraya koenigii (Linn.) Spreng.

SYNONYMS : *Bergera koenigi* Linn.

HABIT : A small tree.

HABITAT : Konkan and Western Ghats.

ENGLISH : Curry-leaf tree.

INDIAN : Kadhilimba, Kadi-patta.

SANSKRTA : Kaiḍarya, Surabhinimba, Saurabhinimba, Kṛṣṇanimba, Kumudikā, Surabhicchada.

CHEMICAL COMPOSITION : Girinimbin from stem bark. Girinimbine, mahanimbine and isomahanimbine from leaves and roots. Essential oil, glucoside koeingin.

Parts used : Leaves and roots.

Ayurvedic properties :
Rasa - Tikta. Vīrya - Śīta.
Guṇa - Laghu. Vipāka - Kaṭu.

कैदर्यः शीतलस्तिक्तः कटुश्च तुवरो लघुः।
दाहार्शः कृमिशुलघ्नः संतापविशनाशनः।।
शोफं कुष्ठं भूतबाधां नाशयेदिति कीर्तितः। (निघण्टु रत्नाकर)

Kaiḍaryaḥ śītalastiktaḥ kaụsca tuvaro saghuḥ,
Dāhārśah krmiśūlaghnah santāpavisanāśanaḥ.
Ṡopham kuṣtham bhūtabādhām nāśayediti kīrtitaḥ.
(Nighanturatnakara)

Actions/Uses : Roots are Uttejaka and Sansrana. Leaves are Grāhī and Pācana.

Therapeutics :

Kaidarya or Surabhinimba is tonic and stomachic. Root is slightly purgative. Bark and root are stimulant, externally used to cure eruptions and bites of poisonous animals. Leaves promote appetite and digestion, destroy pathogenic organisms. It is acrid, bitter, astringent, cool and light and is useful in emaciation or wasting conditions, skin diseases, hemopathy, worm troubles, neurosis and poisons in the system. Green leaves are eaten raw as cure for dysentery; bruised and applied externally to cure eruptions; given in decoction with bitters as febrifuge and used in snake-bite.

Peganum harmala Linn.

HABIT : An bushy herb.

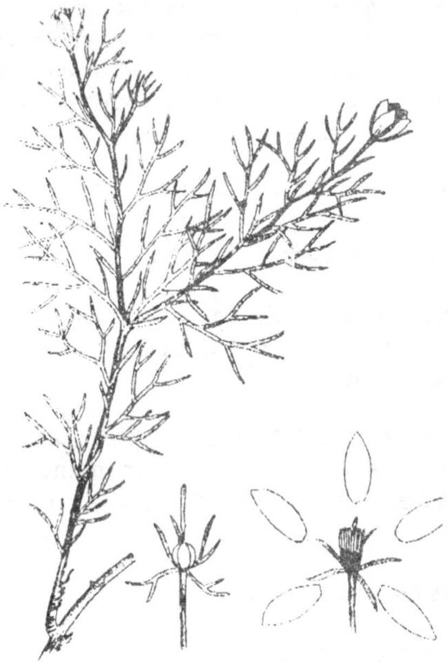

HABITAT : Konkan and Western Ghats, Ladakh and Kashmir and other parts of India.

ENGLISH : Syrian Rue, Rue Sauvage, Foreign henna.

INDIAN : Haramalla, Isband-lahouri.

SANSKṚTA : Haramala, Harmūla, Haramara, Isvanda, Virāti.

CHEMICAL COMPOSITION : Seeds and root contain four alkaloids, harmine, harmaline, harmalol and peganine and quinazoline.

Parts used : Roota, leaves and seeds.

Ayurvedic properties :
Rasa - Tikta, Kaṭu. Vīrya - Ūṣṇa.
Guṇa - Laghu, Rūksa, Tīkṣṇa. Vipāka - Kaṭu.

Actions/Uses : Kaphavātaghna, Pittavardhaka, Vedanāsthāpana, Jantughna, Vātānulomana, Śūlaprasamana, Kṛmighna, Mūtrala, Vṛṣya, Ārtavajanana, Stanyajanana, Garbhāśayasankocaka.

Therapeutics :

Haramala plant is aphrodisiac, alterative, anti-periodic, stimulant, emmenagogue, lactagogue and abortifacient. Root is applied to kill lice. Decoction of leaves is given in rheumatism. Seeds are narcotic, given in fever and colic; used as a remedy for tape-worm.

Ruta graveolens Linn.

ENGLISH : Ruta, Rue.

INDIAN : Sataapa.

SANSKṚTA : Satāpa, Suddāba, Sadapaha, Gucchapatra, Pītapuṣpa, Viṣapaha, Somalatā.

HABIT : A strong-scented, erect, glabrous shrub.

HABITAT : Indeginous in South Europe, cultivated in Indian gardens.

CHEMICAL COMPOSITION : Alkaloids, graveolin; graveolinine, rutamine, dictamine, aeborine and arborinine. Glucoside rutin and essential oil.

Parts used : Whole plant, leaves.

Ayurvedic properties :
Rasa - Kaṭu, Tikta. Vīrya - Ūṣṇa.
Guṇa - Laghu, Tīkṣṇa, Rūkṣa. Vipāka - Kaṭu.

सतपं कटूष्णं परं तिक्तयुक्तं तथैवोग्रगंधी प्रभू ताग्निकारि।
सदाक्षेपशूलाक्रिमिघ्नं प्रयुक्तं रजःस्त्रावकं गर्भपातापकारि।।

Satapam kaṭūṣṇam param tiktayuktam,
Tathaivogragandhī prabhūtāgnikāri,
Sadākṡepaśūlakrimighmam prayuktam,
Rajaḥ srāvakam garbhapātāpakāri.

Actions/Uses : Kaphavātaghna, Pittavardhaka, Dīpana, Anulomana, Vedanāsthāpana, Raktotkleṣakāraka, Jantughna, Uttejaka, Garbhāśayasankocaka, Ārtavajanana, Svedajanana.

Therapeutics :

Satapa, is resolvent, diuretic, emmenagogue, antispasmodic, stimulant, acro-narcotic, poison, irritant and abortifacient. It is useful in hysteria and amenorrhoea. Juice of the herb relieves earache and toothache. Leaves are used in rheumatic pains, in treating hysteria, worms, colic and atonic amenorrhoea and menorrhagia. Herb and the oil act as stimulants, their influence being chiefly directed to the uterine and nervous systems. It is anthelmintic and in excessive doses it is acro-narcotic. Oil is used externally as rubefacient.

HOMEOPATHIC USE : Ruta graveolens is a Homeopathic remedy. The chief symptoms noted are pains in bones, joints and cartilages. It is used in the treatment of rheumatism especially of the wrist and ankle; and to avert abortion or to stop bleeding during pregnancy. It is indicated in many eye troubles.

Toddalia asiatica (Linn.) Lam.

SYNONYMS : *Toddalia aculeata* Pers.; *Paullinia asiatica* Linn.; *Limonia oligandra* Dalz.

ENGLISH : Lopez root tree, Wild orange tree, Forest-pepper.

INDIAN : Jangli-kalli-mirchi, Limri, Manger.

SANSKṚTA : Kancana, Dahana.

HABIT : A sub-scandent shrub or a large tree. Two varieties with red and white flowers are in use.

HABITAT : Konkan and Maharashtra.

CHEMICAL COMPOSITION : Essential oil. Alkaloid berberine. Roots contain a poisonous resin which produced abortion in guinea pigs and in large doses paralysis and death. Leaves contain toddaline and toddalinine, toddalolactone, pimpinellin, isopimpinellin, resins and glycosides. Two new coumarins- norbraylin, 5,7,8-trimethoxycoumarin isolated.

Parts used : Root, bark, leaves and fruit.

Ayurvedic properties :
Rasa - Tikta, Kaṭu. Vīrya - Uṣṇa.
Guṇa - Laghu, Rūkṣa. Vipāka - Kaṭu.

रक्तस्तु काञ्चनः शीतः सरो ह्यग्नि प्रदीपनःऽच।
तुवरो ग्राहकः प्रोक्तः कफपित्तव्रणकृर्मान्।।
गण्डमाला रक्तपित्तकुष्ठ वातांश्च नाशयेत्।
गुदभ्रंशं रक्तपित्तं नाशयेत्पुष्पमस्य च।।
शीतलं तुवरं रुक्षं संग्राहि मधुरं लघु।
पित्तं क्षयं च प्रदरं कासं रक्तरुजं हरेत्।।

Raktastu kāñcanaḥ śītaḥ saro hyagni pradipanaḥ,
Tuvaro grāhakaḥ proktaḥ kaphapittavraṇakṛmīn.
Gaṇḍamālā raktapittakuṣṭha vātañśca nāśayet,
Gudabhrmśam raktapittam nāśayet puṣpamasya ca.
Śītalam tuvaram rūkṣam sangrāhi madhuram laghu,
Pittam kṣayam ca pradaram kāsam raktarujam haret.
(Nighanturatnakara)

श्वेतस्तु काञ्चनो ग्राहि तुवरो मधुरः स्मृतः।
रुच्यो रुक्षः श्वासकासपित्तरक्तविकारहा।।
क्षतप्रदरनुत्प्रोक्तो गुणाश्चान्ये तु रक्तवत्।

Śvetastu kāñcano grāhi tuvaro madhuraḥ smṛtaḥ,
Rucyo rukṣaḥ śvāsakāsa pittarakta vikārahā.
Kṣatapradaranut prokto guṇāścānye tu raktavat.
(Nighanturatnakara)

Actions/Uses : Vātahara, Svedajanana, Dīpana

Therapeutics :
Root bark of Kancana is bitter, aromatic, tonic, stimulant, antiperiodic, diaphoretic, stomachic; given in weak unfusion useful in constitutional debility and convalescence after febrile and other exhausting diseases. An infusion of fresh root bark is stimulating tonic and carminative. Plant is used as a febrifuge.

Zanthoxylum americanum Mill.

SYNONYMS : *Zanthoxylum clava-nerculis* Linn.; *Xanthoxylum fraxineum* Willd.; *Xanthoxylum carolinianum* Lamb.

ENGLISH : Prickly Ash.

SANSKṚTA : Tumbaru-1, Tikta.

HABIT : A tree.

HABITAT : Canada and U.S.A.

CHEMICAL COMPOSITION : It contains Pipevine.

Parts used : Root, bark, leaves and fruit.

Ayurvedic properties :
Rasa - Tikta, Madhura. Vīrya - Ūṣṇa.
Guṇa - Laghu, Rūkṣa. Vipāka - Madhura.

तुंबरुर्मधुररित्तक ऊषण्डश्चाग्निदोपनः।
वीर्योष्णः कृच्छ्रानुद्रूक्षस्तीक्ष्णो रुचिकरो लघुः।।
विदाही हृद्यः कफनुद्वात गुल्मोदरापहः।
शूलाध्मानकृमीन्नेत्रकर्ण मस्तक रोगनुत्।।
कण्ठरोगं वान्तिकुष्ठ प्लीहा श्वासारुचिर्हरेत्।
अपतन्त्रकनामानं वातं चैव विनाशयेत्।।
(निघण्टु रत्नाकर)

Tumbarurmadhurastiktah ūṣaṇa ścāgnidīpanaḥ,
Vīryoṣṇaḥ kṛcchranu drūkṣa stikṣṇo rucikaro laghuḥ.
Vidāhī hṛdyaḥ kaphnudvata gulmodarāpahaḥ,
Śūlādhmāna kṛmīnnetrakaṛnamastaka roganut.
Kaṇṭharogam vāntikuṣṭha plīha śvāsārucirharet,
Apatantrakanāmānam vātam caiva vināśayet.
(Nighanturatnakara)

Actions/Uses : Vātahara, Svedajanana, Dīpana

Therapeutics :

Tumbaru-1 is stimulant, alterative, tonic and diaphoretic. Berries are more active and are also carminative and antispasmodic. Found useful in rheumatism and skin diseases.

HOMEOPATHIC USE : Xanthophyllum is a Homeopathic remedy. It acts upon skin and female sexual organs. Bearing down sensation; vulva inflamed, with furious iitching. Increased sexual desire, with ovarian and uterine pains and leucorrhoea. Muscular lameness. Erythema on skin, with vesicles and intense itching, stinging and burning, dermatitis around knees are chief symptoms.

Zanthoxylum rhesta (Roxb.) DC.

SYNONYMS : *Zanthoxylum budrunga* Wall. *ex DC.*; *Zanthoxylum limonella (Dennst.)* Alston.; *Fagara rhetsa* Roxb.; *Fagara budrunga* Roxb.

INDIAN : Chiraphalla, Tiraphalla, Tessul.

SANSKṚTA : Tejovatī, Aśvaghra, Tumbaru-2.

HABIT : A middle sized tree.

HABITAT : Konkan and Western Ghats and other parts of India.

CHEMICAL COMPOSITION : Quinazolone alkaloids- rhetin, chelerythrine, rhetsine, evodiamine and rhetsinine from trunk bark. Bark contains budrungaine and budrungainine. Seed bark, lupeol. Essential oil, d-terpinene, 4-carene, β-pinene etc. Fruit rind contains an antibacterial agent. Essential oil from fruits showed maximum local anaesthetic activity.

Parts used : Fruit and root-bark and essential oil.

Ayurvedic properties :
Rasa - Madhura, Tikta.   Vīrya - Ūṣṇa.
Guṇa - Laghu, Rūkṣa.   Vipāka - Kaṭu.

तेजोवती कटुस्तिक्ता रुच्या दीपन - पाचनी।
उष्णा वात - कफ - श्वास - कास - हिध्मास्यरोगनुत्।।
(कै. नि.)

Tejovati kaṭustikttā rucyā dīpana pācanī,
Ūṣṇā vātakaphaśvāsakāsa hidhmāsyaroganut.
(Kaiyadevanighantu)

Actions/Uses : Dīpana, Pācana, Grāhī, Kṛmighna, Rucikar.

Therapeutics :

Root bark of Tejovati is considered a purgative of the kidneys. Fruit is aromatic, astringent, stomachic; prescribed in dyspepsia, arising from atrabilis and in some form of diarrhoea and given in honey for rheumatism.

Family : LXXXVIII-Salvadoraceae

Salvadora indica Wight.

SYNONYMS : *Salvadora persica* Linn. *var. Wightiana (Planch. ex Thw.)* Verdc.; *Salvadora persica auct. (non* Linn.*).*

ENGLISH : Toothbrush Tree, Mustard, Saltbush.

INDIAN : Khaankanna, Mirajollee, Khakhin, Miraj, Jhak, Pilva, Kharjal, Rhakhan, Thorapilu.

SANSKRTA : Pīlu, Angāhavā, Sītasahā, Stranvsi, Dagari, Gudaphala, Virekaphala, Śākhī, Karabhavallabha, Bṛhatpilu, Karabhapriya, Dhānī, Śyāma, Sahasrāngī, Sahasrānśī, Tīkṣnataru.

HABIT : A large, much-branched, evergreen shrub or small tree.

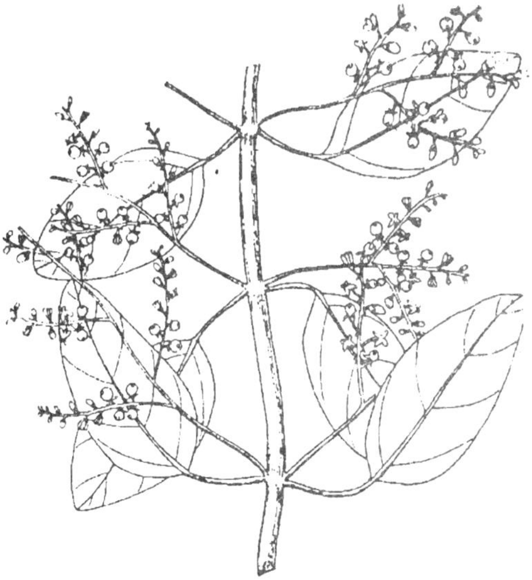

HABITAT : Konkan and Maharashtra, drier parts of India including Punjab and Rajasthan.

CHEMICAL COMPOSITION : Alkaloid trimethylamine, β-sitosterol and elementaly-monoclinic sulphur from roots. Plant extract antibacterial.

Parts used : Root-bark, stem-bark, fruit, flowers, seed, leaves, and oil.

Ayurvedic properties :
Rasa - Tikta, Madhura, Kaṭu. Vīrya - Ūṣṇa.
Guṇa - Laghu, Snigdha, Tīkṣṇa. Vipāka - Madhura.

पीलु श्लेष्मसमीरघ्नं पित्तलं भेदि गुल्मनुत्।
स्वादु तिक्तंच यत्पीलु तन्नात्युष्णं त्रिदोषहृत्।।
(भावप्रकाश)

Pīlu śleṣmasamīraghnam pittalam bhedi gulmanut,
Svādu tiktam ca yatpīlu tannātyuṣṇam tridoṣahṛt.
(Bhavaprakasa)

अङ्गाहवः कटुकः पिलुः कषायो मधुराम्लकः ।
सरः स्वादुश्च गुल्मार्शः शमनः दीपनः परः।।
(राजनिघण्टु)

Aṅgāhavaḥ kaṭukaḥ piluḥ kaṣāyo madhurāmlakaḥ,
Saraḥ svāduśca gulmārśaḥ śamanaḥ dīpanaḥ paraḥ.
(Rajanighantu)

Actions/Uses : Kaphavātaśāmaka, Pittakara, Anulomana, Recana, Vedanāsthāpana, Śothahara, Mūtrala, Ārtavajanana, Aśmarighna, Raktapittapraśamana.

Therapeutics :

Root-bark of Pilu is acrid and is vesicant when applied locally. Leaves possess antiscorbutic and astringent properties. Their juice is given in scurvy. A decoction of leaves is used in asthma and cough and a poultice is applied to painful tumours and piles. They are used as external application in rheumatism. Shoots and leaves are pungent and used as antidote to poisons of all sorts. Decoction of stem-bark is useful in fevers and is used as stimulant and tonic in amenorrhoea. Flowers are laxative and stimulant are beneficial in the treatment of gonorrhoea, leprosy. They are applied externally in painful rheumatic conditions. Fruits are alterative, carminative, diuretic, stomachic, deobstruent and lithontriptic. They are useful in the treatment of billiousness, spenomegaly, tumor and rheumatism. Seeds are bitter, purgative, diuretic and tonic. Seed oil is applied on skin in rheumatism. Decoction of stem bark is used in low fever and as stimulant and tonic in amenorrhoea.

Family : LXXXIX-Santalaceae

Santalum album Linn.

SYNONYMS : *Syrium myrtifolium* Roxb.

ENGLISH : Sandlewood tree.

INDIAN : Chandana.

SANSKṚTA : Candana, Śrikhanda, Gandhaśara, Malayaja, Bhadraśrī, Candradyutī, Mangalya, Sarpāvāsa, Śītala, Bhogivallabha, Gośīrṣa, Mahāruha, Śvetacandana, Tilaparṇa, Gandhādhya, Pāvana, Śiśira, Śri, Hima, Tailaparṇa, Pātīra, Ekānga, Rājayogya, Sarvapriya.

HABIT : A small or medium sized evergreen semi-parasitic tree.

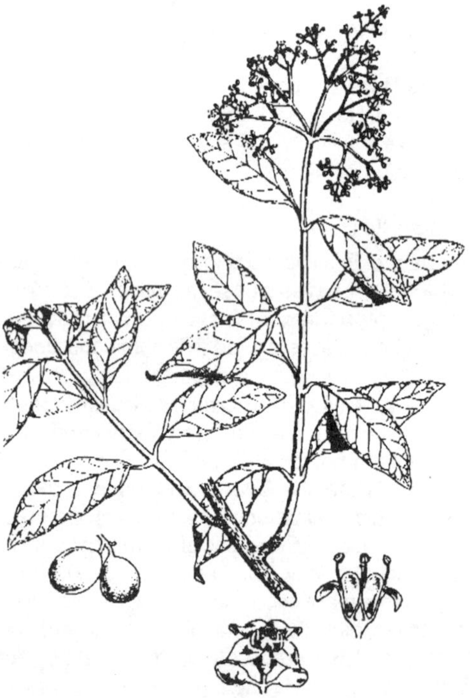

HABITAT : Western Peninsula; cultivated elsewhere.

CHEMICAL COMPOSITION : Leaves, fruits and seeds contain santalbic acid,palmitic acid,oleic acid, linoleic acid and sugars. Fruits yield betulic acid and B-sitosterol and a fatty oil. Betulinic acid, β-sitosterol, glucose, fructose and sucrose from leaves. Heartwood yields essential oil which contains α-, α- and β-santalenes, santenol, teresantanol.

Parts used : Wood and oil

Ayurvedic properties :
Rasa - Tikta. Vīrya - Śīta.
Guṇa - Laghu, Rūksa. Vipāka - Kaṭu.

चन्दनं शीतलं रुक्षं तिक्तमल्हादनं लघु।
श्रमशोषविषश्लेष्मतृष्णापित्तास्त्रदाहनुत्।।
(भावप्रकाश)

Candanam śītalam rūkṣam tiktamlhādanam laghu,
Śramaśoṣaviṣaśleṣmatṛṣṇāpittāsradahanut.
(Bhavaprakasa)

Actions/Uses : Kaphaghna, Pittaghna, Vātakar, Hṛdya, Dāhaśamana, Tṛsnānigrahaṇa, Kaṇḍūghna, Yakṛtotejaka, Raktaprasādana, Varṇya, Viṣghna, Jantughna, Jvaraghna.

Therapeutics :

Both Candana and oil are cooling, diaphretic, diuretic and expectorant. Sandal wood is chiefly employed in the treatment of chronic mucous affections such as inflammation of bladder and other diseases. Ground up with water into a paste it is applied to temples in headache, during fevers as diaphoertic and for burns and local inflammations. It is used in skin diseases to allay heat and pruritus. Decoction of wood, mixed with that of ginger is beneficial in hemorrhoids. Sandal-wood oil is recommended as a remedy in gonorrhoea. It is as equal and frequently superior to Copaiba and Cubebs.

Family : XC - Sapindaceae

Allophylus serratus Radik.

SYNONYM1 : *Allophylus cobbe (Linn.)* Rausch.

INDIAN : Tiphana; Tipani.

SANSKṚTA : Tripuṭa.

HABIT : A low ferrunginous shrub sometimes a climber or small tree.

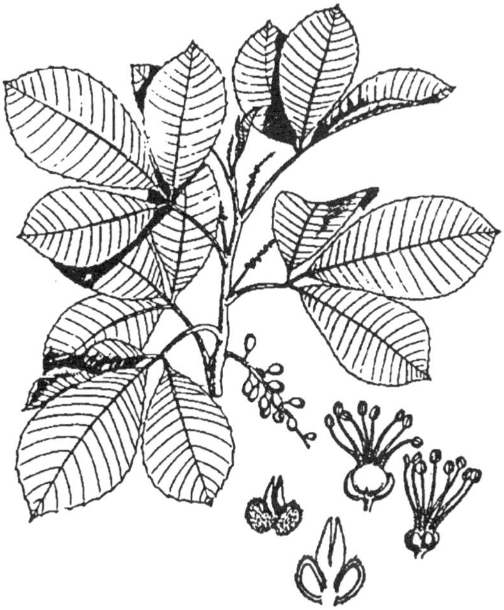

HABITAT : Eastern and Western Peninsula; Assam.

CHEMICAL COMPOSITION : Benzylamide from leaves. Leaves, Ph-acetamide and alkaloids.

Parts used : Whole plant, roots, fruits.

Ayurvedic properties :
Rasa - Mdhura. Vīrya - Śīta.
Guṇa - Laghu. Vipāka - Madhura.

कषायतिक्तमधुरं समूलं शोफनाशनम्।।
अस्थिभङ्गहरं प्रोक्तभङ्गदर्दव्रणापहम्।
फलं स्वादु हिमं रुच्यं मूलं दीपनपाचनम्।।
(स्व)

Kaṣāyatiktamadhuram samūlam śophanāśanam,
Asthibhaṅgaharam
proktamaṅgamarddavraṇāpaham,
Phalam svādu himam rucyam mūlam dīpana pācanam.
(Sva.)

Actions/Uses : Dīpan, Pācana, Rucya, Śothahara.

Therapeutics :

Plant is astringent, bitter, sweet, antinflammatory, vulnerary, digestive, carminative and constipating. It is useful in bone fractures, dislocation, ulcers and wounds. Root is astringent, employed to check diarrhoea; for treating piles and in nose bleeding. Leaves are eaten to induce lactation. Decoction given in colic. Fruits are sweet, cooling and nourishing tonic. Fruit juice is used against tapeworms.

Cardiospermum halicacabum Linn.

ENGLISH : Baloon vine, Blister creeper.

INDIAN : Kaanaphuti, Sabni, Kapaala-phodi, Bodh, Naphat.

SANSKṚTA : Indravalli, Karavi, Karṇasphota, Jyotiśmati, Tejovati.

HABIT : Annual or perennial scrambling shrub climbing by means of tendrillar hoods.

HABITAT : Common throughout the plains of India in waste lands, hedges and thickets.

CHEMICAL COMPOSITION : Seeds, cyanogenic glycoside; seed oil ester. Essential oil from seeds.

Parts used : Entire herb, roots, leaves and seeds.

Ayurvedic properties :
Rasa - Kaṭu, Tikta. Vīrya - Ūṣṇa.
Guṇa - Tīkṣṇa, Snigdha, Sara. Vipāka - Kaṭu.

इन्द्रवल्ली ज्वरहरा शोफपाण्डुहरा स्मृता।
वातघ्नी मूत्रळा केश्या वृद्धिशूलापहारिणी।।
(स्व)

Indravallī jvraharā śophapāṇḍuharā smṛtā,
Vātaghnī mūtralā keśyā vṛddhiśūlāpahāriṇī.
(Svayarnkrti)

दाहप्रदा दीपनकृश्च मेध्या।
प्रज्ञाश्च पुष्णाति तथा तेजोवती।।
(राजनिघण्टु)

Dāhapradā dīpanakrcca medhyā,
Prajnāñca puṣṇāti tathā tejovatī.
(Rajanighantu)

Actions/Uses : Vātaghna, Jvaraghna, Vṛddhināṣaka.

Therapeutics :
Indravalli is used in rheumatism, stiffness of limbs and snake bite. Plant juice is used as a cure for earache. Alcoholic extract of plant is spasomlytic. Root is diaphoratic, diuretic, aperient, laxative, stomachic, alterative, rubifacient, and emmenagogue; occasionaly used for rheumatism, lumbago and nervous diseases, also used for abortion. Leaves are rubifacient, useful as poiutice for rheumatism. Leaves and roots are useful in nervous diseases, piles, chronic bronchitis, fevers, hydrocele, amenorrhoea, sprains and oedema. Powder of leaves is used externally for healing wounds.

Sapindus trifoliatus Linn.

SYNONYMS : *Sapindus emarginatus* Vahl.; *Sapindus laurifolius var. emarginatus* Cooke.; *Sapindus laurifolius* Vahl.

NOTE : *Sapindus mukorossi Gaertn. is Arista witch is chota ritha and S. trifoliatus is Bara-ritha. Arista is cultivated throughout North-West India, West Bengal and Assam.*

ENGLISH : Soapnut tree.

INDIAN : Ritthaa, Rinthi, Ardal, Pitha, Ringin.

SANSKṚTA : Riṭhākaranja, Ariṣṭa, Gucchala, Gucchaphala, Phenila, Gucchapuṣpaka, Mangalya, Kumbhabījaka, Raktābija, Prakīrya, Riṭha, Somavalka, Garbhapātī, Rakśabīja.

HABIT : A handsome medium-sized to large, deciduous tree, or shrub.

HABITAT : Commonly cultivated in villages of South India and West Bengal.

CHEMICAL COMPOSITION : Saponins-sapindoside A and sapindoside B. Nuts contain kaempferol, quercetin and β-sitosterol. Saponin emarginatoside from fruits.

Parts used : Fruit.

Ayurvedic properties :
Rasa - Tikta, Kaṭu. Vīrya - Ūṣṇa.
Guṇa - Laghu, Tīkṣṇa. Vipāka - Kaṭu.

रीठाकरञः तिक्तोष्णः कटुः स्निग्धश्च वातजित्।
कफघ्नः कुष्ठकण्डूतिविषविस्फोटनाशनः।।
(राजनिघण्टु)

Rithākaranjaḥ tiktoṣṇaḥ kaṭuḥ snigdhaśca vātajit,
Kaphaghnaḥ kuṣṭhakaṇḍūtiviṣavisphoṭanāśanaḥ.
(Rajanighantu)

अरिष्टः कटुकः पाके तीक्ष्णश्चोष्णश्च लेखनः।
गर्भपातकरः प्रोक्तो लघुः स्निग्धः त्रिदोषहा।
ग्रहपीडादाहशूलनाशनश्च प्रकीर्तितः।।
(द्रव्यगुण)

Ariṣṭaḥ kaṭukaḥ pāke tīkṣṇaścoṣṇaśca lekhanaḥ,
Garbhapātakaraḥ prokto laghuḥ snigdhaḥ tridoṣāhā,
Grahapīḍādahāśūlanāśanaśca prakīrttitaḥ.
(Dravyaguna)

Actions/Uses : Kaphapittaghna, Vātanāśaka, Vāmaka, Kṛmighna, Recana, Viṣaghna, Śothahara, Vedanāsthāpana, Raktaśudikara, Lekhana, Kaphanissāraka, Garbhāśayasankocaka.

Therapeutics :
Arista fruit is sharp, hot, tonic, alexipharmac, given internally as expectorant, emetic, aphrodisiac, abortifacient, purgative and nauseant; useful in chronic dysentery, diarrhoea, cholera, and tubercular glands. As an errhine used in epilepsy, asthma, hysteria and hemicrania, paralysis of limbs, lumbago, epileptic fits of children; externally it is detergent. It cures tridosha and is sedative to uterus, allays uterine pain. Pessaries made of kernels are used to stimulate uterus during childbirth. Used as fish poison.

Schleichera oleosa (Lour) Oken.

SYNONYMS : *Pistacia oleosa* Lour.; *Schleichera trijiga* Willd. & *Klein.*, *Schleichera oleosa (Lour)* Merr.

ENGLISH : Ceylon Oak, Lac tree, Macassar tree.

INDIAN : Kosimba, Kusum, Gausam, Kohan.

SANSKṚTA : Kośāmra, Ghanaskandha, Vanāmra, Jantupadāpa, Raktāmra, Kśūdrāmra, Lākṣāvṛukṣa, Suraktaka, Sukośa, Kṛumitaru, Kṛmivṛkṣaka, Jantudruma.

HABIT : A large or medium-sized deciduous tree.

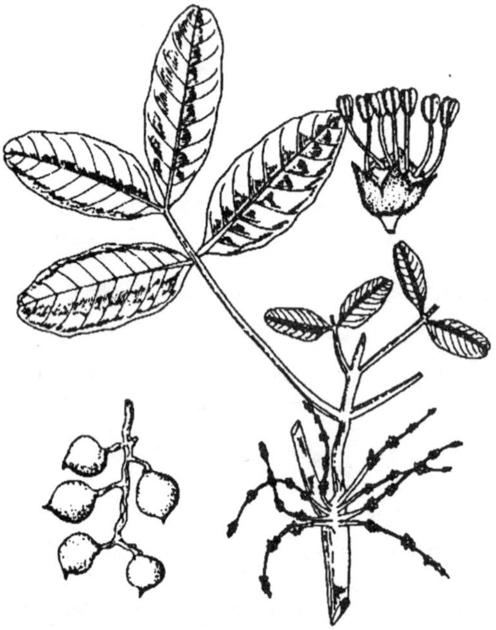

HABITAT : Common in dry forests of India.

CHEMICAL COMPOSITION : Seeds contains phenols and polyene pigments. Kusum oil contains cyanolipid. Leaves contain alcohols, alkanes, stigmasterol, β-sitosterol, campesterol, cholesterol, glucose, fructose, tartartaric acid and oxalic acid.

Parts used : Seeds, Bark and oil.

Ayurvedic properties :
Rasa - Amla. Oil : Kaṭu, Tikta, Kaṣāya. Vīrya - Uṣṇa. Guṇa - Gurū. Vipāka - Kaṭu.

कोषाम्रोऽम्लः कटुः पाके वीर्योष्णोऽथानिलापहः।
कफपित्तकरो रुच्यः कुष्ठघ्नो रक्तशोधनः॥
(ध. नि.)

Koṣāmro amlaḥ katuḥ pāke vīryoṣṇo athānilapāhaḥ,
Kaphapittakaro rucyaḥ kuṣṭhaghno raktaśodhanaḥ.
(Dhanvantari)

Actions/Uses : Fruit : Kaphavātaghna.

Bark : Kaphapittaghna, Dīpana, Rucivardhaka, Virecana, Stambhana, Keṣya, Kṛmighna, Kaṇḍūghna, Vraṇaśodhana, Grāhī, Vedanāsthāpana.

Therapeutics :

Bark of Kosamra is astringent. It is used to cure kapha. Rubbed up with oil it is used as a cure for itch, leprosy, skin diseases, inflammation, ulcers. Oil is efficacious in alopecia, also used for acne, itch and for massage in rheumatism. Unripe fruit is used for vaata. Powder of seeds is applied to ulcers of animals and for removing maggots.

Family : XCI - Sapotaceae

Madhuca indica J. F. Gmel.

SYNONYMS : *Bassia latifolia* Roxb.; *Madhuka longifolia* Koen.; *Madhuca latifolia (Roxb)* Macbride.; *Madhuca longifolia (Koen.)* Macbride., *Bassia longifolia* Koen.

ENGLISH : Butter tree.

INDIAN : Maahua, Moha, Mahwa, Mohwra, Mauwa.

SANSKRTA : Madhūka, Madhūlaka, Gudapuṣpa, Madhudruma, Madhuṣṭīla, Rodhrapuṣpa, Vānaprastha, Dīrghapatra, Madhusrava, Gauravaśāka, Tīkṣṇasāra, Dholāphala.

HABIT : A middle sized to large deciduous tree.

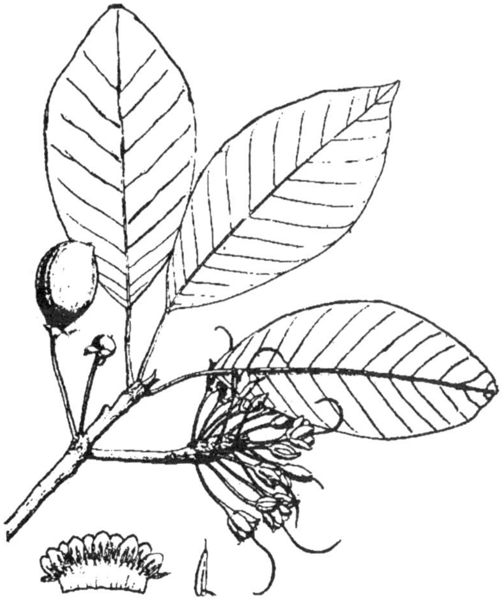

HABITAT : Forests of Western India and Konkan.

CHEMICAL COMPOSITION : Seed kernels yield mahua oil. Fruit on steam distillation yield a volatile oil. Sucrose, β-sitosterol and a sterol glucoside from nuts. Lupeol acetate, β-amyrin acetate, betulinin and oleanolic acids from bark.

Parts used : Flowers, leaves, bark and seeds.

Ayurvedic properties :
Rasa - Madhura, Kaṣāya. Vīrya - Śīta.
Guṇa - Gurū, Snigdha. Vipāka - Madhura.

मधुकं मधुरं शीतं पित्तदाह श्रमापहम्।
वातलं जन्तुदोषघ्नं वीर्यपुष्टिविवर्द्धनम्।।
(राजनिघण्टु)

Madhukam madhuram śītam pittadāha śramāpaham,
Vātalam jantudoṣaghnam vīryapuṣṭivivarddhanam.
(Rajanighantu)

रक्तपित्तहराण्याहुर्गुरुणि मधुराणि च।
बृंहणीयमद्यंच मधुककुसुमं गुरु।
वातपित्तोपशमनं फलं तेनोपदिश्यते।।
(सु. सू.)

Raktapittaharaṇyāhurguruṇi madhurāṇi ca,
Bṛnhaṇīyamaddyañca madhūkakusumam guru,
Vatapittopaśamanam phalam tenopadiśyate.
(Susruta Sutra 46)

Actions/Uses : Pittahara, Vātakara, Balya, Vṛṣya, Śvāsaśamana, Rasāyanī, Brihmana, Tṛṣṇāsamana, Śukrakara, Dāhaśamana, Mūtrala, Stanyajanana, Raktapittahara, Grahaṇī, Śirorogaghna, Vraṇahara.

Therapeutics :

Bark of Madhuka is used for rheumatism, ulcers, itches, bleeding and spongy gums, tonsillitis and diabetes mellitus. A decoction of bark is used as astringent and emolient, also as a remedy for itch. Flowers are cooling, demulcent, laxative, tonic, stimulant, anthelmintic; used in coughs, colds, bronchitis, snake-bite and as fish-poison. Seeds are galactagogue, and laxative in habitual constipation and piles. Oil from seeds has emolient properties and is good for skin diseases. It is used in rheumatism and headache. It is laxative, useful in piles and haemorrhoids, also used as emetic. Honey obtained is used for eye diseases. Gummy juice is used in rheumatism. Fresh juice of Madhuka is alterative and the spirit distilled from the flowers is a powerful diffusible stimulant and an astringent. Mahua cake is insecticidal and pesticidal; used with shikakai for hairwash.

Mimusops elengi Linn.

INDIAN : Bakulla, Maulsari, Ovalli.

SANSKṚTA : Bakula, Madhugandha, Sinhakakesara, Sthirapuṣpa, Kesara, Sīdhugandha, Strimukhamadhudāhada, Bhramarānanda, Sīdhusamjnya, Cirapuṣpa, Dhanvī, Śaradika, Karaka, Gudhapuspa, Viśārada, Sinha, Madyāmoda, Madana, Sinhakesaraka, Śīerṣaka, Gandhāḍhya, Dohalī, Maulasirī, Bolasarī, Magādāma, Pagādāmānu.

HABIT : A small to large evergreen tree.

HABITAT : Cultivated throughout Western Peninsula.

CHEMICAL COMPOSITION : Quercitol, ursolic acid, glucose, a triterpene alcohol, quercetin, dihydroquercetin and β-sitosterol glucoside from fruits and seeds. A mixture of saponins in bark. D-mannitol, β-sitosterol and its glucoside from flowers. Essential oil.

Parts used : Bark, flowers, fruits, seeds and leaves.

Ayurvedic properties :
Rasa - Katu, Kaṣāya. Vīrya - Śīta.
Guṇa - Gurū, Snigdha, Viṣada. Vipāka - Kaṭu.

वकुलस्तुवरोनुष्णः कटुपाकरसो गुरुः।
कफपित्तविषश्वित्रक्रिमिदन्तगदापहः।।
मधुरञ्च कषायञ्च स्निग्धं संग्राहि वा कुलम्।
स्थिरीकरञ्च दन्तानां विशदं फलमुच्यते।
(भावप्रकाश)

Bakulastuvaro anuṣṇaḥ kaṭupākaraso guruḥ,
Kaphapittaviṣaśvitrakrmidantagadāpahaḥ.
Madhurānca kaṣāyañca snigdham sangrāhi vākulam,
Sthirīkarañca dantānam viśadam phalamucyate.
(Bhavaprakasa)

Actions/Uses : Kaphapittaśāmaka, Grāhī, Hṛdya, Dantarogahara, Stambhana, Jantughna, Dantya, Kṛmighna, Viṣahara.

Therapeutics :

Flowers, fruit and bark of Bakula are astringent. Bark is astringent, tonic, increases fertility in women and is useful in fevers. Infusion or decoction of bark is used as a gargle in dieases of gums and teeth. Leaves are used in snake-bite. Pulp of ripe fruit is astringent, used in curing chronic dysentery. Powder of dry flowers is used as snuff to relieve headache. Both flowers and fruits together with other astringents are used in lotion for wounds and ulcers. Seeds are purgative. They are bruised and locally applied within anus of children in cases of constipation.

Family : XCII - Scrophulaceae

Bacopa monniera (Linn.) Wettst.

SYNONYMS : *Bacopa monnieri (Linn.)* Pennel.; *Herpestis monniera (Linn.) H. B. & K.; Moniera cuneifolia* Michx.; *Gratiola monniera* Linn.; *Lysimachia monniera* Linn.

ENGLISH : Thyme-leaved graticula.

INDIAN : Neera-Braamhee, Braahmi, Jalavaveri.

SANSKṚTA : Brāmhi, Sarasvatī, Saumyā, Suraśreṣṭhā, Suvarcalā, Kapotavegā, Vaidhātrī, Divyatejā, Mahauṣadhī, Surejyā, Jalanimba, Bramhakanyakā, Surasā, Mandukamālā, Matsyākṣī, Laghubrahmi, Vīra, Medhyā, Bhāratī, Parameśṭhinī, Divyā, Śāradā, Somā, Kapotavankā, Tvaṣṭā, Bramhasuvarcasā, Vayasthā, Smaraṇī, Bramhacarini, Maṇḍukī, Satyanāmā, Bramhasomā, Munikā, Mandukacadā, Tvāṣṭrī, Satyavatī, Kankolaparṇyakā, Lāvanyā, Śankhadharā, Jalajabrāhmi, Satsyākṣī, Ḋhirā.

HABIT : A small, creeping glabrous or succulent annual herb.

HABITAT : Throughout India in wet, damp and marshy areas.

CHEMICAL COMPOSITION : Alkaloid brahmine; its therapeutic action resembles strychnine, but is less toxic. Three bases isolated, B1 oxalate, B2 oxalate, B3 chloroplatinate and a sterol. Contains alkaloid herpestine. Plant saponins, bacoside A & B; monniern, hersaponin, betulic acid, d-mannitol, stigmasterol, β-sitosterol and stigmastanol. Hersaponin possesses cardiotonic, sedative and spasmodic properties.

Parts used : Stem and leaves

Ayurvedic properties :
Rasa - Tikta, Kaṣāya, Madhura. Vīrya - Śīta.
Guṇa - Laghu, Snigdha. Sara. Vipāka - Madhura.

ब्राह्मी हिमा सरा तिक्ता लघुर्मेध्या च शीतला।
कषाया मधुरा स्वादुपाका युष्या रसायनी।
स्वर्या स्मृतिप्रदाकुष्ठपाण्डुमेहास्त्रकासजित्।
विषशोथ ज्वरहरी (तद्वत् मण्डुक पर्णिनी)।।
(भावप्रकाश)

Brāhmī himā sarā tiktā laghurmedhyā ca śītalā,
Kaṣāya madhurā svādupakāyuṣyā rasāyanī.
Svaryā smṛtipradā kuṣṭhapāṇḍumehāsrakāsajit,
Viṣaśothajvaraharī (tadvat maṇḍukaparṇinī).
(Bhavaprakasa)

Actions/Uses : Tridoṣaghna, Vātaśāmaka, Pittaśāmaka, Kaphaghna, Dīpana, Pācana, Medhya, Svarya, Stambhana, Śothahara, Viṣaghna, Kuṣṭha-kaṇḍūhara, Jvaraghna, Apasmārahara, Mehaghna, Vedanāśāmaka, Rasāyanī.

Therapeutics :
Bramhi, is astringent, bitter, cooling, pungent, heating, emetic, laxative, and improves intellect, useful in bed ulcers, tumours, ascites, enlargement of spleen. It is a potent nerve tonic, cardionic and diuretic, found very effective in cases of anxiety neurosis. It is an anti-anxiety agent having adaptogenic effect. It is aphrodisiac and aperient; used in the treatment of asthma, hoarseness, insanity and epilepsy. It is indicated against dermatosis, anaemia, diabetes, cough, dropsy, fever, arthritis, anorexia, dyspepsia and emaciation.

The entire plant constitutes the well-known drug Brahmi. Stem and leaves are brain tonic which sharpens dull memory; and are used in catarrhal complaints and in snake-bite; also as a safe cardiac tonic. Leaves are used as diuretic and aperient. The juice of leaves is given to children for relief in bronchitis and diarrhoea. Paste of leaves is used as a remedy for rheumatism. Decoction of leaves in cough. Used in rheumatism. It is also capable of imparting youthful vitality and longevity. It forms an important ingredient of Ayurvedic preparations, such as Brahmigrihitam, Brahmirasayanm, etc.

Digitalis purpurea Linn.

SYNONYMS : *Digitalis tomentosa Link & Hoffm.*

ENGLISH : Foxglove, Digitalis.

INDIAN : Ghantaävinna, Tilapushpi.

SANSKṚTA : Hṛtpatrī, Tilapuṣpī.

HABIT : A tall biennial (annual in subtropical climate) herbaceous plant upto a height of 1.8 m.

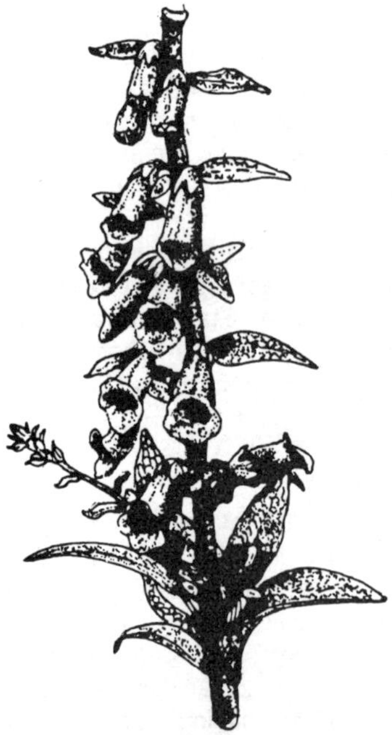

HABITAT : Native to British Isles and Western Europe. Cultivated in Kashmir, Darjeeling distric and Nilgiris.

CHEMICAL COMPOSITION : Leaves contain approximately 30 glycosides with a total glycoside content of 0.1-0.63 %. Three main glycosides used in medicine are digitoxin, gitoxin and gitalin. Other important glycosides are digitoxigenin, glucodigitoxigenine-bis-digitoxoside, glucogitaloxigenin-bis-digitoroside, glucoevatromonoside, glucolanadoxin, verodoxin and stropeside. The leaves also contain anthraquinones.

Parts used : Leaves

Ayurvedic properties :
Rasa - Tikta. Vīrya - Śīta/Uṣṇa.
Guṇa - Laghu, Rūksa. Vipāka - Kaṭu.

Actions/Uses : Kaphavātaśāmaka, Pittavardgaka, Hṛdayapoṣaka, Vājikaraṇa, Ākṣepaśāmaka, Garbhāśayasankocaka, Jvaraghna.

Therapeutics :

Hrtpatri is a sedative of heart's action and a diuretic. Leaves are used in cardiac complaints arising from kidney diseases; also in dropsy and urinary suppression. It acts as an acro-narcotic poison in large doses. It is a valuable remedy in dropsy, especially when it is connected with affections of heart. Digitalis products are cardiotonics, increase contractability of the heart and improve tone of the heart muscles. It also stimulates vagus nerve. These lead to more forceful contraction of the heart without increasing the rate. In congestive heart failures it improves cardiac output and thus relieves congestion. It has well established use in treatment of heart ailaments like arrhythmia, auricular fibrillation, and supra-ventricular trachardia. Digitoxin is used as cardiotonic to improve tone of cardiac muscles causing the heart to be emptied more effectively. Gitalin is employed in congestive heart failure.

HOMEOPATHIC USE : Digitalis is a Homeopathic

remedy. It is seldom called for in diseases in which are absent its characterristic heart and pulse symptoms which are of prime importance. Digitalis disorders are very often traceable to cardiac troubles characterised by slow heart, the pulse being irregular, intermittent and something like 50. It can never be replaced by any medicine for its pulse symptoms.

Family : XCIII-Simaroubaceae

Ailanthus excelsa Roxb.

SYNONYMS : *Pongelion wightii van Tiegh.*

ENGLISH : Tree of heaven.

INDIAN : Mahaarukha, Limbado, Mahanimba.

SANSKṚTA : Aralu, Mahānimba, Nimbākāradala, Viśvaka, Bhalluka, Panktipatraka, Bhangara, Asārā, Adūśa.

HABIT : A lofty deciduous tree.

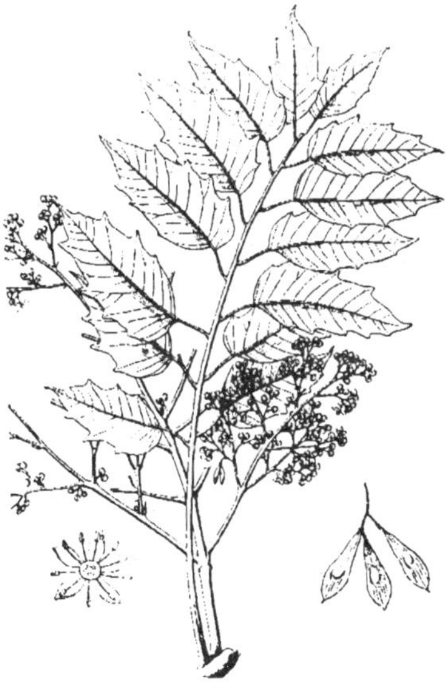

HABITAT : Cultivated in road-sides and garden; also grows wild in some forest of Deccan, Bihar, Western Ghats from N. Kanara and Mysore to Travancore.

CHEMICAL COMPOSITION : Bark contains triacontane, hesatriacontane and a non-glycosidal bitter, excelsin. It yields β-sitosterol, 2,6-dimethoxyhenzoquinone, vitexin, malanthine and melanthin.

Parts used : Bark and leaves.

Ayurvedic properties :
Rasa - Tikta, Kaṣāya. Vīrya - Śīta.
Guṇa - Rūkṣa, Laghu. Vipāka - Kaṭu.

महानिम्बस्तु शिशिरः कषायः कटुतिक्तकः।
अस्त्रदाहवलासघ्नो विषमज्वरनाशनः।।
(राजनिघण्टु)

Mahānimbastu śiśiraḥ kaṣāyaḥ kaṭutiktakaḥ,
Asradahāvalāsaghno viṣamjvaranāśanaḥ.
(Rajanighantu)

अरलुः कफहृद् ग्राहिदीपनः कृमिकुष्ठनुत्।
(शो)

Araluḥ kaphahṛd grāhi dīpanaḥ kṛmikuṣṭhanut.
(Sho.)

Actions/Uses : Kaphapittaghna, Vraṇaśodhana, Dīpana, Pācana, Sangrāhī, Bastirogahara, Raktaśuddhikara, Kuṣṭhaghna, Sandhānīya, Arśoghna, Kṛmighna, Lekhana, Raktastambhana, Kāsahara.

Therapeutics :

Leaves of Aralu are specially useful in asthma, bronchitis, dyspepsia and in the treatment of weakness after child birth. Paste of leaves is applied as polutice in erysipelas, goitre, swellings and ulcers. Bark is aromatic, bitter tonic, febrifuge, expectorant and antispasmodic; used for dyspeptic complaints and as astringent in diarrhoea and dysentery and given in chronic bronchitis and asthma. Bark is used as tonic especially in debility after child-birth.

Balanites roxburghii Planch.

SYNONYMS : *Balanites aegyptiaca (Linn.)* Delile.; *Balanites ingudi.; Ximenia aegyptiaca* Linn.

ENGLISH : Desert date.

INDIAN : Hinganbet, Ingudi, Hingoli.

SANSKṚTA : Ingudī, Hingūpatra, Viṣakaṇta, Anilāntaka, Vyavahārī, Gaura, Suputra, Śulārī, Tāmasadruma, Kroṣṭuphala, Angāravṛukṣa, Bhallaka, Tinguda, Pūtikanṭaka, Angavṛkṣa, Korakavṛkṣa, Tikta, Tiktamajjā, Tāmasataru, Vṛudhakanṭaka, Ingula, Tilvaka.

HABIT : A spiny, evergreen tree.

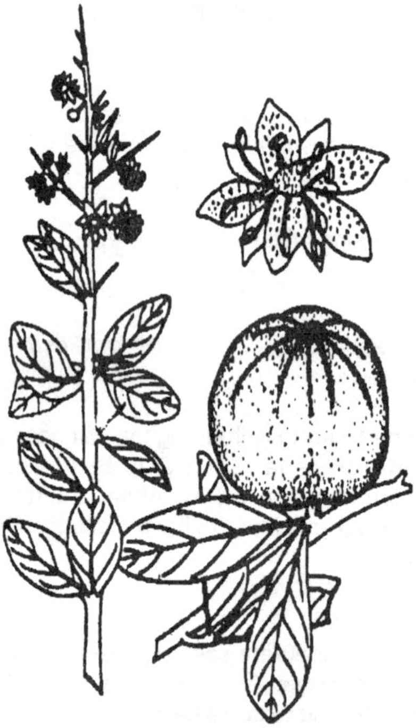

HABITAT : Common in open sandy plains of Indian peninsula, western Rajasthan, West Bengal, Maharashtra and drier parts of India.

CHEMICAL COMPOSITION : Seed kernels contain a steroidal saponin- balanitesin composed of diogenin, glucose, xylose and rhamnose. Nitogenin glucoside isolated. Diosgenin and stigmasterol from leaves.

Parts used : Bark, fruit, seed, leaves and seed oil.

Ayurvedic properties :
Rasa - Tikta, Kaṭu. Vīrya - Ūṣṇa.
Guṇa - Laghu, Snigdha. Vipāka - Kaṭu.

इङ्गुदी मदगन्धी स्यात् कटूष्णा फेनिला लघुः।
रसायनी हन्ति जन्तुवातामयकफव्रणान्।।
(राजनिघण्टु)

Ingudi madagandhi syat katsna phenila laghuh,
Rasayani hanti jantuvatanayakaphavranan.
(Rajanighantu)

इङ्गुदः कुष्ठभूतादिग्रहव्रणविष क्रिमीन्।
हन्त्युष्णः श्वित्रशूलघ्नस्तिक्तकः कटुपाकवान्।।
(भावप्रकाश)

Iṅgudaḥ kuṣṭhabhūtadigraha vraṇāviṣakṛimīn,
Hantyuṣṇaḥ śvitraśūlaghnastiktakaḥ kaṭupākavān.
(Bhavaprakasa)

Actions/Uses : Kṛmighna, Kaphavāataghna, Jantughna, Keṣya, Vraṇaropaṇa, Dīpana, Sramsana, Raktaśudhikara, Mūtrala.

Therapeutics :
Unripe fruits, seeds, leaves and bark of Ingudi are purgative and anthelmintic. Fruit is edible, useful in boils, leucoderma and other skin diseases, pulp in whooping cough. Fruit pulp mixed with goat milk is rubbed on chest to cure pneumonia in children. Seeds possess expectorant properties and useful in cough, whooping cough and colic. Powdered seed is used for easy childbirth. Decoction of root is emetic. Plant is used in snake-bite. It is used in Sudan and Egypt as tonic, laxative and stomachic, also in asthma, renal diseases and hypertension.

Family : XCIV - Solanaceae

Atropa belladonna Linn.

SYNONYMS : *Atropa sciminsts Royle ex Lindl.; Atropa belladona C.B. Clarke (FBI) in part non Linn.*

Similar : Atropa acuminata Royle ex Lindle. (Indian Belladonna).

ENGLISH : Belladonna, Deadly Nightshade.

INDIAN : Soochi, Angurshef, Limuuna, Sag-angur, Girbuti.

SANSKRTA : Sūci, Sāga-angura, Anguraśaphā, Giravūtī, Zalākaphala.

HABIT : A tall, erect, perennial herb.

HABITAT : A. belladonna is Native of central and southern Europe, indigenous in Great Britain. Cultivated in Kashmir at Baramulla, Darang and Yarikh. A. acuminata occurs naturally in north western Himalay in Jammu & Kashmir and Himachal Pradesh.

CHEMICAL COMPOSITION : Leaf and root contain total alkaloids from 0.3 % to 1 %. About three-fourth of these is present in the form of l-hyoscyamine, the remainder being hyoscine (Scopolamine) and atropine. Tropain alkaloids (0.2 -0.5 %), Scopolamine, Apoatropine, Belladonine.

Parts used : Root, leaves, berries and flowering tops.

Ayurvedic properties :
Rasa - Tikta, Kaṭu. Vīrya - Ūṣṇa.
Guṇa - Laghu, Rūkṣa. Vipāka - Kaṭu.

सूची तिक्ता कटुर्लघ्वी रुक्षोष्णा मदकारिणी।
कफवातहरा पित्तकरा धातुप्रशोषणी।।
लालाप्रसेकशमनी शूलप्रशमनी भृशम्।
प्रलाप जननी हृद्या वेदनास्थापनी तथा।।
कासे श्वासे व्रणे शोथे वातव्याधी हृदामये।
अग्निमान्द्ये विषे शूले प्रस्वेदे च प्रशस्यते।
(द्रव्यगुण विज्ञान)

Sūcī tiktā kaṭurlaghvī rukṣoṣṇā madakāriṇī,
Kaphavātaharā pittakarā dhātupraśoṣaṇī,
Lālāprasekaśamanī sūlaprasamanī bhṛśam,
Pralāpajananī hṛdyā vedanāsthāpanī tathā.
Kāse śvāse vraṇe śothe vātavyādhau hṛdāmaye,
Agnimāndye viṣe śūle prasvede ca praśasyate.
(Dravyaguna)

Actions/Uses : Kaphavātahara, Pittavardhaka, Viṣakara, Avasādaka, Kāsahara, Tārakavikāsi, Sankocavikāsapratibandhaka, Hṛdayabalya, Nāḍiśaithilyakara, Śothapratibandhaka, Śvāsahara, Raktapratibandhaka.

Therapeutics :
Suci is used as an anodyne in coughs, whooping cough, febrile conditions, night-sweats, etc., also for spermatorrhoea. It is antispasmodic, antimuscarinic, antisialagogue, cerebral sedative. Various preparations of belladonna are employed in medicine mainly because of anticholinergic action of tropane alkaloids. Berries are highly poisonous; root and leaves are narcotic, diuretic, sedative, mydriatic. In the diseases of respiratory tract like asthma, it is used to check glandular secretion and spasms of bronchial tube. Similarly, for the same purpose it is used in intestinal disor-

ders, peptic ulcers, colic pain, pain of urinary bladder and kidney stones to check spasms and pain as well as glandular secretions. It is also used in epileptic conditions, Parkinson's disease, night-sweats and bradycardia. In suitable doses it is used as a sedative and in the form of plaster and ointments it is used as an antiinflammatory agent. It is often used as antidote for opium type of poisoning. Externally it is used in gouty and rheumatic inflammations.

Capsicum annuum Linn.

SYNONYMS : *Capsicum longum DC.; Capsicum grossum* Willd.; *Capsicum annum* Linn. *var grossum (Willd.)* Sdendt; *Capsicum condiforme* Mill.

ENGLISH : Red Chilli, Sweet-pepper, Paprika.

INDIAN : Lal-mirchi.

SANSKṚTA : Kaṭuvīrā, Lankā, Raktamarica.

HABIT : An annual erect bushy herb.

HABITAT : Cultivated throughout India.

CHEMICAL COMPOSITION : Capsicin, a volatile alkaloid, capsacin, a crystalline substance, solanine, a volatile oil; fixed oil; fatty acid and a resin. Ascorbic acid (0.1 - 0.5 %).

Parts used : Fruit.

Ayurvedic properties :
Rasa - Kaṭu. Vīrya - Ūṣṇa.
Guṇa - Laghu, Rūkṣa, Tīksna. Vipāka - Kaṭu.

कटुवीरो कटुस्तीष्णो गुरुष्णः कफनाशनः।
दीपनो रोचनो रुच्यः तृष्णादाहकरस्तथा।।
रोगिणां नैव पथ्योऽसौ, सोवितस्वतियोगतः।
कुर्यात् स्वस्थनरस्यापि मूत्रकृच्छ्रदिकान् गदान्।। (स्व)

Kaṭuvīro kaṭustīkṣṇo gurūṣṇaḥ kaphanāśanaḥ,
Dīpani rocano rucyaḥ tṛṣṇādāhakarastathā.
Rogināṃ naiva pathyo asau, sevitasvatiyogataḥ,
Kuryāt svasthanarasyāpi mūtrakṛcchādikān gadan.
(Svayamkrti)

लङ्का तीक्ष्णा कटुष्णातिलालास्त्रावकरी मता।
विदोहजननी पित्तकारिणी कफवातहृत्।।
(द्रव्यगुण विज्ञान)

Lañkā tīkṣṇākaṭūṣṇātilālāsrāvakari matā,
Vidāhajananī pittakāriṇī kaphavātahṛt.
(Dravyaguna)

Actions/Uses : Kaphavātaśāmaka, Pittaprakopī, Anulomana, Lekhana, Dīpana, Pācana, Raktotkleśakāraka, Hṛdayottejaka.

Therapeutics :
Katuvira is carminative, nerve stimulant, increases capillarity of blood vessels. Fruit has a bitter sharp taste, analgesic, expectorant, stomachic, tonic, carminative, a poweful local irritant, heart and general stimulant. It is useful in atonic dyspepsia. It enriches blood, lessens inflammation and pain. It is a source of Vitamin C. It acts as an acid stimulant and externally as a rubefacient.

Capsicum frutescens Linn.

SYNONYMS : *Capsicum minimum* Roxb.; *Capsicum fastigiantum* Blume.

HABIT : A small spreading shrub.

HABITAT : Cultivated throughout India.

ENGLISH : Cayenne Pepper. Bird Chillies.

INDIAN : Lal-mirchi, Mirchi.

CHEMICAL COMPOSITION : Capsaicin and dihydrocapsaicin in fruit.

Therapeutics :

Fruit causes a burning sensation, increases appetite, useful in indigestion, diarrhoea, chronic ulcers, loss of consciousness and delirium. Crude juice is antimicrobial, used in stomach ache and backache and for chest trouble and cough. Leaves are used in headache, night blindness, pain, adenoids, soress, dysduria, and bronchitis.

HOMEOPATHIC USE : It affects mouth, tongue, throat, stomach and anus very deeply. Inflammation and burning of all the mucous membranes of these organs. Secretions from mucous membranes are sometimes even bloody. Useful in ague; it has febrile rigor in evening, with thirst, heat and at the same time shivering. Chill, fever and thirst, all three at a time is a characterisitic symptom.

Datura metel Linn.

SYNONYMS : *Datura alba* Nees.; *Datura fastuosa* Linn. *var. alba C. B. Clarke.*

ENGLISH : Indian Datura.

INDIAN : Kaallaa Dhotraa.

SANSKṚTA : Dhuṣṭura, Dhattūra, Dhūrta, Kanaka, Śivapriya, Kitava, Madana, Kanakakauṭhaphala, Mātula, Mohana.

HABIT : An sub-glabrous spreading perennial herb.

HABITAT : Cultivated in India, often found wild.

CHEMICAL COMPOSITION : The principal alkaloid is scopolamine. Hyosciamine, atropine and norhyosciamine are usually small. The seeds contain a fixed oil. Fastunine, fastudine and fastusidine are isolated from seeds. Leaves contain Vitamin C.

Parts used : Root, leaves, flowers, and seeds.

Ayurvedic properties :
Rasa - Kaṣāya, Tikta, Madhura. Vīrya - Uṣṇa.
Guṇa - Laghu, Rūkṣa, Vikāsi, Vyāvayī.
Vipāka - Kaṭu.

धस्तुरः कटुरुष्णश्च कान्तिकार री व्रणार्तिनुत्।
त्वग्दोषखर्ज्जूक कडूतिज्वरहरी भ्रमप्रदः।।
(राजनिघण्टु)

Dhasturaḥ kaṭurūṣṇaśca kāntikāri vraṇārttinut,
Tvagdoṣa kharjjū kaṇḍūtijvaraharī bhramapradaḥ.
(Rajanighantu)

Actions/Uses : Kaphavatasamaka, Vedanasthapana, Svasahara, Sothahara, Jvarapratibandhaka, Kustaghna.

Therapeutics :

Dhastura is bitter, acrid, astringent, germicidal, anodyne, antiseptic, antiphlogistic, narcotic and sedative. It gives good complexion, improves digestion, cures skin diseases such as itching, scabies, ulcers and leprosy, dandruff, fever, dysuria, piles, anaemia and inflammatory swellings. It is also useful in respiratory ailments, rheumatism, elephantiasis, insanity, ear ache and eye diseases. Seeds, leaves and roots are considered useful in insanity, fever with catarrhal and cerebral complications, diarrhoea, skin diseases, lice etc. Root is boiled in milk and administered with clarified butter and treacle in insanity. In traditional medicine, it is a reputed drug in the treatment of rabid dog-bites and poisonous insect bites. The dried leaves and flowering tops are known for their narcotic and anti-spasmodic properties. They are used for the same purposes as leaves of belladonna and stramonium.

Datura stramonium Linn.

SYNONYMS : *Datura tatula* Linn.

ENGLISH : Stramonium; Thorn Apple.

INDIAN : Dhotraa.

SANSKṚTA : Dhattūra, Kitava, Dhūrta, Unmatta, Śivaśekhara, Śāma, Kanaka, Śaṭha, Mātulaka, Madana, Kharjūghna, Śaiva, Kāhalāpuṣpa, Khala, Kanṭaphala, Mohana, Kalabha, Haravallabha, Matta, Devatā, Kali, Madakāra, Turī, Mātula, Tarala, Dhuṣṭara, Mahāmohī, Mātulaputraka, Śivapriya.

HABIT : An glabrous or farinose annual herb, 50-200 cm. in height.

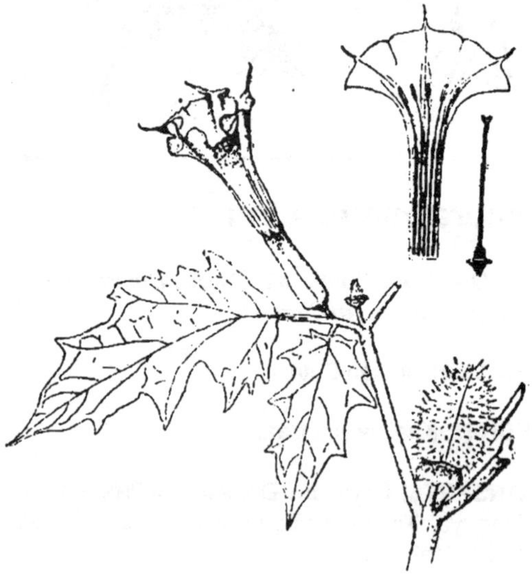

HABITAT : Essentially a temperate plant but is found growing in the vicinity of cultivation, on rank soil, where refuse is deposited, in all parts of world.

Himalays, hilly districs of Cental and South India.

CHEMICAL COMPOSITION : The major constituent are tropane alkaloids which vary from 0.25-o.4 %, two-third of which is hyoscyamine and one-third is hyoscine. A number of withanolides have also been reported from the plant.

Parts used : Leaves, flowers, seeds and root.

Ayurvedic properties :
Rasa - Kaṣāya, Madhura, Tikta. Vīrya - Uṣṇa.
Guṇa - Laghu, Rūkṣa, Vikāsi, Vyāvayi.
Vipāka - Kaṭu.

धत्तूरो मदवर्णाग्निवातकृत् ज्वरकुष्ठनुत्।
कषायो मधुरस्तिक्तो यूकालिक्षाविनाशकः।
उष्णो गुरुर्व्रणश्लेष्मकण्डुक्रिमिविषापहः।।
(भावप्रकाश)

Dhattūro madavarṇāgni vātakṛt jvarakuṣṭhanut,
Kaṣāyo madhurastikto yūkālikṣavināśakaḥ,
Uṣṇo gururvraṇaśleṣmākaṇḍukrimiviṣāpahaḥ.
(Bhavaprakasa)

Actions/Uses : Kaphavātaśāmaka, Vedanāsthāpana, Śvāsahara, Śothahara, Jvarapratibandhaka, Kuṣṭhaghna.

Therapeutics :

Dhattura is toxic, narcotic, aphrodisiac; applied topically it removes pain of tumours and piles. It is narcotic, anodyne, and anti-spasmodic and is used chiefly to relieve the spasm of the bronchioles in asthma. Leaves are used in the treatment of parkinsonism. Ointment is used in the treatment of haemorrhoids. Leaves after roasting are applied locally to releive eye pain. They are applied to boils, sores and juice of flowers is used for earache. The juice from fruits is applied to scalp for curing dandruff and falling hair. Kanaka Asava, an Ayurvedic preparation is used as demulcent, expectorant, anti-spasmodic and anodyne in coughs, asthma and phthisis. The drug is mostly used in crude form as anticholinergic agent in case of asthma and intestinal disorders.
Ayurvedic properties and uses are the same as that of D. metal and English names for both species are same.

HOMEOPATHIC USE : Stramonium is similar to belladonna in the symptoms produced by it, and in its general physiological and therapeutic action. It has incomparable curative action in the peculiar mind affections and in convulsive ailments. It is supereminent for some spasmodic muscular movements and suppressed secretions and excretions; in such cases, absence of pain is a prominent symptom.

Hyoscyamus niger Linn.

SYNONYMS : *Hyoscyamus agrestis* Kit.; *Hyoscyamus pallidus* Kit.

ENGLISH : Henbane, Hogbean, Hyoscyamus.

INDIAN : Khoraasani owaa, Khurasaniajvayan.

SANSKṚTA : Pārasayikayāvanī, Khurāsānīyavānī, Turuśkā, Tīvrā, Pārasikaya, Dipya, Madakāriṇī, Yavāṇikā, Yavāhvā.

HABIT : An erect more or less hairy herb 30 cm. to 1 m. tall.

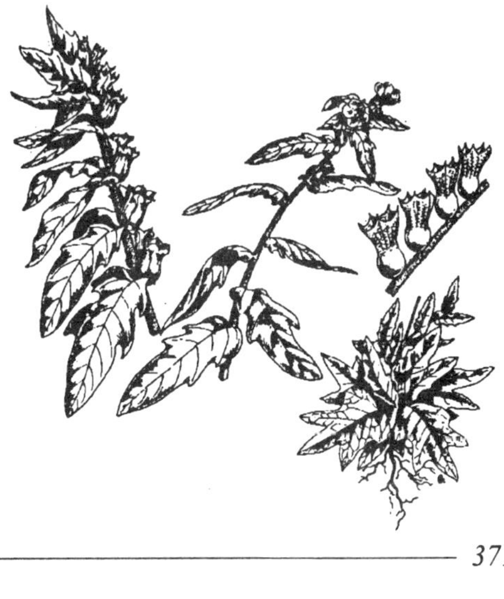

HABITAT : A native of Scandinavia, Europe and Great Britain. Cultivated in USA and in India from Kashmir to Garhwal.

CHEMICAL COMPOSITION : Leaves and flowering tops contain 0.1 % to 0.15 % total alkaloids. The important alkaloids are hyoscyamine, hyoscine and scopolamine.

Parts used : Leaves, flowers and seeds.

Ayurvedic properties :
Rasa - Tikta, Kaṭu, Kaṣāya. Vīrya - Ūṣṇa.
Guṇa - Rūkṣa. Vipāka - Kaṭu.

खूरासानी यमानी तु कटुः रुक्षा च पाचिका।
ग्राहिकोष्णा मादिका च गुर्वी वातकरी तथा।
कफनाशकरी प्रोक्ता गुणास्तन्ये यमानीवत्।।
(द्रव्यगुण)

Khurāsānī yamānī tu kaṭuḥ rukṣā ca pācikā,
Grāhikoṣṇā mādikā ca gurvī vātakarī tathā.
Kaphanāśakarī proktā guṇāstvānye yamānīvat.
(Dravyaguna)

पारसिकयवानी तु यवानीसदृशा गुणैः।
विशेषात् पाचनी रुच्या ग्राहिणी मादिनी गुरुः।।
(भावप्रकाश)

Pārasikayavānī tu yavānisadrśā guṇaiḥ,
Viśeṣāt pācanī rucyā grāhiṇī mādinī guruḥ.
(Bhavaprakasa)

Actions/Uses : Kaphaghna, Vātaghna, Pittavardhaka Śothahara, Grāhī, Vedanāsthāpana, Kāmavāsādaka, Ākṣepahara, Nidrājanana, Śūlaghna.

Therapeutics :

Parasikaya plant is used in indigenous medicine along with other ingredients for diabetes. It is used to a limited extent as tincture or extract for the same diseases where belladonna is employed, or other atropane alkaloids are used. It is mainly used in diseases of respiratory and intestinal tracts as well as urinary bladder pain to control spasms and secretion. Leaves have anodyne, narcotic, antispasmodic, antimuscarinic, antiscialagogue, cerebral sedative and mydriatic properties. They are useful in asthama, nervous affections, whooping cough. Extract or tincture is specially prescribed in convulsions, epileptic mania and hiccup. Seeds have a sharp pungent taste, tastey, heating, stomachic, astringent, toxic; cause vata, cure kapha. They ae principally employed as a sedative in nervous affections and irritable conditions, such as asthma and whooping cough and is substituted for opium in cases where the latter is inadmissible. It is used to counteract griping action of purgatives and relieve spasms in the urinary tract. Plasters, poultice, and medicated oil prepared from leaves and seeds are employed for external application to inflammatory swellings.

HOMEOPATHIC USE : Hyoscyamus acts very prominently on brain and it is indicated for from simple mania to complete madness including delirium in fevers. Alternation of calmness and fury. Extremly furious and naked he passes day and night without sleep and crying out. Whenever there is diarrhoea in typhoid fever Hyos. must be considered. For hydrophobia when there is dryness in throat with thirst, impedement to deglutition Hyos. is useful. Throat is so contracted and dry that a mouthful of tea almost chokes him.

Lycopersicon esculentum Mill.

SYNONYMS : *Lycopersicon lycopersicum (Linn.) Farwell.; Solanum lycopersicum* Linn.; *Lycopersicon licopersicum (Linn.) Karsten. Lycopersicon licopersicum (Linn.) Britton & Brown.*

HABIT : An unarmed, spreading, pubescent sub-erect annual herb.

HABITAT : Cultivated in many parts of India.

ENGLISH : Tomato.

INDIAN : Taamaatu, Taambetaa, Tamatar, Vel Vangi, Vilaayati Baingan.

SANSKRTA : Raktavṛtanaka.

CHEMICAL COMPOSITION : Ripe and green fruits contain minerals and vitamins. Calcium, phosphorus, carotene, thiamine, riboflavin, niacin amd vitamin C. Crude tomatine.

Parts used : Fruits.

Ayurvedic properties :

Rasa - Amla, Madhura. Vīrya - Śīta.
Guṇa - Laghu. Vipāka - Madhura.

Actions/Uses : Dīpana, Pācana, Raktaśodhana, Sāraka.

Therapeutics :

Raktavrtanaka or tomatoes are the richest of all foods in vitamins. They are effective blood cleanser. They are richest of all vegetables in the natural health acids which keep our stomachs and intenstines in condition. They are the most extraordinary corrective for kidneys, a gentle, natural stimulant, which helps to wash away poison which cause disease and contaminate our system. Tomato pulp and juice is an appetizing and nourishing beverage. They are mild aperient, blood purifier, cholagogue, digestive. They promote gastric secretions, stimulate torpid liver. They are useful in asthma, bronchitis, atonic dyspepsia and apthous stomatitis.

Nicotiana tabacum Linn.

SYNONYMS : *Nicotiana macrophylla* Lehm.; *Nicotiana auriculata* Bert.

ENGLISH : Tobacco, Tabacum.

INDIAN : Tambaakhoo, Tamaaku.

SANSKṚTA : Tāmraparṇa, Dhumrapatra.

HABIT : A stout viscid annual with a thick erect stem.

HABITAT : Cultivated throughout India.

CHEMICAL COMPOSITION : Nicotine, seed oil. Phenol, polyphenol and tannin.

Parts used : Leaves.

Ayurvedic properties :
Rasa - Tikta, Kaṭ . Vīrya - Ūṣṇa.
Guṇa - Laghu, Tīkṣṇa, Vyāvayi, Vikāsi.
Vipāka -Kaṭu.

धूम्रपत्रा रसे तिक्ता शोफघ्नी क्रिमिनाशिनी।
उष्णा कासहरा चैव रुच्या दीपनकारिणी।।
(राजनिघण्टु)

Dhūmrapatrā rasetiktā śophghnī krimināśinī,
Ūṣṇā kāsaharā caiva rucyā dīpanakāriṇī.
(Rajanighantu)

शिरोगच्छित् क्षवनः कलञ्जो वम्यो विषं विश्वविषस्य हन्ता।
कलञ्जसंवेष्टितधूमपानात् स्याद्दन्तशुद्धिर्मुख रोगहानिः।।
(आ.वि.)

Śirogacchit kṣavanaḥ kalañjo,
Vamyo viṣam viśvavisạsyahantā,
Kalānjasamveṣtitadhūmapānāt
Syāddantaśudhirmukha rogahāniḥ.
(Ayurvedvijnana)

Actions/Uses : Kaphavātaśāmaka, Pittavardhaka, Vātānulomana, Śothahara, Vedanāsthāpana, Bodhaka, Kṛmighna, Svasanottejaka, Mādaka, Chardikara.

Therapeutics :
Leaves of Dhumrapatra are laxative, tonic, emetic, narcotic, carminative, antelmintic, antiseptic, antispasmodic and sedative. They have a sharp bitter taste, and are useful in bronchitis, asthma, caries of teeth, skin diseases. They cause bad eyesight. Decoction of leaves is locally applied for muscle relaxation in dislocation, strangulated hernia and orchitis. It relieves pain and irritation in rheumatic swelling. It is useful in skin diseases and syphilitic nodes. Ointment, made by simmering leaves in lard has been used in curing old ulcers and painful tumours. Oil is useful in arthralgia, gout, lumbago and rheumatism. As a medicine, tobacco owes its value to the properties of alkaloid nicotine which is a most energetic poison, a powerful sedative and antispasmodic.

Solanum dulcamara Linn.

SYNONYMS : *Solanum persicum* Willd. *ex R. & S.; Dulcamara flexuosa* Moench.

ENGLISH : Bittersweet, Dulcamara.

INDIAN : Anabesaalib, Kaakamaachi.

SANSKṚTA : Kākajanghā, Kākamāci-viśeṣa.

HABIT : A climbing shrub.

HABITAT : From Kashmir to Garhwal and Sikkim to Baluchistan.

CHEMICAL COMPOSITION : Leaves, stems and fruits contain glycoside alkaloid solanine, alkaloid solanidine. Herb contains bitter principle dulcamarin and two other saponins dulcamarinic acid and dulcamaretinic acid. The plant contains alkaloids α-, β-, and y-soladulcine.

Parts used : Roots, stems, twigs and berries.

Ayurvedic properties :

Rasa - Madhura, Tikta.	Vīrya - Śīta, Anūṣṇa.
Guṇa - Laghu	Vipāka - Kaṭu.

काकजङ्घा च तिक्तोष्णा रक्तपित्तज्वरापहा।
कृमिदोषहरा वर्ण्या विषदोषहरा मता।।

Kākajaṅghā ca tikttoṣṇā raktapittajvarāpahā,
Kṛmidoṣaharā varṇyā viṣadośaharā matā.
(Dhanvantari)

Actions/Uses : Svedajanana, Mūtrajanana, Raktaśuddhikara.

Therapeutics :

Berries of Kakajangha are alterative, diuretic, diaphoretic; useful in skin diseases, psoriasis, lepra, syphilitic affections, chronic rheumatism, enlargement of liver and as hydragogue cathartic. Twigs are diuretic, resolvent, narcotic; promote all secretions and are used in rheumatism, obstinate cutaneous eruptions, scrofula etc. Stems are antirheumatic and are applied in the form of decoction for carbuncles and boils. Root is good for skin diseases.

Solanum indicum Linn.

SYNONYMS : *Solanum sodomenum L. nom ambig.; Solanum anguivi* Lam; *Solanum violaceum* Ortega.

ENGLISH : Indian night shed, Poison-berry.

INDIAN : Dorle, Ringani, Barhanta, Birhatta.

SANSKṚTA : Bṛhatī, Kṣudrabhandākī, Sinhī, Mahatī, Sthūlakaṇtakā, Vrihati, Bhanṭaki, Kanṭaphala, Duśpradhrśā, Candākī, Krāntā, Vārtākī, Ḍoralī, Ākulī, Rāśtrikā, Mahotikā, Bahuputrī, Kanṭatanu, Kanṭālu, Varavṛuntakī, Kāntā, Viśadā, Sinhikā, Sthūlabhanṭākinī, Hingulī, Vanavrintaki.

HABIT : A biennial erect spiny herb.

HABITAT : Cultivated throughout tropical India.

CHEMICAL COMPOSITION : Enzyme in fruits. Alkaloids solanine, solanidine in roots and leaves. Fruits contain 1.8 % of alkaloids and can form a good source material for cortisone and sex hormone preparations.

Parts used : Root, leaves, fruits and Seeds.

Ayurvedic properties :
Rasa - Tikta, Kaṭu. Vīrya - Ūṣṇa.
Guṇa - Laghu, Rūksa. Tīkṣṇa. Vipāka - Kaṭu.

वृहती ग्राहिणी हृद्या पाचनी कफवातजित्।
कटुतिक्तास्यवैरस्यमलारोचकनाशिनी।।
उष्णा कुष्ठज्वरश्वासशूलकासाग्नि मान्घजित्।।
(भावप्रकाश)

Bṛhatī grāhiṇī hṛdyā pācanī kaphavātajit,
Kaṭutiktāsyavairasya malārocakanāśinī,
Ūṣṇā kuṣṭhajvaraśvāsa śūlakāsāgnimāndyajit.
(Bhavaprakasa)

Actions/Uses : Kaphaghna, Vātaghna, Vedanāsthāpana, Uttejaka, Kaṇḍūghna, Keṣya, Dīpana.

Therapeutics :

Bṛhati plant and root, both are pungent, bitter, stimulant, digestive, astringent, anthelmintic, carminative, diaphoretic and expectorant. They are beneficial in catarrhal affections, asthma, dry cough, dropsy, colic, dysuria, flatulence, cardiac troubles, leucoderma, worm complaints, fevers, difficult parturition, toothache and ischuria. It removes foulness of mouth. Plant is used as aphrodisiac. Juice of leaves with fresh juice of ginger is taken to stop vomiting. Leaves and fruit rubbed up with sugar used as external application for itch. Fruits are digestive and laxative and their juice is beneficial in alopecia. Decoction of seeds is used in dysuria and vapour from seeds is useful in otalgia.

Solanum melongena Linn.

ENGLISH : Bringal, Egg plant.

INDIAN : Waange, Baingan, Bhanta, Badanjan.

SANSKṚTA : Vārtākī, Bhanṭaki, Vatigama, Jukutam, Hingoli.

HABIT : A herbaceous prickly or sometimes unarmed perennial.

HABITAT : Widely cultivated in India.

CHEMICAL COMPOSITION : Solasodine from fruits. Arginine, aspartic acid.

Parts used : Root, leaves, fruit and seeds.

Ayurvedic properties :

Rasa - Kaṭu.	Vīrya - Uṣṇa.
Guṇa - Laghu, Tīkṣṇa.	Vipāka - Kaṭu.

वार्त्ताकी कटुका रुच्या मधुरा पित्तनाशनी।
बलपुष्टिकरी हृद्या गुरुर्वातेषु निन्दिता।।
(राजनिघण्टु)

Vārtākī kaṭukā rucyā madhurā pittnāśanī,
Balapuṣṭikarī hṛdyā gururvāteṣu ninditā.
(Rajanighantu)

Actions/Uses : Vātahara, Kaphahara, Agnidīpana, Śūkrajanana.

Therapeutics :

Vartaki root is antiasthmatic and stimulant. Pounded and applied to nasal ulcers. Juice of root is used in otitis. Leaves are narcotic and sialagogue, beneficial in asthma, bronchitis and dysuria. Fruit is hypnotic, costive and excitive of wind. Long sized fruits are phlegmatic and generative of phthisis, cough and loss of appetite. Tender fruits are antiphlegmatic, alleviative of wind and ripe ones are billious. Unripe fruit is bitter, pungent, heating, sweetish, saltish, tasty; improves appetite, aphrodisiac; enriches blood; beneficial in vata and kapha; increases asthma and bronchitis. Seeds are stimulant, costive and cause indigestion.

Solanum nigrum Linn.

SYNONYMS : *Solanum subrum* Mill *ex* Wight.; *Solanum incertum* Dunal *ex* Graham.

ENGLISH : Black night shade.

INDIAN : Kaangannee, Kaamonnee, Makoi.

SANSKṚTA : Kākamācī, Dhyānksamācī, Vāyasāhvā, Vāyasī, Sarvatiktā, Bahuphalā, Kaṭphalā, Rasāayanī, Kākamātā, Gucchaphalā, Svādupākā, Sundarī, Vidrāviṇī, Matsyākṣī, Kuṣṭhanāśanī, Tiktā, Bahutiktā, Kākāhvā, Varā, Rasāyanavarā.

HABIT : A herbaceous or suffrutescent weed.

HABITAT : Common in waste places and road-sides throughout India.

CHEMICAL COMPOSITION : Solasomine and solamargine in leaves and glucoalkaloids from immature fruits.

Parts used : Whole plant, root-bark, leaves, flowers and fruit.

Ayurvedic properties :
Rasa - Tikta. Vīrya - Anūṣṇa.
Guṇa - Laghu, Snigdha. Vipāka - Kaṭu.

काकमाची कटुस्तिक्ता रसोष्णा कफनाशनी।
शूलार्शः शोथदोषघ्नी कुष्ठकण्डूतिहारिणी।।
स्वरदा शुक्रदा चैव चक्षुष्या च रसायनी।।
(राजनिघण्टु)

Kākamācī kaṭustikttā rasoṣṇa kapha nāśanī,
Śūlārśaḥ śothadoṣaghnī kuṣṭhakaṇḍūtihāriṇī,
Svaradā śukradā caiva cakṣuśyā ca rasāyanī.
(Rajanighantu)

काकमाची त्रिदोषघ्नी स्निग्धोष्णा स्वरशुक्रदा।
तिक्ता रसायनी शोथकुष्ठार्शोज्वरमेहजित्।।
कटुर्नेत्रहिता हिक्काच्छर्दिहृद्रोगनाशिनी।।
(भा. प्र.)

Kākamāci tridoṣaghnī snigdhoṣṇā svaraśukradā,
Tiktā rasāyanī śothakuṣṭhārśojvaramehajit,
Kāturnetrahitā hikkāccharddihṛdroganāśinī.
(Bhavaprakasa)

Actions/Uses : Tridoṣāśāmaka, Śothahara, Vraṇaśodhana, Dīpana, Vedanāsthāpana, Balya, Savarniekarana, Yakṛtottejaka, Virecana, Pittasāraka, Mūtrala, Raktaśodhana, Jvaraghna, Śvāsahara, Plīhāhara.

Therapeutics :
Kakamaci plant is emolient, diuretic, laxative, alterative, sedative, diaphoretic, diuretic, hydrrogogue and expectorant; locally anodyne and its decoction is antispasmodic and narcotic. Herb has antiseptic and antidysenteric properties and is given internally for cardalgia and gripe. Infusion of plant is used as an enema for infants having abdominal upsets. It is a household remedy for anthrax pustules and applied locally. Root bark is laxative; useful in diseases of ears, eyes and nose; good for ulcers on neck, burning of throat, inflammation of liver, chronic fever. Leaves are used as polutice over rheumatic and gouty joints and also as a remedy for skin diseases. Fresh juice produces dilatation of pupils. Hot leaves are applied with benefit to painful and swollen testicles. Flowers

are prescribed in cough and cold. Berries are bitter, tonic, diuretic, laxative, alterative, aphrodisiac; improve appetite and taste; useful for itch, dysentery, hiccough, vomiting, asthma, bronchitis, in fever diseases of heart and eye, pains, piles, inflammation, leucoderma and urinary discharges.

Solanum tuberosum Linn.

ENGLISH : White flower potato, Potato.

INDIAN : Aalu, Bataataa.

SANSKṚTA : Āluka, Āruka, Golālu.

HABIT : A much-branched, bushy herb.

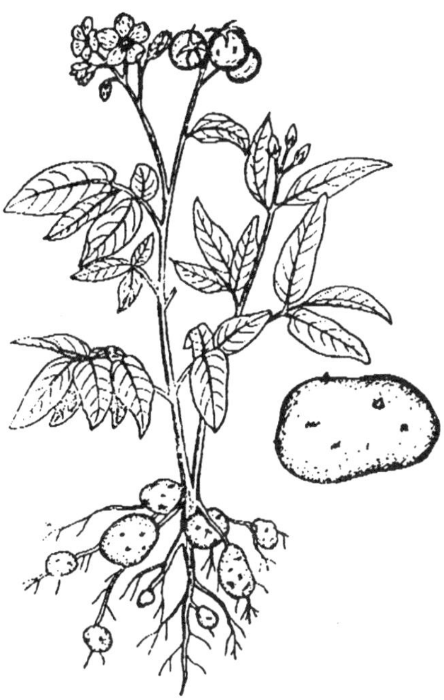

HABITAT : Cultivated throughout India excluding drier regions.

CHEMICAL COMPOSITION : Nitrogen content of different varieties of potatoes varies from 1.2 to 2.0 per cent. Less than half of this is present as protein, 35-60 per cent consists of free amino acids, amides and nitrogenous bases. Principal protein is globulin. Potatoes are rich in a variety of enzymes.

Parts used : Leaves and tubers.

Ayurvedic properties :
Rasa - Madhura. Vīrya - Śīta.
Guṇa - Rūkṣa, Śīta, Gurū. Vipāka - Madhura.

आलुकं शीतलं सर्वं विष्टम्भि मधुरं गुरु।
सृष्टमूत्रमलं रुक्षं दुर्जरं रक्तपित्तनुत्।।
कफानिलकरं वल्यं वृष्यं स्तन्यविवर्द्धनम्।।
(द्रव्यगुण)

Ālukam śītalam sarvam viṣṭambhi madhuram guru,
Sṛṣṭamutramalam rukṣam durjaram raktapittanut,
Kaphānilkaram balyam vṛṣyam
stanyavivarddhanam.
(Dravyaguna)

Actions/Uses : Balya, Vṛṣya, Malamūtranissāraka, Viṣṭambhī.

Therapeutics :
Aluka leaves in the form of extract are employed as antispasmodic in chronic cough. Tubers are antiscorbutic. They are employed as aperient, diuretic, galactagogue, nervous sedative and stimulant in gout; ground into a paste they are applied as plaster to burns caused by fire.

Solanum xanthocarpum Schrad. & Wendl.

SYNONYMS : *Solanum surattense* Burm. *f.*; *Solanum maccanni* Sant.; *Solanum virginianum* Linn.

INDIAN : Kaate ringannee, Bhueeringannee.

SANSKRTA : Kaṇṭakārī, Dusparśā, Vyāghrī, Draviṇī, Nindigdhikā, Kṣudrakaṇṭakā, Kṣudraphalā, Duh praharśiṇī, Pracodanī, Bahukaṇṭā, Bahugudākulī, Vārtākī, Raśṭrikī.

HABIT : A perennial prickly prostrate herb.

HABITAT : Common in waste places, road sides throughout India.

CHEMICAL COMPOSITION : Crude plant extract caused hypotension. Plant powder is anti-tussive and is beneficial in bronchial asthma and non-specific cough.

Parts used : Whole plant, root, leaves, stem, flowers, fruit and seeds.

Ayurvedic properties :
Rasa - Tikta, Kaṭu. Vīrya - Uṣṇa.
Guṇa - Laghu, Rūkṣa. Vipāka - Kaṭu.

कण्टकारी कटूष्णा च दीपनी श्वासकासजित्।
प्रतिश्यायार्तिदोषघ्नी कफवात ज्वरार्तिनुत्।।
(राजनिघण्टु)

Kaṇṭākarī kaṭūṣṇā ca dīpanī śvasakāsajit,
Pratiśyāyārtidoṣaghnī kaphvāta jvarārtinut.
(Rajanighantu)

कण्टकारी सरा तिक्ता कटुका दीपनी लघुः।
रुक्षोष्णा पाचनी कासश्वासज्वर कफानिलान्।
निहन्ति पीनसश्वासपार्श्वपीडाहृदामयान्।।
(भा.प्र.)

Kaṇṭakāri sarā tiktā kaṭukā dīpanī laghuḥ,
Rūkṣoṣṇā pācanī kāsāśvāsajvara kaphānilān,
Nihanti pīnasaśvāsapārśvapīḍahṛdāmayān.
(Bhavaprakasa)

Actions/Uses : Kaphaghna, Vātaghna, Vedanāsthāpana, Śothahara, Svedajanana, Śūkravirecana, Garbhaśṭhapana, Raktaśudhikara, Kṛmighna, Āmapācana.

Therapeutics :
Kaṇṭakari plant is astringent, stimulant, aperient, diuretic, pungent, bitter, digestive, expectorant, febrifuge, laxative, and cardiotonic. It is used in fever, cough, asthma and costiveness. It is an important therapeutic agent for dislodging tenacious phlegum and is extensively used in asthma, cough, and bronchitis. It is useful in cases of influenza, enteric fever and allied conditions, used against difficult urination, bladder stones, rheumatism, sore throat, enlargement of liver and spleen. A decoction of plant is used in gonorrhoea. It promotes conception in females. Root is pungent, bitter, heating, appetiser, laxative, stomachic, anthelmintic and aphrodisiac, useful in asthma, bronchitis, fever, lumbago, pains, piles, urinary concretions and diseases of heart. It is an effective diuretic, expectorant and febrifuge. Roots are one of the constituents of "Sashamula Asava". Leaves are anodyne. Their juice in combination with black pepper is prescribed

in rheumatism. Stems, flowers and fruits are bitter and carminative and are prescribed in burning of feet in cases associated with a vesicular and watery eruption. Juice of fruits is beneficial in sore throat. Fine powder of berries mixed with honey is given to children in chronic cough. Vapour of burning seeds is beneficial as an expectorant in asthma and cough and cures toothache.

Withania coagulans Dunal.

ENGLISH : Vegetable rennet.

INDIAN : Ashwagandhaa, Aaskand, Punir.

SANSKṚTA : Aśvagandhā, Vārahakarṇī, Baladā, Kuṣṭhagandhinī, Havī, Turakī, Gandhāntā, Vājināmā, Varadā, Vājigandhā, Kambukāṣṭhā, Varāhikā, Vanajā, Vājinī, Puṣṭidā, Hayagandhā, Puṇyā, Pīvarā, Palāśaparṇī, Vātaghnī, Śāmalā, Kāmarūpiṇī, Valyā, Kālapriyā, Gandhapatrī, Vārāhaputrī, Vājikarī, Kancukyā, Gokarṇī, Vṛṣā, Turangi-gandha.

HABIT : An erect, evergreen, tomentose shrub.

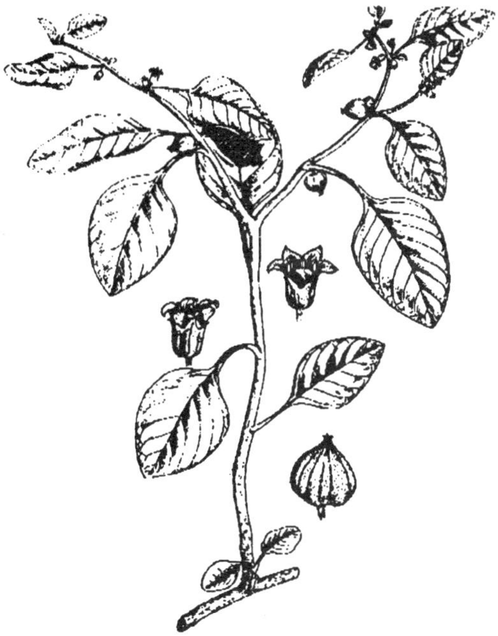

HABITAT : Punjab, Satlej valley, Sind and Baluchistan.

CHEMICAL COMPOSITION : Roots contain several pyrazole alkaloids. Withasomnine and steroidal lactones, withaferin A and withanolides. They also contain starch, reducing sugars, hentriacontane, glycosides, dulcitol, withaniol, an acid and a neutral compound. Withaferin is a bacteriostatic and antitumerous agent.

Parts used : Root, leaves, fruits and seeds.

Ayurvedic properties :
Rasa - Tikta, Kaṣāya. Vīrya - Ūṣṇa.
Guṇa - Laghu, Snigdha. Vipāka - Kaṭu.

अश्वगन्धा कटूष्णा स्यात्तिक्ता च मदगन्धिका।
वल्या वातहरा हन्ति कासश्वासक्षयव्रणान्।।
(राजनिघण्टु)

Aśvagandhā katūṣṇāsyāt tikatācamadagandhikā,
Balyā vātaharā hanti kāsaśvāsa kṣaya vraṇān.
(Rajanighantu)

अश्वगंधानिलश्लेष्मश्वित्रशोथक्षयापहा।
बल्या रसायनी तिक्ता कषायोष्णाऽतिशुक्रला।।
(भा. प्र.)

Aśvagandhānilaśleṣmaśvitrasothakṣayāpahā,
Balyā rasāyanī tiktā kaṣāyoṣṇā atiśukralā.
(Bhavaprakasa)

Actions/Uses : Vātaghna, Kaphaghna, Śothahara, Vātānulomana, Vedanāsthāpana, Dīpana, Kṛimighna, Śvāsahara, Mūtrala, Rasāyana, Raktaśuddhikara, Raktabhāraśāmaka, Brihana, Balya.

Therapeutics :

Asvagandha plant is sedative, tonic, stimulant, aphrodisiac and helps toning up of uterus of women. Internally it is given in marasmus in children. Externally it is used in the treatment of inflammatory conditions, ulcers and scabies. Root is adaptogenic, alterative, aphrodisiac, deobstruent, diuretic and tonic. It is useful in cough, dropsy, hiccup, leucorrohea and menstrual troubles. It restores loss of memory and is used in cases of nervous exhaustion, spermatorrhoea and senile debility. Powder of root mixed with equal parts of ghee and honey is beneficial in impotency or seminal debility. Decoction boiled with milk and ghee promotes nutrition. Root and leaves are used for emphysematous dyspnea. Leaves are bitter, antipyretic and anthelmintic. Bruised leaves and ground root is locally applied in carbuncles, scabied, painful swellings and ulcers. Fruit is sweet, applied to wounds, and is used in asthma, biliousness and strangury. Ripe fruits are anodyne or sedativre. Round capsular fruit is used in fresh state as an emetic and when dried is used as a stomachic. In small doses it is a remedy in dyspepsia and flatuent colic. It coagulates milk. Seeds are emmenagogue, diuretic, useful in lumbago, ophthalmia. They lessen inflammation of piles.

Family : XCV - Sterculiaceae

Abroma augusta Linn. f.

SYNONYMS : *Ambroma augusta* Linn. *f.*

ENGLISH : Devil's Cotton.

INDIAN : Pishaacha-karpaasa, Ullatakambal, Kumal, Ulat-kambal.

SANSKṚTA : Pīvarī, Piśācakārpāsa, Yośinī.

HABIT : A large, spreading, quick-growing hairy shrub or a small tree.

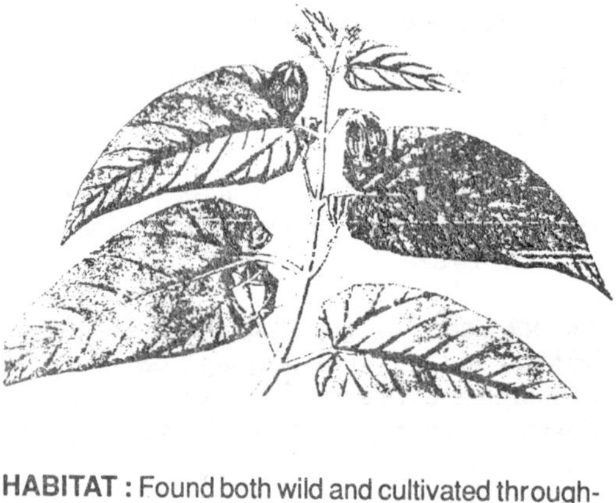

HABITAT : Found both wild and cultivated throughout hot and moist parts of India.

CHEMICAL COMPOSITION : Stem-bark contains B-sitosterol and friedelin. Seeds contain fixed oil (20.2%) composed of linoleic, oleic, hexadecenoic, palmatic and stearic acids. Root contains a fixed oil, resins, an alkaloid in minute quantity and water soluble bases.

Parts used : Root, root-bark, stem and leaves.

Ayurvedic properties :
Rasa - Kaṭu, Tikta. Vīrya - Uṣṇa.
Guṇa - Laghu, Rūkṣa, Tīkṣṇa. Vipāka - Kaṭu.

पीवरी योषिनी सा स्यात् योनिव्यापद्विनाशिनी।
वजोदोषप्रशमर्नां प्रदरार्शोनिवारिणी।। (द्रव्यगुण)

Pīvarī yoṣinīsāsyād yonivyapadvināśinī,
Rajodoṣaśamanī pradarārśonivāriṇī.
(Dravyaguna)

Actions/Uses : Kaphavātaghna, Pittavardhaka, Ārtavajanana, Garbhāśayottejaka, Vedanāsthāpana.

Therapeutics :

The root and root-bark of Pivari are uterine tonic and emmenagogue. They contract the uterus and are used for treatment of congestive and nervous dysmenorrhoea, ammenorrhoea, sterility and other menstrual disorders. Powdered roots act as an abortifacient and anti-fertility agent. Leaves useful in treating uterine disorders, diabetes, rheumatic pains of joints, and headache with sinusitis. Leaves and stem are demulcent and an infusion of fresh leaves and stems in cold water is very efficacious in gonorrhoea.

Helicteres isora Linn.

ENGLISH : East Indian screw tree.

INDIAN : Murudashenga, Marorphali, Varkati, Jonkaphal, Kewan.

SANSKṚTA : Āvartanī, Mṛga-śinga.

HABIT : A sub-deciduous shrub or small tree.

HABITAT : Dry forests throughout Central and Western India.

CHEMICAL COMPOSITION : Bark contains chloroplast pigments, phyto-sterol, a hydroxy-carboxylic acid, a crystalline colouring matter, saponins, sugars, phlobotannins and lignin.

Parts used : Root, bark, fruits and seeds.

Ayurvedic properties :
Rasa - Kaṣāya. Vīrya - Ūṣṇa.
Guṇa - Laghu, Rūkṣa. Vipāka - Kaṭu.

आवर्त्तकी कषायाम्ला शीतला पित्तहारिणी।
(राजनिघण्टु)

Āvartakī kaṣāyāmla śītalā pittahāriṇī.
(Rajanighantu)

आवर्तनी लघुः शीता कषाया त्वतिसारनुत्।
बलासपित्तशूलास्त्रकृमिरोगविनाशिनी।।

Āvartanī laghuḥ śīta kaṣāyā tvatisāranut;
Balāsapittaśūlāsrakṛmirogavināśinī.

Actions/Uses : Kaphapittaghna, Raktastambhana, Vraṇaropaṇa, Stambhana, Śūlapraśamana, Kṛmighna.

Therapeutics :
Root and stem barks of Avartani are demulcent, expectorant, astringent and antigalactagogue. Juice of root is beneficial in empyema and stomach affections and used in diabetes. Bark is used for diarrhoea, dysentery and billiousness. Dried fruit is useful in intestinal complaints and prescribed for colic, flatulence and diarrhoea.

Pterospermum suberifolium Lam.

SYNONYMS : *Pterospermum canescens* Roxb.

INDIAN : Muchakunda.

SANSKṚTA : Mucakunda, Kṣatravṛkṣa, Supuṣpa, Citraka, Bahupatra, Prativiṣṇuka, Sudala, Harivallabha, Arghyārha, Lakṣmaṇaya, Vasu, Raktaprasava.

HABIT : A small to medium sized tree.

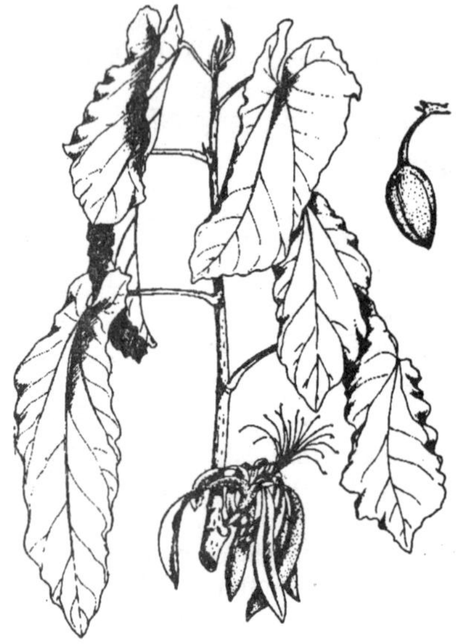

HABITAT : In forests of Karnataka and Tamil Nadu. Planted in Maharashtra.

CHEMICAL COMPOSITION : β-sitosterol and a mixture of fatty acids containg myristic, palmitic, stearic, arachidic, oleidc and linoleic acids from flowers.

Parts used : Flowers, leaves and bark.

Ayurvedic properties :
Rasa - Kaṣāya, Kaṭu, Tikta. Vīrya - Alpa Ūṣṇa.
Guṇa - Picchila. Vipāka - Kaṭu.

मुचकुन्दः कटुतिक्तः कफकासविनाशानश्च कण्ठकरः।
त्वग्दोषशोफशमनो व्रणपामाविनाशनश्चैव ॥
(राजनिघण्टु)

Mucukundaḥ kaṭutiktaḥ kaphakāsavināśanaśca kaṇṭhakaraḥ,
Tvagdoṣaśophaśamano vraṇapāmāvināśanaścaivā.
(Rajanighantu)

Actions/Uses : Triḍosaghna, Vedanāsthāpāna, Raktastambhana, Viṣaghna, Kuṣṭhaghna, Kaṇṭhya.

Therapeutics :

Flowers of Muchakunda are bitter and render water mucilaginous. They are made into a paste with rice and vinegar for application in hemicrania. A paste from leaves is used in headache. Bark and flowers are charred and mixed with kamala powder for use on small-pox eruptions. Fruits are made into a jam.

Family : XCVI - Strychnaceae.

Ignatia amara Linn.

SYNONYMS : *Ignatiana philippinica* Lour.; *Strychnos ignatii* Berg.; *Strychnos maingayi Cl. var. fruticosa Cl.*

ENGLISH : Ignatia, St. Ignatius Beans.

INDIAN : Papeetaa.

SANSKṚTA : Papītā.

HABIT : A large, woody forest vine.

HABITAT : Native to the Philippine Islands. Grows in Singapur.

CHEMICAL COMPOSITION : Detection of strychnine, brucine, pseudostrychnine, pseudobrucine, N-cyano-sec.-pseudostrychnine, N-cyano-sec.-pseudocolubrine.

Therapeutics :

Papītā is stimulant and tonic. Its properties are similar to those of Nux Vomica. It is used in functional nervous disorders, neuralgia and debility.

HOMEOPATHIC USE : Ignatia is a Homeopathic remedy, especially suitable to nervous and hysterical persons, either male or female. But it is generally called a female remedy and is much employed and needed frequently for them because they are more hysterical and more nervous. The emotional disposition if patients for whom Ignatia is serviceable, differs widely from those for whom Nux Vomica is of use. Ignatia is for those who are subject to rapid alternations of gaity and disposition to weep, or in those we notice the other emotional states of Ignatia. It is not suitable for patients in whom anger, eagerness, or violence is prominent. It is a main remedy in cases of vexation.

Strychnos nux vomica Linn.

SYNONYMS : *Strychnos colubrina* Wight.

HABIT : A middle sized deciduous tree.

HABITAT : Indigenous to India but growing in Burma, China and Australia.

ENGLISH : Poison Nut, Nux vomica, Crow gig, Kachita.

INDIAN : Kuchalaa, Kaajaraa, Bailewa, Chibbige, Jharakatachura.

SANSKṚTA : Kupīlu, Kāraskara, Viṣatinduka, Kākapiluka, Kālapīlu, Mūṣakaviṣa, Kucalā, Vidāravṛkṣa, Nirbhedin, Kuncavṛkṣa, Kulaka, Markaṭatinduka, Viṣamuṣṭi.

CHEMICAL COMPOSITION : Glucoside loganin from fruits. Brucine, strychnine, vomicine and methoxystrychnine from leaves and root bark. C-mavacurine in roots. Isostrychnine from seeds. Vanillic, p-hydroxybenvoic, 2-hydroxy-4-methoxybenzoic, sinapic and syringic acids and kaempferol, quercetin and 3'-O-methylquercetin in plant. A new alkaloid- protostrychnine along with normacusine B and 4-hydroxy-3-methoxystrychnine isolated. Fixed oil 2-4 %.

Parts used : Root, leaves, seeds, bark and wood.

Ayurvedic properties :
Rasa - Tikta, Kaṭu. Vīrya - Ūṣna.
Guṇa - Laghu, Śīta, Rūkṣa, Tīkṣṇa. Vipāka - Kaṭu.

कुपीलु शीतलं तिक्तं वालश्च मदकृल्लघुः।
परं व्यवहारं ग्राहि कफपित्तस्त्रनाशनम्।।
मूत्रप्रवर्त्तनं वल्यं वह्णिकृत् कामदीपनम्।
शूलमेकाङ्ग रोगञ्चशुक्रमेहमपस्मृतिम्।
सर्वाङ्ग कम्पनं दौर्बल्यं न चिरणे विनाशयेत्।।
सारमेयविषोन्मादहरो मदकरः सरः।। (द्रव्यगुण)

Kupīlu śītalam tiktaṃ vālañca madakṛllaghu,
Param vyathahāram grāhi kaphapittāsranāśanam.
Mūtrapravarttanam balyam vahiṇakṛt kāmadīpanam,
Śūlamekāṅga rogānca śukramehamapasmṛtim.
Grāhaṇīmatisārañca gudabhraṃśam madātyayam,
Sarvāṅga kampanam daurbalyam na cireṇa vināśayet.
Sārameya viṣonmādaharo madakaraḥ saraḥ.
(Dravyaguṇa)

कारस्करः कटूष्णश्चं तिक्तः कुष्ठविनाशनः।
वातामयास्त्रकण्डूतिकफामार्शो व्रणापहः।।
(राजनिघण्टु)

Kāraskaraḥ kaṭūṣṇaṣcya tiktaḥ kuṣṭhavināśanaḥ,
Vātāmayāsrakaṇḍūtikaphāmārśo vraṇāpahaḥ.
(Rājanighaṇṭu)

Actions/Uses : Kaphaghna, Vātakara, Pittaghna, Vedanāsthāpana, Āmavātśamana, Hṛdayottejaka, Raktabhāravardhaka.

Therapeutics :

Kupīlu is highly toxic to man and animals, producing stiffness of muscles and convulsions. Root is bitter, useful in intermittent fevers. Root bark, ground up into fine paste with lime juice and made into pills are used for cholera. Leaves are applied as polutice to sloughing wounds and ulcers, specially in cases when maggots have formed. Seeds are bitter, atonic, nervine tonic, stomachic, antidiarrhoeal, antidysenteric, antispasmodic, emetic, febrifuge; spinal, respiratory and cardiac stimulant, aphrodisiac. They are used as a general tonic, mostly in combination with other remedies, for neuralgia, dyspepsia, debility, impotence and in chronic constipation, as it increases peristalsis. With aromatics they are given in colic. Wood is used in dysentery, fevers and dyspepsia.

HOMEOPATHIC USE : Nux Vomica is a polychrest Homeopathic remedy whose symptoms correspond in similarity with the symptoms of the commonest and most frequent of human diseases arising out of the conditions of modern life such as mental labour of an engrossing character; sedentary life at desk; high living; strenuous life which irritates temper and excites desire for dissipation at night; consequent late hours; gastric derangement; use of tonics, liver pills, cathartics, drug mixtures; sparkling liquors ets. Nux Vom. in its action is an admirable combination of all these causations which correspond to the typical life of the sterner sex of the age, and that is why it has gained reputation of being preeminently a male remedy.

Strychnos potatorum Linn. f.

HABIT : A middle sized glabrous tree.

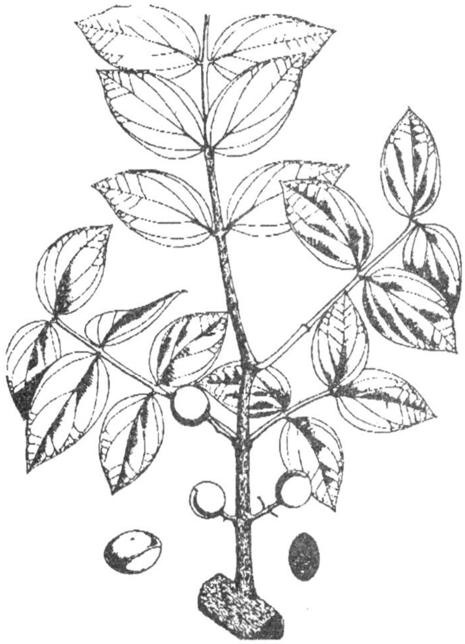

HABITAT : Found in deciduous forests of Central India. Konkan, Maharashtra; N. Kanara; Karnatak.

ENGLISH : Clearing nut.

INDIAN : Niwallee, Nirmallee, Gajrah, Nelmal.

SANSKṚTA : Nirmaḷī, Kataka, Payaḥprasādī, Cakṣuśya, Lekhanātmaka, Ambuprasāda, Tiktaphala, Chedanīya, Gucchaphala, Ślaṣṇa, Kata, Rucya, Netravikārajit, Vāriprasādana, Dṛḍhaphala, Chedī, Kātya, Śodhana, Ruciṣṇa.

CHEMICAL COMPOSITION : Brucine; diaboline; β-sitosterol and stigmasterol; oleanolic acid and its 3B-acetate and a saponin, galactose and mannose from seeds. Isomotiol; mixture of sitosterol, stigmasterol and campesterol from leaves and bark. Quercetin and vanillic, syringic, 2-hydroxy-4methoxybenzoic, chlorogenic and sinapic acids in plant.

Parts used : Root, fruits and seeds.

Ayurvedic properties :
Rasa - Madhura, Kaṣāya, Tikta. Vīrya - Śīta.
Guṇa - Laghu, Viśāda. Vipāka - Madhura.

कतका कटुतिक्तोष्णः चक्षुष्यो क्रिमिदोषनुत्।
रुचिकृत् शूलदोषघ्नो वीजमम्बु प्रसादनम्।।
(राजनिघण्टु)

Katakaḥ kaṭutiktoṣṇaḥ cakṣuṣyo krimidoṣanut,
Rucikṛt śūladoṣaghno vījamambu prasadānam.
(Rājanighaṇṭu)

कतकं शीतलं प्राहुस्तृष्णा विषविनाशनम्।
नेत्रोत्थरोगविध्वंस्सेसि विधिनाऽञ्जनयोगतः।।
कतकस्य फलं तिक्तं चक्षुष्यं शीतलं मृदु।
वारिप्रसादनं कृच्छ्र शर्कराश्मरी जयेत्।। (ध. नि.)

Katakam śītalam prāhustṛsṇāviṣavināśanam,
Netrot tharogavidgvansi vidhina anjanayogataḥ.
Katakasya phalam tiktam cakṣuṣyam śitalam mṛdu,
Vāriprasādanam kṛccham śarkarāśmarīm jayet.
(Dhanvantarī)

Actions/Uses : Kaphavātaśāmaka, Rucivardhaka, Chedana, Dīpan, Pācana, Stambhana, Mūtrala, Atisārnāṣaka, Pūyameha, Aśmarighna, Mūtrakricch, Netrabhiṣyanda, Gulmaghn.

Therapeutics :
Kataka is acrid, bitter, alexiteric, anthelmintic, increases appetite and improves taste; good in troubles of eyes, thirst, burning sensation, tumours, pains and urinary discharges. Leaves are used as poultice over maggot-infested ulcers. Powdered bark, mixed with lime juice, is given in cholera. Fruits are antidiabetic, antidysenteric, emetic. Pulp is expectorant and a good substitute for Ipecacuanha as antidysenteric. Seeds are tonic, alterative, stomachic, demulcent and emetic; are used in acute diarrhoea, diabetes, gonorrhoea etc. and as a local application in eye diseases; rubbed with honey and little camphor, the mixture applied to eyes in lachrymation and conjunctivitis.

Family : XCVII - Styraceae.

Styrax benzoin Dryand.

SYNONYMS : *Styrax paralleloneurum* Perkins.; *Laurus benzoin* Houtt.; *Benzoin officinale* Hayne.

ENGLISH : Benzoin.

INDIAN : Ooda, Luban, Loban.

SANSKṚTA : Ūda, Kapardaka-ūda, Sayāma-dhūpa.

HABIT : A shrub or small tree.

HABITAT : Siam Java and Sumatra, Malacca and Malaya.

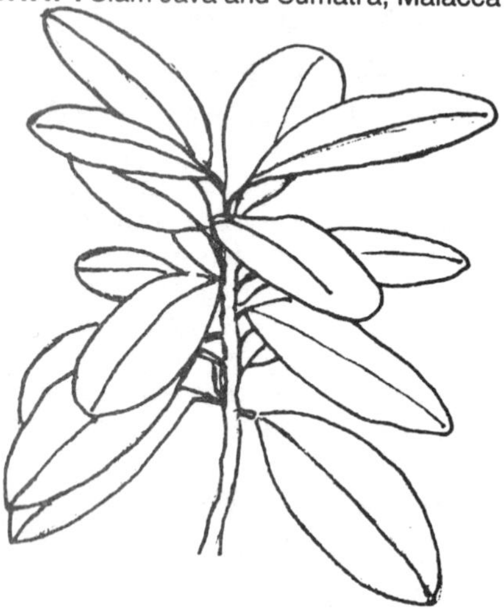

CHEMICAL COMPOSITION : Benzoic and cinnamic acids and their esters, Siaresinolic acid and Sumaresinolic acid.

Parts used : Niryaas.

बलासवातग्रहवान्तिहिक्कां शिरोर्तिशौर्थल्यभयं निमित्तम्।
भेत्तुः क्षमः स्निग्धवलक्षर्ताष्णो मया प्रयुक्तः खलु लोहबाणः।।
(सिद्धभैषज्यमणिमाला)

Balāsavātagrahavānti hikkām
Śirortiśaithilyabhayam nimitam,
Bhettuḥ kṣamaḥ snigdhavalakṣartaṣṇo,
Meyā prayuktaḥ khalu lohabāṇaḥ.
(Nighaṇtuadarsa)

Actions/Uses : Pūtihara, Durgandhināṣaka, Vraṇaśodhan, Vraṇaropaṇa, Śonitasthāpan, Śleśmagna, Uttejaka, Kaphaghna, Mūtrajanana.

Therapeutics :

Uda gum is antiseptic, stimulating, expectorant, diuretic, irritating and carminative, forms an ingredient of inhalations in the treatment of catarrh of upper respiratory tract. As diuretic it is useful in calculous disorders from phosphatic deposits in urine. Useful in coughs, bronchitis and externally applied to wounds, sores etc.

Family : XCVIII-Symplocaceae.

Symplocos racemosa Roxb.

ENGLISH : Lodhra tree, Cinchona.

VERNACULAR : Lodhra, Lodh.

SANSKṚTA : Lodhra, Sthūlavalkala, Tilva, Tirita, Śābara, Gālava, Kramuka, Jīrṇapatra, Bṛhatpatra, Pattī, Lākṣāprasādana, Rodha, Bhillataru, Cillaka, Kāṭhakīlaka, Lodhraka, Hastilodhraka, Śūka, Kāṇḍahīna, Hemapuṣpaka, Santatodbhava, Mārjana, Kānīna, Śāmbar, Tirītaka, Tilvakataru, Śavarapādapa.

HABIT : A middle-sized evergreen tree or shrub.

HABITAT : Throughout N.E. India. Occasionally in Konkan along river sides and hill slopes.

CHEMICAL COMPOSITION : Bark contains two alkaloids, loturine and colloturine which are chemically related to harmine. Two new monomethyl pelargonidin glucosides and pelarggonidin-3-O-glucoside isolated from trunk bark.

Ayurvedic properties :

Rasa - Kasāya. Vīrya - Sīta.
Gunna - Laghu, Rūksa. Vipāka - Katu.

Parts used : Bark and wood.

लोध्रो ग्राहि लघुः शीतः चक्षुष्यः कफपित्तनुत्।
कषायोरक्तपित्तासृग्ज्वरातीसारशोथहृत्।।
(भा. प्र.)

Lodhro grāhi laghuḥ śītaḥ cakṣuṣyaḥ kaphapittanut,
Kaṣāyo raktapittāṣrg jvarātisāraśothahṛt.
(Bhavaprakasa)

Actions/Uses : Kaphapittaśāmaka, Śothahara, Sandhānīya, Kuṣṭhaghna, Raktastambhana, Vraṇaropaṇa, Śoṇitasthāpana, Garbhāśayaśotha-nāśana, Srāvaṇāśana.

Therapeutics :

Lodhra is a specific remedy for uterine complaints, vaginal and menstrual disorders. Bark is light, cooling, mild astringent, antidiarrhoeal, antidysenteric, aphrodisiac; useful in dropsy, elephantiasis, filaria, liver complaints, bowel complaints, eye diseases, ulcers, menorrhagia and leucorrhoea. Decoction of bark or wood is used as gargle for giving firmness to spongy and bleeding gums. Paste of wood is externally applied to boils to promote suppuration and discharge of pus.

Family : XCIX-Thymelaeaceae.

Aquilaria agallocha Roxb.

SYNONYMS : *Aquilaria malaccenisis* Lamk.

ENGLISH : Agarwood; Aloewood; Malacca eaglewood.

INDIAN : Agara, Agaru.

SANSKṚTA : Aguru, Loha, Kṛumija, Kṛmijaghna, Bhṛngaja, Keṣya, Agnikāṣṭha, Sudhūpya, Gandhadhūmaja, Kālāgaru, Kāleya, Kālīyaka, Laghucandana, Vanadruma, Mallikāgandha, Rājārha, Vanśika, Śīrṣa, Śṛngāra, Vallabha, Jongāka, Kākatunḍaka, Malinakālaśāra, Puram, Tailāgaru, Gandharājaka, Malligandhi.

HABIT : A large evergreen tree.

HABITAT : Eastern Himalayas, Bhutan, Assam and hill forests of Khasia, Nagaland, Manipura, Tripura.

CHEMICAL COMPOSITION : Agarwood on distillation, yields an essential oil known as AGAR OIL. The yield 0.12-3.6 % dry basis. It is highly priced and has a large potential market. A sequiterpenoid -agarospiroi - isolated from oil. Holocellulose, lignin and pentosans.

Parts used : Fragrant, resinous wood and oil.

Ayurvedic properties :
Rasa - Tikta, Kaṭu. Vīrya - Ūṣṇa.
Guṇa - Laghu, Rūkṣa, Tīkṣṇa. Vipāka - Kaṭu.

अगुरूष्णं कटु त्वच्यं तिक्तं तीक्ष्णं च पित्तळम्।
लघु कर्णाक्षिरोगघ्नं शीतवातकफप्रणुत् ।।
कृष्णं गुणाधिकं ततु लोहवद्वारी मज्जति।
अगुरुप्रभवः स्नेहः कृष्णागुरुसमः स्मृतः।।
(भा. प्र.)

Aguruṣṇam kaṭu tvacyam tiktam tīkṣṇam ca pittalam,
Laghu Karṇākṣirogaghnam śīta vāta kaphapraṇut.
Kṛṣṇam guṇādhikam tattu lohavadvārī majjati,
Agaruprabhavaḥ snehaḥ kṛṣṇāgurusamaḥ smṛtaḥ.
(Bhavaprakasa)

Actions/Uses : Kaphaghna, Vātaghna, Pittavardhaka, Uttejaka, Anulomana, Śītapraśamana, Śvāsahara, Śīrovirecana, Jantughna, Kothapraśamana, Durgandhahara, Vraṇaśodhana, Vedanāsthāpana, Śothahara,

Therapeutics :

Wood of Aguru is acrid, bitter, stimulant, thermogenic, digestive, carminative, deodorant, sudorific, anodyne, anti-inflamatory, cardiotonic. tonic, aprhodisiac, antiasthamatic, antidysenteric, astringent in diarrhoea and vomiting. Useful in gout, rheumatism and paralysis and as a linament in various skin diseases. It is used by traditional vaidyas as contraceptive. It is used as a perfume in the form of powder and internally as stimulant, cholagogue and deobstruent. Leaves boiled in oil, taken for removing fish spines from throat.

Daphne mezerium Linn.

SYNONYMS : *Daphne oleoides* Schreb.; *Mezereum officinarum C.A.* Meyer.

ENGLISH : Mezereum, Mezereon.

INDIAN : Mezeriyana.

SANSKṚTA : Mezeriyuna.

HABIT : A hardy shrub.

HABITAT : Native of Great Britain and Northern countries. Growing in Moscow.

CHEMICAL COMPOSITION : Bark yields poisonous daphnetoxin. Leaves, β-sitosterol and luteolin. Seeds, glococide daphnoside and Mezerein, an antiinflammatory and anticarcinogenic.

Parts used : Bark, root and root-bark.

Ayurvedic properties :

Rasa - Tikta, Madhura. Vīrya - Śīta.
Guṇa - Śīta. Vipāka - Kaṭu.

Actions/Uses : Mūtrajanana, Svedajanana, Śoṇitasthāpana, Tvagrogahara.

Therapeutics :

As an internal remedy Mezeriyuna it is stimulant, alterative, diaphoretic and diuretic. It has been given in syphilitic, scrofulous and cutaneous affections and in chronic rheumatism. Externally used as lotion to blistered surfaces and indolent ulcers.

HOMEOPATHIC USE : Its main use, always, has been made for skin diseases. And for the diseases which come after suppressed eruptions by ointments. Very obstinate itching all over the body. Peeling off of the skin. Severely itching rash on the nape, back and thighs, always worse and more gnawing after scratching. Severe itching on the head. The scales of dandruff are whiter. Eczema, itching intolerably, copious serous exudation. Scabs thick, bloody secretion beneath. Scurflike fish-scales on back, chest and thighs and scalp. Head covered with thick leathery crust, under which pus collects.

Family : C - Tiliaceae.

Triumfetta rhomboidea Jacq.

SYNONYMS : *Triumfetta bartramia* Linn.; *Bartramia indica* Linn.; *Triumfetta angulata* Lamk.

ENGLISH : Burbush, Burweed.

INDIAN : Jhinjhira, Nichaardi, Chikti.

SANSKRTA : Jhinjhirīṭā, Kaṇtaphala, Jhirpaṭa.

HABIT : A herbaceous perennial.

HABITAT : Throughout tropical and subtropical India.

CHEMICAL COMPOSITION : Leaves contain proteins and seeds a greenish yellow fatty oil. 4-Hydroxyisoxazole and triumbodin.

Parts used : Fruits, flowers, leaves, bark and root.

Ayurvedic properties :

Rasa - Kaṭu, Kasāya. Vīrya - Śīta.
Guṇa - Śīta. Vipāka -Kaṭu.

झिञ्झिरीटा कटुः शीता कषायाचातिसारजित्।
वृष्या सन्तर्पणी वष्या महिषीक्षीररवर्द्धिनी।।
(राजनिघण्टु)

Jhinjhirīṭā kaṭuḥ śīta kaṣāyāḥ cātisārajit,
Vṛṣyā santarpaṇī balyā mahiṣīkṣīravarddhinī.
(Rajanighantu)

Actions/Uses : Balya, Atisārajit, Vṛṣya.

Therapeutics :

Root of Jhinjhirita is bitter and diuretic. Root is used in dysentery and bark and fresh leaves are used in diarrhoea. Pounded roots are given for intestinal ulcers and a hot infusion is taken to facilitate childbirth or hasten inception of parturition when it is delayed. Leaves and flowers are used against leprosy. Flowers, fruits and leaves are mucilagious, demulcent, astringent, used with sugar during gonorrhoea.

Family : CI - Trapaceae.

Trapa bispinosa Roxb.

SYNONYMS : *Trapa natans* Linn. *var. bispinosa (Roxb.)* Makino.; *Trapa quadrispinosa* Wall.

ENGLISH : Water chestnut, Caltrops, Sunghara nut.

INDIAN : Shingaaddaa.

SANSKṚTA : Śṛṅgaṭaka, Jalaphala, Śṛṅgaruha, Jalavallī, Jalāśrayā, Śṛṅgakanda, Śṛṅgamula, Visani, Sṛṅghva, Saptanādīka, Jalakanṭa, Trikanṭa, Kaseruka, Ambukanda, Trika, Paniphala, Triconaphala, Lālaphala.

HABIT : An aquatic floating herb.

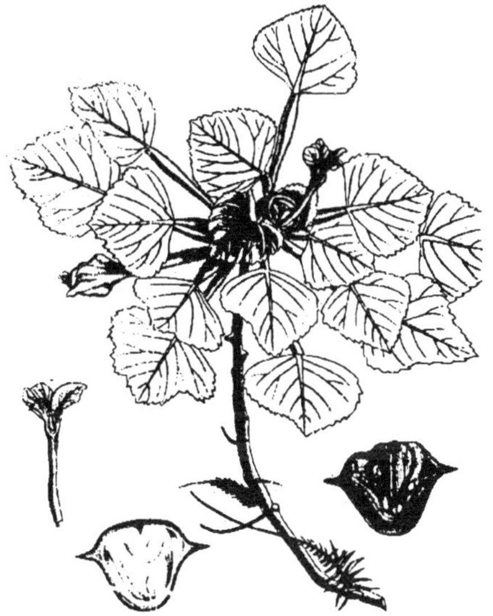

HABITAT : Throughout India. Cultivated in ponds in different parts, particularly West Bengal, Bihar, Orissa, U.P. and M.P.

CHEMICAL COMPOSITION : Manganese in plant and kernel.

Parts used : Seeds and stem.

Ayurvedic properties :
Rasa - Madhura, Kasāya. Vīrya - Sīta.
Guṇa - Gurū, Rūksa. Vipāka - Madhura.

शृंङ्गाटकं हिमं स्वादु गुरु वृष्यं कषायकम्।
ग्राहि शुक्रानिलश्लेष्मप्रदं दाहास्त्रपित्तनुत्।।
(भावप्रकाश)

Śṛgāṭakam himam svādu gurū vṛṣyam kaṣāyakam,
Grāhi śukrānilaśleṣmapradam dāhāsrapittanut.
(Bhavaprakasa)

Actions/Uses : Kaphakara, Vātakara, Pittasāmaka, Stambhana, Raktapittasāmaka, Garbhasthāpana, Dāhahara, Balya, Vrsya.

Therapeutics :

Śṛṅgataka fruit is sweetish, aphrodisiac, appetiser, antipyretic; useful in chronic fevers,lumbago, pain, sore throat, biliousness and bronchitis. Nuts are used as food, considered cooling, useful in diarrhoea and billious affections. They are given in powder form with milk in nervous and general debility, leucorrhoea and seminal weakness. Juice of stem is beneficial in eye diseases and its poultice acts as an agent for resolution of tumor.

Family : CII - Verbenaceae.

Callicarpa macrophylla Vahl.

SYNONYMS : *Callicarpa nudiflora* Hook. & *A.*; *Callicarpa revesii* Wall. *ex* Schaucer.; *Callicarpa acuminata* Roxb.

INDIAN : Piyunga, Gavhalaa, Daya.

SANSKṚTA : Priyaṅgu, Phalinī, Priyā, Śyāmā, Vṛttā, Priyavallī, Phalapriyā, Gaurigovandinī, Kārambhā, Kangu, Kangunī, Viṣvakṣenā, Bhangurā, Subhagā, Parṇabhedinī, Śubhā, Pītā, Mangalyā, Śreyasī, Varṇabhedinī,. Striphalī, Gandhapriyangu, Mahilā, Nandinī, Vanitā, Latā, Gandhaphalī, Angaṇapriyā, Gundrā, Kāntā, Priyaka, Kṛsāṅgī, Puśkaraprṇī, Nārivallabhā, Draviḍodbhavā.

HABIT : An erect shrub or a small tree.

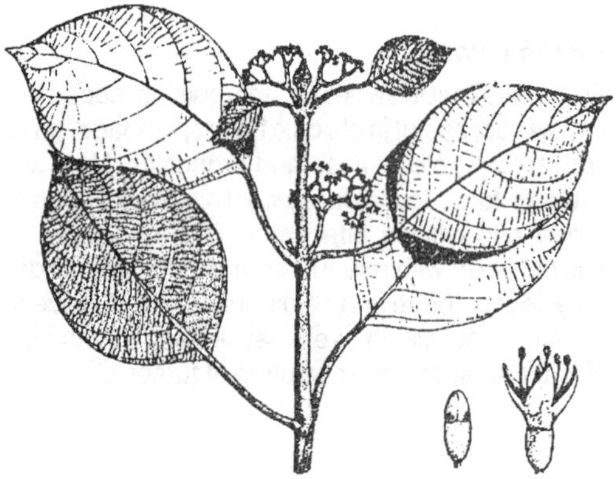

HABITAT : Upper Gangetic Plain and W. Himalayas.

CHEMICAL COMPOSITION : 2-tetracyclic diterpenoid, calliterpenone and its mono- OAc. Leaves, sitosterol, calliterpenone; luteolin, apigenin, ursolic acid, crategolic acid etc.

Parts used : Roots, leaves, flowers, fruits and wood.

Ayurvedic properties :

Rasa - Tikta, Kaṣāya.	Vīrya - Śīta.
Guṇa - Gurū, Rūksa.	Vipāka - Kaṭu.

प्रियङ्गुः शीतळा तिक्ता तुवरानिलपित्तहृत्।
रक्तातिसारदौगन्ध्य स्वेद द्वाह ज्वरा पहा।।
गुल्मतृड्विषमेहघ्नी तद्वगन्धप्रियङ्गुका।
तत्फलं मधुरं रुक्षं कषायं शीतलं गुरु।।
विबन्धा ध्मानबलकृत् सङ्गाहि कफपित्तजित्।
(भा. प्र.)

Priyaṅguḥ śītalā tiktā tuvarānilapittahṛt,
Raktātisāradaurgandhya svedadāhajvarāpahā.
Gulmatṛḍ viṣamehaghnī tadvad gandhapriyaṅgukā.
Tatphalam madhuram rukṣam kaṣāyam śītakam gurū,
Vibandhādhmanabalakṛt saṅgrāhi kaphapittajit.
(Bhavaprakasa)

प्रियङ्गु शीतळा तिक्ता मोहदाहविनाशिनी।
ज्वरवान्तिहरा रक्तमुद्रिक्तं च प्रसादयेत्।।
(ध. नि.)

Priyaṅgu śītalā tikṭā mohadāhavināśinī,
Jvaravāntiharā raktamudriktanca prasādayet.
(Dhanvantarinighantu)

Actions/Uses : Tridoṣaśāmaka, Vātapittaghna, Dīpana, Dāhapraśamana, Vedanāsthāpana, Durgandhanāśana, Vātānulomana, Stambhana, Raktaśodhana, Raptapittaśāmaka, Mutravirajaniya.

Therapeutics :

Priyangu plant is used against dysentery; wood paste in mouth and tongue sores. Root or oil from root is aromatic and stomachic. Leaves are useful in rheumatism and gout. They are heated and applied to rheumatic joints and smoked to relieve headache. Flowers and fruits are bitter, sweet, astringent, anodyne, digestive, constipating and febrifuge. They are useful in rheumatoid arthritis, dyspepsia, colic, diarrhoea, dysentery, diabetes, fever and general weakness. Seed paste used in stomatitis, useful in leprosy and as diuretic. Fruit paste in treatment of boils and blisters on tongue.

Clerodendrum serratum (Linn.) Moon.

SYNONYMS : *Volkameria serrata* Linn.

INDIAN : Bhaarangee.

SANSKRTA : Bhārangī, Bramhaṇayaṣṭikā, Kharasaka, Padmā, Barbara, Gardabhaśāka, Kāsajit, Vātāri, Bhṛngajā, Kāsaghni, Bhramareṣṭā, Śakramātā, Phanjī, Angāravallarī, Surupā, Bhargakāgraṇī, Pālindī, Bhūmijumbū, Dvijā, Māngalyavallikā, Varca, Gandhaparvaṇī, Bhārlī.

HABIT : A blue flowered low shrub.

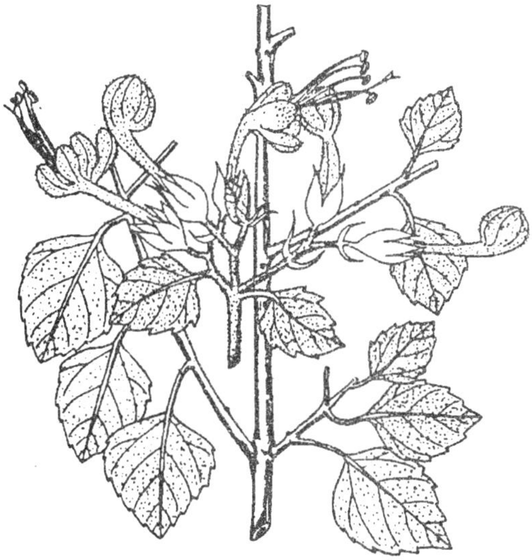

HABITAT : More or less throughout India.

CHEMICAL COMPOSITION : Plant contains saponin and mannitol. Root-bark yields saponin, manitol and stigmasterol.

Parts used : Root, leaves and seeds.

Ayurvedic properties :
Rasa - Tikta, Kaṭu, Kaṣāya. Vīrya - Ūṣṇa.
Guṇa - Laghu, Rūkṣa. Vipāka - Kaṭu.

भाङ्‌र्गी रुक्षा कटुस्तिक्ता रुच्योष्णा पाचनी लघुः।
दीपनी तुवरा गुल्मरक्तनुन्नाशयेद् ध्रुवम्।
शोधकासकफश्वासपनिसज्वरमारुतान्।।
(भा. प्र.)

Bhāṅrgī rūkṣā kaṭustiktā rucyoṣṇā pācanī laghuḥ,
Dīpanī tuvarā gulmaraktanunnāśayed dhrrvam,
Śothakāsakaphaśvāsapīnasajvaramārutān.
(Bhavaprakasa)

Actions/Uses : Kaphaghna, Vātaghna, Śothahara, Raktaśuddhīkara, Vraṇapācana, Agnidīpana, Āmapācana, Garbhāśayaśodhana, Anulomana.

Therapeutics :

Root of Bhangri has a pungent, bitter and acrid taste, and is antispasmodic, carminative, expectorant, febrifuge and tonic and is useful in anasarca, coryza, cough, dyspnoea, catarrhal affections, epilepsy. It increases appetite, lessens expectoration, useful in inflammations, bronchitis, asthma and is used for fevers, rheumatism and dyspepsia. Decoction of roots is effective in catarrhal bronchitis. Leaves are used for fevers and boiled with oil and butter made into an ointment are used as external applications in cephalgia and ophthalmia. Seeds are slightly aperient and are used for dropsy. The drug dispels kapha and is a specific remedy for respiratory diseases.

Gmelina arborea Linn.

SYNONYMS : *Gmelina rheedii* Hook.

ENGLISH : Cashmere tree.

INDIAN : Shivanna, Kambhari.

SANSKṚTA : Kāśamarī, Gambhārī, Śrīparṇi, Kāśhmirī, Mṛdi, Hīra, Madhuparṇikā, Kṛṣṇavṛintā, Bhadraparṇi, Kumbhāri, Saphala, Mahī, Pītarohiṇī, Bhadrā, Gopabhadrā, Madhumatī, Kṛṣṇā, Mahābhava, Kaṭphalā, Sarvatobhadrikā, Svabhadrā, Sthūlatvacā, Mudikāmahi, Sudṛḍhatvacā, Mahākumuda, Vidāriṇī, Kṣīriṇī.

HABIT : A large deciduous tree.

HABITAT : Found in road-sides and throughout Deccan Peninsula.

CHEMICAL COMPOSITION : Drupes contain butyric acid, tartaric acid in traces and resinous and saccharine substances.

Parts used : Whole plant, root, leaves, bark and fruit.

Ayurvedic properties :

Rasa - Tikta, Kaṣāya.	Vīrya - Ūṣṇa.
Guṇa - Gurū.	Vipāka - Kaṭu.

काश्मरी तुवरा तिक्ता वीर्योष्णा मधुरा गुरुः।
दीपनी पाचनी मेध्या भेदिनी भ्रमशोषजित्।।
तत्फलं बृंहणं वृष्यं गुरुं केश्यं रसायनम्।
वातपित्ततृषारक्तक्षयमूत्रविबंधनुत्।।
(भा. प्र.)

Kāśmarī tuvarā tiktā vīryoṣṇā madhurā guruḥ,
Dīpanī pacani medhyā bhedinī bhramaśoṣajit.
Tatphalam bṛnhaṇam vrsyam gurum keśyam rasāyanam,
Vātapittatṛṣāraktakṣayamūtravibandhanut.
(Bhavaprakasa)

कश्मरी कटुका तिक्ता गुरुष्णा कफशोफनुत्।
त्रिदोषविषदाहार्ति ज्वरतृष्णास्यदोषजित्।।
(राजनिघण्टु)

Kaśmarī kaṭukā tiktā gurūṣṇā kaphaśophnut,
Tridoṣaviṣadāhārti jvaratṛṣṇāsyadoṣajit.
(Rajanighantu)

Actions/Uses : Vātakaphaghna, Pittaghna, Bhedana, Bhramahara, Śoshahara, Dīpana, Pāchana, Tṛṣṇāghna, Āmadoṣahara, Śūlaghna, Arśoghna, Viṣaghna, Dāhaghna, Jvaraghna.

Therapeutics :

Kasamari root an important ingredient of dashamula. It is astringent, bitter tonic, stomachic, digestive, cardiotonic, laxative, galactagogue, pulmonary and nervine tonic. It improves memory, overcomes giddiness and is useful in burning sensation, fever, thirst, emaciation, heart diseases, nervous disorders and piles. Pulverised root is applied locally for gout. The drupes are sweetish and bitter and are used as an astringent of refrigerant decoctions for fevers and bilious affections. The tender leaves are demulcent. A paste of the leaves is applied to the head for the relief of headache in fevers. The leaf juice is used as a wash for foul ulcers. Flowers are given in blood diseases. Fruits are bitter, cooling, tonic and overcome thirst, pitta, vatarakta and useful in pleural and lung diseases.

It produces corpulency, promotes sexual power and is good for growth of hair. It is useful in difficult urination, vitiation of blood and rheumatism. In the form of infusion or decoction it is prescribed in indigestion, fevers and anasarca. Bark is a bitter tonic and stomachic and is useful in fever and indigestion.

Lippia nodiflora Mich.

SYNONYMS : *Phyla nodiflora (Linn.)* Greene.; *Verbena nodiflora* Linn.

ENGLISH : Wild Sage.

INDIAN : Jala pimpallee, Bhui-okra, Ratoliya.

SANSKRTA : Jalapippalī, Vaśira, Śakulādanī, Agnijvala, Lāngali, Citrapatri, Maharaṣṭri.

HABIT : A creeping much-branched herb.

HABITAT : Common in hedges, waste places throughout India.

CHEMICAL COMPOSITION : Two glucosides, nodiflorin-A and nodiflorin-B from dried plant. Plant also contains traces of a non-glucoside bitter substance, an essential oil, resin and a large mount of potassium nitrate.

Ayurvedic properties :

Rasa - Kaṭu, Kaṣāya, Tikta.	Vīrya - Śīta.
Guṇa - Laghu, Rūkṣa.	Vipāka - Kaṭu.

Parts used : Whole plant and leaves.

जलपिप्पली, शारदी, शकुलादनी, मत्स्यगंधा
मत्स्यादनी मत्स्यगन्धा लाङ्लीत्यपि कीर्तिता।
जलपिप्पलीका हृद्या चक्षुष्या शुक्रला लघुः।।
संग्राहिणी हिमा रुक्षा रक्तदाहव्रणापहा।
कटुपारसा रुच्या कषाया वह्निवर्द्धिनी।।

Jalapippalīkā hṛdyā cakṣuṣyā śukralā laghuḥ,
Sangrāhiṇī himā rūkṣā raktadāha vraṇāpahā,
Kaṭupākarasā rucyā kaṣāyā vanhivardhinī.
(Bhavaprakasa)

Actions/Uses : Dīpana, Rocana, Grāhī, Mūtral, Hṛdya, Śūkrala, Agnivardhaka, Vraṇahara, Dāhaghna, Vedanāhara, Raktaprasādana.

Therapeutics :

Jalapippali or Vasira plant possesses cooling, diuretic and febrifuge properties and is used in stoppage of bowels and pain in knee-joints. It is stomachic, antiseptic, antispasmodic, carminative, diaphoretic, aphrodisiac, astringent, vulnerary and anthelmintic. Useful in diseases of heart, eye and for piles. It promotes healing of wounds and is good in ulcers, wounds, burning sensation, asthma, bronchitis. A paste or poultice from fresh plant is applied as suppurant for boils, swollen cervical glands, erysipelas, fistula and chronic indolent ulcers. Decoction is laxative and is useful in malaria and rheumatixsm. Infusion of leaves or tender stalks is useful in indigestion in children. In combination with cumin it is useful in gonorrhoea. Extract of leaves is antibacterial.

Premna integrifolia Linn.

SYNONYMS : *Premna serratifolia* Linn.; *Premna latifolia* Dalz. & Gibs.; *Cornutia corymbose* Burm.*f.*; *Premna obtusifolia R. Br.; Premna corymbosa sensu* Hook.*f. (non Rottl. & Willd.).*

INDIAN : Aaranna, Narawela, Agetha, Ustabunda, Chamari.

SANSKṚTA : Agnimantha, Tarkārī, Maṭha, Jayā, Śrīparṇa, Gaṇikārikā, Nādevī, Araṇī, Āsāvarikā, Vātaghnī, Vaijayantikā, Jayanti, Ketu.

HABIT : A large, thorny, deciduous shrub or tree.

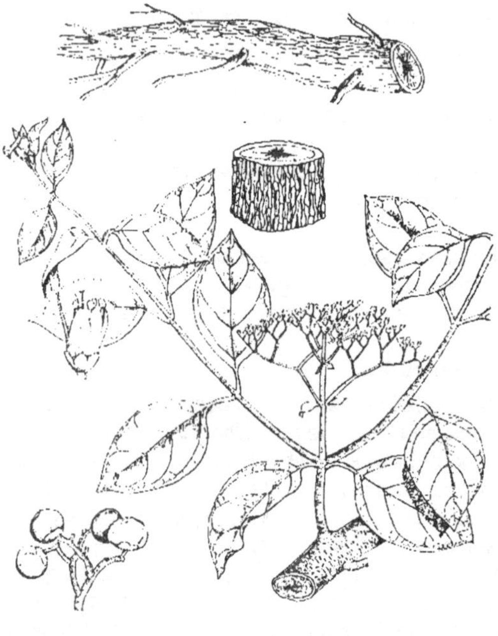

HABITAT : Konkan near the sea from Bombay to Malacca, Ceylon and the Andamans.

CHEMICAL COMPOSITION : Chief alkaloids are premnine, ganikarine and ganikarine.

Parts used : Whole plant, roots, root-bark and leaves.

Ayurvedic properties :
Rasa - Kaṭu, Tikta, Kaṣāya, Madhura.
Vīrya - Ūṣṇa.
Guṇa - Laghu, Rūkṣa. Vipāka - Kaṭu.

तर्क्कारी कटुरुष्णा च तिक्तानिलकफापहा।
शोफश्लेष्माग्निमान्द्यार्शो विट्वन्धाध्मान नाशनी।।
(राजनिघण्टु)

Tarkārī kaṭurūṣṇā ca tiktānilakaphāpahā,
Śophaśleṣmāgnimāndyārśo
viṭvandhādhmānanāśanī.
(Rajanighantu)

अग्निमन्थः श्वयथुनुद्वीर्योष्णः कफवातहृत्।
पाण्डुनुत् कटुकस्तिक्तस्तुवरो मधुरोऽग्निदः।।
(भा. प्र.)

Agnimanthaḥ śvayathunudvīryoṣṇaḥ kaphavātahṛt,
Pāṇḍunut kaṭukastiktastuvaro madhuro agnidaḥ.
(Bhavaprakasa)

Actions/Uses : Kaphaghna, Vātaghna, Agnidipana, Sāraka, Āmapācana, Anulomana, Raktaśodhana, Pānduhara, Vibandhahara, Śothahara, Vedanāsthāpana, Śleśmāhara, Jvaraghna, Avasādaka for Garbhāṣaya.

Therapeutics :

Agnimantha is a constituent of Dashamula, used for obstinate fevers. Root is acrid, bitter, astringent, cardiotonic, laxative, stomachic, cordical and tonic; useful in constipation, fever, heart diseases, neurological diseases and rheumatism. It improves digestion and is good for liver complaints. Leaves are carminative and galactagogue and are used as a soup given as a stomachic. Decoction of leaves used in the treatment of colic and flatulence, and that of tender parts for rheumatism and neuralgia. Fresh bark is at first sweetish but bitter and astringent latter. Aqueous extracts of the plant has a powerful action on uterus and gut of experimental animals.

Tectona grandis Linn. f.

ENGLISH : Teak.

INDIAN : Saagawaana, Sagun, Saaga.

SANSKṚTA : Śāka, Saka.

HABIT : A tall deciduous tree with rounded crown.

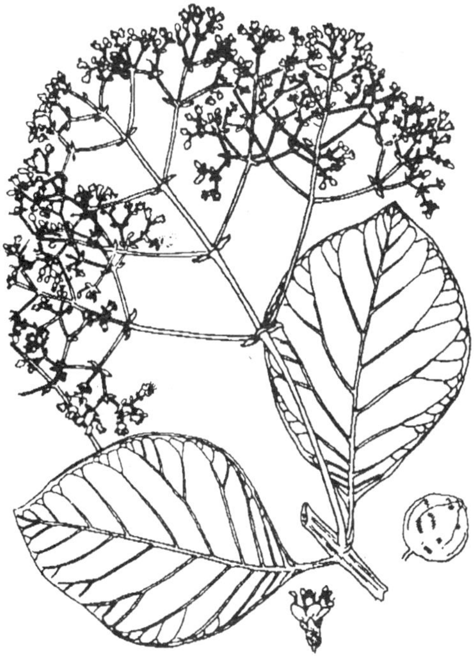

HABITAT : Konkan and Western Ghats, Karnatak and Madhyapradesh.

CHEMICAL COMPOSITION : Wood contains in its cavities white crystellinee deposits of calcium phosphate, silica and ammonium and magnesium phosphate, also a resin. Seed contains a bland fatty oil.

Parts used : Flowers, seeds, seed-oil and bark.

Ayurvedic properties :
Rasa - Kaṣāya. Vīrya - Śīta.
Guṇa - Sīta. Vipāka - Kaṭu.

शाकः कषायाशिशिरो रक्तपित्तप्रसादनः।
कुष्ठश्लेष्मानिलहरो गर्भसन्धानस्थैर्यकृत्।।
शाकपुष्पं प्रमेहघ्नं रुक्षं तुवरतिक्तकम्।
कफपित्तहरं वातकोपनं विशदं लघु।।
(कैयदेवनिघण्टु)

Śākaḥ kaṣāyāh śiśiro raktapittaprasadanaḥ,
Kuṣṭhasleśmānilaharo garbhasandhānasthyiryakṛt.
Śakapuṣpam pramehaghna rukṣam tuvartiktam,
Kaphapittaharam vātakopanam viśadam laghu.
(Kaiyadevanighantu)

Actions/Uses : Raktapittaśāmaka, Kaphapittahara, Mūtrjanana, Grāhī, Garbha-sthiryakara.

Therapeutics :

Roots of Saka are given in anuria and retention of urine. Flowers are acrid, bitter and diuretic, useful in bronchitis, biliousness, urinary discharges. Wood paste is astringent, diuretic, hepatic, stimulant, local refrigerant and sedative. Wood brayed in water is used as a local application to relieve headache, toothache, and to subdue inflammation and irritation of skin. Ashes are applied to swollen eyelids. Oil of nuts is used in scabies and to promote growth of hair.

Vitex negundo Linn.

SYNONYMS : *Vitex bicolor* Willd.; *Vitex trifolia* Graham.; *Vitex nirgundi trifolia.*

ENGLISH : Three-leaved chaste-tree.

INDIAN : Nirguddi, Sambhalu, Shivari, Nisida, Nigudi.

SANSKṚTA : Nirguḍī, Sinduvāraka, Śephālī, Suvahā, Śvetakusuma, Nīlasinduka, Puṣpanīlaka, Anilamanjirī, Śephālikā, Bhūtakeśī, Sinduka, Śītabhīru, Vanaka.

HABIT : A large aromatic shrub or small tree.

HABITAT : Throughout india in warmer zone.

CHEMICAL COMPOSITION : Leaves contain two alkaloid nishindine and hydrocotylene. Fresh leaves yield a pale greenish yellow oil.

Parts used : Whole plant, root, leaves, bark, flowers, fruits and seeds.

Ayurvedic properties :
Rasa - Tikta, Kaṭu, Kaṣāya. Vīrya - Uṣṇa.
Guṇa - Laghu, Rūkṣa. Vipāka - Kaṭu.

सिन्धुवारः कटुस्तिक्तः कफवातक्षयापहः।
कुष्ठकण्डूतिशमनः शूलहृत् कासशुद्धिदः।।
(राजनिघण्टु)

Sindhuvarāḥ kaṭustikaḥ kaphavātakṣayāpahaḥ,
Kuṣṭhakaṇḍūtiśamanaḥ śūlahṛt kāsaśuddhidaḥ.
(Rajanighantu)

निर्गुंडी स्मृतिदा तिक्ता कषाया कटुका लघुः।
केश्या नेत्रहिता हन्ति शूलशोथाममारुतान्।।
कृमीकुष्ठारुचिश्लेष्मज्वरान्नीला हि तद्विधा।
सिन्दुवारदलं जन्तुवातश्लेष्महरं लघु।।
(भा. प्र.)

Nirguḍī smṛtidā tiktā kaṣāya katuka laghuh,
Keśyā netrahitā hanti śūlaśothāmamārutān.
Kṛmīkuṣṭhāruciśleṣmajvarānnilā hi tadvidha,
Sinduvāradalam jantuvātaśleṣmaharam laghu.
(Bhavaprakasa)

Actions/Uses : Kaphaghna, Vātaghna, Pittakar, Vedanāsthāpana, Śothahara, Vraṇaśodhana, Ropaṇa, Medhya, Jantughna, Agnidipana, Keṣya, Āmapacana, Kāsahara, Ārtavajanana.

Therapeutics :

Nirgudi plant is astringent, bitter, cephalic and stomachic. It has germicidal properties. It is easily digestible and cures cough, asthma, fever, eye diseases, inflammatory, glandular and rheumatic swellings, intestinal worms, ulcers, skin diseases, nervous disorders and leprosy. Roots are tonic, anodyne, febrifuge, expectorant and diuretic. They are used in dyspepsia and rheumatism and also for boils. Powdered root is prescribed as an anthelmintic and as a demulcent in dyentery. Tincture of root-bark is useful in rheumatism and irritable bladder. Leaves are alterative, effective in gonorrhoeal epididymitis, orchitis and as vermifuge. Externally they are anodyne and antiparasitic. Rheumatic patients are benefited by bathing in water boiled with the leaves. Smoke from burning dried leaves relieves catarrh and headache. Juice of leaves is used externally for foetid discharge and maggots in ulcers. Flowers are astringent and are used in fever, diarrhoea and liver complaints. Fruits are cephalic, emmenagogue and nervine tonic. They are prescribed in headache, catarrh and watery eyes and dried fruit are veremifuge.

Family : CIII - Violaceae.

Viola odorata Linn.

SYNONYMS : *Viola suavis* Bieb.; *Viola imberbis* Leighton.; *Viola alba* Besser.

ENGLISH : Sweet Violet.

INDIAN : Banafshah, Kaamapushpa, Bagabanosa.

SANSKRTA : Banapsikā, Jvarāpaha, Nīlapuṣpa, Sukṣmapatra, Vanapsa.

HABIT : A perennial herb.

HABITAT : Widely found in Europe, Asia and British Isles. Kashmir; planted in many hill stations.

CHEMICAL COMPOSITION : Flowers contain an emetic principle violin which is acrid and bittter, a volatile oil, rutin, cyanin, a colourless chromogen, a glycoside of methyl salicylate and sugar. Leaves contain an essential oil, an alkaloid, colouring matter, friedelin, β-sitosterol and an alcohol. Rootstock contains saponins, a glycoside, an essential oil and an alkaloid, odoratine.

Parts used : Whole plant, flowers.

Ayurvedic properties :
Rasa - Kaṭu, Tikta. Vīrya - Ūṣṇa.
Guṇa - Laghu, Snigdha. Vipāka - Kaṭu.

बनप्सा कटुतिक्तोष्णा शीतज्वरनिवारणी।
कासश्वासहरा त्वच्या सरा वातकफापहा।।

Banapsā kaṭutiktoṣṇā śītajvaranivaraṅī,
Kāsaśvāsaharā tvacyā sarā vātakaphāpahā.

Actions/Uses : Vātapittaghna, Vedanāśāmaka, Śothahara, Jantughna, Pittahara, Virecana, Vamana, Raktaśuddhikara.

Therapeutics :
Banapsika is valued as an expectorant, diuretic, diaphoretic, and antipyretic used as a laxative in bilious affections. It is used alone or in mixture with other herbs for catarrhal and pulmonary troubles and for calculous affections. The herb is well known for its medicinal virtues since olden times and is used for several diseases in Ayurvedic and Unani medicine. Drug "banafshah" was imported from Iran. Now it is collected in Kashmir, Kangra and Chamba. Root and seeds are emetic and purgative. Flowers are slightly laxative, emollient and demulcent and are used for sherbet preparation which is useful for coughs, sore throat, hoarseness and ailments of infants. Herbaseous parts of this and of V.tricolor have been employed for their mucilaginous, demulcent and expectorant properties.

HOMEOPATHIC USE : Useful in treatment of diseases of skin and eyes and for relief from pain in the ear.

Family : CIV - Vitaceae.

Cissus quadrangularis Linn.

SYNONYMS : *Vitis quadrangularis* Wall.; *Cissus edulis* Dalz.; *Saelanthus quadragonus* Forsk., *Cissus quadrangularis* Wall.

ENGLISH : Edible-stemmed vine, Bone-setter.

INDIAN : Chaudhari, Kaandawela, Haadajoda, Harasankari. Naller.

SANSKṚTA : Vajravallī, Asthiśṛnkhalā, Granthimāna, Amara, Asthisanhāra, Vajrāngī, Caturdhārā, Kāṇḍavallī.

HABIT : A climber with stout fleshy quadrangular stem.

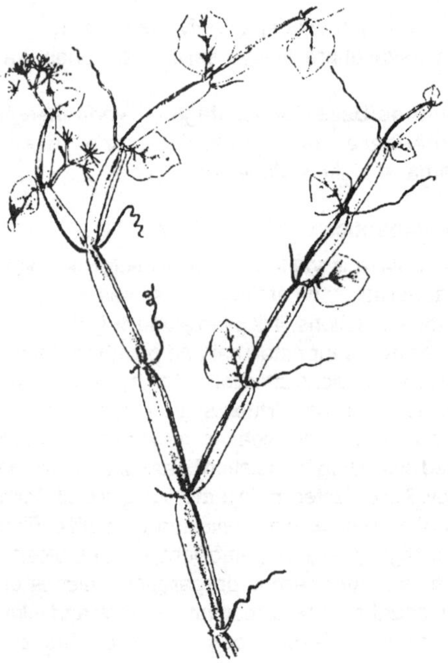

HABITAT : Common and planted in gardens throughout the hotter parts of India.

CHEMICAL COMPOSITION : Plant contains proteins, 12.8; fat and wax, 1.0; fibre. 15.6; carbohydrates, 36.6; mucilages and pectins, 1.2 % dry basis. A yellow wax and tartaric acid and the acid potassium salt are present. The plant is remarkably rich in vitamin C. Calcium oxalate crystals account for the irritating action of fresh stems.

Parts used : Root, stem and leaves.

Ayurvedic properties :
Rasa - Madhura. Vīrya - Uṣṇa.
Guṇa - Laghu, Rūkṣa, Sara.
Vipāka - Madhura, Kaṭu.

अस्थिसंहारक : प्रोक्तो वातश्लेष्महरोऽस्थियुक्।
उष्ण: सर: कृमिघ्नश्च दुर्नामघ्नाअक्षिरोगजित्।
रुक्ष: स्वादु: लघुर्वृष्य: पाचन: पित्तल: स्मृत:।।
(भावप्रकाश)

Asthisanhārakaḥ prokto vātaśleṣmaharo asthiyuk,
Uṣṇaḥ saraḥ kṛmighnaśca durnāmaghno akṣirogahrt.
Rūkṣah svādurlaghurvṛṣyaḥ pācanaḥ pittalaḥ smṛtaḥ.
(Bhavaprakasa)

Actions/Uses : Kaphavātaghna, Pittavardhaka, Sandhānīya, Dīpana, Pācana, Stambhana, Anulomana, Kṛmighna, Raktaśodhanī, Vṛṣya, Raktastambhanī, Netrarogahara, Arśoghna, Balaprada, Apasmāraghna, Vātarogaśāmanī, Bhagnasandhānakara.

Therapeutics :

Vajravalli is light, sweet, hot, alterative, carminative, anthelmintic, aphrodisiac and stomachic. It is useful in dyspsia, indigestion, piles, worms, asthma and vaatarakta. It has ability to rejoin broken bones. Juice of plant is beneficial in scurvy. Infusion of plant is purgative. A paste of the stem is plastered over the fracture and swellings. Powdered dry shoots are used in digestive troubles. It is also applied in cases of otorrhoea and epistaxis and used as alterative. The stem beaten into a paste is given in asthma and powdered root is a specific in the treatment of fractured bones. Fresh shoots are pounded and applied for burns and wounds.

Vitis vinifera Linn.

ENGLISH : Grape.

INDIAN : Draaksha, Angura, Dakh, Abai.

SANSKRTA : Drākṣa, Cāruphalā, Kṛṣṇā, Priyālā, Rasālā, Himā, Tāmasapriyā, Gucchaphalā, Amṛtaphalā, Phalottamā, Madhurasā, Gostanī, Hemavatī, Dīrghaphalā, Kapilaphalā, Mṛdvī, Amṛtarasā, Uttarapathika, Kasiri, Satavirya, Madhuli, Suphala, Harhura, Guda, Yakṣmaghni.

HABIT : A large, deciduous climber with long bifid tendrils.

HABITAT : Cultivated in many parts of India.

CHEMICAL COMPOSITION : Main constituent is grape-sugar, gum, tannin, tartaric, citric, racemic and malic acids; chlorides of potassium and sodium, sulphate of potash, tartarate of lime, mangnesia, alum, iron, some albumin.

Parts used : Fruits, leaves and young branch.

Ayurvedic properties :
Rasa - Madhura. Vīrya - Śīta.
Guṇa - Snigdha, Mṛudu. Vipāka - Madhura.

द्राक्षातिमधुराम्ला च शीता पित्तार्तिदाहजित्।
मूत्रदोषहरा रुच्या वृष्या सन्तर्पणी परा।।
(राजनिघण्टु)

Drākṣāti madhurāmlā ca śīta pittārti dāhajit,
Mūtradoṣaharā rucyā vṛṣyā santarpaṇī parā.
(Rajanighantu)

मांसमेवाश्नतो युक्त्या मार्द्वीकं पिबतोऽनुच।

Mānsamevāśnato yuktyā mārdvīkam pibato anuca.

तेषां द्राक्षा सरा स्वर्या मधुरा स्निग्धशीतला।
रक्तपित्तज्वरश्वास तृष्णादाहक्षयापहा।। (च.सू.४७)

Teṣām drākṣā sarā svaryā madhurā snigdhaśītalā,
Raktapittajvaraśvāsa tṛṣṇādāhakṣayāpahā.
(Caraka Sutra 47)

Actions/Uses : Vātapittaghna, Balya, Anulomana, Tṛṣṇānigrahaṇa, Sandhānīya, Raktaprasādana, Raktapittaśāmaka, Jīvanīya, Vṛṣya, Mūtrala, Garbhasthāpana, Snehana, Bṛnhana.

Therapeutics :
Fresh Draksa or grapes are cooling, demulcent, diuretic laxative, refrigerant, stomachic and tonic. They are useful in chronic bronchitis, dyspepsia, dysuria, haemorrhagia, heart diseases and strangury. Juice is highly efficacious in severe cold, fever, jaundice. Useful for thirst and constipation during teething in children. Juice of unripe grapes is astringent, and is useful in throat affections. Dried grapes are cooling, laxative, demulcent, stomachic and expectorant. They are useful in catarrh, cough, consumption, thirst, hoarseness, jaundice, enlarged liver and spleen. They are considered as attentuant, suppurative, nutritious and blood-purifier. Leaves are antidiarrhoeal and astringent. Expressed juice of young branch is useful in skin diseases.

Family : CV - Zingiberaceae.

Costus speciosus (Koen.ex Retz.) Sm.

SYNONYMS : *Banksea speciosa* Koenig.

INDIAN : Kemuka, Pewa, Kushtha, Pushkarmula.

SANSKṚTA : Kebuka, Kuṣṭha, Ruk, Pakala, Kāśmīraja, Vaśya, Peu, Vāpya, Utpala, Rogāhvaya, Thāpya, Pāribhāvya, Puṣkara, Candā, Padmapatraka, Bramhatīrtha, Kāsārī, Puṣkarajatā, Śiphā, Śūlahara, Cirantana, Vīra, Svāsāri, Sukhasambhava, Sthalapadminikanda, Puṇyasāgara, Vātyāhva, Sugasndhika, Puṣkarādyā, Vṛkṣaruha.

HABIT : An erect succulent herb with root stocks and tuberous rhizome, stem spirally twisted growing in marshy palaces and shades.

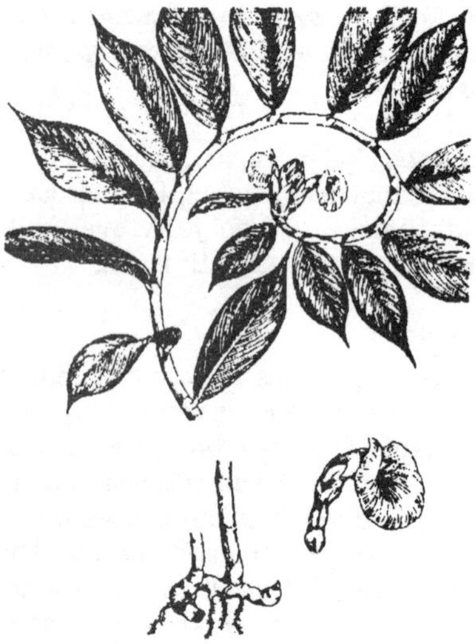

HABITAT : Throughout India, particularly common in Konkan and Bengal. Often cultivated as an ornamental plant.

CHEMICAL COMPOSITION : Rhizomes and stems yield diosgenin and tigogenin.

Parts used : Root and tuber.

Ayurvedic properties :
Rasa - Tikta, Kaṭu, kasāya. Virya - Ūṣṇa, Śīta.
Guṇa - Laghu, Rūkṣa. Vipāka - Kaṭu.

केबुकं कटुकं पाके तिक्तं ग्राहि हिमं लघू।
दीपनं पाचनं हृद्यं कफपित्तज्वरापहम्।
कुष्ठकासप्रमेहास्त्रनाशनं वातलं कटू।।
(भा. प्र.)

Kebukam kaṭukam pāke tiktam grāhi himam laghū,
Dīpanam pācanam hṛdyam kaphapittajvarāpaham,
Kuṣṭhakāsapramehāsranāśanam vātalam kaṭū.
(Bhavaprakasa)

Actions/Uses : Kaphapittaśāmaka, Vātavardhaka, Grāhī, Dīpana, Pācana, Raktaśodhana, Kṛmghna, Medohara, Śothahara, Kāsaghna, Śūkrala, Garbhāśayasankocaka, Pramehaghna, Kuṣṭhaghna, Śvāsaghna, Kaṇḍūghna, Viṣaghna, Visarpahara.

Therapeutics :

Kemuka root is bitter, astringent, stimulant, digestive, depurative and used as tonic, aphrodisiac and anthelmintic. It is useful in cattarrhal fevers, coughs, dyspepsia, worms and skin diseases. Rhizome is edible and is rich in starch. It has purgative and tonic properties.

Curcuma amada Roxb.

ENGLISH : Mango Ginger.

INDIAN : Ambehallada, Am haldi.

SANSKṚTA : Āmraharidrā, Sugandhī, Karpūraharidā.

HABIT : An erect herb with rhizome.

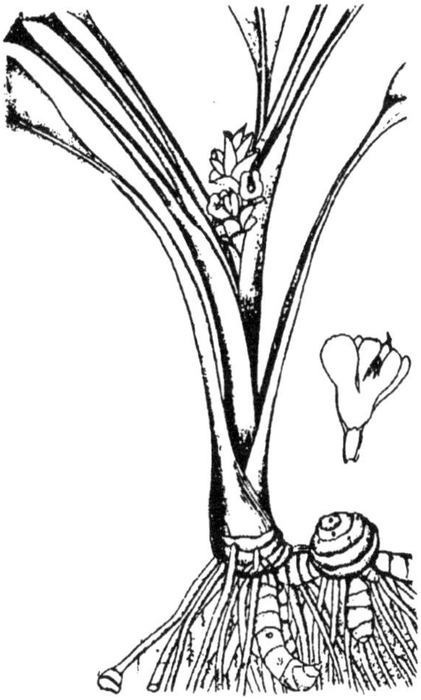

HABITAT : Wild in parts of Bengal, Koṇkan and Tamilnadu.

CHEMICAL COMPOSITION : Essential oil, resin, sugar, gum, starch, albuminoids and organic acids. Clinical trials prove that it reduces serum cholesterol level considerably in hypercholesterolemic rats. Dried ginger is effective in treatment of "grahaniroga. The drug significantly improved body weight, appetite and haemoglobin percentage and controlled number of motions.

Parts used : Rhizome

Ayurvedic properties :
Rasa - Madhura, Tikta. Vīrya - Śīta.
Guṇa - Laghu, Rūkṣa. Vipāka - Kaṭu.

आम्रगन्धिहरिद्रा या सा शीता बातला मता।
पित्तहृन्मधुरा तिक्ता सर्वकण्डूविनाशिनी।।
(भा. प्र.)

Āmragandhiharidrā yā sā śītā vātalā matā,
Pittahṛnmadhurā tiktā sarvakaṇḍūvināśinī.
(Bhavaprakasa)

Actions/Uses : Kaphapittaśāmaka, Vātakar, Vātānulomana, Dīpana, Pācana, Vedanāsthāpana, Śothahara, Sarvakaṇḍūvināśinī.

Therapeutics :
Root of Amraharidra has a bitter sharp taste; diuretic, maturant, emollient, expectorant, antipyretic, appetiser; useful in inflammations, diarrhorea and gleet. Rhizome is acrid, hot, anodyne, antirheumatic, carminative, cooling, aromatic, bitter, stomachic, diuretic, aphrodisiac and astringent. It promotes digestive power, cleans throat and tongue, dispels cardiac disorders and cures vomiting, ascites, cough, dyspnoea, anorexia, fever, anaemia, flatulence, colic, constipation, dysuria, swelling and elephantiasis. It has specific action in rheumatism and inflammation of liver. Tubers are useful in prurigo. They are used externally in the form of paste as an application for bruises and skin diseases.

Curcuma aromatica Salisb.

ENGLISH : Wild turmeric.

INDIAN : Raana hallada, Jangli-haldi.

SANSKṚTA : Araṇyakaharidrā, Vana-haridrā.

HABIT : An erect herb with rhizome.

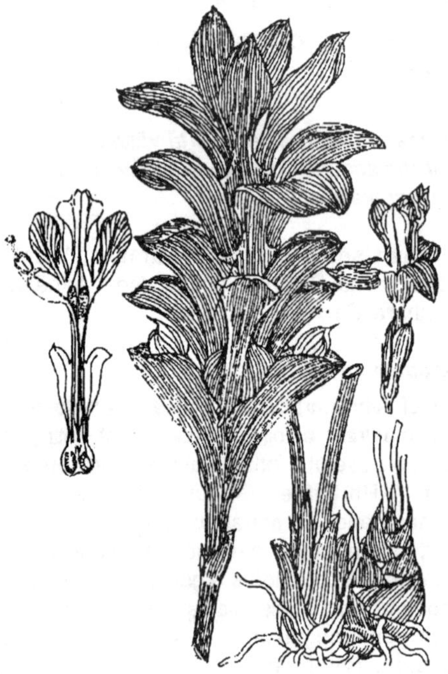

HABITAT : Wild throughout India and cultivated chiefly in Bengal and Trvancore.

CHEMICAL COMPOSITION : A volatile oil, resin, starch, mucilage, sugar, gum, albuminoids and curcumin.

Parts used : Tuber or rhizome.

Ayurvedic properties :
Rasa - Tikta, Kaṭu. Vīrya - Ūṣṇa.
Guṇa - Laghu, Rūkṣa. Vipāka - Kaṭu.

वनहरिद्रा :

आरण्यकहरिद्रानुकटुकामघुरामता।
रुच्याग्नि दीपनी तिक्ताकुण्ठ वातत्रिदोषहा।।
रक्तदोषं विषंश्वासंकासंहिक्कांचनाशयेत्।।
(निघण्टु रत्नाकर भाग १)

Āraṇyakaharidrānu kaṭukā madhurāmatā,
Rucyāgnidīpanī tiktākuṣṭhavāta tridoṣahā.
Raktadoṣam viṣam śvāsam kāsam hikkāñca nāśayet.
(Nighanturatnakar)

Actions/Uses : Kaphaghna, Pittavirecaka, Pittaśāmaka, Varṇya, Tvagrogahara, Mehaghna, Raktadoṣahara, Kaṇḍughna, Pāṇḍughna, Vraṇahara, Viṣaghna, Viśodhinī, Pinasanāśinī, Śoṣaghna.

Therapeutics :

Rhizome of Aranyakaharidra is used and its uses are similar to those of Curcuma longa. It is tonic, stimulant and carminative useful in leucoderma and diseases of blood. Dried rhizome is used as an aromatic adjunct to other medicines used in skin diseases and impurities of blood. Boiled in oil it is used externally as an application to sprains and bruises.

Curcuma longa Linn.

SYNONYMS : *Curcuma domestica* Valenton.; *Curcuma rotunda* Linn.; *Amomum curcuma* Jacq.

ENGLISH : Turmeric.

INDIAN : Hallada, Haridraa.

SANSKṚTA : Haridrā, Kāmini, Niśā, Varṇini, Gaura, Yośitpriya, Kāñcani, Varavarṇini, Haladi, Rajani, Kṛimighnā, Haṭtavilāsini, Harita, Jayanti, Varṇadāyi.

HABIT : An erect perennial herb with rhizome.

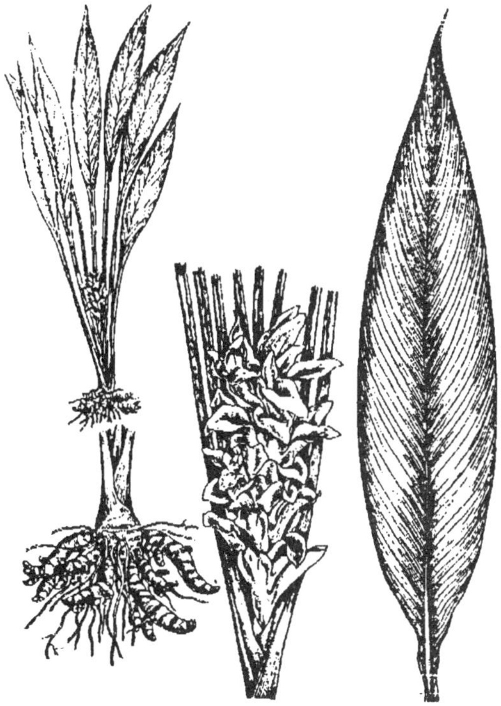

HABITAT : Cultivated throughout India.

CHEMICAL COMPOSITION : Campesterol, stigmasterol, β-sitosterol, cholesterol and fatty acids from rhizome. An essential oil, resin, an alkaloid, curcumin, turmeric oil. Aqueous extract suppressed carrageenin-induced oedema and showed very potent activity in granuloma pouch test.

Parts used : Tubers and Rhizome.

Ayurvedic properties :
Rasa : Tikta, Kaṭu. Vīrya : Ūṣṇa.
Guṇa - Laghu, Rūkṣa. Vipāka - Kaṭu.

हरिद्रा कटुकातिक्ता रुक्षोष्णा कफपित्तनुत्।
वर्ण्या त्वग्दोषमेहास्त्रशोथपाण्डुव्रणापदा।।
(भा. प्र.)

Haridrā kaṭukātiktā rūkṣoṣṇā kaphapittanut,
Varṇyā tvagdoṣamehāsraśothāpāṇḍuvraṇāpahā.
(Bhavaprakasa)

हरिद्रा लेखनिया, कुष्ठघ्नी, विषघ्नी च।
(चरक)

Haridrā lekhanīyā kuṣṭhaghnī viṣaghnī ca.
(Caraka)

Actions/Uses : Kaphaghna, Pittavirecaka, Pittaśāmaka, Varṇya, Ṭvagrogahara, Mehaghna, Raktadoṣahara, Kaṇḍughna, Pāṇḍughna, Vraṇahara, Viṣaghna, Viśodhinī, Pināsanaśinī, Śosaghna.

Therapeutics :
Due to strong antiseptic properties Haridra is useful for all kinds of poisonous affections, ulcers and wounds. It purifies blood by destroying pathogenic organisms. It is useful in cold, cough, bronchitis, conjuctivitis and liver affections. Root of Haridra is usefully administered in intermittent fevers. Rhizome is aromatic, pungent, bitter, laxative, anthelmintic, stimulant, tonic, emollient, maturant, diuretic and carminative. Oral admisnistration of powdered rhizome gives relief in cases of asthma and cough. Internally, juice is anthelmintic. Juice of fresh rhizome is applied to recent wounds and bruises.

Curcuma zedoaria Rosc.

ENGLISH : Zedoary.

INDIAN : Kachoraa.

SANSKṚTA : Karcūra.

HABIT : An erect herb with rhizome.

HABITAT : Cultivated throughout India. Wild in moist deciduous forests.

CHEMICAL COMPOSITION : An essential oil containing curzerenone and curcumol, a bitter soft resin, organic acids, gum, starch, resins, sugar, curcumin arabins, albuminoids. A yellowish oil from rhizome.

Parts used : Rhizome.

Rasa - Kaṭu, Tikta.	Vīrya - Ūṣṇa.
Guṇa - Laghu.	Vipāka - Kaṭu.

कर्चूरो दपिनो रुच्यः कटुकस्तिक्त एव च।
सुगन्धिः कटु-पाकः स्यात्कुष्टार्शो-व्रण-कास-नुत्॥
उष्णो लघुर्हरेच्छ्वास - गुल्म - वात - कफ - कृमीन्॥
(भा. नि.)

Karcūro dīpano rucyaḥ kaṭukstikta eva ca,
Sugandhiḥ kaṭupakaḥ syāt kuṣṭhārśo vraṇakāsanut.
Ūṣṇolaghurhare cchvāsa gulma vāta kapha kṛmīn.
(Bhavaprakasa)

Actions/Uses : Rucya, Kaphahara, Vātahara, Mūtrajanana, Kṛmighna.

Therapeutics :

Karcura plant is stimulant, diuretic, demulcent, carminative, expectorant and rubefacient. Root is aromatic, cooling and diuretric. It is useful in flatulence and dyspepsia and as a corrector of purgatives. Fresh root checks lecorrhoeal and gonorrhoeal discharges. Rhizome possesses stimulant and carminative properties. A decoction of the rhizome administered along with pepper, cinnamon and honey is beneficial for colds. It is stomachic.

Elettaria cardamomum Maton.

SYNONYMS : *Amomum repens* Sonnerat.; *Amomum cardamomum* White *non* Linn.; *Alpine cardamomum* Roxb.

ENGLISH : Cardamom.

INDIAN : Weladoddaa, Welachi.

SANSKṚTA : Elā, Triputa, Treti, Suksmailā, Drāviḍī, Himā, Candrā, Candralatā, Candrabhāgā, Bahulagandhā, Niśhkuti, Kapotavarṇi, Gaurangī, Aindrī, Balavatī, Upakuncī, Sāgaragāminī, Garbhāri, Gandhaphalikā, Kāyasthā, Bālā, Candrasambhavā, Korangī.

HABIT : A tall herbaceous perennial, with branching subterranean rootstock.

HABITAT : Malabar, Western Ghats, wild or cultivated.

CHEMICAL COMPOSITION : A volatile oil 2 to 8 % from seeds and a fixed oil; limonene, borneol, L-terpineol, potassium salts, starch, nitrogenous mucilage. a-pinene, sabinene, myrcene, limonene, cineol, cymane, Me heptenine, linalool, linalys acetate etc. from essential oil.

Parts used : Fruit and seed.

Ayurvedic properties :
Rasa - Kaṭu, Madhura. Vīrya - Śīta.
Guṇa - Laghu, Snigdha, Sugandhī, Sūkṣma.
Vipāka - Madhura.

रसे तु कटुका शीता लघ्वी वातहरी मता।
एला सूक्ष्मा कफश्वासकासार्शो मूत्रकृच्छ्रहृत्।।
(भा. प्र.)

Rase tu kaṭukā śitā lagvī vātaharī matā,
Elā sūkṣmā kaphaśvāsakāsārśo mūtrakṛcchahṛt.
(Bhavaprakasa)

सूक्ष्मैला मूत्रकृच्छघ्नी श्वासकासक्षयेहिता।
सूक्ष्मैला शीतला स्वादी हृद्यारोचन दीपनी।।
(ध. नि.)

Śukṣmailā mūtrakṛcchaghnī śvāsakāsakṣayehitā,
Sūkṣmailā śītalā svādī hṛdyārocana dīpanī.
(Dhanvavtarinighantu)

Actions/Uses : Kaphaghna, Vātaghna, Pittaghna, Dīpana-Pācana, Mukhadurgandhihara, Chardinigrahaṇa, Hallasanigrahaṇa, Hṛdya, Mūtrajanana, Śūkravrddhikara, Dāhaśāmaka.

Therapeutics :

Ela has carminative and stimulant properties, but it is rarely prescribed alone, but commonly either as adjuvants or correctives of cordial, tonic and purgative medicines; also used as a spice and masticatory. The effects of Cardamom are those of agreeable aromatic.

Hedychium spicatum Buch.-Ham. ex Smith.

ENGLISH : Spiked ginger lily.

INDIAN : Kapurkaachari.

SANSKṚTA : Sati-2, Kārṣa, Grandhamūla, Drāviḍā, Kacūra, Gandha, Vedhamukhya, Durlabha, Gandhasāra, Jatila, Sthūlakanda, Kālpaka, Śatati, Śadgranthā, Gandhamūlikā, Palāśī, Suvratā, Gandhārikā, Kapurakacali, Gandhasati.

HABIT : A perennial rhizomatous herb or a tall tree with root stock.

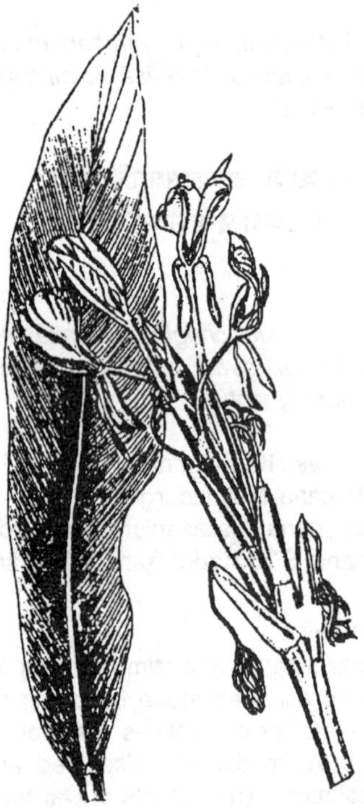

HABITAT : China, Subtropical Himalayas, Nepal, Kumaon.

CHEMICAL COMPOSITION : Rhizomes contain an essential oil, starch, organic acids including resinic acid, a glycoside and ash. Oil showed tranquillising activity of short duration.

Parts used : Rhizome.

Ayurvedic properties :
Rasa - Kaṭu, Tikta, Madhura. Vīrya - Uṣṇa.
Guṇa - Laghu, Tīkṣṇa. Vipāka - Kaṭu.

शटी स्यात् कटुतीक्ष्णोष्णा सन्निपातज्वरापहा।
कफास्त्रव्रणकासघ्नी वक्त्रशुद्धिविधायिनी।।
(ध. नि.)

Śatī syāt kaṭutīkṣṇoṣṇā sannipātajvarāpahā,
Kaphāsravraṇakāsaghnī vaktraśudhividhāyinī.
(Dhanvantari nighantu)

Actions/Uses : Kaphavātaśāmaka, Vedanāsthāpana, Dīpana, Pācana, Śothahara, Durgandhanāśana, Kesya, Uttejaka, Śūlapraśamana, Raktaśodhana, Śvāsahara, Hikkānigrahaṇa, Vraṇahara, Hidmahara, Grāhī, Asyāmalanāśinī.

Therapeutics :

Root-stock of Sati is acrid, bitter, pungent and astringent; useful in inflammations, asthma, pains, foul breath, bronchitis, hiccough, vomiting and diseases of blood. It is laxative, tonic to brain, emmenagogue, good for liver complaints. Rhizomes are stomachic, carminative, stimulant and tonic and are used in dyspepsia in the form of powder or decoction.

Kaempferia galanga Linn.

ENGLISH : Black thorn.

INDIAN : Chandramula, Kapurkaachari (2), Kachri.

SANSKṚTA : Karpurakacorī, Candramūlikā.

HABIT : An erect herb with root stocks.

HABITAT : Cultivated more or less throughout India.

CHEMICAL COMPOSITION : Rhizome contains an essential oil. It contains p-Methoxycinnamic acid and its methyl and ethyl esters.

Parts used : Rhizome.

Ayurvedic properties :
Rasa - Kaṭu, Tikta. Vīrya - Śīta.
Guṇa - Laghu, Tīkṣṇa. Vipāka - Kaṭu.

कापूरकचरी
स सुगन्धः कर्चुरकस्तीष्णो दाही कटुः स्मृतः।
तिक्तश्च तुवरश्चैव शीतवीर्यो लघुः स्मृतः।।
किंचित्पित्तं कोपयति कासश्वासज्वरापहः।
शूलहिक्कागुल्मरक्तरुग्वातविलनाश नः।।
मुखवैरस्यदौर्गंध्यव्रणामच्छर्दिहिध्महा।
(निघण्टु रत्नाकर)

Sa sugandhaḥ karcurakastīkṣṇo dāhi kaṭuḥ smṛtaḥ,
Tiktaśca tuvarścaiva śītavīryo laghuḥ smṛtaḥ.
Kincit pittam kopayati kāsāśvasa jvarāpahaḥ,
Śūlahikkā gulma raktarugvātavila nāśanaḥ.
Mukhavairasya daurgandhya vraṇāmacchardi hidhmahā.
(Nighanturatnakara)

Actions/Uses : Mutrajananna, Uttejaka, Vātahara.

Therapeutics :

Rhizome of Karpuraka is stomachic, anti-inflammatory. In the form of powder or ointment it is applied to wounds and bruises to reduce swellings Rhizomes are stimulating, expectorant, carminative and diuretic. They improve complexion and cure burning sensation, mental disorders and insomnia. They are used in the preparation of gargles and administered with honey in cough and pectoral affections. Decoction of rhizomes is used for dyspepsia, headache and malaria. Roasted rhizomes are applied hot in rheumatism and for hastening the ripening of inflammatory tumors. They are used as a wash in dandruff and for relieving irritation produced by stinging caterpillars. Leaves are used in lotions and poultices for sore eyes, sore throat, swellings, rheumatism and fevers.

Kaempferia rotunda Linn.

INDIAN : Bhuee chaaphaa, Bhuichampaa.

SANSKṚTA : Bhūcampaka, Bhūmicampā.

HABIT : An erect herb with tuberous rhizomes.

HABITAT : Throughout India.

CHEMICAL COMPOSITION : Crotepoxide from tubers. A compound, mp. 149 C. from tubers which yielded benzoic acid or hydrolysis. Rhizomes yield a light yellow volatile oil. Oil contains cineol and probably methyl chavicol.

Parts used : All parts, tubers, leaves.

Ayurvedic properties :

Rasa - Kaṭu, Tikta. Virya - Uṣṇa.
Guṇa - Laghu, Tikṣṇa. Vipaka - Kaṭu.

भूमिजश्चम्पकश्चोष्णः कटुः शोथरुजापहः।
गलगण्डं व्रणं चैव नाशयेदिति कीर्तितः।।
(निघण्टु रत्नाकर)

Bhūmijaścampakaścoṣṇaḥ kaṭuḥ śotharujāpahaḥ,
Galagaṇḍam vraṇam caiva nāśayediti kīrtitaḥ.
(Nighanturatnagara)

Actions/Uses : Śothaghna, Vraṇaropaṇa.

Therapeutics :

Tubers of Bhucampaka are used as local application for tumors, swellings and wounds. They are stomachic and given in gastric complaints. They help to remove blood clots and other purpulent matter in the body. Juice of tubers is given in dropsical affections of hands and feet, and for effusions in joints. Herb is used in ointments for wounds.

Zingiber officinale Rosc.

SYNONYMS : *Amomum zingiber* Linn.

ENGLISH : Ginger.

INDIAN : Aale, Adaraka.

SANSKṚTA : Ādraka, Śuṇṭhī, Nāgara, Viśva, Śoṣaṇa, Viśvabheṣaja, Viśvauṣadha, Kaṭugranthi, Kaṭubhadrā, Ūṣṇa, Kaṭūṣaṇa, Sauparṇa, Śṛuṅgabera, Kaphāri, Kaṭutikta.

HABIT : An erect herb with rhizome.

HABITAT : Widely cultivated in India.

CHEMICAL COMPOSITION : An aromatic volatile oil having a characteristic flavour and containing terpenes, cineol, borneol, citral, camphene, phelandrene, gingerol, shogaol, zingerfone, zinziberin.

Parts used : Dry and fresh rhizome.

Ayurvedic properties :
Rasa - Kaṭu. Vīrya - Ūṣṇa.
Guṇa - Guru, Rūksa, Tīkṣṇa, Laghu, Snigdha.
Vipāka - Kaṭu, Madhura.

अर्द्रिका भेदिनी गुर्वी तीक्ष्णोष्णा दीपनी मता।
कटुका मधुरा पाके रुक्षा वातकफापहा।
ये गुणाः कथिताः शुण्ठ्यास्तेऽपि सन्त्यार्द्रकेऽखिला।।
(भा. प्र.)

Ardrikā bhedinī gurvi tīkṣṇoṣṇā dīpanī matā,
Kaṭukā madhurā pāke rūkṣā vātakaphāpahā,
Ye guṇāḥ kathitāḥ śuṇṭhyāste api santyārdake akhilā.
(Bhavaprakasa)

Actions/Uses : Kaphaghna, Vātaghna, Pittakara, Vātānulomana, Agnidipana, Āmapācana, Śūlapraśamana, Hṛdayottejaka, Rucikara, Rasasamvahana.

Therapeutics :
Ādraka possesses stimulant, aromatic and carminative properties when taken internally; and when chewed it acts as a sialagogue. Externally applied it is rubefacient. It is of much value in atonic dyspepsia, especially if it is acompined with much flatulence; and as an adjunct to purgative medicines to correct griping. When chewed it is serviceable in relaxed conditions of uvulva and tonsils. As a rubefacient it relieves headache and headache.

Family : CVI-Zygophyllaceae.

Fagonia arabica Linn.

SYNONYMS : *Fagonia cretica* Linn.; *Fagonia bruguieri DC.*; *Fagonia mysorensis* Roth.

INDIAN : Dhamaasaa, Ustar-khar, Dumaso, Hinguna, Usturgar.

SANSKṚTA : Durālabha, Dhanvayāsa, Uṣṭrabhakṣa, Yāsa, Marudbhava, Dhanvayavaśa, Mūla, Samudrānta, Niṣkanṭaka, Yāvasā, Ajabhaksya, Tamramuli, Bālapatra, Kacchura, Harivigraha, Gāndhāri, Duhsparṣa.

HABIT : A woody perennial spiny undershrub.

HABITAT : Punjab, Konkan and various parts of India.

CHEMICAL COMPOSITION : Saponin.

Parts used : Leaves, twigs and juice.

Ayurvedic properties :
Rasa - Tikta, Kaśāya, Madhura, Kaṭu.
Vīrya - Śīta.
Guṇa - Laghu, Rūkṣa. Vipāka - Kaṭu, Madhura.

दुरालभा कटुस्तिक्ता मधुरा रक्तशुद्धिकृत्।
शीता चोष्णा विसर्पघ्नी विषमज्वर नाशिनी।।
तृट्छर्दिमेहगुल्मघ्नी मोहरक्तरुजा पहा।
वातपित्तं कफं कुष्ठं ज्वरं चैव विनाशयेत्।।

Durālabhā kaṭustiktā madhurā raktaśuddhikṛt,
Śīta coṣṇā visarpighnī viṣamajvara nāśinī.
Tṛtūcchardimeha gulmaghnī moharakta rujāpahā,
Vātapittam kapham kuṣṭham jvaram caiva vināśayet.
(Nighanturatnagara)

Actions/Uses : Dāhapraśamana, Viṣamajvaraghna, Tritprasamana, Chardipraśamana, Pramehaghna, Bhramaghna, Gulmahara, Kuṣṭhaghna, Stambhana, Vraṇaropaṇa, Kāsaghna, Raktaśuddhikara, Jvaraghna, Vamanapraśamana, Visarpaghna.

Therapeutics :

Duralabha is bitter astringent, cooling, tonic and febrifuge. It removes thirst, fever, and cures typhoid, asthma, tumors and delirium. Plant has antiseptic properties and is applied both in the form of paste or polutice to tumours and abscesses, wounds and scrofulous glands; it is also given as a prophylactic against small pox. It is employed in the preparation of Kumari Asava, an indigenous medicinal preparation known for its stimulative, laxative and alterative properties. Twigs are used as tooth-brushs and bark in scabies.

Guaicum officinale Linn.

SYNONYMS : *Guaicum sanctum* Linn.; *Lignum vitae*; *Lignum sanctum* Linn.

ENGLISH : Pockwood tree, Brazil wood, Gum Guaiacum.

INDIAN : Lohalakkada, Chobahayaata.

SANSKṚTA : Jīvadāru, Lohakāṣṭha, Vṛdhamitra, Amrtadāru, Guhyākṣa.

HABIT : A small to middle-sized tree.

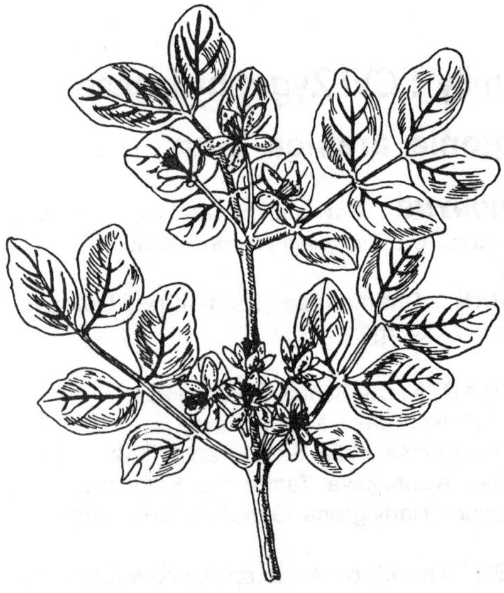

HABITAT : A native of the West Indies and South America, grows in Indian gardens.

CHEMICAL COMPOSITION : Guaiacum or guaiac

resin from tree. The stem-bark yields another resin distinct from but analogous to that of wood. Alkaloids free of harman, harmine and harmol obtained from wood. Furoguaiacin, O-methylfuroguaiacin and furoguaiacidin from hearwood.

Parts used : Wood and latex.

Ayurvedic properties :

Rasa - Kaṭu. Vīrya - Ūṣṇa.
Guṇa - Rūkṣa. Vipāka - Kaṭu.

Actions/Uses : Dīpana, Pācana, Ānulomic, Svedajanana, Mūtrala, Vedanāsthāpana, Śothahara, Āartavajanana.

Therapeutics :

Jivadaru increases appetite. When given for a long period it improves metabolic activity and activity of kidneys and skin. It suits old people more. Gum is a mild laxative, diaphoretic, alterative. A valuable remedy in gout, chronic rheumatism. In the form of lozenges it is employed in the treatment of tonsillitis and pharyngitis with rheumatism.

HOMEOPATHIC USE : Guaiacum is a Homeopathic remedy. It acts on mucous membranes, muscles, joints and bones and causes contraction of tendons with resulting deformity. Gouty nodosities on joints. It is antipsoric and is best known as remedy in gout and rheumatism.

REFERENCES

AUTHOR	TITLE
Akhtar, Husain.	Medicinal Plants And Their Cultivation.
Apte, M. V.	Vanashreesrushtee, 2 Vols. (In Marathi).
Arya Vaidya Sala.	Indian Medicinal Plants. 4 Vols.
Asolkar, L.V., Kakkar, K.K., Chakre.	Glossary of Indian Medicinal Plants with Active Principles. Part-I (A-K).
Baapaalaala, G. Vaidya.	Nigantu Aadarsha. (In Hindi). 2 Vols.
Bor. Dr., Raizada	Some Beautiful Indian Climbers and Shrubs.
Chatterjee, Asima., Pakrashi, Satyesh Chandra.	Medicinal Plants, The Treatise on Indian. 4 Vols.
Chopra, R.N., Nayar S.L., Chopra I.C.	Glossary of Indian Medicinal Plants, Supplement To.
Chopra, R.N., Nayar S.L., Chopra I.C.	Glossary of Indian Medicinal Plants.
Chunekar, K.C.	Bhavaprakakasa-nighantu of Shri Bhavamisra.(In Hindi).
Clarke, John Henry.	A Dictionary of Practical Materia Medica. 3 Vols.
Coats, Alice M.	The Book of Flowers
Desai, Vaman Ganesha.	Aushadheesangraha. (In Marathi).
Deshpande, A.P., Javalgekar, R.R. and Ranade, Subhasha.	Dravyagunavidnyana. (In Marathi).
Dwivedishaastri, Vishwanaatha.	Bhavaprakasha nighantu. (In Hindi).
Ghose, Dr. S.C.	Drugs of Hindustan.
Gogate, Vishnu Mahadeva.	Dravyagunavidnyana. (In Marathi).
Gupta, Dr.B.Prasad.	Encyclopaedia of Homoeopathy.
Hamilton, Edward., M.D., F.L.S.	Flora Homoeopathica.
Hooker, J.D.	The Flora of British India.
Indian Council of Medicinal Research.	Medicinal Plants of India.
Kirtikar, K.R. & Basu, B.D.	Indian Medicinal Plants, 4 Vols.
Mathura, Dattaraam Shrikrushnalaal.	Nighantu Ratnaakara. (In Hindi).
Nadkarni, K.M.	The Indian Materia Medica.
Navre, Krishnasastri R.	Nighantu Ratnaakara. (In Hindi).
Novak, F. A.	Pictorial Encyclopedia of Plants and Flowers
Oza, Zarakhande.	Dhanvantari nighantu. (In Hindi).
Pal, B.P.	Beautiful Climbers of India.
Pande, Gangaa Sahaaya, and Chunekara, Krushnachadra.	Bhavaprakasha nighantu. (In Hindi).
Rastogi, Ram, P., Mehrotra B.N.	Indian Medicinal Plants, Compendium of. 2 Vols.
Satyavati, G.V., Raina, M.K., Sharma, M.	Medicinal Plants of India 2 Vols.
Savanta, Sadashiva Yashavanta.	Maharaashtraatila Divya Vanaushadhee. (In Marathi).
Shaligram, Vaishyavarya.	Shaligrama nighantu. (In Hindi).
Sharma, Acharya Priyavrata.	Dhanvantari nighantu. (In Hindi).
Sharma, Acharya Priyavrata. and Sharma Dr. Guruprasada.	Kaiyadeva nighantu. (In Hindi).
Shivarajan, and Balachandran I.	Ayurvedic Drugs and their Plant Sources.
Tripaathi, Indradeva.	Raajanighantu. (In Hindi).
Udayalaal, Mahatmaa.	Dhanwantari, Vanaushadhi Visheshaanka (In Hindi). 10 Vols.
Vartak, V.D.	Enumeration of Plants from Gomantaka, India.
Wealth of India.	Wealth of India. 3 Vols.
Wren, R.C., F.L.S.	Potter's New Cyclopaedia of Botanical Drugs & Preparations.
Yadavjee, Trikamjee.	Dravyagunavidnyanum, (In Hindi). 2 Vols.

INDEX OF BOTANICAL NAMES

INDEX OF ENGLISH & INDIAN NAMES

Raantaakalla	121
Raasnaa	84
Raayaaawallaa	178
Radish	106
Rakat rohan	263
Raktachandana	208
Rakta-chitra	312
Raktakanchan	114
Raktapushpa	62
Raktavasu	295
Raldhup	110
Ram's horn	65
Ramturai	249
Ran tulasi	229
Rangoon creeper	141
Rangunachi veli	141
Ranjani	330
Ran-limba	342
Rataalle	148
Rati	190
Ratoliya	397
Rauvolfia	47
Red cedar	264
Red Chickweed	325
Red Chilli	368
Red creeper	332
Red onion	11
Red quash gourd	159
Red sanders	208
Red Sandlewood	208
Revand-chini	323
Rhakhan	355
Rhubarb	323
Ribbed gourd	163
Riber eboney	173
Ridge Gourd	163
Ringani	376
Ringin	359
Ringworm weed	117
Rinthi	359
Ritthaa	359
Rohan	263
Rohini	186
Rohitaka	260
Rojmari	71
Rose bay	45
Rose Laurel	45
Rue	351
Rue Sauvage	350
Ruee	63
Ruhin	263
Rui	252
Rus	1
Ruta	351
Saaga	399
Saagaragotaa	115
Saagawaana	399
Saalaee	109
Saalawanna	196
Saapasana	61
Saayidevi	89
Sabni	358
Sabzaa	228
Sacred bash	230
Sacred basil	227, 230
Sadaru	141
Sadodi	89
Sadphali	238
Safed jeera	34
Safed safari	293
Safed-ak	64
Safedaranda	184
Safed-musallee	69
Safetadamara	172
Safflower	78
Saffron	219
Safron tree	99
Sag-angur	367
Sage-leaved alangium	10
Sagun	399
Sahadevi	89
Sahadevi	89
Sahstrapatree	71
Sain	146
Saj	146
Sakkar teti	156
Salpan	196
Salphullie	109
Salsa	66
Saltbush	355
Sambhalu	400
Samudraphalla	237
Samudraphalla-2	238
Samudulan	82
Sandlewood tree	356
Sanmadat	141
Santha	246
Santonica	75
Santre	343
Saona	96
Sapashi	61
Saptaparnna	40
Saraso	104
Sareevaa	43

INDEX OF SANSKRIT NAMES

Bhārangī 395
Bhāraṭī 363
Bhārgavī 317
Bhārlī 395
Bhāskarapriya 208
Bhargakāgraṇī 395
Bharodvāha 141
Bhavābhīṣṭha 108
Bhavarakta 219
Bhayadā 187
Bhedinī 30
Bhikṣu 86
Bhilla 190
Bhillataru 389
Bhinnadalī 330
Bhīrū 70
Bhiṣagvarā 144
Bhiṣakpriya 268
Bhiṣaṅgmātā 1
Bhogivallabha 356
Bhṛngaja 390
Bhṛngajā 395
Bhṛngamohī 247
Bhṛṅgābhīṣṭa 23
Bhṛngara 81
Bhṛngarāja 81
Bhṛungavallabhā 337
Bhṛungeṣṭā 337
Bhṛahmadanti 303
Bhramarānanda 362
Bhramarātithi 247
Bhramarapriya 127
Bhramareṣṭā 395
Bhrangurā 328
Bhrngeṣṭā 239
Bhujangāksī 47
Bhunimba 216
Bhūcampaka 412
Bhūcaṇaka 191
Bhūdhātrī 187
Bhūkṣāra 106
Bhūmiari 118
Bhūmicampā 412
Bhūmijā 191
Bhūmijumbū 395
Bhūmikūṣmānda 149
Bhūmimudga 191
Bhūmināgaranga 262
Bhūmyāmalak 187
Bhūnimba 2
Bhūrdiśam 195
Bhūripuṣpā 28
Bhūstha 191
Bhūsūtā 10
Bhūtaghna 12, 289
Bhūtahara 108
Bhūtakeśī 400
Bhūtavṛukṣa 102
Bhūtika 315
Bhūtī 315
Bhupadma 118
Bhustraṇa 315
Bhutaghni 230
Bhutakusum 179
Bhutali 218
Bhutavasa 142
Bhuthala bhairi 179
Bibhītaka 142
Bijaka-2 146
Bilva 341
Bilvapatra 127
Bimbī 154
Bimbora 64
Binducita 315
Bindupatrī 187
Bījadhānya 33
Bījaka 207
Bījaka-1 145
Bījāhva 346
Bījāmla 138
Bījapūra 346
Bījapuraka 346
Bījarecanī 180
Bījasar 207
Bījasneha 192
Bisaja 297
Bodha 282
Bodhidru 282
Bodhivṛukṣa 282
Bolasarī 362
Bṛhatī 376
Bṛhatpallī 79
Bṛhatpatra 389
Bṛhatphala 152, 278
Bṛhatpilu 355
Bṛmhaṇī 202
Brahiṣṭa 322
Bramhacarini 363
Bramhadaṇḍī 88
Bramhadandi 80
Bramhadārū 283
Bramhaghnī 239

Gokśūra	305
Golālu	379
Golī	82
Golomī	52, 317
Gopabhadrā	396
Gopakanyā	66
Gopavalli	330
Gopavallī	66
Gopimulam	66
Gopī	66
Gorakhamuṇḍī	86
Gorakh-cinca	100
Gorakṣagañjā	15
Gorakṣakarkaṭī Gangerukīe	100
Gorakṣī	100
Gośīrṣa	356
Gośṛnga	270
Gostanī	403
Govindaphala	126
Gṛhakanyā	239
Gṛhakumarī	239
Gṛndhranakhī	332
Gṛunjana	35
Grahinī	67
Grandhamūla	410
Granthila	126, 214
Granthimāna	402
Granthinī	286
Grāhiphala	347
Gridhrani	60
Gucchadantikā	286
Gucchala	359
Gucchapatra	56, 351
Gucchaphala	359,387
Gucchaphalā	286,378,403
Gucchapuṣpa	20
Gucchapuṣpaka	40,205,259
Gucchapuspaa	246
Gucchasangha	246
Guḍāśaya	220
Guḍatvak	234, 235
Guḍatvaka	236
Guḍuci	268
Guda	403
Gudajaghna	53
Gudāmayahara	53
Gudaphala	332, 355
Gudapuṣpa	361
Gudhapatra	10
Gudhapuspa	362
Guggul̥a	108
Guggulu	108
Guhā	213
Guhyākṣa	414
Guṇasū	253
Gundrā	394
Gunjaka	35
Gunjā	190
Gunjavallī	190
Guptasneha	10
Gurupatrā	124
Gurutwak	235, 236
Guṭiā	27
Guvāka	55
Haimamatī	144
Haladi	407
Halinī	240
Halī	240
Haliyuna	69
Hangsavati	9
Hansapadi	95
Hansapādi	9
Hārya	142
Hapuṣā	50
Hapuśa	310
Haramala	350
Haramara	350
Haravallabha	370
Harhura	403
Haricandana	208
Haridrā	407
Harimantha	194
Haripriya	50, 58
Hariśṛngārapuṣpaka	300
Harita	129
Harita	317, 407
Haritakī	144
Haritamanjari	174
Haritā	209
Haritākṣi	281
Haritaparṇa	106
Haritaparṇī	290
Haritaśāka	284
Haritasāraka	284
Harivallabha	384
Harivāsa	282
Harivigraha	413
Harmūla	350
Harṇya	339
Harśanṇī	125
Hastidanta	106
Hastidantī	153

Laghukāśmarī	287
Laghuparṇī	330
Laghupathā	266
Laguda	44
Lajjālu	277
Lakṣmaṇaya	384
Lakṣmīphala	341
Lakṣmīvān	282
Lalana	22
Lalanapriya	322
Lalukarṇikā	330
Lambakarṇa	10
Lanalī	57
Lankā	368
Lākṣāprasādana	389
Lākṣāvṛukṣa	360
Lāksā	337
Lālaphala	393
Lāmajjaka	316
Lāngali	397
Lāngalinī	240
Lāngalī	202, 213, 240
Lāvanyā	363
Laśuṇa	12
Latā	317, 394
Latākaranja	115
Latākastūrikā	250
Latākastūrī	250
Lataparṇa	235, 236
Lavali	26
Lavalī	178
Lavaliphala	178
Lavaṇī	131
Lavanī	26
Lekhanātaka	102
Lekhanātmaka	387
Lekhyapatra	56
Linginī	161
Lobhanīyā	86
Locamostaka	29
Lodhra	389
Lodhraka	389
Lodhrapuṣpiṇī	246
Loha	390
Lohakāṣṭha	414
Lohita	208
Lohitacandana	219
Lohitānga	186
Lohitapuṣpaka	326
Lohita-sūraṇa	54
Lomaśā	52
Lomaśapuṣpaka	275
Lunga	346
Lūtārī	85
Macchaka	18
Mada	203
Madagandha	40
Madagandhā	241
Madakāra	370
Madakāriṇī	371
Madana	10339, 362, 369, 370
Madanakūlakā	175
Madanamasta	54
Madanāhva	220
Madanaphala	339
Madanī	246
Madanīyā	246
Madaśaunda	288
Madayantikā	245
Madhubīja	326
Madhudruma	361
Madhudūta	23
Madhudūtī	98
Madhudvatī	98
Madhugandha	362
Madhukā	199
Madhukarapriya	23
Madhukarkaṭa	129
Madhukarkaṭi	345
Madhuli	403
Madhulikā	330
Madhulī	23
Madhumajjā	220
Madhu-mālati	69
Madhumatī	330, 396
Madhunāśīnī	65
Madhūka	361
Madhūlaka	361
Madhupāka	156
Madhuparṇa	199
Madhuparṇikā	18
Madhuparṇikā	396
Madhuparṇī	268
Madhupippalī	283
Madhupuṣpa	123, 175, 275
Madhupuṣpī	179
Madhurakāṣṭha	199
Madhuramlaka	25
Madhuramūlika	199
Madhurā	28,39,70,241
Madhurasā	28, 330, 403
Madhurasrava	199

Mangalyā 52, 394
Manjari 230
Manjiṣṭhā 340
Manmatha 23
Manmathālaya 23
Manojnya 23
Manojnyā 298
Manojyna 119
Mansaruhā 263
Manthāna 120
Mācikā 254
Mādanī 125
Mādhavadruma 23
Mādhavī 39, 248
Māgadhā 39
Māgadhī 309
Māghadhī 28
Mālākaṇṭa 14
Mālatī 298
Mālatiphala 288
Mālatisuta 288
Mālavi 266
Mālukāpatra 113
Mālūra 341
Māmsala 293
Māngalya 57, 282
Māngalyārhā 257
Māngalyavallikā 395
Mānsapācanī 129
Mānsaraktā 263
Mānsarohiṇī 263
Māriṣa 16
Mārjana 389
Mārjāragandhikā 203
Mārkanḍikā 118
Mārkava 81
Māṣa 204
Māṣaparṇī 204
Māsapatrikā 203
Mātā 239
Mātrukā 1
Mātula 339, 369, 370
Mātulaka 346, 370
Mātulanī 125
Mātulaputraka 370
Mardaka 122
Marica 310
Marjara 35
Markātamra 25
Markaṭa-pippalī 14
Markaṭatinduka 386
Markaṭī 14, 202
Marubaka 231
Marudbhava 413
Marudurvā 253
Marudviṣta 108
Marūbaka 339
Marusambhava 153
Maruvaka 231
Maśakī 281
Maskara 314
Masruṇā 241
Masūra 200
Maṭha 398
Matidā 131
Matsyabhedī 331
Matsyākṣī 363, 378
Matsyapittā 331
Matta 370
Maulasirī 362
Mayūrahāśikhā 9
Mayūrajangha 96
Mayūraka 14
Mayūri 29
Mayūr-śikhā 17
Medhyā 131, 144, 363
Medravalli 65
Meghamedinī 294
Meghanāda 16
Mekhalā 213
Mendikā 245
Meṣakusuma 122
Meṣalocana 122
Meṣāhva 122
Meṣaṣṛiṅgī 65
Meṣaviṣāṇikā 65
Methikā-1 201
Methikā-2 212
Methinī 212
Mezeriyuna 391
Mimbataru 198
Miṣreyā 28
Miṣṭanimbuphala 347
Miśi 28, 39
Miśra 106
Miśreyā 39
Mlencchakandaka 12
Mocaka 284
Mocā 101, 286
Moda 23
Modākhya 23
Mogarā 299

Pittapāpadā 6
Pittapriya 81
Pittavallī 68
Pittavṛkṣa 25
Piyala 22
Plavangaka 275
Plihāri 211
Plihāśatrū 211
Plīhāri 260
Pophaḷa 55
Potakī 92
Pṛṣṇiparṇī 213
Pṛthakabīja 24
Pṛthakbīja 20
Pṛthakparṇī 330
Pṛthakpatra 40
Pṛthaktvāca 293
Pṛthāja 141
Pṛthu 239, 330
Pṛthusimbi 96
Pṛthvī 295
Pṛthvīka 128
Prabhadra 198, 261
Prabhātā 240
Pracandā 317
Pracetasī 287
Pracodanī 380
Pradīpa 240
Pragraha 120
Praharavallī 263
Prakīrya 115, 359
Prakīryā 205
Pramadanī 246
Prācina 266
Prācīnā 267
Prājakta 300
Prāṇadā 144
Prānśū 56
Prāvṛśāyanā 202
Prāvṛuṣya 41
Prapayyā 144
Prapunnāta 122
Prapuṣpa 122
Prasādaka 324
Prasāraṇī 257
Pratanini 266
Pratāpī 258
Pratihāsa 44
Pratīyasa 63
Prativiṣṇuka 384
Pratyakapuspa 14
Pratyakśreṇī 175
Pravālaphala 208
Priya 129
Priyajīva 96
Priyaka 207, 394
Priyaka sauri 207
Priyaṅgu 394
Priyā 394
Priyāla 22
Priyālā 403
Priyāmbū 23
Priyavadā 298
Priyavallī 394
Pruthakparṇī 213
Pudinā 225
Pulaka 137
Puṇyā 381
Puṇyasāgara 404
Puṇyaṭruṇa 318
Punarnavā-Rakta 295
Punarnavā-Śveta 295
Punḍarīka 76
Punnāga 136
Punyalatha 159
Pūga 55
Pūra 108
Pūranī 101
Pūta 318
Pūtadhānya 306
Pūtadru 192
Pūtanā 144
Pūtā 317
Pūtaphala 278
Pūti 315
Pūtibarbara 135
Pūtihā 226
Pūtika 115
Pūtikanṭaka 366
Pūtikaranja 115
Pūtikītā 135
Pūtimāruta 341
Pūtimayūrikā 135
Pūtinaśa 225
Pūtiphala 206
Pūtivāha 341
Pūtodakam 57
Puram 390
Purāmla 138
Puruṣa 136
Puṣkara 297, 404
Puṣkarajatā 404

Vanodbhava 306
Vansa 1
Vanśadalā 338
Vanśapatī 338
Vanśika 390
Vanyajīraka 79
Vājidanta 1
Vājigandhā 381
Vājikara 125
Vājikarī 381
Vājikhurā 195
Vājināmā 381
Vājinī 381
Vākerī 116
Vāla 322
Vālaka 256
Vālungī 158
Vāmana 76
Vāmanī 253
Vāmāvartā 119
Vāmaviṣānikā 119
Vānaprastha 361
Vānti 87
Vāntiśodhinī 30
Vāpya 404
Vārahakarṇī 381
Vāraka 137
Vāraṇā 286
Vāraṇāmbusā 286
Vārāhāngī 175
Vārāhaputrī 381
Vārida 170
Vāriprasādana 387
Vārtākī 376,377,380
Vāruṇa 313
Vāsaka 1
Vāsanī 253
Vāsantī 248
Vāsikā 1
Vāstuka 131, 324
Vāstuka-2 134
Vāsudevī 70
Vāṭyā 258
Vātaghnī 196,381,398
Vātala 209
Vātaloma 225
Vātāda 20, 334
Vātāma 334
Vātāri 24,53,395
Vātārī 188,289
Vātapotha 192
Vātavairī 334
Vātyāhva 404
Vāyasāhvā 378
Vāyasī 378
Vaṛnini 407
Vara 139
Varada 128
Varadā 381
Varadruma 253
Varagada 279
Varakāṣthikā 258
Varamanjarī 123
Varā 378
Varāhikā 381
Varāhitā 202
Varatikta 41,65,261
Varavāhinī 258
Varavarṇini 407
Varavṛuntakī 376
Varayonika 219
Varca 395
Vardhamāna 188
Vareṇya 219
Varhiṇī 11
Varitara 322
Varṇabhedinī 394
Varṇadāyi 407
Varṇaka 35, 186
Varṣaketu 295
Varṣā 30
Varṣapuṣpā 257
Varṣika 125
Varsabhava 218
Vartala 126, 160
Varuṇa 127
Varvara 228
Vasanta 142
Vasantadūta 23
Vasantapādapa 23
Vasapuspa 105
Vasati-tikta 267
Vaśira 397
Vaśirā 14
Vastraranjaka 78
Vastraranjinī 340
Vastuphala 57
Vasu 384
Vasuka 63
Vaśya 404
Vaṭa 279
Vaṭyālikā 258